BIOLOGICAL AND MEDICAL PHYSICS, BIOMEDICAL ENGINEERING

Shiyi Shen · Jack A. Tuszynski

Theory and Mathematical Methods for Bioinformatics

With 47 Figures and 59 Tables

Prof. Shiyi Shen
Nankai University
College of Mathematical Sciences
Tianjin 300071
China
syshen@nankai.edu.cn

Prof. Jack A. Tuszynski
University of Alberta
Department of Physics
Edmonton T6G 2G7, Alberta
Canada
jtus@phys.ualberta.ca

ISBN 978-3-540-74890-8 e-ISBN 978-3-540-74891-5

DOI 10.1007/978-3-540-74891-5

Biological and Medical Physics, Biomedical Engineering ISSN 1618-7210

Library of Congress Control Number: 2007939206

Cover Design: eStudioCalamar S.L., F. Steinen-Broo, Girona, Spain

Shen & Tuszynski: Theory & Mathematical Methods for Bioinformatics

First Indian Reprint 2010

ISBN 978-81-8489-625-1

This edition is published by Springer (India) Private Limited, (a part of Springer Science+Business Media), Registered Office: 3rd Floor, Gandharva Mahavidyalaya, 212, Deen Dayal Upadhyaya Marg, New Delhi – 110 002, India

Printed in India by Rakmo Press (P) Ltd., New Delhi.

Preface

Bioinformatics is an interdisciplinary science which involves molecular biology, molecular chemistry, physics, mathematics, computational sciences, etc. Most of the books on biomathematics published within the past ten years have consisted of collections of standard bioinformatics problems and informational methods, and focus mainly on the logistics of implementing and making use of various websites, databases, software packages and serving platforms. While these types of books do introduce some mathematical and computational methods alongside the software packages, they are lacking in a systematic and professional treatment of the mathematics behind these methods.

It is significant in the field of bioinformatics that not only is the amount of data increasing exponentially, but collaboration is also both widening and deepening among biologists, chemists, physicists, mathematicians, and computer scientists. The sheer volume of problems and databases requires researchers to continually develop software packages in order to process the huge amounts of data, utilizing the latest mathematical methods. The intent of this book is to provide a professional and in-depth treatment of the mathematical topics necessary in the study of bioinformatics.

Although there has been great progress in bioinformatics research, many difficult problems are still to be solved. Some of the most well-known include: multiple sequence alignment, prediction of 3D structures and recognition of the eukaryote genes. Since the Human Genome Project (HGP) was developed, the problems of the network structures of the genomes and proteomes, as well as regulation of the cellular systems are of great interest. Although there is still much work to be done before these problems are solved, it is our hope that the key to solving these problems lies in an increased collaboration among the different disciplines to make use of the mathematical methods in bioinformatics.

This book is divided into two parts: the first part introduces the mutation and alignment theory of sequences, which includes the general models and theory to describe the structure of mutation and alignment, the fast algorithms for pairwise and multiple alignment, as well as discussion based on the output given by fast multiple alignment. Part I contains a fairly advanced treatment, and it demonstrates how mathematics may be successfully used in bioinformatics. The success achieved using fast algorithms of multiple alignment illustrates the important role of mathematics.

Part II analyzes the protein structures, which includes the semantic and cluster analysis based on the primary structure and the analysis of the 3D structure for main chains and side chains of proteins. The wiggly angle (dihedral angle) was used when analyzing the configuration of proteins, making the description of the configuration more exact. Analyzing the configuration differs from predicting the secondary or 3D structures. We collect all pockets, grooves, and channels in a protein as configuration characteristics, analyze the structure of these characteristics, and give the algorithms to compute them.

Parts I and II offer independent treatments of biology and mathematics. This division is convenient, as the reader may study both separately. In each part we include results and references from our own research experiences. We propose some novel concepts, for example, the modulus structures, alignment space, semantic analysis for protein sequences, and the geometrical methods to compute configuration characteristics of proteins, etc. Study of these concepts is still in its infancy and so there is much to still be explored. It is our hope that these issues continue to be examined mathematically so that they remain at the forefront, both in mathematics and bioinformatics.

We recognize the importance of considering the computational aspect while introducing mathematical theories. A collection of computational examples have been included in this book so that our theoretical results may be tested, and so that the reader may see the corresponding theories illustrated. Additionally, some of these examples have implications which may be applied to biology directly, and may be downloaded from the website [99] as the data is too large to include in this book. (As an example, when examining the alignment of the HIV gene, the size is $m = 405, n = 10{,}000$ bp.)

An understanding of the fundamentals of probability, statistics, combinatorial graph theory, and molecular biology is assumed, as well as programming ability. In order that the reader may solidify their understanding, problems have been added at the end of each chapter, with hints to help get started, It is our hope that this book will be useful in the field of bioinformatics both as a textbook and as a reference book.

Acknowledgements

This book is the product of a collaboration between the Chinese and Canadian groups. In addition to the two authors, many members of both groups made important contributions. Shiyi Shen's assistants Gang Hu, Kui Wang, Liuhuan Dong, Hua Zhang (doctoral candidate), Zhonghua Wu (doctoral candidate), and Guogen Shan helped him to collect and analyze the data, along with other works in order to create the Chinese edition of this work. Dr. Jishou

Ruan and his graduate students provided the initial translation of the Chinese edition into non-native-speaker English, and later checked the native-speaker English edition to confirm that the meaning of the original was retained. These students are Guangyue Hu, Tuo Zhang, Guoxiang Lu, Jianzhao Gao, Yubing Shang, Shengqi Li, and Hanzhe Chen. The manuscript was rendered into native-speaker English by Jack Tuszynski's assistant, Michelle Hanlon. The authors acknowledge and are grateful to all involved for the hard work that went into this project.

Shen's group would like to thank Chun-Ting Zhang (Fellow of the Chinese Academy of Sciences) who encouraged Shiyi Shen's work on bioinformatics. The group also wishes to thank the Liuhui Center for Applied Mathematics, the Tianjin University and Nankai University joint program, the NSFC (grant NSFC10271061), and the Sciences and Technology program of Tianjin (STPT 04318511-17) for providing funding for this project. We would also like to thank the Chern Institute of Mathematics, which offers much assistance, including computers and offices.

Both authors would like to thank Angela Lahee of Springer-Verlag for her guidance and encouragement.

Tianjin, China
Edmonton, Canada

Shiyi Shen
Jack A. Tuszynski
June 2007

Contents

Outline

This book discusses several important issues including mathematical and computational methods and their applications to bioinformatics.

This book contains two parts. Part I introduces sequence mutations and the theory of data structure used in alignment. Stochastic models of alignment and algebraic theory are introduced as a means to describe data structure. Dynamic programming algorithms and statistical decision algorithms for pairwise sequence alignment, fast alignment algorithms, and the analysis and application of multiple sequence alignment output are also introduced. Part II includes the introduction of frequency analysis, cluster analysis and semantic analysis of the primary sequences, the introduction of the 3D-structure analysis of the main chains and the side chains, and the introduction of configuration analysis.

Many mathematical theories are presented in this monograph, such as stochastic analysis, combinational graph theory, geometry and algebra, informatics, intelligent computation, etc. A large number of algorithms and corresponding software packages make use of these theories, which have been developed to deal with the large amounts of biological and clinical data that play such an important role in the fields of bioinformatics, biomedicine, and so on.

This book has three main goals. The first is to introduce these classical mathematical theories and methods within the context of the current state of mathematics, as they are used in the fields of molecular biology and bioinformatics. The second is to discuss the potential mathematical requirements in the study of molecular biology and bioinformatics, which will drive the development of new theories and methods. Our third goal is to propose a framework within which bioinformatics may be combined with mathematics.

Within each chapter, we have included results and references from our own research experience, to illustrate our points. It is our hope that this book will be useful as a textbook for both undergraduate and graduate students, as well as a reference for teachers or researchers in the field of bioinformatics or mathematics, or those interested in understanding the relation between the two.

Part I

Mutations and Alignments

1

Introduction

1.1 Mutation and Alignment

1.1.1 Classification of Biological Sequences

The term "biological sequence" is generally used to refer to DNA sequences, RNA sequences and protein sequences. In the sense of molecular biology, a biological sequence is composed of many macromolecules, all with specific biological functions under certain conditions. Furthermore, a macromolecule itself can be divided into a large number of functional micromolecules in a certain way. Typically, a DNA sequence (or RNA sequence) is based on four nucleotides, while a protein is based on 20 amino acids. If we consider the nucleotides of a DNA sequence or the amino acids in a protein to be basic units, then a biological sequence is simply a combination of these basic units.

Definitions and Notations for Biological Sequences

There are many ways in which to represent the structure of a biological sequence. The most popular is to describe the primary, secondary, and tertiary (or three-dimensional) structures. For a protein, the primary structure describes the combination of amino acids making up the protein. In a DNA (or RNA) sequence, the primary structure specifies the component nucleotides. We generally use the following description of a biological sequence:

$$A = (a_1, a_2, \cdots, a_{n_a}), \quad B = (b_1, b_2, \cdots, b_{n_b}), \quad C = (c_1, c_2, \cdots, c_{n_c}), \tag{1.1}$$

where the capital letters A, B, C represent the sequences, and a_i, b_i, c_i represent the basic units of the sequence, at positions i, whose elements are obtained from the set $V_q = \{0, 1, \cdots, q-1\}$. Typically, $q = 4$ and $V_4 = \{a, c, g, t\}$ (or $\{a, c, g, u\}$) if A, B and C are DNA (or RNA) sequences. $q = 20$ and V_q is the set of the 20 most common amino acids if A, B, and C are protein sequences. In (1.1), n_a, n_b, n_c are the lengths of sequences A, B, C.

Generally, a multiple sequence (or group of sequences) would be denoted as:

$$\mathcal{A} = \{A_1, A_2, \cdots, A_m\} , \tag{1.2}$$

in which each A_s is a separate sequence defined on V_q, and its complete expression is

$$A_s = (a_{s,1}, a_{s,2}, \cdots, a_{s,n_s}) , \quad s = 1, 2, \cdots, m , \tag{1.3}$$

where n_s is the length of the sequence A_s, and m is the number of sequences in each group.

Classification of Biological Sequences

The primary structure of a biological sequence specifies its component nucleotides or amino acids. The tertiary or three-dimensional structure of a biological sequence describes the three-dimensional arrangement (position coordinates) of the constituent atoms in the molecule. The secondary structure of a biological sequence describes its local properties. For example, the secondary structure of a protein denotes the special structures (motifs) of each protein segment, where a helix, strand or other structure might exist. Super-secondary structure is also frequently used to describe an intermediate state between the secondary structure and the tertiary structure, which consists of some larger compact molecular groups (domains).

Modern molecular biology tells us that DNA (or RNA) sequences and protein sequences are the basic units involved in special biological functions. Their functional characteristics not only involve their primary structure, but also their three-dimensional shapes. For example, the binding pockets of a protein play an important role in controlling its functions. Thus, the shape formed by the amino acid sequences in three-dimensional space can become highly relevant to the clinical treatment involving a serious genetic mutation present in a disease. We will use the configuration of a protein to replace the shape of the protein in three-dimensional space. Since the mutation of a biological sequence changes its configuration and therefore may affect its function, and since alignment is the most popular method for scanning the mutation positions, we begin by discussing the basic characteristics of mutations, as well as alignment methods for biological sequences.

1.1.2 Definition of Mutations and Alignments

The success of cloning demonstrates that a DNA sequence contains the complete information regarding the construction of a life form. However, there are many complex processes that must occur when building structures from DNA to RNA, RNA to protein, protein to organelle, organelle to cell and, finally, from cell to organism. Some of these processes include transcription, translation, and duplication. Within these processes, the mechanisms of recognition, regulation, and control are still not entirely clear. There remain a great number of biological phenomena which cannot be explained at this time. For example, the mechanism of mutation within biological sequences is yet to be fully explored. Mutations can lead to the growth and death of cells, and may also lead to disease. Sequence alignment is an important tool in the analysis of the positions and types of mutations hidden in biological sequences, and allows an exact comparison. The earliest evidence that mutations may cause tumor growth was found in 1983, when it was shown that cancer is the result of uncontrolled cell growth in an organism and that this growth is often due to a mutation. Sequence alignment is also important in that it can be used to research genetic diseases and epidemics. For example, it is possible to determine the origin, variety, variance, diffusion, and development of an epidemic, and then find the viruses and bacteria responsible and appropriate medication. Thus, sequence alignment is very important in the fields of both bioinformatics and biomedicine due to its powerful predictive function. In order to obtain better high-level alignment algorithms, more mathematical theories are required.

Sequence alignment has many applications in bioinformatics besides its direct use in comparing the structures of proteins. Some typical applications are listed as follows:

Gene Positioning and Gene Searching

For a given gene in a certain organism, we must consider whether that same gene (or a similar gene) may be found in another organism. These are the basic problems of gene searching and gene positioning based on the GenBank database. Gene positioning and gene searching is the basis of gene analysis. A better method of gene searching would allow the development of a more accurate alignment algorithm. Many alignment software packages have been developed based on the principles of gene searching and gene positioning, and are used frequently in bioinformatics, such as BLAST [3,67] and FASTA [73–75]. According to some reports, BLAST is visited by more than 100,000 visitors per day, lending credence to the statement that sequence alignment is widely used in the study of bioinformatics.

Gene Repeating and Gene Crossing and Splicing

Gene repeating and gene splicing frequently occur within the same organism's genome, which has become obvious as the genomes of various organisms (including humans) are sequenced. Gene repeating refers to the repetition of a long DNA segment within the same genome. The length of some segments may be in the millions of base pairs (bp), and the number of repetitions may also be in the millions. These repetitions may not be identical, but are typically similar. They may therefore be found through alignment algorithms.

A gene usually is composed of segments of several different genes. These segments are called exons, and the intervals are called introns. When a gene translates into a protein, the introns are cut off, and the exons array in reverse order. This phenomenon is known as gene crossing, and its analysis also depends on alignment algorithms.

Genome Splicing

In the process of gene sequencing, a long sequence of a chromosome is first cut into pieces, the individual DNA segments are then sequenced independently, and the segments will be assembled together. That is, the nucleotides of the entire chromosome are not sequenced simultaneously. To assemble the segments properly, many copies of a genome are cut into random segments, which are then sequenced independently. The common information (found through alignment) is then used to asemble a complete genome.

Other Applications

It is difficult to identify and search the introns and exons of a eukaryote while identifying and searching the regulation genes. As a result, many identification methods have emerged. Among these, alignment-based methods are very significant.

In summary, it is important to note that mutations and alignments are not used simply to study biological evolution but may also be used to study the relationships among genes, proteins and biological macromolecular structure in living systems.

1.1.3 Progress on Alignment Algorithms and Problems to Be Solved

Researchers today are performing sequence alignments and database searches billions of times per day. Due to their importance, alignment methods should be familiar to all biologists and researchers in the field of bioinformatics. Alignment methods must also be continually updated to address new challenges as they arise in the life sciences. We now briefly review the progress made in sequence alignment algorithms, as well as the challenges involved.

Progress in Pairwise Sequence Alignment

The method of dynamic programming-based alignment was first proposed by Needleman and Wunsch [69]. It involves drawing a graph with one sequence written across the page and the other down the left-hand side, scoring matches between the sequences (or penalties for mismatches) and linking with the inserted virtual symbol. The alignment with the highest (or lowest) possible score is defined as the optimal alignment. In 1980s, Smith and Waterman [95] developed an important modification of the dynamic programming-based algorithm, referred to as the local optimal alignment or the Smith–Waterman algorithm. Segments of this local optimal alignment can be determined independently, and a global optimal sequence alignment is obtained by connecting the segments. Although both approaches are dynamic programming-based algorithms, the Smith–Waterman algorithm greatly simplifies the Needleman–Wunsch algorithm.

Following the development of the Smith–Waterman algorithm, sequence alignment became a topic of great importance in bioinformatics. Many papers were published, which not only improved the Smith–Waterman algorithm, but also adapted it to apply to protein sequences. As a result, many types of applied software based on the alignment algorithm were developed, and exist today as powerful bioinformatics tools. The alignment of protein sequences is more complex than the alignment of gene sequences because it is much more difficult to develop scoring matrices (which quantify the similarities between the sequences) for protein sequences. Researchers have proposed many types of scoring systems to produce these scoring matrices, such as the PAM system and the BLOSUM system. For the scoring matrix of the PAM system, the probability of mutations based on the evolution time of the homologous family is determined first, followed by the development of the scoring matrix. The scoring matrix of the BLOSUM system finds the probability of mutations based on the conservative regions of the homologous family, then develops a scoring matrix. Therefore, depending on their requirements, users can choose their scoring matrix based on either the PAM system or the BLOSUM system, then combine the scoring matrix with a dynamic programming algorithm to calculate the highest scoring functions. We will go into more detail later regarding the scoring matrices of the PAM system and the BLOSUM system. The reader is also referred to the literature [24, 40] for more information on this topic.

Besides being adapted for use with proteins, there are many other applications of alignment algorithms. Nowadays, over ten types of software packages exist for the purpose of database searching. Among them, BLAST and FASTA are the most popular of those available as free downloads.

A dynamic programming-based algorithm needs to be aligned along both the vertical axis and the horizontal axis. One must first assign the penalty scores (or matching scores) at the crossed entries intersected by both the vertical axis and horizontal axis, and the links optimized. Therefore, the computational complexity of this algorithm can not be less than $O(n^2)$ (where n is the length of the aligned sequence). For longer sequences, alignment and searching are difficult tasks, although they are easily realized using the methods of computational science. For example, these alignment algorithms cannot currently be performed on a PC if the length of the sequence exceeds 100 kbp. For lengths exceeding 10 Mbp, the alignment algorithms cannot be performed by any computers currently in existence. In 2002, a probability and statistics-based alignment algorithm was proposed by the Nankai University group, called the super pairwise alignment algorithm (SPA algorithm for short) [90]. For homologous sequences, the computational complexity of SPA is only $O(n)$, that is, linearly proportional to the length of sequence. This makes the algorithm run much faster, and makes possible the alignment and searching of super-long sequences.

It may seem as if the problems inherent in the method of pairwise alignment have all been addressed by dynamic programming-based algorithms and statistical decision-based algorithms. However, there is much room for improvement, and for more applications to be developed. For example, the SPA algorithm is a suboptimal algorithm, although it is able to deal with super-long sequences. It still has the potential to be further improved, because its accuracy is lower than that of the optimal solution within 0.1–1%. Additionally, in order to process super-long sequences (i.e., when the length exceeds 100 Mbp), an "anchor" must be incorporated into the algorithm.

Multiple Alignment Algorithms

Compared to pairwise alignment, multiple alignment is much more difficult. The optimal solution of this problem was regarded as one of the unsolved problems of computational biology and bioinformatics during the period between 2000 and 2002. It is sometimes referred to in the literature as the "NP-complete problem" or the "NP-hard problem" [15,36,46,104,106]. The importance of multiple alignment has driven the development of software packages that are able to handle multiple alignment algorithms. These software packages do not search for the optimal solution theoretically; rather, they make comparisons based on some specific indices. In Chap. 5, we will examine the following indices of multiple alignment:

1. **The scope of multiple alignment.** The same type of sequences can be aligned using multiple alignments, i.e., nucleotide sequences are compared with other nucleotide sequences, and amino acid sequences are compared with other amino acid sequences. It is generally expected that there is some further similarity among the sequences to be compared, as multiple alignments are used to compare homologous sequences.
2. **The scale of the multiple alignment.** Let (m, n) denote the length and number of the sequences to be aligned. The maximum size (m, n) permitted by the multiple alignments is then referred to as the scope of the multiple alignment. We will list several software packages concerned with the scale of multiple alignment in Chap. 5.
3. **The computational speed.** The time required to determine the multiple alignment of sequences of scale (m, n) is referred to as the computational speed.
4. **The optimizing indices.** In the literature, some optimizing indices for multiple alignments such as the "SP-scoring function" and "entropy function" are frequently mentioned. In Chap. 5, we introduce two new indices: "similarity" and "rate of insertion." We will discuss these indices in more detail at that time.

We will also introduce the super multiple alignment (SMA) [89] method in Chap. 5. The computational complexity of SMA is just $O(m \cdot n)$, and its scale, speed, and indices are superior to those of HMMER [78]. Based on HIV-1, we compare the SMA method with both SMA and HMMER methods for all indices and show the final results in Tables 5.1 and 5.2. HIV-1 is the known genome of the AIDS virus, and according to GenBank, its scale is $(m, n) = (706, 10{,}000)$. It takes 40 min to perform multiple alignment on a PC using SMA, which is better and much faster than HMMER.

Analysis and Application of Multiple Alignment Results

As the results of a pairwise alignment or multiple alignment are produced, the central problem both in theory and practice becomes the analysis and use of those results. The most pressing problem is the analysis of multiple alignment results, which we explain here.

We have mentioned that the purpose of multiple alignment is to determine the relationships among mutational sequences. Thus, it is first necessary to find common conservation regions, which is simplified by the use of multiple alignment. Correlation among the mutations in different sequences then becomes the key problem. The traditional method of analysis for determining the mutual relationship is called clustering analysis. The most classical method is the "system evolutionary tree" or "minimum distance tree" (which will be discussed in Chap. 6). The "system evolutionary tree" or "minimum distance tree" method is a clustering relation established by the mutation distance determined by different aligned sequences. Thus, its structure is measurable, and partly reflects the degree of being "far" or "near." However, it is not comprehensive, and some useful information is missed if the analysis is based only on the "system evolutionary tree" and the "minimum distance tree."

Currently, in order to analyze the results of multiple alignment, we propose the "multiple sequence mutation network theory" ("mutation network" for short). The theory is that we can replace the "topologically metric structure" by a "modulus structure" based on multiple alignment. This is an effective method to describe the mutations, and is introduced in Chap. 3. Determination of the modulus structure involves a series of algebraic operations. The "mutation network" is a combination of the "topological graph" and "modulus structure," and as such, it comprises a complete description of the alignment. We can then endow the mutation network with operational laws such as decomposition, combination and so on. As a result, the mutation network theory is an important tool for analyzing alignment results.

1.1.4 Mathematical Problems Driven by Alignment and Structural Analysis

With respect to modeling, computation, and analysis, pairwise alignment and multiple alignment can be seen as typical mathematical problems, rather than biological problems. Many mathematical theories and methods are involved, some of which are listed below:

Stochastic Analysis

Stochastic analysis is the basis of the probability and statistics analysis and stochastic processes used in alignment modeling, the creation of fast algorithms, and analysis of the results. Following from the mechanism of mutation, we know that mutations at each site in a sequence obey a Poisson flow. Thus, the structure of different types of mutations should be a renewal process. We can also say that type-I mutated structures obey a Poisson flow based on observations of many sequences.

Stochastic analysis is the basis of mutation structure analysis, and it allows us to understand the overall data character, based on all given sequences. It also plays an important role in the development of the alignment algorithm and computation of the indices (such as complexity estimation, error estimation and the values of optimizing indices).

Algebraic Structure

To describe the structural character of mutation and sequence alignments, in Chap. 3 we propose algebraic operations for the molecular structure. This theory defines the types of various modulus structures, the equivalent representation and the algebraic operations. The algebraic operations of modulus structures are key in the development of fast multiple alignment algorithms and analysis of the results.

Combinational Graph Theory

Combinatorial graph theory is an important tool, both when building fast multiple alignment algorithms and when analyzing the alignment output. Using combinatorial graph theory, cluster analysis is made possible. This is the basis for analyzing multiple alignment outputs to construct the systemic tree and minimum distance tree, and also to construct the mutation networks.

Alignment Space Theory

Alignment space is a data space theory based on mutation error, and is a new concept proposed by the Nankai University group. Alignment space is a nonlinear metric space, and is the theoretical basis for alignment. As a result, its applications will be far-reaching.

The mathematical problems mentioned above are essential for the construction of alignment algorithms and analysis of the output. This combination of mathematics and biology is still in its infancy, and many problems await deeper discussion. There is obviously much space for future development.

1.2 Basic Concepts in Alignment and Mathematical Models

For the sake of simplicity, we confine our discussion to DNA (or RNA) sequences unless otherwise specified. For protein sequences, we need only replace V_4 by V_{20} and then follow a similar argument. Let us begin by introducing the basic problems and mathematical models that will be used in alignment and the analysis of mutations.

1.2.1 Mutation and Alignment

Classification of Mutations

In molecular biology, some small molecules' mutation of its sequence A will cause A to change into sequence B. Sequence B is then referred to as the mutation (sequence) of A. The mutation of DNA sequences can be classified into four types as follows:

Type I: a mutation caused by a nucleotide changing from one into another, i.e., "a" changing into "g."

Type II: a mutation caused by a nucleotide segment permuting its position, i.e., the segment "accgu" permutes into the segment "guacc."

Type III: a mutation caused by inserting a new segment into an existing sequence, i.e., inserting "aa" into the middle position of segment "gguugg" so that it becomes a new segment "gguaaugg."

Type IV: a mutation caused by a segment of nucleotides being deleted from an existing sequence, i.e., deleting the nucleotides "ag" from the segment "acaguua," we are left with the segment "acuua."

Since types I and II do not change the positions of all the nucleotides, these mutations are called substitution mutations. Types III and IV change the positions of all the nucleotides, and so these mutations are called displacement mutations. The basic problem of alignment is to search the mutated sites and decide which regions are conserved and which have been changed. The evolutionary relationship and the changes of both structure and function in the evolution process can then be determined. Alignment is obviously an important tool in this bioinformatics process.

Definition 1. *If sequence B is a mutation sequence of sequence A, and they have the same biological meaning, then they are homologous sequences.*

In sequence analysis, if we know that B is a mutation sequence of A but we do not know whether or not they have the same biological meanings (i.e., whether the differences are caused by a metrical error), then we say that the two sequences are mutually similar. The terms "homologous sequences" and "similar sequences" are frequently used in the discussion of sequence analysis, and note should be taken of the distinction.

Definition of Alignment

To confirm the relationship between the mutations, a common approach is to compare the differences within a family of sequences, which can be viewed as operations in the mathematical sense. This is referred to as sequence alignment or alignment for short. The key to sequence alignment is deciding on the displacement mutation. Let A, B be the two sequences defined in (1.1). Inserting the virtual symbol "$-$" into A, B so that they become two new sequences A', B', the elements of A', B' are then in the range of $V_5 = \{0,1,2,3,4\} = \{a,c,g,t,-\}$ where, V_4 and V_5 are called the quaternary set and the five-element set, respectively.

Definition 2.

1. *Sequence A' is the virtual expansion sequence (expansion, for short) of sequence A, if the rest of A' is just the old sequence A with the insertion symbol "$-$" added.*
2. *Sequences $A' = (a'_1, a'_2, \cdots, a'_{n'_a})$, $B' = (b'_1, b'_2, \cdots, b'_{n'_b})$ are called the expansions of double sequences A, B, if A', B' are the expansions of A, B, respectively.*
3. *Sequence group $\mathcal{A}' = \{A'_s,\ s = 1, 2, \cdots, M\}$ is called the expansion of multiple sequence $\mathcal{A}$, if each A'_s of $\mathcal{A}'$ is the expansion of A_s.*
4. *$\mathcal{A}$ is called the original sequence of $\mathcal{A}'$ if the multiple sequence $\mathcal{A}'$ is an expansion of $\mathcal{A}$. We then denote the sequence in $\mathcal{A}'$ by*

$$A'_s = \left(a'_{s,1}, a'_{s,2}, \cdots, a'_{s,n_s}\right), \quad a'_{s,j} \in V_5 . \tag{1.4}$$

   In the expansion $\mathcal{A}'$ of $\mathcal{A}$, the data value 4 corresponds to the virtual insertion data (or symbol) and the data values 0, 1, 2, 3 correspond to the nucleotide data appearing in sequence $\mathcal{A}$.
5. *Sequence group $\mathcal{A}'$ is called the alignment sequence group (alignment for short) of $\mathcal{A}$ if the lengths of the sequences in $\mathcal{A}'$ are the same, if they are expansions of $\mathcal{A}$ and if $a'_{1,j}, a'_{2,j}, \cdots, a'_{M,j}$ do not simultaneously equal 4 at each position j.*

The definitions of decompression and compression of sequences will be given in Sect. 3.1.

Optimizing Principles of Alignment

The aim of sequence alignment is to find the expansion $\mathcal{A}'$ of a given group $\mathcal{A}$ so that all sequences in $\mathcal{A}'$ have lower "difference" or higher "similarity." In bioinformatics, "difference" is usually quantified using a "penalty matrix" or "scoring matrix."

The basis of the penalty function is the penalty matrix. It stands for the degree of difference of each molecular unit (such as a nucleotide or amino acid) in a biological sequence. It is usually expressed in matrix form as follows:

$$D = (d(a,b))_{a,b \in V_5} \ . \tag{1.5}$$

In bioinformatics, the penalty matrix of DNA sequence alignment is usually fixed by the Hamming matrix or the WT-matrix. The Hamming matrix on V_5 is defined as follows:

$$d_{\mathrm{H}}(a,b) = \begin{cases} 0\,, & \text{if} \quad a = b \in V_5\,, \\ 1\,, & \text{otherwise}\,, \end{cases} \tag{1.6}$$

while the WT-matrix is

$$d_{\mathrm{W}} = [d_{\mathrm{W}}(a,b)]_{a,b \in V_5} = \begin{pmatrix} 0 & 0.77 & 0.45 & 0.77 & 1 \\ 0.77 & 0 & 0.77 & 0.45 & 1 \\ 0.45 & 0.77 & 0 & 0.77 & 1 \\ 0.77 & 0.45 & 0.77 & 0 & 1 \\ 1 & 1 & 1 & 1 & 0 \end{pmatrix} \ . \tag{1.7}$$

The value of the scoring matrix is a maximum if $a = b$. Generally, the scoring matrix is denoted by $G = [g(a,b)]_{a,b \in V_5}$. The entries in the scoring matrix are opposite in value to the corresponding values in the penalty matrix. For example, the scoring matrix of the Hamming matrix is $g(a,b) = 1 - d_{\mathrm{H}}(a,b)$, $a, b \in V_5$.

The penalty matrix (or scoring matrix) is used to optimize the results of the alignment. Thus, both matrices are referred to as the optimizing matrix and are denoted by $W = [w(a,b)]_{a,b \in V_5}$.

The optimizing function measures the optimal value of the two sequences. If

$$A' = (a'_1, a'_2, \cdots, a'_{n'}) \ , \quad B' = (b'_1, b'_2, \cdots, b'_{n'}) \ ,$$

are two sequences on V_5, then the optimizing function is defined as

$$w(A', B') = \sum_{j=1}^{n'} w\left(a'_j, b'_j\right) \ , \tag{1.8}$$

where $w'(A', B') = \frac{1}{n'} w(A', B')$ is the average optimal rate of (A', B'). In future, we will not distinguish between the optimizing function and average optimal rate, and the reader is expected to discern which one is implied according to the context.

The most frequently used optimizing function in multiple alignment is the SP-function. If $\mathcal{A}$ are multiple sequences given by (1.2), and $\mathcal{A}'$ is the expansion of $\mathcal{A}$ given in Definition 2, then the SP-function is defined by:

$$w_{\mathrm{SP}}(\mathcal{A}') = \sum_{s=1}^{m-1} \sum_{t>s} w\left(A'_s, A'_t\right) = \sum_{s=1}^{m-1} \sum_{t>s} \sum_{j=1}^{n'} w\left(a'_{s,j}, a'_{t,j}\right) . \tag{1.9}$$

Then, $w_{\mathrm{SP}}(\mathcal{A}')$ denotes the optimizing function, or optimizing measurement, to align the multiple sequences $\mathcal{A}'$.

Definition 3. *Optimal alignment of multiple sequences is the situation where, for given multiple sequences $\mathcal{A}$, the expansion $\mathcal{A}'_0$ satisfies the optimizing function SP-function in (1.9). Alternatively, if we find the expansion $\mathcal{A}'_0$ of $\mathcal{A}$ such that*

$$\begin{cases} w_{\mathrm{SP}}(\mathcal{A}'_0) = \min\{w_{\mathrm{SP}}(\mathcal{A}') \colon \mathcal{A}' \textit{ is the multiple sequence of } \mathcal{A} \\ \qquad\qquad \textit{while } W \textit{ is the penalty matrix}\}, \\ w_{\mathrm{SP}}(\mathcal{A}'_0) = \max\{w_{\mathrm{SP}}(\mathcal{A}') \colon \mathcal{A}' \textit{ is the multiple sequence of } \mathcal{A} \\ \qquad\qquad \textit{while } W \textit{ is the scoring matrix}\}. \end{cases} \tag{1.10}$$

$\mathcal{A}'_0$ is then called the optimal alignment of $\mathcal{A}$.

The optimal alignment $\mathcal{A}'_0$ determined by the SP-function is called the SP-optimal solution or the SP-function-based optimal solution. The process to find this SP-optimal solution is known as the SP-method.

Pairwise alignment is the simplest case of multiple alignment. We will discuss the optimal criteria for multiple alignment in Chap. 7.

Example 1. We discuss the following sequences:

$$\begin{cases} A = (00132310322), \\ B = (1323210322), \\ A' = (400441323103422), \\ B' = (144323421044322), \\ A'_0 = (001323410322), \\ B'_0 = (441323310322). \end{cases}$$

We can see that B is a mutated sequence of A, and A', B' are expansions of sequences A, B, respectively. Then

$$d_{\mathrm{H}}(A', B') = 12 > d_{\mathrm{H}}(A'_0, B'_0) = 3 .$$

Therefore, the penalty of (A'_0, B'_0) is smaller than that of (A', B').

For the alignment of protein sequences, we usually adopt a scoring matrix. Since the gene sequence of a protein is complex, the PAM and BLOSUM matrices are used to obtain the required scoring matrix. We will discuss this in the corresponding chapters for protein sequence alignment.

We should note that the optimal alignments may not be unique for a given sequence (A, B) under a given optimal matrix W. This is demonstrated in Example 2.

Example 2. If

$$\begin{cases} A = (000132)\,, \\ B = (00132)\,, \end{cases}$$

then

$$\begin{cases} A' = (000132)\,, \\ B' = (400132)\,, \end{cases} \qquad \begin{cases} A_1' = (000132)\,, \\ B_1' = (040132)\,, \end{cases} \qquad \begin{cases} A_2' = (000132) \\ B_2' = (004132) \end{cases}$$

are the alignments of (A, B) with the minimum penalty scores. Since their penalty scores are the same

$$d_{\rm H}(A', B') = d_{\rm H}(A_1', B_1') = d_{\rm H}(A_2', B_2') = 1\ ,$$

it is obvious that we can not find an alignment with a smaller penalty score.

For simplicity, in the following, the term alignment always refers to the minimum penalty-based alignment unless otherwise specified. Obviously, the corresponding conclusions also apply to the maximum scoring-based alignment.

1.3 Dynamic Programming-Based Algorithms for Pairwise Alignment

1.3.1 Introduction to Dynamic Programming-Based Algorithms

Dynamic programming-based algorithms represent the usual method for solving the optimal problem and are broadly applied in many fields. The validity of the dynamic programming-based algorithm depends on whether or not the problem to be solved has an optimal substructure. That is, it depends on whether or not the problem satisfies the optimizing principle. The so-called optimal substructures are those substructures for which every optimal solution of the entire optimal problem (restricted to these substructures) is also an optimal solution. In the optimal problem for the alignment, this substructure exists. For example, let

$$A' = (a_1', a_2', \cdots, a_{n'}')\ , \quad B' = (b_1', b_2', \cdots, b_{n'}')$$

be the optimal alignment of the given sequence pair

$$A = (a_1, a_2, \cdots, a_{n_a}) , \quad B = (b_1, b_2, \cdots, b_{n_b}) .$$

Then the penalty score is defined (see (1.8))

$$w(A', B') = \sum_{i=1}^{n'} w(a'_i, b'_i)$$

as a minimum, where $w(a'_i, b'_i)$ is the penalty score of a'_i and b'_i given by the penalty matrix. Typically, for a fixed position n_0, the penalty is given by

$$w(A', B') = \sum_{i=1}^{n_0} w(a'_i, b'_i) + \sum_{i=n_0+1}^{n'} w(a'_i, b'_i) .$$

Therefore, the pair of subsequences

$$A'_{(0,n_0)} = (a'_1, a'_2, \cdots, a'_{n_0}) , \quad B'_{(0,n_0)} = (b'_1, b'_2, \cdots, b'_{n_0})$$

also represent an optimal alignment. Otherwise, the optimality of A' and B' would not be true. Thus, we may use the dynamic programming-based algorithm to search for the optimal alignment.

Dynamic programming-based algorithms have been successfully used in bioinformatics to perform alignment for a long time. In 1970, Needleman and Wunsch proposed the global alignment algorithm [69]. In 1981, Smith and Waterman gave the mathematical proof [95] and improved the algorithm to apply to local alignment. The time complexity and space complexity of both are $O(n^2)$. Although the time complexity still cannot be reduced, many improved algorithms have been proposed [16–19] that may greatly reduce the space complexity, from $O(n^2)$ to $O(n)$.

1.3.2 The Needleman–Wunsch Algorithm, the Global Alignment Algorithm

The Needleman–Wunsch algorithm is a global alignment algorithm for a pair of sequences. Its procedure is as follows:

Arrange the Two Sequences in a Two-Dimensional Table

If the sequences are

$$A = (a_1, a_2, \cdots, a_n) , \quad B = (b_1, b_2, \cdots, b_m)$$

then the two-dimensional table is constructed as in Table 1.1, in which the element $s(i, j)$ in the two-dimensional table is calculated in step 2.

Table 1.1. Two-dimensional table of sequences A, B

		a_1	a_2	...	a_n
	$s(0,0)$	$s(1,0)$	$s(2,0)$	...	$s(n,0)$
b_1	$s(0,1)$	$s(1,1)$	$s(2,1)$	...	$s(n,1)$
b_2	$s(0,2)$	$s(1,2)$	$s(2,2)$	...	$s(n,2)$
...	...	...	...	...	...
b_m	$s(0,m)$	$s(1,m)$	$s(2,m)$	...	$s(n,m)$

Calculate the Elements $s(i,j)$ of the Two-Dimensional Table

Each element $s(i,j)$ of the two-dimensional table is determined by the three elements; $s(i-1,j-1)$ in the upper left corner, $s(i-1,j)$ on the left side and $s(i,j-1)$ on top. First of all, we determine the marginal scores $s(i,0)$ and $s(0,j)$. For simplicity, we assume that the penalty score of a string of virtual symbols is $d \times |\text{virtual symbol}|$ if the penalty score of a virtual symbol is d, where, $|\text{virtual symbol}|$ is the length of the string of virtual symbols. Thus, $s(0,j) = -j \times d, s(i,0) = -i \times d$, letting $s(0,0) = 0$.

Then, we calculate $s(i,j)$ using the formula:

$$s(i,j) = \max\{s(i-1,j-1) + s(a_i,b_j), s(i-1,j) - d, s(i,j-1) - d\} . \tag{1.11}$$

Figure 1.1 illustrates the computation of $s(i,j)$.

While calculating $s(i,j)$, we should also store the three neighbors of $s(i,j)$, which will be used to produce the backward pathway of the traceback algorithm in the next step.

Traceback Algorithm

The last value $s(n,m)$ is the maximum score of sequences (A,B) after being aligned and $s(n,m)$ is the starting point of the backward pathway. For

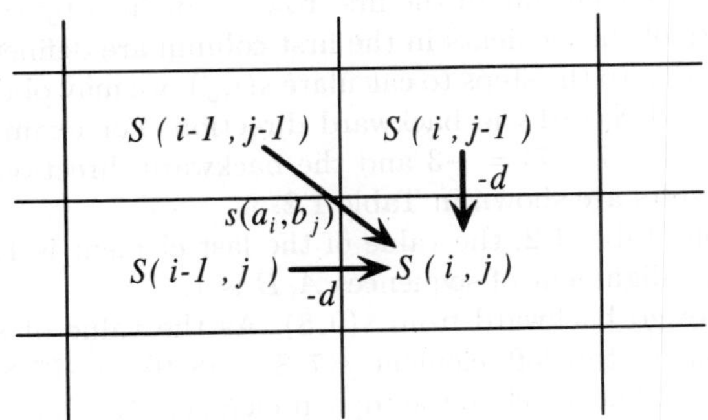

Fig. 1.1. Calculation of $S(i,j)$

each $s(i,j)$, the backward pathway is recorded in the process of calculating Table 1.1. For example, if $s(i,j) = s(i-1,j-1)+s(a_i,b_j)$, then the backward pathway is $(i,j) \longrightarrow (i-1,j-1)$. Proceeding from $s(n,m)$ back to the end $s(0,0)$, we find the backward pathway. We may then recover the alignment of the sequences according to the backward pathway as follows: For the element $s(i,j)$ on the backward pathway:

1. Record the corresponding pairs of nucleic acids a_i, b_i if the backward direction is from a_i, b_i to its upper left corner.
2. Insert a virtual symbol in the vertical sequence and record $(a_i, -)$ if the direction is horizontal.
3. Insert a virtual symbol in the horizontal sequence and record $(-, b_i)$ if the direction is vertical.
4. Finally, we obtain the optimal alignment of the two sequences.

The reader should note that sometimes the backward pathway may not be unique since the backward method itself may not be unique. In fact, it is possible to have several optimal alignments with the same optimal score.

Example 3. Consider the sequences

$$\begin{cases} A = \text{aaattagc}\,, \\ B = \text{gtatatact}\ . \end{cases}$$

We will use the dynamic programming-based algorithm to obtain the alignment. If the penalty score is 5 for matching, -3 for not matching, and -7 for inserting a virtual symbol, that is,

$$s_{a_i,b_j} = \begin{cases} 5\,, & \text{if } a_i = b_j\,, \\ -3\,, & \text{otherwise}\,, \end{cases}$$

and $d = 7$, then:

1. Build a two-dimensional table and calculate the value of each element. The values of the elements in the first row are defined by $s(i,0) = -i \times d$ and the values of the elements in the first column are defined by $s(0,j) = -j \times d$. According to the steps to calculate $s(i,j)$, we may obtain the values of all $s(i,j)$ and record the backward direction. For example, $s(1,1) = \max(0\text{-}3, -7-7, -7-7) = -3$ and the backward direction is $(1,1) \longrightarrow (0,0)$. The results are shown in Table 1.2.
   Following from Table 1.2, the value of the last element is 1, so the score of the optimal alignment of sequences A, B is 1.
2. Traceback: We go backward from $s(9,8)$. As the value of $s(8,9) = 1$ is obtained from its top left element $s(7,8)$, $s(8,9) = s(7,8) + s(c,t) = 4-3 = 1$, we first backtrack to the top left element $s(7,8)$, $(8,9) \longrightarrow (7,8)$. This is repeated until the backtracking path reaches $s(0,0)$.

Table 1.2. Two-dimensional table formed by the sequence A, B

		a	a	a	t	t	a	g	c
	0	–7	–14	–21	–28	–35	–42	–49	–56
g	–7	–3	–10	–17	–24	–31	–38	–37	–44
t	–14	–10	–6	–13	–12	–19	–26	–33	–40
a	–21	–9	–5	–1	–8	–15	–14	–21	–28
t	–28	–16	–12	–8	4	–3	–10	–17	–24
a	–35	–23	–11	–7	–3	1	2	–5	–12
t	–42	–30	–18	–14	–2	2	–2	–1	–8
a	–49	–37	–25	–13	–9	–5	7	0	–4
c	–56	–44	–32	–20	–16	–12	0	4	5
t	–63	–51	–39	–27	–15	–11	–7	–3	1

		a	a	a	t	t	a	g	c
	0	–7	–14	–21	–28	–35	–42	–49	–56
g	–7	**–3**	–10	–17	–24	–31	–38	–37	–44
t	–14	–10	**–6**	–13	–12	–19	–26	–33	–40
a	–21	–9	–5	**–1**	–8	–15	–14	–21	–28
t	–28	–16	–12	–8	**4**	–3	–10	–17	–24
a	–35	–23	–11	–7	**–3**	1	2	–5	–12
t	–42	–30	–18	–14	–2	**2**	–2	–1	–8
a	–49	–37	–25	–13	–9	–5	**7**	0	–4
c	–56	–44	–32	–20	–16	–12	0	**4**	5
t	–63	–51	–39	–27	–15	–11	–7	–3	**1**

Fig. 1.2. Backtracking path

The backward pathway is $(8,9) \longrightarrow (7,8) \longrightarrow (6,7) \longrightarrow (5,6) \longrightarrow (4,5) \longrightarrow (4,4) \longrightarrow (3,3) \longrightarrow (2,2) \longrightarrow (1,1) \longrightarrow (0,0)$.
Figure 1.2 shows a schematic representation of the backtracking path.
According to the backward pathway, we can recover the result of the alignment as follows:

$$A' = (\text{aaat-tagc}) ,$$
$$B' = (\text{gtatatact}) .$$

1.3.3 The Smith–Waterman Algorithm

In bioinformatics, the role played by global alignment is limited because of the complexity of biological sequences. Since global optimizing algorithms always ignore local properties, we sometimes are concerned not with global properties but with whether or not the two sequences have similar conservation regions.

For example, two sequences with low global similarity may have domains which are highly homologous. Therefore, finding alignment algorithms that target these "domains" with a minimal penalty score would be more useful in practice.

The Smith–Waterman algorithm is a type of local alignment algorithm. Although it may simply seem to be an improvement of the dynamic programming-based algorithm which fits local alignment, it is widely useful in bioinformatics. For example, BLAST, a well-known software package, has been developed based on this algorithm. The two aspects of the Smith–Waterman algorithm which may still be improved are stated as follows.

Calculation of the Values in a Two-Dimensional Table

The Smith–Waterman algorithm adds a 0 while calculating $s(i,j)$. Thus, a negative score will never occur in the Smith–Waterman algorithm. The advantage of this will become clear when constructing the backward pathway.

$$s(i,j) = \max \begin{cases} 0\,, \\ s(i-1,j-1) + s(x_i, y_j)\,, \\ s(i-1,j) - d\,, \\ s(i,j-1) - d\,. \end{cases} \tag{1.12}$$

Traceback Algorithm

The start and end points of the backtrace of the Smith–Waterman algorithm are different from the global alignment algorithm. The starting point can be chosen arbitrarily in theory and we usually choose elements with a higher score. The end point is the first element with the value 0 in the process of backtrace. If the purpose of alignment is to find the optimal alignment of two sequences, the Smith–Waterman algorithm should backdate from the element with the maximum score and should not end at the first element where the value 0 appears rather than $s(0,0)$. The starting point with the maximal score guarantees the maximal score of local sequence alignment, and the end point is the first element with a value 0, ensuring that segment is not exceeded. At this time, the segment corresponding to the backward pathway is the segment with maximum penalty.

We use the same example

$$\begin{cases} A = \text{aaattagc}\,, \\ B = \text{gtatatact}\,, \end{cases}$$

to find the optimal alignment subsequences. The penalty score is 5 for matching, −3 for mismatching and $d = 3$. The construction of the two-dimensional table and calculation of the alignment of two sequences is given as shown in Fig. 1.3.

		a	a	a	t	t	a	g	c
	0	0	0	0	0	0	0	0	0
g	0	0	0	0	0	0	0	5	0
t	0	0	**0**	0	5	5	0	0	0
a	0	5	5	↖ **5**	0	2	10	3	0
t	0	0	2	2	↖ **10**	5	3	7	0
a	0	5	5	7	↑ **3**	7	10	3	4
t	0	0	2	2	12	↖ **8**	4	7	0
a	0	5	5	7	5	9	↖ **13**	6	4
c	0	0	2	2	4	2	6	10	11
t	0	0	0	0	7	9	2	3	7

Fig. 1.3. Backtracking path

The maximum score in this table is 13. Thus, we begin at the corresponding element $s(6,7)$ and stop at $s(2,2)$ which is the first element with a value 0. We then obtain the backward pathway as follows:

$$(8,9) \longrightarrow (6,7) \longrightarrow (5,6) \longrightarrow (4,5) \longrightarrow (4,4) \longrightarrow (3,3) \longrightarrow (2,2) .$$

According to this backward pathway, we obtain the following alignment of segments with maximal penalty score:

at-ta
atata

Discussion of Dynamic Programming-Based Algorithms

Some notes about the dynamic programming-based algorithms are given below:

1. Different penalty matrices produce different alignments. So the choice of an appropriate penalty matrix is very important to the dynamic programming algorithm. Some penalty matrices are appropriate to global alignment and some to local alignment. In the extreme case where there is no negative penalty and the virtual symbol also gives no penalty (choosing the Hamming penalty matrix), the result of the local alignment algorithm is almost the same as that of the global alignment algorithm.
2. For the pair of sequences whose lengths are n and m, respectively, we find that the space complexity and time complexity are both $O(nm)$. If the lengths of the two sequences are approximately equal, then these complexities are close to $O(n^2)$. Therefore, the complexity of computation is permissible if the sequences are shorter. However, for longer sequences, such

as genome sequences, this computational complexity makes the problem computationally intractble for present-day computers. If many pairwise alignments must be performed while doing multiple alignments, the scope of applications of the dynamic programming-based algorithm is restricted. The fact that the time complexity would not be reduced is a huge disadvantage of the dynamic programming-based algorithm, although many improved algorithms [16–19] may reduce the space complexity to as low as $O(n)$.

3. One of the purposes of this book is to show how to create an alignment algorithm using stochastic analysis, so that the time complexity may be reduced to as little as $O(n)$ for pairwise alignment. Therefore, we will not discuss dynamic programming-based algorithms further. The reader is referred to the relevant literature for further insights.

1.4 Other Notations

There are some notations that arise frequently in this book when discussing alignment, and we will address them specifically now.

1.4.1 Correlation Functions of Local Sequences

Local Sequences

Let A, B, C be the three sequences given in (1.1), and let n_a, n_b, n_c be their lengths, respectively. Let $N_a = \{1, 2, \cdots, n_a\}$ be a set of integers which is the set of the subscripts of A. The subscript $i \in N_a$ of N_a is a subscript (or position for short) of sequence A. If the subset of N_a is represented by α, β, then

$$\alpha = \{i_1, i_2, \cdots, i_k\} \tag{1.13}$$

is a subset of N_a arranged from the largest to the smallest number, $1 \leq i_1 < i_2 < \cdots < i_k \leq N$. Then,

$$a_\alpha = \{a_{i_1}, a_{i_2}, \cdots, a_{i_k}\} \tag{1.14}$$

is a subsequence of A in the region α.

If N_a and α are both given, then we denote $\alpha^c = N_a - \alpha$ as the complement of set α and α^c as the subset of set N_a. Thus, a_{α^c} is a subsequence of A and $A = (a_\alpha, a_{\alpha^c})$ is referred to as the decomposition of A. In the special case, let $\alpha = [i, j]$, or $(i, j), [i, j), (i, j]$, in which case,

$$[i, j] = (i, i+1, \cdots, j), \quad (i, j) = (i+1, i+1, \cdots, j-1),$$

$$[i, j) = (i, i+1, \cdots, j-1), \quad (i, j] = (i+1, i+1, \cdots, j).$$

These are the closed interval, open interval or half-open interval of N_a, respectively. The corresponding vectors are then

$$a_{[i,j]} = (a_i, a_{i+1}, \cdots, a_j), \quad a_{(i,j)} = (a_{i+1}, a_{i+1}, \cdots, a_{j-1}),$$

$$a_{[i,j)} = (a_i, a_{i+1}, \cdots, a_{j-1}), \quad a_{(i,j]} = (a_{i+1}, a_{i+1}, \cdots, a_j). \qquad (1.15)$$

The subsequences of (1.15) are referred to as the local vectors defined on the intervals $[i,j], (i,j)$ or $[i,j), (i,j]$. We use the following notation for the local vectors.

$$\bar{a}_i = \boldsymbol{a}_i = (a_{i+1}, a_{i+2}, \cdots, a_{i+k}), \qquad (1.16)$$

where i denotes the first position of vector $\bar{a}_i$ and k denotes the length of $\bar{a}_i$. For simplicity, we consider these three symbols $\bar{a}, \boldsymbol{a}$ and $a^{(k)}$ to be equivalent. The length of vectors $\bar{a}$ and $\boldsymbol{a}$ is always k unless otherwise specified.

Correlation Functions

The local correlation function of sequences A, B based on a penalty matrix w is defined as follows:

$$w(A,B;i,j,n) = \sum_{k=1}^{n} w(a_{i+k}, b_{j+k}), \quad i+n \le n_a, \quad j+n \le n_b. \qquad (1.17)$$

In the case $B = A$, the local correlation function in (1.17) becomes the local autocorrelation function of A.

1.4.2 Pairwise Alignment Matrices Among Multiple Sequences

We have mentioned above that the minimum penalty algorithm for multiple sequence alignment is an unsolved problem in bioinformatics, although the fast algorithm of pairwise alignment has been determined. Therefore, we discuss the pairwise alignment within multiple sequences before moving on to multiple alignments. Let

$$\mathcal{B} = \{B_{s,t},\ s,t = 1,2,\cdots,M\} = (B_{s,t})_{s,t=1,2,\cdots,m} \qquad (1.18)$$

be the sequence matrix, in which, each $B_{s,t} = (b_{s,t;1}, b_{s,t;2}, \cdots, b_{s,t;n_{s,t}})$ is a five-dimensional vector. That is, for any s,t,j, there is a $b_{s,t;j} \in V_5$.

Definition 4.

1. *The matrices $\mathcal{B}$ in (1.18) are referred to as the pairwise expansion of multiple sequence $\mathcal{A}$, if $(B_{s,t}, B_{t,s})$ is the expansion of the sequence pair (A_s, A_t) for each s,t.*

2. *Matrix* $\mathcal{B}^o = (B^o_{s,t})_{s,t=1,2,\cdots,m}$ *of (1.18) is referred to as the pairwise minimum penalty alignment matrix for the multiple sequence* $\mathcal{A}$*, if* $\mathcal{B}^o$ *is the pairwise expansion matrix of* $\mathcal{A}$ *and* $(B^o_{s,t}, B^o_{t,s})$ *is the minimum penalty alignment sequence of* (A_s, A_t) *for all* s, t*. Here,*

$$w(B^o_{s,t}, B^o_{t,s}) = \min\{w(B_s; B_t) \colon (B_s, B_t) \text{ is the alignment of } (A_s, A_t)\} \tag{1.19}$$

is tenable and

$$w(B_{s,t}, B_{t,s}) = \sum_{j=1}^{N'_{s,t}} w(b_{s,t;j}, b_{t,s;j}) \ . \tag{1.20}$$

Following the fast pairwise alignment algorithm, we can determine the pairwise minimum penalty alignment matrices $\mathcal{B}^o$ for the multiple sequence $\mathcal{A}$. One of the purposes of this book is to demonstrate how to use $\mathcal{B}^o$ to construct the minimum penalty alignment of $\mathcal{A}$.

1.5 Remarks

The mathematical methods introduced in this book are suited for bioinformatics and computational biology students. Additionally, this book also refers to some important databases, such as GenBank [10], PDB [13], PDB-Select [41], Swiss-Prot [8, 33]. The reader is assumed to be familiar with these databases in the following aspects:

1. Know the Web sites that provide the databases and know the updating situation of these databases.
2. Know the content of these databases. For example, representations of primary structure, secondary structure, and 3D structure may be found in the PDB database; representations of genes, introns, and exons are in the GenBank, etc.
3. Know how to obtain the data required when using computers for analysis, e.g., know how to use the computer to obtain the primary structure, secondary structure, and space coordinates of a given atom based on the PDB database.
4. Know how to use the corresponding databases for other requirements [99].

Besides databases, the reader should also know how to use some popular software packages, for example, BLAST [3], FASTA [73–75] and other specialized software packages (such as the software package for multiple alignment) that will be referred to later. For visual software, we recommend RASWIN [83], which may be used to find the 3D configurations of proteins. Its function is superior to other packages in some aspects such as rotating, moving and marking objects. It is available as a free download from its Web site [76].

1.6 Exercises, Analyses, and Computation

Exercise 1. For the RNA sequences E.co and B.st given in Table 4.5, use the dynamic programming-based algorithm to compute their minimal penalty alignment based on the Hamming penalty matrix.

Exercise 2. For the RNA sequences Mc. vanniel and Mb. tautotr given in Table 4.5, use the dynamic programming-based algorithm to determine the optimal alignment sequences based on the following requirements:

1. Compute the minimal penalty alignment based on the Hamming penalty matrix and the WT-penalty matrix, respectively.
2. If $d_{\mathrm{H}}(a,b)$, $a,b \in V_5$ is the Hamming penalty matrix, then $g_{\mathrm{H}}(a,b) = 1 - d_{\mathrm{H}}(a,b)$, $a,b \in V_5$ is the corresponding scoring matrix. Compute the maximal score alignment.
3. Compare the computational results of the minimal penalty alignment and the maximal score alignment.

Exercise 3. Continuing from Exercise 2, for RNA sequences Mc. vanniel and Mb.tautotr, compute the optimal alignment by using the dynamic programming-based algorithm, based on the three criteria of the Hamming penalty matrix, the WT-penalty matrix and the Hamming scoring matrix, respectively. Compare the corresponding results.

Exercise 4. For an arbitrary pair of sequences with different lengths, compute the optimal alignment using the dynamic programming-based algorithm, based on the penalty matrix and the scoring matrix, respectively.

Exercise 5. Test your dynamic programming-based algorithm for the optimal pairwise alignment according to the following indices:

1. For data with length 1 kbp, such as the sequences Mc.vanniel and Mb.tautotr given in Table 4.5, visually check whether they arrive at the target.
2. For data with length 1–100 kbp (such as sequences 1–10 given in Table 4.4), use your dynamic programming-based algorithm to test the relationship between the length of the sequence and the CPU time required for computation.
3. Analyze the relationship between similarity and the CPU time required for computation.

Hint

1. Data for the two sequences Mc.vanniel and Mb.tautotr given in Table 4.5 may be downloaded from the Web site [99].

2. The dynamic programming algorithm for the optimal alignment of sequence pairs should be done following Sect. 1.3. If this proves too difficult, the algorithm from our Web site [99] may be downloaded, and your program may be based on it.

2

Stochastic Models of Mutations and Structural Analysis

The mathematical methods used to study mutation and alignment fall mainly into three groups – stochastic analysis, modulus structure and combinational graph theory. In this chapter we introduce stochastic analysis and show how this theory is used to understand the characteristics of biological structures. Stochastic analysis of mutations includes three aspects: the randomness of sequence structure, the randomness with which mutation flows happen, and the randomness with which each type of mutation occurs.

2.1 Stochastic Sequences and Independent Sequence Pairs

At first glance, a DNA sequence may seem disorderly and unsystematic, and the nucleotides at each position (or a group of positions) are not fixed. That is to say that a biological sequence is a stochastic sequence. Statistically, we may find that the frequency of observing small molecules or segments changes based on a different benchmark dataset of biological sequences. Therefore, we may use stochastic models to describe biological sequences.

2.1.1 Definitions and Notations of Stochastic Sequences

Random Vectors, Stochastic Sequences, and Stochastic Processes

Random variables, random vectors, stochastic sequences and stochastic processes are the fundamental concepts of probability theory and stochastic processes theory. For convenience, we briefly summarize their definitions, notations, and properties as follows.

$$\tilde{\xi} = (\xi_1, \xi_2, \cdots, \xi_{n_\xi}), \quad \tilde{\eta} = (\eta_1, \eta_2, \cdots, \eta_{n_\eta}) \tag{2.1}$$

or

$$A^* = \left(a_1^*, a_2^*, \cdots, a_{n_a}^*\right), \quad B^* = \left(b_1^*, b_2^*, \cdots, b_{n_b}^*\right)$$

denote stochastic sequences, in which ξ_j, η_j or a_j^*, b_j^* are random variables whose range is the integral set V_4, and n_ξ, n_η, n_a, n_b are the lengths of the corresponding sequences. In this book, these two kinds of symbols will be used interchangeably.

In (2.1), $N_\xi = \{1, 2, \cdots, n_\xi\}$, $N_\eta = \{1, 2, \cdots, n_\eta\}$ are sets of subscripts or positions of sequences $\tilde{\xi}$ and $\tilde{\eta}$, respectively. If n_ξ, n_η are large, we consider $N_\xi = N_\eta = \{1, 2, 3, \cdots\}$ to be the set of all natural numbers. Then $\tilde{\xi}$ and $\tilde{\eta}$ given in (2.1) are considered to be infinite stochastic sequences and we rewrite them as follows:

$$\tilde{\xi} = (\xi_1, \xi_2, \xi_3, \cdots), \quad \tilde{\eta} = (\eta_1, \eta_2, \eta_3, \cdots) . \tag{2.2}$$

Biological sequences are generally very long. Thus, the symbols defined in (2.1) and (2.2) are frequently considered to be equivalent. In addition to the stochastic sequence notation of $\tilde{\xi}$ in (2.2), we use the symbols $\{\xi_1, \xi_2, \xi_3, \cdots\}$, and therefore the sequence $\tilde{\xi}$ can be regarded as an ordered set.

Frequently, the DNA sequence to be processed is very long, e.g., the length of the human genome is 3.2×10^9 bp. The sequence $\tilde{\xi}$ can then be regarded as an isochronously sampled sequence drawn from a continuous stochastic process with the same identified distribution. Therefore, the mother stochastic process $\tilde{\xi}$ is denoted by

$$\tilde{\xi} = \{\xi_t, t \in (0, \infty)\} . \tag{2.3}$$

If ξ_i is a component of the sequence in (2.2), then the corresponding random variable in (2.3) is $\xi_t = \xi_{i/n_0}$. In this book, we will use (2.1)–(2.3) interchangeably. We can consider the other related notations in a similar fashion, and hence we do not repeat them here.

The Family of Probability Distributions of a Stochastic Process

The structural characteristics of a stochastic sequence are determined by its family of probability distributions. For an arbitrary k, $\xi^{(k)} = (\xi_1, \xi_2, \cdots, \xi_k)$ is a random vector that is a segment of $\tilde{\xi}$. For any given vector $a^{(k)} = (a_1, a_2, \cdots, a_k), a_i \in V_4$, the probability of $\xi^{(k)} = a^{(k)}$ is defined as: $p_{\xi^{(k)}}(a^{(k)}) = P_r\{\xi^{(k)} = a^{(k)}\}$. We denote the probability of $\xi^{(k)}$ by

$$P_{\xi^{(k)}} = \left\{p_{\xi^{(k)}}\left(a^{(k)}\right) : a^{(k)} \in V_4^{(k)}\right\} \tag{2.4}$$

in which $V_4^{(k)}$ is the set of $a^{(k)}$. Following from probability theory, we have $p_{\xi^{(k)}}(a^{(k)}) \geq 0$ for all $a^{(k)} \in V_4^{(k)}$ and $\sum_{a^{(k)} \in V_4^{(k)}} p_{\xi^{(k)}}(a^{(k)}) = 1$.

As $k = 1, 2, 3, \cdots$ is changing, we can obtain the probability distribution family

$$\mathcal{P}_{\tilde{\xi}} = \{P_{\xi^{(k)}}, k = 1, 2, 3, \cdots\} \tag{2.5}$$

which is referred to as the probability distribution family determined by the stochastic sequence $\tilde{\xi}$.

It is a necessary condition that the probabilities within the probability distribution family determined by a stochastic sequence must satisfy compatibility. That is, the probability distribution $P_{\xi^{(k)}}$ is a marginal distribution $P_{\xi^{(k')}}$ for any positive integer $k' > k$. It is equivalent to

$$p_{\xi^{(k)}}\left(a^{(k)}\right) = \sum_{b^{(k'-k)} \in V_4^{(k'-k)}} p_{\xi^{(k')}}\left(a^{(k)}, b^{(k'-k)}\right) , \tag{2.6}$$

for all $a^{(k)} \in V_4^{(k)}$, in which $(a^{(k)}, b^{(k'-k)}) = a^{(k')} \in V_4(k')$.

Based on the famous Kolmogorov theorem, we may construct a stochastic sequence derived from given probability distributions $\mathcal{P}$, with compatibility conditions. Moreover, we can build different kinds of stochastic sequences or stochastic processes using this Kolmogorov theorem. The most typical stochastic sequences or stochastic processes are: independently and identically distributed (i.i.d.) sequences, Markovian sequences, Poisson processes and renewal processes. In this book, we begin by discussing i.i.d. sequences. A DNA sequence is frequently approximated as an i.i.d. sequence, if the sequence is sufficiently long.

2.1.2 Independently and Identically Distributed Sequences

Definition of i.i.d Sequences

Let $\tilde{\xi}$ be a stochastic sequence. Then $\tilde{\xi}$ is referred to as:

1. An independent sequence if the probability distribution of the stochastic vector $\xi^{(k)}$ satisfies

$$p_{\xi^{(k)}}\left(a^{(k)}\right) = P_r\left\{\xi^{(k)} = a^{(k)}\right\} = \prod_{i=1}^{k} p_i(a_i) , \tag{2.7}$$

where $p_i(a) = P_r\{\xi_i = a\}$.
2. An identically distributed sequence if $p_i(a) = p(a)$, $a \in V_q$ is the same for all k.
3. An i.i.d. sequence if it is both an independent sequen[illegible] an identically distributed sequence, in which $p(a) = P_r\{\xi_i = a\}$ is [illegible]robability distribution of ξ_i. For independently and identically dist[illegible]ted sequences, (2.7) becomes

$$p\left(a^{(k)}\right) = P_r\left\{\xi_i^{(k)} = a^{(k)}\right\} \tag{2.8}$$

and $p(a)$, $a \in V_4$ is called the generation distribution of ξ. If $p(a) = 1/q$ for all $a \in V_q$, then $p(a)$, $a \in V_q$ is a uniform distribution.

Definition 5. *If $\tilde{\xi}$ is an i.i.d. sequence and its generating distribution $p(a)$ is a uniform distribution, then $\tilde{\xi}$ is referred to as a perfectly stochastic sequence.*

Therefore, a stochastic sequence $\tilde{\xi}$ is a perfectly stochastic sequence if and only if $p_{\xi^{(k)}}(a^{(k)}) = \frac{1}{4^k}$ holds for all $k = 1, 2, 3, \cdots$, and $a^{(k)} \in V_4^{(k)}$.

Penalty Functions

For a given penalty matrix $W = w(a, b)$, $a, b \in V_4$ defined in (1.5), we define two indices for this matrix as follows:

$$\begin{cases} w_0 = \dfrac{1}{4} \sum_{a \in V_4} w(a, a) , \\ w_1 = \dfrac{1}{16} \sum_{a,b \in V_4} w(a, b) . \end{cases} \tag{2.9}$$

Definition 6. *A penalty matrix $\boldsymbol{W}$ is called strongly symmetric if $w(a, b) = w(b, a)$, and $w(a, a) = 0$ holds for any $a, b \in V_4$, if every row is the permutation of another row, and if each column is the permutation of another column.*

Obviously, the Hamming matrix and the WT-matrix are both strongly symmetric, with average penalties defined as follows:

$$\begin{cases} (w_0, w_1) = (0, 3/4) , & \text{if } \boldsymbol{W} \text{ is the Hamming matrix} , \\ (w_0, w_1) = (0, 0.4975) , & \text{if } \boldsymbol{W} \text{ is the WT-matrix} . \end{cases} \tag{2.10}$$

Theorem 1. *If the penalty matrix W is strongly symmetric, then $w_0 = 0$ and $P_r\{w(a, \zeta) = c\} = \rho(c)$ for any uniformly random variable ζ defined on V_4.*

Proof. $w_0 = 0$ follows directly because the matrix W is strongly symmetric. To prove that $P_r\{w(a, \zeta) = c\} = \rho(c)$, we consider the number of elements of

$$D_{a,c} = \{b \in V_4 : w(a, b) = c\} .$$

It is invariant for different a due to symmetry. Consequently,

$$P_r\{w(a, \zeta) = c\} = \frac{\| D_{a,c} \|}{q}$$

is constant if ζ is a uniform distribution on V_4.

Locally Random Correlation Functions

Let $\tilde{\xi}, \tilde{\eta}$ be two stochastic sequences and let

$$\xi_i^{(n)} = (\xi_{i+1}, \xi_{i+2}, \cdots , \xi_{i+n}) , \quad \eta_j^{(n)} = (\eta_{j+1}, \eta_{j+2}, \cdots , \eta_{j+n}) .$$

be their local vectors. Their local correlation function is then defined as follows:

$$w\left(\tilde{\xi};\tilde{\eta};i,j;n\right) = w\left(\xi_i^{(n)},\xi_j^{(n)}\right) = \sum_{k=1}^{n} w\left(\xi_{i+k},\xi_{j+k}\right) . \quad (2.11)$$

Since this $w(\tilde{\xi},\tilde{\eta};i,j;n)$ is a random variable, it is called the locally random correlation function. Specifically, it is known as the locally random autocorrelation function if $\tilde{\xi}=\tilde{\eta}$.

If $\tilde{\xi}$ is a perfectly stochastic sequence, then its locally random autocorrelation function satisfies:

$$E\{w(\tilde{\xi};i,j;n)\} = E\{w(\xi_i,\xi_j)\} = \begin{cases} \mu_0 , & \text{if } i=j , \\ \mu_1 , & \text{otherwise} . \end{cases} \quad (2.12)$$

Equation (2.12) is implied by the definitions of perfect sequences and penalty functions, so we omit the proof. An important characteristic of a perfect sequence is that expressed by (2.12).

2.1.3 Independent Stochastic Sequence Pairs

Since independent stochastic sequence pairs are defined with respect to both mutation and the structural analysis of alignment, we will discuss them in more detail here.

Definition 7. *Let* $\tilde{\xi}=(\xi_1,\xi_2,\cdots,\xi_{n_\xi})$, $\tilde{\eta}=(\eta_1,\eta_2,\cdots,\eta_{n_\eta})$ *be two sequences defined on* V_4. *These are referred to as independent stochastic sequence pairs, if the following conditions are satisfied:*

1. $\tilde{\xi}$ *and* $\tilde{\eta}$ *are two perfectly stochastic sequences.*
2. $\{(\xi_1,\eta_1),(\xi_2,\eta_2),(\xi_3,\eta_3),\cdots\}$ *is an i.i.d. two-dimensional sequence.*
3. *There is a fixed* $\epsilon>0$ *such that for every pair* (ξ_j,η_j), *the joint probability distribution satisfies the following:*

$$P_r\{\xi_j \neq \eta_j\} = \epsilon , \quad or \quad P_r\{\xi_j = \eta_j\} = 1-\epsilon . \quad (2.13)$$

If the two sequences are both mutated sequences resulting from type-I mutation, then $(\tilde{\eta},\tilde{\xi})$ is an independent sequence pair. Specifically, if $\epsilon=0$, then $\tilde{\eta}$ and $\tilde{\xi}$ are two identical sequences, and $(\tilde{\eta},\tilde{\xi})$ is an independent sequence pair. The following are basic characteristics of independent sequence pairs, and are defined in the context of structure analysis.

Theorem 2. *If* $(\tilde{\xi},\tilde{\eta})$ *is an independent sequence pair and* $w(a,b)$ *is a function defined on* V_4^2 *which satisfies the strong symmetry condition, then*

$$\zeta_k = w(\xi_{i+k},\eta_{j+k}) , \quad k=0,1,2,\cdots \quad (2.14)$$

is an i.i.d. sequence for all $i,j\geq 0$.

Proof. If $i = j$, the proof can be deduced from condition 2 of Definition 3 and the characteristic of independent random variables (a function of an independent random variable is itself an independent random variable).

Here we discuss the special condition of $i \neq j$.

We begin with the probability distribution $\zeta_k = w(\xi_{i+k}, \eta_{j+k})$. Let

$$\begin{aligned} p_k(z) &= P_r\{\zeta_k = z\} = P_r\{w(\xi_{i+k}, \eta_{j+k}) = z\} \\ &= \sum_{a\in V_4} P_r\{\xi_{i+k} = a\} P_r\{w(\xi_{i+k}, \eta_{j+k}) = z | \xi_{i+k} = a\} \\ &= \frac{1}{4} \sum_{a\in V_4} P_r\{w(a, \eta_{j+k}) = z\} = \frac{1}{4} \sum_{a\in V_4} \rho(z) = \rho(z) . \end{aligned} \quad (2.15)$$

The fourth equation is deduced from the strongly symmetric character of $w(a, b)$ and Theorem 1. The third equation is deduced by $i \neq j$ and ξ_{i+k} is independent of η_{j+k}. Consequently, the probability distribution of ζ_k does not depend on k.

Now, we prove the independence of the sequence

$$\zeta_k = w(\xi_{i+k}, \eta_{j+k}) ,$$

where $k = 0, 1, 2, \cdots$. To do this, we calculate the probability distribution of $\bar{\zeta} = (\zeta_1, \zeta_2, \cdots, \zeta_n)$. For simplicity, we only discuss the situation where $i = 1, j = 2, n = 4$, so that

$$\zeta_1 = w(\xi_1, \eta_2) , \quad \zeta_2 = w(\xi_2, \eta_3) , \quad \zeta_3 = w(\xi_3, \eta_4) , \quad \zeta_4 = w(\xi_4, \eta_5) .$$

Note that the random variables ξ_1, $\{\xi_2, \xi_3, \xi_4, \eta_2, \eta_3, \eta_4\}$, η_5 are independent. We denote

$$p(\bar{z}) = P_r\{\bar{\zeta} = \bar{z}\} = P_r\{\bar{w}^* = \bar{z}\} ,$$

in which $\bar{w}^* = (w_1^*, w_2^*, w_3^*, w_4^*)$ and $w_j^* = w(\xi_j, \eta_{j+1}), j = 1, 2, 3, 4$. If we denote

$$\bar{\xi} = (\xi_2, \xi_3, \xi_4) , \quad \bar{\eta} = (\eta_2, \eta_3, \eta_4) , \quad \bar{d} = (a, b, c) \in V_4 ,$$

then

$$p(\bar{z}) = \sum_{\bar{d}\in \mathbf{Z}_4^3} P_r\{\bar{\xi} = \bar{d}\} P_r\{\bar{w}^* = \bar{z} | \bar{\xi} = \bar{d}\} = \frac{1}{64} \sum_{\bar{d}\in \mathbf{Z}_4^3} P_r\{\bar{w}^* = \bar{z} | \bar{\xi} = \bar{d}\}$$

$$= \frac{1}{64} \sum_{\bar{d}\in \mathbf{Z}_4^3} P_r\{w(\xi_1, \eta_2) = z_1, w(a, \eta_3) = z_2, w(b, \eta_4) = z_3, w(c, \eta_5) = z_4 | \bar{\xi} = \bar{d}\}$$

$$\begin{aligned} = \frac{1}{64} \sum_{c\in V_4} \sum_{b\in V_4} \sum_{a\in V_4} [&P_r\{w(\xi_1, \eta_2) = z_1 | \xi_2 = a\} P_r\{w(a, \eta_3) = z_2 | \xi_3 = b\} \\ &P_r\{w(b, \eta_4) = z_3 | \xi_4 = c\} P_r\{w(c, \eta_5) = z_4\}] . \end{aligned} \quad (2.16)$$

In (2.16), the last equation is obtained from the independence of η_5 and (ξ_2, ξ_3, ξ_4). The strong symmetry of the $w(a, b)$ function turns (2.16) into

$$\begin{aligned}
p(\bar{z}) &= \frac{\rho(z_4)}{64} \sum_{\bar{d} \in \mathbf{Z}_4^3} [P_r\{w(\xi_1, \eta_2) = z_1, w(a, \eta_3) = z_2, w(b, \eta_4) = z_3 | \bar{\xi} = \bar{d}\}] \\
&\overset{\text{i}}{=} \frac{\rho(z_4)}{16} \sum_{\bar{a}, b \in V_4} [P_r\{w(\xi_1, \eta_2) = z_1, w(a, \eta_3) = z_2, \\
&\qquad w(b, \eta_4) = z_3 | \xi_2 = a, \xi_3 = b\}] \\
&\overset{\text{ii}}{=} \frac{\rho(z_4)}{16} \sum_{\bar{a}, b \in V_4} [P_r\{w(\xi_1, \eta_2) = z_1, w(a, \eta_3) = z_2 | \xi_2 = a, \xi_3 = b\} \\
&\qquad P_r\{w(b, \eta_4) = z_3\}] \\
&\overset{\text{iii}}{=} \frac{\rho_{z_3} \rho(z_4)}{16} \sum_{\bar{a}, b \in V_4} P_r\{w(\xi_1, \eta_2) = z_1, w(a, \eta_3) = z_2 | \xi_2 = a, \xi_3 = b\} \\
&\overset{\text{iv}}{=} \frac{\rho_{z_3} \rho(z_4)}{4} \sum_{\bar{a} \in V_4} P_r\{w(\xi_1, \eta_2) = z_1, w(a, \eta_3) = z_2 | \xi_2 = a\} \\
&\overset{\text{v}}{=} \frac{\rho_{z_3} \rho(z_4)}{4} \sum_{\bar{a} \in V_4} P_r\{w(\xi_1, \eta_2) = z_1 | \xi_2 = a\} p_r\{w(a, \eta_3) = z_2\} \\
&\overset{\text{vi}}{=} \frac{\rho_{z_2} \rho_{z_3} \rho(z_4)}{4} \sum_{\bar{a} \in V_4} P_r\{w(\xi_1, \eta_2) = z_1 | \xi_2 = a\} \\
&\overset{\text{vii}}{=} \rho_{z_2} \rho_{z_3} \rho(z_4) P_r\{w(\xi_1, \eta_2) = z_1\} \overset{\text{viii}}{=} \rho_{z_1} \rho_{z_2} \rho_{z_3} \rho(z_4)
\end{aligned} \tag{2.17}$$

in which equations i, iv, vii are obtained using the relationship of joint probability distribution and marginal distribution respectively; equations ii, v are obtained from the independence of η_4, η_3 and $(\xi_2, \xi_3), \xi_2$; and equations iii, iv, viii are obtained from

$$P_r\{w(\xi_\tau, \eta_{\tau+1} = z_\tau\} = \rho(z_\tau), \quad \tau = 3, 2, 1 .$$

Consequently, $p(\bar{z}) = \rho(z_1) \cdot \rho(z_2) \cdot \rho(z_3) \cdot \rho(z_4)$ holds. Since the joint probability distribution of $\bar{\zeta} = (\zeta_1, \zeta_2, \cdots, \zeta_n)$ is the product of the marginal distributions of $\zeta_1, \zeta_2, \cdots, \zeta_n$, it implies that $\bar{\zeta}$ is an i.i.d. sequence. The theorem is thus proved.

From the proof of this theorem, we may find that independent stochastic sequence pairs can be broadened. Typically, under the conditions of Definition 3, if requirement 1 were deleted, the conclusion of Theorem 2 holds as well.

2.1.4 Local Penalty Function and Limit Properties of 2-Dimensional Stochastic Sequences

Based on Theorem 2, we can give the local penalty function and the limit properties of two-dimensional stochastic sequences.

Definition 8. *For a two-dimensional stochastic sequence $(\tilde{\xi}, \tilde{\eta})$ and a fixed penalty function $w(a, b)$, we define the local penalty function by*

$$w(\tilde{\xi}, \tilde{\eta}, i, j, n) = \frac{1}{n} w(\bar{\xi}_i, \bar{\eta}_j) = \frac{1}{n} \sum_{k=0}^{n-1} w(\xi_{i+k}, \eta_{j+k}) . \tag{2.18}$$

It is the average of the penalty of local vectors $\xi_{[i,i+n-1]}$ and $\eta_{[j,j+n-1]}$, and is simply referred to as the local penalty function.

Lemma 1. *If $(\tilde{\xi}, \tilde{\eta})$ is a double independent stochastic sequence, then the local penalty function is*

$$\mu_{ij} = E\{w(\tilde{\xi}, \tilde{\eta}, i, j, n)\} = \begin{cases} \mu_0 , & \text{if } i = j , \\ \mu_1 , & \text{otherwise} , \end{cases} \tag{2.19}$$

in which

$$\begin{cases} \mu_0 = E\{w(\xi_1, \eta_1)\} = \sum_{a,b \in V_4} w(a, b) p_{\xi_1, \eta_1}(a, b) , \\ \mu_1 = E\{w(\xi_1, \eta_2)\} = \frac{1}{16} \sum_{a,b \in V_4} w(a, b) . \end{cases} \tag{2.20}$$

If $w(a, b) = d_{\mathrm{H}}(a, b)$ is the Hamming function, then $\mu_0 = \epsilon, \mu_1 = 3/4$.

Theorem 3. *If $(\tilde{\xi}, \tilde{\eta})$ is an independent stochastic sequence pair and the penalty function $w(a, b)$ is strongly symmetric, then the following limit theorems hold.*

1. *As $n \to \infty$, following from the law of large numbers, we have the limit formula as follows:*

$$w\left(\tilde{\xi}, \tilde{\eta}, i, j, n\right) \longrightarrow \mu_{ij} , \quad \text{a.e.,} \tag{2.21}$$

in which μ_{ij} is given in (2.19) and a.e. is the abbreviation of almost everywhere convergence.

2. *Central limit theorem: If n is large enough, then*

$$\frac{1}{\sqrt{n}\,\sigma_{ij}} \sum_{k=0}^{n-1} [w(\xi_{i+k}, \eta_{j+k}) - \mu_{ij}] \sim N(0, 1) , \tag{2.22}$$

in which $N(0,1)$ denotes a standard normal distribution and

$$\sigma_{ij}^2 = E\left\{\left[w(\tilde{\xi},\tilde{\eta},i,j,n) - \mu_{ij}\right]^2\right\}$$
$$= \begin{cases} \sum_{a,b\in V_4} [w(a,b) - w_0]^2 p(a,b)\,, & \text{if } i = j\,, \\ \frac{1}{16}\sum_{a,b\in V_4} [w(a,b) - \mu_1]^2\,, & \text{otherwise}\,. \end{cases} \tag{2.23}$$

Proof. Following from Theorem 2, Lemma 1, the Kolmogorov law of large numbers, and the Levy–Lindberg central limit theorem, when $w(a,b) = d_{\mathrm{H}}(a,b)$ is the Hamming function, then from Definition 3 of an independent sequence pair, we calculate σ_{ij} as follows:

1. While $i = j$,

$$\sigma_{ij}^2 = \sum_{a,b\in V_4} [w(a,b) - w_0]^2 p(a,b) = (1-\epsilon)\epsilon^2 + (1-\epsilon)^2\epsilon = (1-\epsilon)\epsilon\,.$$

2. While $i \neq j$,

$$\sigma_{ij}^2 = \frac{1}{16}\sum_{a,b\in V_4} [w(a,b) - w_1]^2 = \frac{1}{16}\left(\frac{9}{4} + \frac{12}{16}\right) = \frac{3}{16}\,.$$

2.2 Stochastic Models of Flow Raised by Sequence Mutations

A biological sequence is composed of many small molecular units, and sequence mutations most often result from the mutation of some of these units. In stochastic processes, the structure of this type of mutation is usually discussed in two stages. First, we discuss how the mutations occur, and then discuss the effects of these mutations. It is similar to describing the arrival of customers at a shop, and then describing how much these customers spend. These two stages happen randomly. The stochastic process which describes the occurrence of these mutations is referred to as the random flow of mutations or simply the flow of mutations. There are four kinds of mutation effects, which may be described by the corresponding stochastic processes. Combining the results from the two stages, we obtain the stochastic model of sequence mutations.

2.2.1 Bernoulli Processes

Usually, random flow is described by the Bernoulli process or the Poisson process in probability theory. The common characteristic of these two processes is that a random flow is composed of many small probability independent events.

For example, customers may arrive in a shop at different times, and the numbers of customers arriving at a given time are independent. Generally, this characteristic coincides with the characteristic of mutation flow in a biological sequence. Consequently, mutation flow can be described by the Bernoulli process or the Poisson process. The corresponding symbols definitions and properties are given below.

Definition and Notations of the Bernoulli Process

The Bernoulli process is a stochastic sequence whose time and state are discrete. If we denote it by $\tilde{\zeta} = (\zeta_1, \zeta_2, \zeta_3, \cdots)$, then it has the following characteristics:

1. $\tilde{\zeta}$ is an independent sequence and each element ζ_j is determined according to the Bernoulli experiment. Here, each ζ_j has only the value 0 or 1, which denotes whether or not the event happens at time j.
2. The probability distribution of ζ_j is

$$P_r\{\zeta_j = 1\} = \epsilon_j \,, \quad P_r\{\zeta = 0\} = 1 - \epsilon_j \,. \tag{2.24}$$

Here, $\tilde{\epsilon} = (\epsilon_1, \epsilon_2, \epsilon_3, \cdots)$ is called the strength sequence of the Bernoulli process. If the strength sequence $\tilde{\epsilon}$ in (2.24) is a constant series ϵ, then $\tilde{\zeta}$ is called an homogeneous Bernoulli process and ϵ is the strength of the Bernoulli process.

For simplicity, we will discuss only homogeneous Bernoulli processes. Since the nonhomogeneous and nonindependent situations follow similar arguments, we will omit them in this book.

Counting Process and Dual Renewal Processes Associated with Bernoulli Processes

Let $\tilde{\zeta}$ be a given homogeneous Bernoulli process, and let $v_n^* = \sum_{j=1}^{n} \zeta_j$ be the total number of mutations happening in region $[1, n]$. So $\tilde{v}^* = \{v_1^*, v_2^*, v_3^*, \cdots\}$ is then referred to as the counting process. $\tilde{v}^*$ is also referred to as the renewal process of $\tilde{\zeta}$ in the sense of stochastic processes. The following characteristics hold:

1. Let $\tilde{v}^*$ be a binomial, homogeneous, and independent increment process and

$$P_r\{v_{n'}^* - v_n^* = k\} = b(n' - n, k; \epsilon) = \frac{(n' - n)!}{k! \cdot (n' - n - k)!} \epsilon^k (1 - \epsilon)^{n' - n - k} \,, \tag{2.25}$$

hold for all $0 \le n \le n'$ and all $k = 0, 1, 2, \cdots$; here $v_0^* \equiv 0$. The probability distribution

$$b(n, k; \epsilon) = \frac{n!}{k! \cdot (n - k)!} \epsilon^k (1 - \epsilon)^{n-k} \,, \quad k = 0, 1, 2, \cdots, n$$

is then a binomial distribution.

2. $\tilde{v}^*$ generates the dual renewal process of $\tilde{v}^*$:

$$j_k^* = \sup\{j : v_j^* \le k\} , \tag{2.26}$$

in which j_k^* denotes the time at which the kth renewal happens in renewal process theory. In this book, j_k^* denotes the position of the kth mutation. Hence

$$\tilde{j}^* = (j_0^*, j_1^*, j_2^*, \cdots) \tag{2.27}$$

is referred to as the mutation point sequence. It is equal to the sequence of the renewal position in the renewal process, in which $j_0^* \equiv 0$. A detailed discussion about renewal sequences (or processes) and dual renewal sequences (or processes) may be found in [25].

Characteristics of Dual Renewal Processes

The relationships and characteristics of the renewal process $\tilde{v}^*$ and its dual renewal process $\tilde{j}^*$ are stated as follows:

1. The sequence $\tilde{j}^*$ is a homogeneous and independent incremental process. If we let $\ell_j^* = j_k^* - j_{k-1}^*$, then $\tilde{\ell}^* = \{\ell_1^*, \ell_2^*, \ell_3^*, \cdots\}$ is an i.i.d. sequence and each ℓ_j^* has a geometric distribution $e_\epsilon(n)$. That is,

$$P_r\{j_k^* - j_{k-1}^* = n\} = e_{1-\epsilon}(n) = \epsilon(1-\epsilon)^{n-1} , \quad k, n = 1, 2, 3, \cdots . \tag{2.28}$$

2. Following from the property of geometric distribution, we have that j_k^* is a negative binomial distribution. That is,

$$P_r\{j_k^* = n\} = C_{n-1}^{k-1}\epsilon^k(1-\epsilon)^{n-k} = \frac{(k-1)!}{(n-1)!(k-n)!}\epsilon^k(1-\epsilon)^{n-k} , \quad k \ge n . \tag{2.29}$$

3. Since the sequence $\tilde{v}^*$ is the dual renewal process of $\tilde{j}^*$, it follows that

$$v_n^* = \sup\{k : j_k^* \le n\} \tag{2.30}$$

holds.

Consequently, the four sequences $\tilde{\zeta}$, $\tilde{v}^*$, $\tilde{j}^*$, $\tilde{\ell}^*$ can be determined from each other. Moreover, $\tilde{\zeta}$, $\tilde{\ell}^*$ are i.i.d. sequences that obey the Bernoulli distribution b_ϵ, and geometric distribution $e_\epsilon(k)$, respectively. Furthermore, $\tilde{v}^*$ and $\tilde{j}^*$ are the renewal processes of ζ and ℓ^* respectively, where,

$$v_n^* = \sum_{j=1}^{n} \zeta_j , \quad j_k^* = \sum_{i=1}^{k} \ell_i^* .$$

Following from (2.26) and (2.30), we have that $\tilde{v}^*$ and $\tilde{j}^*$ are the dual renewal processes for each other.

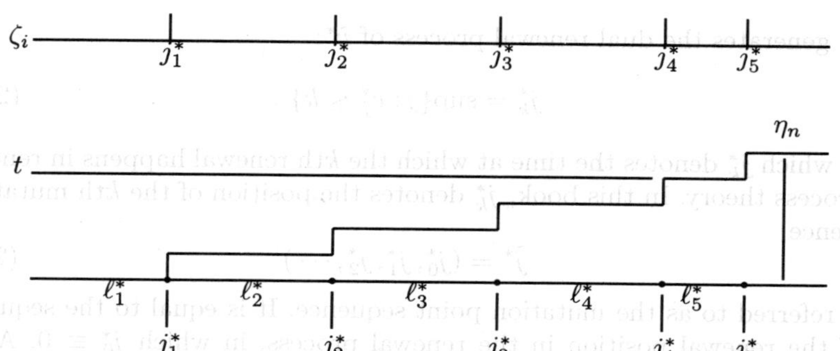

Fig. 2.1. The Bernoulli process, the counting process, and its dual renewal process

These properties also can be found in textbooks on the theory of stochastic processes, for example, in Doob's book [25]. The relationships among sequences $\tilde{\zeta}$, $\tilde{v}^*$, $\tilde{j}^*$, $\tilde{\ell}^*$ are described in Fig. 2.1.

In Fig. 2.1, ζ is a Bernoulli process. The top line denotes the value of ζ_i as 0 or 1, in which, j_s^*, $s = 1, 2, 3, \cdots$ denotes the positions held by $\zeta_j = 1$. The lowest line denotes the relationship between ζ_i and η_n. Following from the definition of the dual renewal process, we have

$$v_s^* = \sup\{v\colon \eta_v \le s\} = j_s^* , \quad s = 1, 2, 3, \cdots$$

and $\ell_s^* = j_s^* - j_{s-1}^*$.

2.2.2 Poisson Flow

The Poisson process is one of the most basic and important stochastic processes describing mutation flows. We recall its notations and properties below.

Definition and Generation of Poisson Flows

Poisson flow (also called the Poisson process) is a stochastic process with sequential time and discrete states. Poisson flow is a stochastic flow composed of an accumulation of many small probability events. It has three main characteristics: the stationary property, the nonbackward property and the sparseness property. Let v_s^* be the number of times that some event happens in interval $(0, s)$, then $v^*(s, s+t)$ and v_s^* have the same distribution for all $t > 0$. Then $v^*(s_1, s_2)$ and $v^*(s_2, s_3)$ are independent if $0 \le s_1 < s_2 < s_3$. The probability of the event happening twice or more in the interval $(0, \Delta s)$ is infinitesimal in higher orders such that Δs is infinitesimal. The probability distribution of $v^*(s)$ is then defined by:

$$p_{\lambda,s}(k) = P_r\{v_s^* = k\} = \frac{(\lambda\, s)^k}{k\,!} e^{-\lambda\, s} , \quad k = 0, 1, 2, \cdots . \tag{2.31}$$

The probability distribution in (2.31) is called a Poisson distribution, in which, $\lambda > 0$ is the strength of the Poisson flow.

Properties of Poisson Flows

1. The mean and variance of a Poisson flow are defined as follows:

$$\begin{cases} E\{v_s^*\} = \sum_{k=0}^{\infty} k p_\lambda(k) = \lambda s \,, \\ D\{v_s^*\} = E\{(v_s^* - \lambda)^2\} = \sum_{k=0}^{\infty} (k-\lambda)^2 p_\lambda(k) = \lambda s \,. \end{cases} \tag{2.32}$$

2. In the interval $(0, s)$, the probability that Poisson flow will happen at least once following from (2.31) is computed as follows:

$$\begin{cases} p_0(s) = P_r\{v_s^* = 0\} = \mathrm{e}^{-\lambda s} \,, \\ p_1(s) = P_r\{v_s^* = 1\} = (\lambda s)\mathrm{e}^{-\lambda s} \,, \end{cases} \tag{2.33}$$

   Therefore, this probability distribution obeys an exponential distribution.

3. If $v_{j,s}^*$, $j = 1, 2, \cdots, k$ are k independent Poisson flows, the mixed Poisson flow is defined as follows:

$$v_s^* = \sum_{j=1}^{k} v_{j,s}^* \,, \quad s \leq 0 \,. \tag{2.34}$$

   This is the total number of times that k Poisson flows happen in the interval $(0, s)$. v_s^* is then also a Poisson flow with the strength parameter $\lambda = \sum_{j=1}^{k} \lambda_j$.

Dual Renewal Process of Poisson Flows

Poisson flow is a kind of renewal process and its dual renewal process is defined by

$$t_k^* = \sup\{t ; v^*(t) \leq k\} \,. \tag{2.35}$$

t_k^* is then the random variable for position renewal, that is, the random variable describing the mutation at the kth position. It follows from (2.35) that the stochastic sequence

$$\tilde{t}^* = \{t_1^*, t_2^*, t_3^*, \cdots\} \tag{2.36}$$

is a homogeneous and independently incremental process. $\tilde{\ell}^* = (\ell_1^*, \ell_2^*, \ell_3^*, \cdots)$ is then an independently and identically distributed process and each ℓ_j^* obeys an exponential distribution, in which $\ell_k^* = t_k^* - t_{k-1}^*$:

$$P_r\{\ell_k^* \leq t\} = 1 - \mathrm{e}^{\lambda t} \,, \quad \lambda, t > 0 \,. \tag{2.37}$$

Here $t_n^* = \sum_{j=1}^{n} \ell_j^*$. Consequently, $\tilde{t}^*$ is the renewal process of $\tilde{\ell}^*$.

Each one of the two processes $\tilde{v}^*$ and $\tilde{t}^*$ is the dual renewal process of the other one, determined by (2.26) and (2.30) if we replace j^* by t^* in the corresponding formula. For this reason, we can get three alternative representations

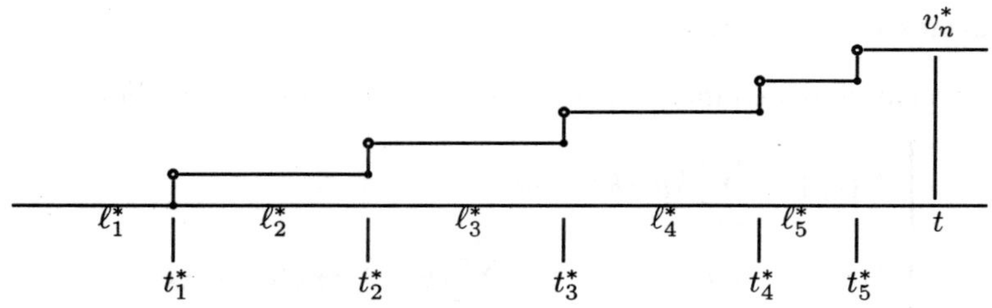

Fig. 2.2. The sample track function for a Poisson process

of Poisson flow: $\tilde{v}^*$, $\tilde{t}^*$, $\tilde{\ell}^*$, in which the random variable $v^*(t)$ of $\tilde{v}^*$ means the time of mutation happening in the position region $(0, t)$. The random variable t_k^* of $\tilde{t}^*$ refers to the position of the kth mutation and the random variable ℓ_k^* denotes the period between the $(k-1)$th and the kth mutations.

The relationships among the three sequences $\tilde{v}^*$, $\tilde{t}^*, \tilde{\ell}^*$ are illustrated in Fig. 2.2.

From Fig. 2.2, we deduce that the sample track function of the Poisson process is similar to the counting process. Its properties are as follows:

1. $v = v_s$, $s \geq 0$ is a jumping and increasing step function with continuous time and discrete values. It is left continuous. That is, $\lim_{t \to s_0 -} v_s = v_{s_0}$ always holds. The left continuous points are denoted by solid dots and the points that are not within the range of the function are denoted as voids.
2. Following from the sparseness property of the Poisson process, the heights at all jump points of the sample track function are at most 1.
3. Let $\{s_1, s_2, s_3, \cdots\}$ be a set of jump points arranged in order, then $\tilde{s} = (s_1, s_2, s_3, \cdots)$ is called the renewal sequence of the function $\tilde{v}$. Generally, the renewal sequence satisfies the following condition:

$$v_{t+} - v_{t-} = \begin{cases} 0, & \text{if } s \text{ does belong to set } \{s_1, s_2, s_3, \cdots\}, \\ 1, & \text{if } s \in \{s_1, s_2, s_3, \cdots\}, \end{cases} \tag{2.38}$$

in which

$$v_{t-} = \lim_{t' \uparrow s_0 -} v_{t'}, \quad v_{t+} = \lim_{t' \downarrow s_0 +} v_{t'} .$$

4. In Poisson flow $\tilde{v}^*$, the probability of all of these samples $\tilde{v}$ satisfying the following conditions is 1: (a) $0 \leq s_1 < s_2 < s_3 < \cdots$ holds and (b) $s_k \to \infty$ if $k \to \infty$.

The Relationship Between the Bernoulli Process and the Poisson Process

We have mentioned that the Bernoulli process and the Poisson process can both describe the mutation of sequences. Now, we further explain their properties and relationships as follows:

1. The Bernoulli process $\tilde{\zeta}$ applies to discrete stochastic sequences and each component $\zeta_i = 1, 0$ designates whether or not a mutation happened at position i. Thus, it is easy to understand it intuitively.
2. The crucial disadvantage of the Bernoulli process is that the probability of its counting sequence $\eta_n = \sum_{i=1}^n \zeta_i$ is difficult to compute directly if n is too large, because in the binomial distribution $b(n, k; \epsilon)$, $C_k^n = \frac{n!}{k!(n-k)!}$ and ϵ^k, and we are unable to calculate it exactly for large numbers. As a result, Poisson flow is typically used to approximate Bernoulli processes.
3. When Poisson flow is used to approximate counting sequences, the changing of positions becomes continuous. A region is chosen with proper length n_0 as its unit. The position n can be replaced by $t = \frac{n}{n_0}$. Thus, the discrete position region $\{1, 2, 3, \cdots\}$ becomes a continuous region $(0, \infty)$. If $n_0\epsilon = \lambda$, then $n\epsilon = \lambda t$ and the binomial distribution
$$b(n, k; \epsilon) = C_k^n \epsilon^k (1-\epsilon)^{n-k} \sim p_{\lambda t}(k) = \frac{(\lambda t)^k}{k!} \mathrm{e}^{-\lambda t}$$
approximates the probability of the Poisson flow. This describes the probability of the number of times mutation occured in the integer region $[1, n]$, or a continuous interval $(0, t)$.

2.2.3 Mutated Flows Resulting from the Four Mutation Types

The four types of mutation in biological sequences are defined in Sect. 1.2.1. Consequently, there are four types of mutated flows corresponding to the four mutation types. All four types of mutated flows can be denoted by Bernoulli processes and Poisson processes.

Representation Using Bernoulli Processes

If we use a Bernoulli process

$$\tilde{\zeta}_\tau = (\zeta_{\tau,1}, \zeta_{\tau,2}, \zeta_{\tau,3}, \cdots), \quad \tau = 1, 2, 3, 4 , \tag{2.39}$$

to represent the mutated flow, then $\tilde{\zeta}_\tau$ is also a Bernoulli process for each $\tau = 1, 2, 3, 4$. $\zeta_{\tau,j}$ denotes the random variable that represents whether or not the mutation type τ happened at position j. Let ϵ_τ denote the strength of the mutation type τ, then

$$P_r\{\zeta_{\tau,j} = 1\} = \epsilon_\tau , \quad P_r\{\zeta_{\tau,j} = 0\} = 1 - \epsilon_\tau . \tag{2.40}$$

Based on many calculations, we know that $\epsilon_2, \epsilon_3, \epsilon_4 \ll 1$ and $\epsilon_1 < 1/2$ within the homologous sequence family. Following from $\tilde{\zeta}_\tau$, $\tau = 1, 2, 3, 4$, we may write the corresponding renewal process as follows:

$$v_{\tau,n}^* = \sum_{j=1}^n \zeta_{\tau,j} , \quad \tau = 1, 2, 3, 4 , \tag{2.41}$$

and

$$j^*_\tau(k) = \sup\{j: v^*_{\tau,j} \le k\}, \quad \tau = 1, 2, 3, 4 . \tag{2.42}$$

Then $\{\tilde{\zeta}_\tau, \tilde{v}^*_\tau, \tilde{j}^*_\tau, \tilde{\ell}^*_\tau\}$ are the four equivalent representations of the mutation type τ, in which

$$\ell^*_{\tau,k} = j^*_{\tau,k} - j^*_{\tau,k-1}, \quad k = 1, 2, 3, \cdots . \tag{2.43}$$

For each fixed $\tau = 1, 2, 3, 4$, the four ζ_τ, $\tilde{v}^*_\tau$, $\tilde{j}^*_\tau$, $\tilde{\ell}^*_\tau$ are determined according to the relationships given in Sect. 2.1.

Representation Using Poisson Flow

The four different types of mutation flows resulting from the four mutation types can also be represented using Poisson flow,

$$\tilde{v}^*_\tau = \{v^*_\tau(t),\ t \ge 0\}, \quad \tau = 1, 2, 3, 4 , \tag{2.44}$$

and the corresponding strength of the Poisson flow is denoted by λ_τ.

For a fixed Poisson flow $\tilde{v}^*_\tau$ resulting from type τ mutation, we denote its dual process by $\tilde{t}^*_\tau$, in which

$$t^*_{\tau,k} = \sup\{t: v^*_\tau(t) \le k\}, \quad \tau = 1, 2, 3, 4 . \tag{2.45}$$

With arguments similar to those used in Sect. 2.1, we find that $\tilde{v}^*$ and $\tilde{t}^*$ are mutually determined for a fixed τ, and satisfy the relationships given in Sect. 2.1.

Synthesis of Mutation Sequences

Based on the four kinds of mutated processes, we can have an addition of mutations if we assume that the four types of mutations happen independently. If we let $\tilde{v}^*_\tau$, $\tau = 1, 2, 3, 4$ denote Poisson flow resulting from the four mutation types, their sum is defined as follows:

$$\tilde{v}^*_0 = \sum_{\tau=1}^{4} \tilde{v}^*_\tau = \left\{ \sum_{\tau=1}^{4} v^*_{\tau,1}, \sum_{\tau=1}^{4} v^*_{\tau,2}, \sum_{\tau=1}^{4} v^*_{\tau,3}, \cdots \right\} . \tag{2.46}$$

The term $\tilde{v}^*_0$ represents the mutated flow resulting from the four mutation types happening within the same sequences. Following the properties of Poisson flow, we have:

1. $\tilde{v}^*_0 = \{v^*_0(t),\ t \ge 0\}$ is also a Poisson flow, in which $v^*(t)$ is the total number of times the four types of mutation happen in the region $(0, t)$ altogether.
2. The strength of the Poisson flow $\tilde{v}^*_0$ is $\lambda_0 = \lambda_1 + \lambda_2 + \lambda_3 + \lambda_4$.
3. Dual process of the Poisson flow $\tilde{v}^*_0$ is $\tilde{t}^*$. They are determined by each other, and satisfy the relationships given in Sect. 2.1.

2.3 Stochastic Models of Type-I Mutated Sequences

In the last section, we introduced the four types of mutated flows resulting from the corresponding four mutation types. Based on these mutated flows, we now discuss how to obtain the stochastic models for these mutated flows. Since the effects of the four mutated flows are different, we analyze these mutation types in detail and give the stochastic models of the mutated sequences resulting from each of these mutation types.

2.3.1 Description of Type-I Mutation

The type-I mutation of a sequence is defined in Sect. 1.2.1, which is the first kind of mutation. In comparison, a sequence $\tilde{\eta}$ mutated from a sequence $\tilde{\xi}$ resulting from a type-I mutation is simply called the type-I mutated sequence (of $\tilde{\xi}$) in the following text. Although the wording of these two terms is similar (type-I mutation and type-I mutated sequence), their meanings are very different. We will discuss how to describe type-I mutated sequences and the corresponding data structure in this subsection. Let $\tilde{\xi}, \tilde{\eta}$ be two stochastic sequences defined on V_4 as given in (2.1). Let

$$(\tilde{\xi}, \tilde{\eta}) = ((\xi_1, \eta_1), (\xi_2, \eta_2), (\xi_3, \eta_3), \cdots) \tag{2.47}$$

be a two-dimensional sequence, in which ξ_j, $\eta_j \in V_4$. This practice problem requires mention of the fact that $\tilde{\eta}$ is the sequence (mutated from the sequence $\tilde{\xi}$) resulting from type-I mutation. For simplicity, $\tilde{\eta}$ is called the type-I mutated sequence of $\tilde{\xi}$ in the following text. To construct the stochastic models of $(\tilde{\xi}, \tilde{\eta})$, we must make the following two assumptions:

I-1 For the two-dimensional sequence $(\tilde{\xi}, \tilde{\eta})$, (ξ_j, η_j), $j = 1, 2, 3, \cdots$ is an i.i.d. sequence of two-dimensional random vectors, in which $\tilde{\xi}$ is a perfectly stochastic sequence.

I-2 For any $j = 1, 2, 3, \cdots$, the joint probability distribution of the two-dimensional random vector (ξ_j, η_j) satisfies the following condition:

$$P_r\{(\xi_j, \eta_j) = (a, b)\} = \begin{cases} \dfrac{1-\epsilon_1}{4}, & \text{if } a = b\,, \\ \dfrac{\epsilon_1}{12}, & \text{otherwise}\,, \end{cases} \tag{2.48}$$

in which ϵ_1 is the strength of the type-I mutation.

If a two-dimensional sequence $(\tilde{\xi}, \tilde{\eta})$ defined by (2.47) satisfies conditions **I-1** and **I-2**, then $\tilde{\eta}$ represents the standard type-I mutated sequence of $\tilde{\xi}$.

Theorem 4.

1. *If there is a stochastic sequence $\tilde{\vartheta} = (\vartheta_1, \vartheta_2, \vartheta_3, \cdots)$ which satisfies the following conditions:*

I-3 *$\tilde{\vartheta}$ is an independently and identically distributed sequence and it is independent of $\tilde{\xi}$.*

I-4 *The probability distribution of each ϑ_j is defined as*

$$P_r\{\vartheta_j = a\} = \begin{cases} 1-\epsilon_1 \,, & \text{if } a = 0 \,, \\ \dfrac{\epsilon_1}{3} \,, & \text{otherwise} \,, \end{cases} \tag{2.49}$$

then $(\tilde{\xi}, \tilde{\eta})$ satisfies the conditions **I-1** *and* **I-2**, *in which sequence $\eta_j = \xi_j + \vartheta_j$, $j = 1, 2, 3, \cdots$.*

2. *If a two-dimensional sequence $(\tilde{\xi}, \tilde{\eta})$ satisfies conditions* **I-1** *and* **I-2**, *then there is a stochastic sequence $\tilde{\vartheta}$ satisfying the conditions* **I-3** *and* **I-4** *such that $\tilde{\eta} = \tilde{\xi} + \tilde{\vartheta}$ holds and $\tilde{\eta}$ is a perfectly stochastic sequence.*

Proof. We prove this theorem in several steps:

1. We first prove proposition 1 of the theorem. If the sequence $\tilde{\vartheta}$ satisfies conditions **I-3** and **I-4** and $\tilde{\eta} = \tilde{\xi} + \tilde{\vartheta}$, then (ξ_j, η_j), $j = 1, 2, 3, \cdots$ is an independently and identically distributed sequence and

$$P_r\{(\xi_j, \eta_j) = (a, b)\} = P_r\{\xi_j = a\} P_r\{\eta_j = b | \xi_j = a\}$$
$$= \frac{1}{4} P_r\{\vartheta_j = b - a\} = \begin{cases} \dfrac{1-\epsilon_1}{4} \,, & \text{if } b = a \,, \\ \dfrac{\epsilon_1}{12} \,, & \text{otherwise} \,. \end{cases}$$

Consequently, (2.48) holds, and proposition 1 of the theorem is proved.

2. If $(\tilde{\xi}, \tilde{\eta})$ satisfies conditions **I-1** and **I-2**, then we prove that $\tilde{\eta}$ is a perfectly stochastic sequence. Alternatively, we prove that η_j is a uniform distribution, such that

$$P_r\{\eta_j = b\} = \sum_{a \in V_4} P_r\{(\xi_j, \eta_j) = (a, b)\}$$
$$= P_r\{(\xi_j, \eta_j) = (b, b)\} + \sum_{a \neq b} P_r\{(\xi_j, \eta_j) = (a, b)\}$$
$$= \frac{1-\epsilon_1}{4} + \sum_{a \neq b} \frac{\epsilon_1}{12} = \frac{1-\epsilon_1}{4} + \frac{\epsilon_1}{4} = \frac{1}{4} \,.$$

Thus, we have proved that η_j is a perfectly stochastic sequence.

3. If $(\tilde{\xi}, \tilde{\eta})$ satisfies conditions **I-1** and **I-2** then we prove that there is a stochastic sequence $\tilde{\vartheta}$ satisfying **I-3** and **I-4** such that $\tilde{\eta} = \tilde{\xi} + \tilde{\vartheta}$ holds. Let $\vartheta_j = \eta_j - \xi_j$, $j = 1, 2, 3, \cdots$; then following from condition **I-1**, we have that $\tilde{\vartheta}$ is an i.i.d. sequence. Thus, we need only prove that (2.49) holds. In fact,

$$P_r\{\vartheta_j = c\} = P_r\{\eta_j - \xi_i = c\} = \sum_{b-a=c} P_r\{(\xi_j, \eta_j) = (a, b)\}$$
$$= \sum_{a \in V_4} P_r\{(\xi_j, \eta_j) = (a, c + a)\} = \begin{cases} 1-\epsilon_1 & \text{if } c = 0 \,, \\ \dfrac{\epsilon_1}{3} \,, & \text{otherwise} \,. \end{cases}$$

That is, (2.49) holds, and the proposition 2 holds. This completes the proof of the theorem.

Theorem 5. *A sequence $\tilde{\vartheta}$ given in Theorem 4 can be decomposed as follows:*

$$\vartheta_j = \zeta_j \cdot \vartheta'_j\,, \quad j = 1, 2, 3, \cdots\,, \tag{2.50}$$

with two components satisfying the following conditions:

I-5 *$\tilde{\zeta}, \tilde{\vartheta}'$ are two i.i.d. sequences, and $\tilde{\zeta}$ and $\tilde{\vartheta}'$ are independent of each other. Furthermore, both $\tilde{\zeta}$ and $\tilde{\vartheta}'$ are independent of $\tilde{\xi}$.*

I-6 *$\tilde{\zeta}$ is a Bernoulli process with strength ϵ_1, and the probability distribution of $\tilde{\vartheta}'$ is:*

$$P_r\left\{\vartheta'_j = a\right\} = \begin{cases} 0\,, & \text{if } a = 0\,, \\ \dfrac{1}{3}\,, & \text{otherwise}\,. \end{cases} \tag{2.51}$$

$(\tilde{\xi}, \tilde{\eta})$ then satisfies the conditions **I-1** *and* **I-2** *if and only if there are two stochastic sequences $\tilde{\zeta}$ and $\tilde{\vartheta}$ satisfying conditions* **I-5** *and* **I-6** *and which satisfy the following:*

$$\eta_j = \xi_j + \zeta_j \vartheta'_j\,, \quad j = 1, 2, 3, \cdots\,. \tag{2.52}$$

Definition 9. *In (2.52), sequence $\tilde{\zeta}$ represents type-I mutated flow, and the sequence $\tilde{\vartheta}'$ represents random additive noise (or random interference).*

In conclusion, a type-I mutated sequence can be decomposed into the sum of a Bernoulli process and the product of a Bernoulli process and a random noise term. We denote the model by

$$\mathcal{E}_1^* = \left\{\tilde{\xi}, \tilde{\zeta}_1, \tilde{\vartheta}_1, \tilde{\eta}\right\}\,, \tag{2.53}$$

in which $\tilde{\xi}, \tilde{\zeta}_1, \tilde{\vartheta}_1$ are three independently stochastic sequences. Any one of the three stochastic sequences is i.i.d., and the three common probability distributions are the uniform distribution, the Bernoulli distribution and the distribution given in (2.51). The mutated sequence is then determined by (2.52).

2.3.2 Properties of Type-I Mutations

Reversibility of the Type-I Mutated Sequence

The so-called reversibility of the sequence is the fact that $\tilde{\eta}$ can be mutated inversely to $\tilde{\xi}$ by type-I mutation if $\tilde{\eta}$ is the type-I mutated sequence of $\tilde{\xi}$.

Theorem 6. *Under type-I mutation, if $\tilde{\eta}$ is the mutated sequence of $\tilde{\xi}$ with strength ϵ_1, then $\tilde{\xi}$ is also the mutated sequence of $\tilde{\eta}$ with strength ϵ_1.*

Proof. For the theorem, we need only prove that sequence $\tilde{\vartheta}$ and sequence $\tilde{\eta}$ are independent. To do this, we consider

$$P_r\{(\eta_j, \vartheta_j) = (b, c)\} = \sum_{a \in V_4} P_r\{(\xi_j, \eta_j, \vartheta_j) = (a, b, c)\}$$

$$\sum_{a \in V_4} P_r\{(\xi_j \vartheta_j) = (a, c)\} P_r\{\eta_j = b | (\xi_j, \vartheta_j) = (a, c)\} .$$

Since $\eta_j = \xi_j + \vartheta_j$, we have

$$P_r\{\eta_j = b | (\xi_j, \vartheta_j) = (a, c)\} = \begin{cases} 1 , & \text{if } b = a + c , \\ 0 , & \text{otherwise} , \end{cases}$$

and

$$\begin{aligned} P_r\{(\eta_j, \vartheta_j) = (b, c)\} &= P_r\{(\xi_j \vartheta_j) = (b - c, c)\} \\ &= \frac{1}{4} P_r\{\vartheta_j c\} = P_r\{\eta = b\} P_r\{\vartheta_j = c\} , \end{aligned} \tag{2.54}$$

in which the last two equations hold due to the fact that ξ_j, η_j both have uniform distributions on V_4.

Following from (2.54), we have that η_j, ϑ_j are two independent random sequences. From the independently and identically distributed property of $(\xi_j, \eta_j, \vartheta_j)$, $j = 1, 2, 3, \cdots$, we show the independence between $\tilde{\eta}$ and $\tilde{\vartheta}$. Therefore, $\tilde{\xi} = \tilde{\eta} - \tilde{\vartheta}$ holds. That is, $\tilde{\xi}$ is the standard type-I mutated sequence of $\tilde{\eta}$ with strength ϵ_1. The theorem is therefore proved.

Local Penalty Function of Type-I Mutated Sequences

Here, we discuss the local penalty function for a type-I mutated sequence. Let $(\tilde{\xi}, \tilde{\eta})$ be a two-dimensional sequence which satisfies conditions **I-1** and **I-2**, and is therefore independent (definition in Sect. 2.1.2). If $w(a, b)$, $a, b \in V_4$ is a strongly symmetric penalty function on V_4, we have the following properties:

1. The mean of $w(\xi_i, \eta_j)$ satisfies

$$w_{i,j} = E\{w(\xi_i, \eta_j)\} = \begin{cases} \dfrac{\epsilon_1 w'}{3} , & \text{if } i = j , \\ \dfrac{w'}{4} , & \text{otherwise} , \end{cases} \tag{2.55}$$

in which $w' = w(0, 1) + w(0, 2) + w(0, 3)$. When $i = j$, we have

$$\begin{aligned} E\{w(\xi_i, \eta_j)\} &= \sum_{a, b \in V_4} w(a, b) P_r\{(\xi_i, \eta_j) = (a, b)\} \\ &= \sum_{a \neq b \in V_4} w(a, b) P_r\{(\xi_i, \eta_i) = (a, b)\} \\ &= \frac{\epsilon_1}{12} \sum_{a \neq b \in V_4} w(a, b) = \frac{\epsilon_1}{3} \sum_{b=1}^{3} w(0, b) = \frac{w' \epsilon_1}{4} , \end{aligned}$$

and when $i \neq j$, we have

$$E\{w(\xi_i, \eta_j)\} = \sum_{a \neq b \in V_4} w(a,b) P_r\{(\xi_i, \eta_j) = (a,b)\}$$
$$= \frac{1}{16} \sum_{a \neq b \in V_4} w(a,b) = \frac{1}{4} \sum_{b=1}^{3} w(0,b) = \frac{w'}{3} .$$

It follows that (2.55) holds.

2. The variance $\sigma_{i,j} = D\{w(a^*_{i+k}, b^*_{j+k})\}$ of $w(\xi_i, \eta_j)$, where

$$\sigma_{i,j} = \sum_{a,b \in V_4} [w(a,b) - w_{i,j}]^2 P_r\{(\xi_i, \eta_j) = (a,b)\} . \tag{2.56}$$

For any fixed (i,j), we calculate the value of $\sigma_{i,j}$ as follows:

$$\begin{aligned} \sigma_{i,j} &= \sum_{a,b \in V_4} [w(a,b) - w_{i,j}]^2 P_r\{(\xi_i, \eta_j) = (a,b)\} \\ &= \sum_{a,b \in V_4} \left[w^2(a,b) - w_{i,j}^2\right] P_r\{(\xi_{ij}, \eta_j) = (a,b)\} \\ &= \sum_{a \neq b \in V_4} w^2(a,b) P_r\{(\xi_i, \eta_j) = (a,b)\} - w_{i,j}^2 , \end{aligned} \tag{2.57}$$

in which

$$w_{i,j}^2 = \begin{cases} \left(\dfrac{w'\epsilon_1}{3}\right)^2 , & \text{if } i = j , \\ \left(\dfrac{w'}{4}\right)^2 , & \text{otherwise} , \end{cases}$$

and

$$\sum_{a,b \in V_4} w^2(a,b) P_r\{(\xi_i, \eta_j) = (a,b)\} = \begin{cases} \dfrac{w''\epsilon_1}{3} , & \text{if } i = j , \\ \dfrac{w''}{4} , & \text{otherwise} , \end{cases}$$

and $w'' = w(0,1)^2 + w(0,2)^2 + w(0,3)^2$. Thus, we have

$$\sigma_{i,j} = \begin{cases} \dfrac{w''\epsilon_1}{3} - \left(\dfrac{w'\epsilon_1}{3}\right)^2 = \dfrac{\epsilon_1}{3}\left[w'' - \dfrac{(w')^2\epsilon_1}{3}\right] , & \text{if } i = j , \\ \dfrac{w''}{4} - \left(\dfrac{w'}{4}\right)^2 = \dfrac{1}{4}\left[w'' - \left(\dfrac{w'}{2}\right)^2\right] , & \text{otherwise} . \end{cases} \tag{2.58}$$

Example 4. If $w(a,b) = d_{\mathrm{H}}(a,b)$ is the Hamming matrix, we have $w' = w'' = 3$. Therefore, we have

$$w_{i,j} = \begin{cases} \epsilon_1 , & \text{if } i = j , \\ \dfrac{3}{4} , & \text{otherwise} , \end{cases} \qquad \sigma_{i,j} = \begin{cases} \epsilon_1(1-\epsilon_1) , & \text{if } i = j , \\ \dfrac{3}{16} , & \text{otherwise} . \end{cases} \tag{2.59}$$

3. If $w(a,b) = d_{\mathrm{WT}}(a,b)$ is the WT-matrix, we can see that: $w' = 1.99$, $w'' = 1.3883$, then we have

$$w_{i,j} = \begin{cases} 0.66 \times \epsilon_1 , & \text{if } i = j , \\ 0.4975 , & \text{otherwise} , \end{cases}$$

$$\sigma_{i,j} = \begin{cases} \epsilon_1(0.4711 - 0.044 \times \epsilon_1) , & \text{if } i = j , \\ 0.1058 , & \text{otherwise} . \end{cases} \tag{2.60}$$

Using the calculation in Example 4, we obtain an overall impression of the local penalty function for a type-I mutated sequence.

Limit Properties of the Local Penalty Function of a Type-I Mutated Sequence

Following from Theorems 3 and 4, we obtain the limit properties of the local penalty function of a type-I mutated sequence.

Theorem 7. *If $\tilde{\eta}$ is a type-I mutated sequence of $\tilde{\xi}$ satisfying conditions* **I-1** *and* **I-2**, *and if W is a strongly symmetric matrix, then the limit properties of the local penalty function of $\tilde{\xi}, \tilde{\eta}$ are as follows:*

1. *As $n \to \infty$, the limitation*

$$w(\tilde{\xi}, \tilde{\eta}; i, j, n) = \frac{1}{n} \sum_{k=1}^{n} w(\xi_{i+k}, \eta_{j+k}) \to w_{i,j} \quad a.e. \tag{2.61}$$

in which $w_{i,j}$ is computed by (2.55).

2. *The central limit theorem: if n is large enough,*

$$\frac{1}{\sigma_{ij}\sqrt{n}} \sum_{k=1}^{n} [w(\xi_{i+k}, \eta_{j+k}) - \mu_{ij}] \sim N(0,1) , \tag{2.62}$$

in which $\sigma_{i,j}$ is given by (2.58). The results of (2.61) and (2.62) follow from the Kolmogorov law of large numbers and the Levy–Lindberg central limit theorem.

2.4 Type-II Mutated Sequences

We continue to discuss stochastic models for type-II, type-III, and type-IV mutations. The description for these is similar to queuing theory – where we consider a service station and the customers arriving as a representation of the mutation flow. The time each customer spends at the station is analogous to the lengths resulting from type-II, type-III, and type-IV mutations. In this subsection, we consider the models that arise from type-II mutations.

2.4.1 Description of Type-II Mutated Sequences

A type-II mutation (defined in Sect. 1.2) refers to the permutation of some segments of a biological sequence $A = (a_1, a_2, \cdots, a_N)$. For example,

$$\begin{cases} A = (00201[332]0110203[01022]23101011[20]3321)\,, \\ B = (00201[20]0110203[332]23101011[01022]3321)\,. \end{cases} \tag{2.63}$$

Then, in sequence A the data [332], [01022], [20] in the square brackets permute and turn into the segments [20], [332], [01022] of sequence B. Data permutation on more disconnected segments is very important in gene or protein analysis. In recent years, bioinformatics has begun to solve these problems. We do not, however, intend to address the subject in this book due to its complexity.

In this book, we confine our discussion to simpler cases. That is, we only discuss data permutation of two coterminous segments. For example,

$$\begin{cases} A = (00201\{[332][0110](00201\{[332][0110]\}20301022231\{[01011][20]\}3321)\,, \\ B = (00201\{[0110][332]\}20301022231\{[20][01011]\}3321)\,. \end{cases} \tag{2.64}$$

The sequence B results from the permutation of the data segments in large brackets $\{[332][0110]\}, \{[01011][20]\}$ of sequence A. After this, the new segments of sequence B are $\{[0110][332]\}, \{[20][01011]\}$, in which each large bracket contains the permutation of two coterminous segments.

2.4.2 Stochastic Models of Type-II Mutated Sequences

The following assumptions are required in order to build models of type-II mutated sequences:

II-1 The mutation sequence $\tilde{\eta}$ is determined by a stochastic sequence $\tilde{\xi}$, $\tilde{\zeta}_2$, and $(\tilde{\ell}_1^*, \tilde{\ell}_2^*)$. The explanation is as follows:

1. $\tilde{\xi}$ is a perfectly stochastic sequence on V_4. It is an initial sequence to be mutated.
2. $\tilde{\zeta}_2$ is a Bernoulli process with strength ϵ_2. It is similar to the sequence defined in (2.38), to describe whether or not type-II mutation happens.
3. $(\tilde{\ell}_1^*, \tilde{\ell}_2^*)$ is a stochastic sequence to describe the permutation length of the type-II mutation, in which

$$\tilde{\ell}_\tau^* = (\ell_{\tau,1}^*, \ell_{\tau,2}^*, \ell_{\tau,3}^*, \cdots)\,, \quad \tau = 1, 2\,, \tag{2.65}$$

are two independently and identically distributed stochastic sequences and each $\ell_{\tau,j}^*$ obeys a geometric distribution:

$$P_r\{\ell_{\tau,j}^* = k\} = e_{p_\tau}(k) = p_\tau(1 - p_\tau)^{k-1}\,, \quad \tau = 1, 2\,. \tag{2.66}$$

Here, $(\ell^*_{1,j}, \ell^*_{2,j})$ represent two segments with lengths $\ell^*_{1,j}, \ell^*_{2,j}$ permutated after the jth mutation. Thus, $(\tilde{\ell}^*_1, \tilde{\ell}^*_2)$ is the length sequence of the permutated segments in type-II mutation.

II-2 Suppose $\tilde{\xi}$, $\tilde{\zeta}_2$, $\tilde{\ell}^*_1, \tilde{\ell}^*_2$ are four independent stochastic sequences. Let $\tilde{a}$, $\tilde{z}$, $\tilde{\ell}_1, \tilde{\ell}_2$ denote the samples of the stochastic sequences $\tilde{\xi}$, $\tilde{\zeta}_2$, $\tilde{\ell}^*_1, \tilde{\ell}^*_2$ in which

$$\begin{cases} \tilde{a} = (a_1, a_2, a_3, \cdots), \\ \tilde{z} = (z_1, z_2, z_3, \cdots), \\ \tilde{\ell}_\tau = (\ell_{\tau;1}, \ell_{\tau;2}, \ell_{\tau;3}, \cdots), \quad \tau = 1, 2 . \end{cases} \tag{2.67}$$

The construction of the type-II mutated sequence $\tilde{\eta}$ produced by $(\tilde{a}, \tilde{z}, \tilde{\ell}_1, \tilde{\ell}_2)$ is then as follows:

1. The renewal processes $\tilde{v}^*_2$ and $\tilde{j}^*_2$ are caused by $\tilde{\zeta}_2$.

$$v^*_{2,n} = \sum_{j=1}^{n} \zeta_{2,j} , \quad j^*_{2,k} = \sup\{n : v^*_{2,n} < k\} , \tag{2.68}$$

and their samples are respectively

$$\begin{cases} \tilde{v}_2 = (v_{2,0}, v_{2,1}, v_{2,2}, \cdots), \\ \tilde{j}_2 = (j_{2,0}, j_{2,1}, j_{2,2}, \cdots), \end{cases} \tag{2.69}$$

in which $v_{2,n}$ denotes the time of the type-II mutation in position region $\{1, 2, \cdots, n\}$. $j_{2,k}$ denotes the kth position occurrence of the type-II mutation, in which $v_{2,0} = j_{2,0} = 0$. $\tilde{v}^*_2, \tilde{j}^*_2$ are then determined by $\tilde{\zeta}_2$. Consequently, $\tilde{v}_2, \tilde{j}_2$ are determined by $\tilde{z}_2$.

2. If $\tilde{a}$, $\tilde{j}_2$, $\tilde{\ell}_1, \tilde{\ell}_2$ are similar to that given in (2.67) and (2.69), then sequence $\tilde{a}$ can be decomposed into several regions as follows:

$$\begin{cases} \delta_{2,0;k} = [j_{2,k-1} + \ell_{1,k-1} + \ell_{2,k-1} + 1, j_{2,k}] , \\ \delta_{2,1;k} = [j_{2,k} + 1, j_{2,k} + \ell_{1,k}] , \\ \delta_{2,2;k} = [j_{2,k} + \ell_{1,k} + 1, j_{2,k} + \ell_{1,k} + \ell_{2,k}] , \end{cases} \tag{2.70}$$

in which $k = 1, 2, 3, \cdots$, $[i, j] = \{i, i+1, \cdots, j\}$ is the set of positive integers. $[i, j] = \Phi$ if $i > j$, and $j_{2,0} = \ell_{1,0} = \ell_{2,0} = 0$.
Formula (2.70) divides a long sequence $V_+ = (1, 2, 3, \cdots)$ into several regions as follows:

$$V_+ = (\delta_{2,0,1}, \delta_{2,1,1}, \delta_{2,2,1}, \delta_{2,0,2}, \delta_{2,1,2}, \delta_{2,2,2}, \cdots , \\ \delta_{2,0,k}, \delta_{2,1,k}, \delta_{2,2,k}, \cdots) . \tag{2.71}$$

Then $\Delta_{2,0}$, $\Delta_{2,1}, \Delta_{2,2}$ represent the non-type-II mutation region and type-II region mutation respectively, in which

$$\begin{cases} \Delta_{2,0} = (\delta_{2,0;1}, \delta_{2,0;2}, \delta_{2,0;3}, \cdots), \\ \Delta_{2,1} = (\delta_{2,1;1}, \delta_{2,1;2}, \delta_{2,1;3}, \cdots), \\ \Delta_{2,2} = (\delta_{2,2;1}, \delta_{2,2;2}, \delta_{2,2;3}, \cdots) . \end{cases} \tag{2.72}$$

Equations (2.71) and (2.72) refer to the region structure of type-II mutated positions. The region in the region structure of type-II mutated positions is also random if $\tilde{\zeta}_2$, $\tilde{\ell}_1$, $\tilde{\ell}_2$ are stochastic sequences. Hence, (2.70) becomes

$$\begin{cases} \delta^*_{2,0;k} = \left[j^*_{2,k-1} + \ell^*_{1,k-1} + \ell^*_{2,k-1} + 1, j^*_{2,k}\right] , \\ \delta^*_{2,1;k} = \left[j^*_{2,k} + 1, j^*_{2,k} + \ell^*_{1,k}\right] , \\ \delta^*_{2,2;k} = \left[j^*_{2,k} + \ell^*_{1,k} + 1, j^*_{2,k} + \ell^*_{1,k} + \ell^*_{2,k}\right] . \end{cases} \tag{2.73}$$

Correspondingly, (2.71) and (2.72) become stochastic regions, such that

$$V_+ = (\delta^*_{2,0;1}, \delta^*_{2,1;1}, \delta^*_{2,2;1}, \delta^*_{2,0;2}, \delta^*_{2,1;2}, \delta^*_{2,2;2}, \cdots) .$$

Similarly,

$$\Delta^*_{2,\tau} = (\delta^*_{2,\tau;1}, \delta^*_{2,\tau;2}, \delta^*_{2,\tau;3}, \cdots) , \quad \tau = 0, 1, 2 .$$

Consequently, these random variables satisfy the relationships $j^*_{2,k-1} < j^*_{2,k}$, $\ell^*_{\tau,k} > 0$ for any $\tau = 1, 2$, $k = 1, 2, 3, \cdots$.

3. Let $\tilde{b} = (b_1, b_2, b_3, \cdots)$, $b_j \in V_4$ denote the samples of $\tilde{\eta}$, so that $\tilde{b}$ is determined by the following steps:
   (a) If $j \in \Delta_{2,0}$, then j is the position in the non-type-II mutated region and $b_j = a_j$ holds. This means the data are invariant.
   (b) If $j \in \Delta_{2,1} \cup \Delta_{2,2}$, then j is the position in the type-II mutated region and the data segments in $\delta_{2,1}$ and $\delta_{2,2}$ are permutated. Let the kth segment of $\Delta_{2,1}$ and $\Delta_{2,2}$ be

$$\begin{cases} \delta_{2,1;k} = \{j_k + 1, j_k + 2, \cdots , j_k + \ell_{1;k}\} , \\ \delta_{2,2;k} = \{j_k + \ell_{1;k} + 1, j_k + \ell_{1;k} + 2, \cdots , j_k + \ell_{1;k} + \ell_{2;k}\} . \end{cases} \tag{2.74}$$

   Then the permutated data of the kth type-II mutation is

$$b_{\delta_{2,0;k}} = a_{\delta_{2,0,k}} , \quad (b_{\delta_{2,1,k}}, b_{\delta_{2,2,k}}) = (a_{\delta_{2,2,k}}, a_{\delta_{2,1,k}}) , \tag{2.75}$$

   the notations of which are defined in (1.13) and (1.14).
4. Following from the above discussions, we will obtain the type-II mutated sequence $\tilde{\eta}$ defined by $\tilde{\xi}, \tilde{\zeta}_2, \tilde{\ell}_1, \tilde{\ell}_2$. Hence, let

$$\mathcal{E}^*_2 = \left\{\tilde{\xi}, \tilde{\zeta}_2, \tilde{\ell}^*_1, \tilde{\ell}^*_2, \tilde{\eta}\right\} \tag{2.76}$$

be the stochastic model of the type-II mutated sequence, in which each stochastic sequence satisfies conditions **II-1** and **II-2** and the supplemental explanations. From (2.75), the type-II mutated sequence $\tilde{\eta}$ can be determined.

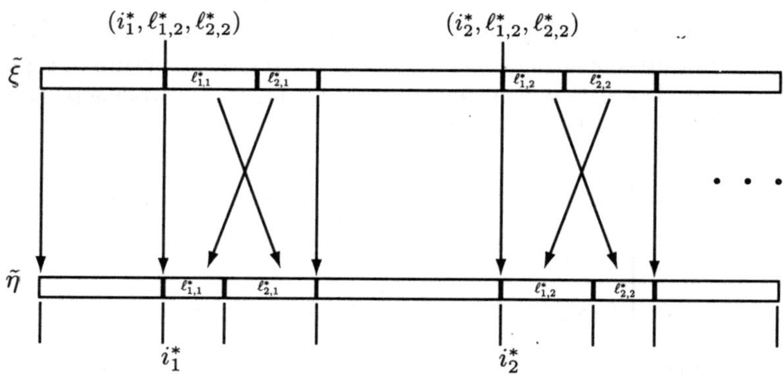

Fig. 2.3. Relationships of type-II mutations

Relationships of the type-II mutated sequences are shown in Fig. 2.3.

In Fig. 2.3, $\tilde{\eta}$ is a type-II mutated sequence of $\tilde{\xi}$, in which, i_k^* is the kth type-II mutated position and $\ell_{1,k}^*, \ell_{2,k}^*$ are the lengths of the permutated segments of the kth type-II mutations.

2.4.3 Error Analysis of Type-II Mutated Sequences

The error analysis of type-II mutated sequences indicates the calculation of penalty scores resulting from comparing a mutated sequence with the initial sequence. If $\tilde{\eta}$ is a type-II mutated sequence of $\tilde{\xi}$, then the definition of the local penalty function is the same as (2.55). Similarly, let

$$w\left(\tilde{\xi}, \tilde{\eta}; i, j, n\right) = \frac{1}{n} w\left(\bar{\xi}_i, \bar{\eta}_j\right) = \frac{1}{n} \sum_{k=1}^{n} w(\xi_{i+k}, \eta_{j+k}) . \tag{2.77}$$

We then discuss how to calculate or estimate the penalty scores.

Estimating the Lengths of Permutated Segments of Type-II Mutated Sequences

To estimate the value of (2.77), we need to estimate the total length of the permutated segments in the type-II mutated sequences. We begin by estimating the length of the permutated segment of type-II mutated sequences at the kth position.

1. Denote the length of the permutated segment at the kth position by $\ell_{0,k}^* = \ell_{1,k}^* + \ell_{2,k}^*$. Following from the i.i.d. property of $\tilde{\ell}_\tau^*$, $\tau = 1, 2$ individually, and the independence between the two sequences, we have that

$$\tilde{\ell}_0^* = \{\ell_{0,1}^*, \ell_{0,2}^*, \ell_{0,3}^*, \cdots\}$$

is an independent and identically distributed sequence, and the common probability distribution of $\ell_{0,k}^*$ can be calculated as follows:

$$\begin{aligned}
P_r\{\ell_{0,k}^* = \ell\} &= P_r\{\ell_{1,k}^* + \ell_{2,k}^* = \ell\} \\
&= \sum_{\ell'=1}^{\ell-1} P_r\{\ell_{1,k}^* = \ell'\} P_r\{\ell_{2,k}^* = \ell - \ell'\} \\
&= \sum_{\ell'=1}^{\ell-1} p_1(1-p_1)^{\ell'-1} p_2(1-p_2)^{\ell-\ell'-1} \\
&= \frac{p_1 p_2}{(1-p_1)(1-p_2)} \sum_{\ell'=1}^{\ell-1} (1-p_1)^{\ell'}(1-p_2)^{\ell-\ell'} .
\end{aligned}$$

When $p_1 \neq p_2$ we have

$$\begin{aligned}
P_r\{\ell_{2,j}^* = \ell\} &= \frac{p_1 p_2 (1-p_2)^\ell}{(1-p_1)(1-p_2)} \sum_{\ell'=1}^{\ell-1} \left(\frac{1-p_1}{1-p_2}\right)^{\ell'} \\
&= \frac{p_1 p_2 (1-p_2)^{\ell-1}}{1-p_2} \left(\frac{1 - \left(\frac{1-p_1}{1-p_2}\right)^{\ell-1}}{1 - \frac{1-p_1}{1-p_2}} \right) \\
&= p_1 p_2 \frac{(1-p_2)^{\ell-1} - (1-p_1)^{\ell-1}}{p_1 - p_2} .
\end{aligned} \tag{2.78}$$

If $p_1 = p_2 = p$, then (2.78) can be simplified as

$$P_r\{\ell_{2,j}^* = \ell\} = (\ell-1)p^2(1-p)^{\ell-2} , \quad \ell = 2, 3, 4, \cdots . \tag{2.79}$$

For simplicity, the probability distribution of $\ell_{0,k}^*$ is denoted by $p_0(\ell) = P_r\{\ell_{0,k}^* = \ell\}$.

Based on the expectation value formula and the variance of geometric distribution given, we calculate the mean and variance of $\ell_{2,j}^*$ as follows:

$$\begin{cases}
E\{\ell_{0,k}^*\} = E\{\ell_{1,k}^*\} + E\{\ell_{2,k}^*\} = \dfrac{1}{p_1} + \dfrac{1}{p_2} , \\
D\{\ell_{0,k}^*\} = D\{\ell_{1,k}^*\} + D\{\ell_{2,k}^*\} = \dfrac{1-p_1}{p_1^2} + \dfrac{1-p_2}{p_2^2} .
\end{cases} \tag{2.80}$$

2. Following from the definition of $\mathcal{E}_2^*$, we have that the counting process of type-II mutated sequences from the Bernoulli process $\tilde{\zeta}_2$ is defined as follows:

$$\tilde{v}_{2,n}^* = \sum_{i=1}^{n} \zeta_{2,i} , \quad n = 1, 2, 3, \cdots .$$

Then, let

$$\psi_{2,n} = \sum_{k=1}^{v_n^*} \ell_{0,k} , \quad n = 1, 2, 3, \cdots \tag{2.81}$$

be the compound renewal sequence of $\tilde{\zeta}_2$ and $\tilde{\ell}_0^*$.

3. Based on the property of the compound renewal process we can determine the limit character of the sequence $\psi_{2,n}$. Here

$$\frac{1}{n}\psi_{2,n} = \frac{1}{n}\sum_{k=1}^{v_n^*}\zeta_{2,k} \sim \epsilon_2\mu_0 \ , \tag{2.82}$$

in which $\mu_0 = \frac{1}{p_1} + \frac{1}{p_2}$, and

$$E\{\psi_{2,n}\} = E\left\{\sum_{k=1}^{v_n^*}\zeta_{2,k}\right\} = n\epsilon_2\mu_0 \ . \tag{2.83}$$

We can see that the central limit property is given by

$$\frac{1}{\sigma_2\sqrt{n}}\left[\psi_{2,n} - n\epsilon_2\mu_0\right] = \frac{1}{\sigma_2\sqrt{n}}\left(\sum_{k=1}^{v_n^*}\zeta_{2,k} - n\epsilon_2\mu_0\right) \sim N(0,1) \ , \tag{2.84}$$

in which

$$\sigma_2^2 = \frac{1}{n}D\{\psi_{2,n}\} = \epsilon_2\left[\frac{2-\epsilon_2-p_1}{p_1^2} + \frac{2-\epsilon_2-p_2}{p_2^2} + \frac{2(1-\epsilon_2)}{p_1p_2}\right] \ . \tag{2.85}$$

Equation (2.85) is obtained by the calculation of $\psi_{2,n}$. We do not present the derivations here.

Estimation of the Penalty Function $w(\tilde{\xi}, \tilde{\eta}; i, j, n)$

To estimate the penalty function of (2.84), the following must be considered:

1. We first discuss the calculation of $w(\xi_i, \eta_j)$. Here $w(a,b)$ is the penalty matrix on V_4 which satisfies the strongly symmetric condition. The values of the type-II mutated sequence η_j in $w(\xi_i, \eta_j)$ are just the components ξ_k of $\tilde{\xi}$. Therefore,

$$E\{w(\xi_i,\eta_j)\} = \begin{cases} 0 \ , & \text{if } \eta_j = \xi_k \ , \quad k = i \\ w_1 \ , & \text{if } \eta_j = \xi_k \ , \quad k \neq i \ , \end{cases} \tag{2.86}$$

in which $w_1 = \frac{1}{4}[w(0,1) + w(0,2) + w(0,3)]$.

2. We estimate the value of $w(\tilde{\xi}, \tilde{\eta}; i, j, n)$ for the condition $i = j$. For universality, we set $i = j = 0$, and denote

$$\Delta_2^*(n) = (\Delta_{2,1}^* \cup \Delta_{2,2}^*) \cap N \ , \tag{2.87}$$

in which the definition of $\Delta_{2,1}^*$, $\Delta_{2,2}^*$ is in (2.72) and $N = \{1, 2, \cdots, n\}$. Here $||\Delta_2^*(n)|| = \psi_{v_{2,n}^*}$ is established. If i is not in $\Delta_2^*(n)$, $w(\xi_i, \eta_i) \equiv 0$

holds. It implies that

$$w(\tilde{\xi},\tilde{\eta};i,j,n)=\frac{1}{n}\sum_{k\in\Delta_2^*(n)}w(\xi_k,\eta_k)\sim\frac{3}{4n}E\{\psi_{v_{2,n}^*}\}=\frac{3\epsilon_2}{4}\left(\frac{1}{p_1}+\frac{1}{p_2}\right) \tag{2.88}$$

holds. Relation (2.88) can be proved by the Markov large number theorem. We can regard the sum of $\{w(\xi_k,\eta_k),\ k\in\delta_n^*\}$ as the sum of a group of independent sequences

$$w_k^*=\sum_{j\in\delta_{2,1;k}^*\cup\delta_{2,2;k}^*}w(\xi_j,\eta_j)\,,\quad k=1,2,3,\cdots$$

so we can use the large number theorem to obtain (2.88).

3. If $i\neq j$, we can also get

$$w(\tilde{\xi},\tilde{\eta};i,j,n)\sim\frac{1}{4}[w(0,1)+w(0,2)+w(0,3)]\,. \tag{2.89}$$

Using the central limit theorem, we can accurately estimate the relationship (2.89). Details of the calculation are omitted here.

2.4.4 The Mixed Stochastic Models Caused by Type-I and Type-II Mutations

We have already described type-I and type-II mutated sequences. Now we consider the combination of these two types.

The Mixed Stochastic Models for Both Type-I and Type-II Mutated Sequences

The mixed stochastic models for both type-I and type-II mutated sequences are described as follows. Here, the mixed stochastic model is denoted by

$$\mathcal{E}_{\mathrm{I,II}}^*=\{\tilde{\xi},\tilde{\vartheta}_1,\tilde{\zeta}_\tau,\tilde{\ell}_\tau^*,\tilde{\eta},\ \tau=1,2\} \tag{2.90}$$

in which:

1. $\tilde{\xi}$ is the initial sequence, which is also a perfect sequence.
2. ζ_τ, $\tau=1,2$ are two Bernoulli processes representing the flows caused by type-I and type-II mutations, respectively.
3. $\tilde{\vartheta}_1$ is the sequence added by all type-I mutated errors given in (2.49).
4. ℓ_τ^*, $\tau=1,2$ are the sequences consisting of the lengths of the permutated segments of type-II mutation, given by (2.65).

All stochastic sequences in $\mathcal{E}_{\mathrm{I,II}}^*$ are therefore independent, and their mutation types can be determined based on the regulations **I-1**, **I-2**, **II-1**, and **II-2**. The mutated sequence $\tilde{\eta}$ is then produced by the following rules.

5. Let $\eta_j = \xi_j$ if $\zeta_{1,j} = \zeta_{2,j} = 0$.
6. Using the rules **I-1** and **I-2** to determine η_j if $\zeta_{1,j} = 1$ and $\zeta_{2,j} = 0$.
7. We follow the rules **II-1** and **II-2** to determine the values of $\eta_{\Delta_{2,0}}$ and $\eta_{\Delta^*_{2,1} \cup \delta^*_{2,2}}$ if $\zeta_{1,j} = 0$ and $\zeta_{2,j} = 1$.
8. We assume that $\zeta_{1,j} = 1$, $\zeta_{2,j} = 1$ do not occur simultaneously.

Error Analysis of Type-I and Type-II Mutations

If $\tilde{\eta}$ is a sequence caused by type-I and type-II mutations, then the local penalty functions $w(\tilde{\xi}, \tilde{\eta}; i, j, n)$ about both $\tilde{\eta}$ and $\tilde{\xi}$ are estimated as follows:

1. If $i = j$, we can see from (2.61) and (2.88) that
$$w(\tilde{\xi}, \tilde{\eta}; i, j, n) \sim \frac{\epsilon_1 w}{3} + \left(\frac{3}{4} - \frac{w\epsilon_1}{3}\right) \epsilon_2 \left(\frac{1}{p_1} + \frac{1}{p_2}\right) , \qquad (2.91)$$
in which ϵ_1, ϵ_2 are the strengths of the Bernoulli processes, and $\tilde{\zeta}_\tau$, $\tau = 1, 2$ and p_1, p_2 are parameters of the geometric distribution of $\tilde{\ell}^*_\tau$, $\tau = 1, 2$ and $w = w(0,1) + w(0,2) + w(0,3)$.
2. If $i \neq j$, from (2.77) and (2.89) we obtain
$$w(\tilde{\xi}, \tilde{\eta}; i, j, n) \sim \frac{w'}{4} . \qquad (2.92)$$

2.5 Mutated Sequences Resulting from Type-III and Type-IV Mutations

Definitions of type-III and type-IV mutations are given in Chap. 1. They are referred to as displacement mutations. We discuss the stochastic models of type-III and type-IV mutated sequences below.

2.5.1 Stochastic Models of Type-III and Type-IV Mutated Sequences

The definition of the stochastic model for type-III and type-IV mutated sequences is as follows:

$$\begin{cases} \mathcal{E}^*_3 = \{\tilde{\xi}, \tilde{\zeta}_3, \tilde{\ell}^*_3, \tilde{\vartheta}_3, \tilde{\eta}\} , \\ \mathcal{E}^*_4 = \{\tilde{\xi}, \tilde{\zeta}_4, \tilde{\ell}^*_4, \tilde{\eta}\} . \end{cases} \qquad (2.93)$$

These satisfy the following conditions:

III-1 For a fixed $\tau = 3, 4$, the three stochastic sequences $\tilde{\xi}, \tilde{\zeta}_\tau, \tilde{\ell}^*_\tau$ are i.i.d. processes, in which:
1. $\tilde{\xi}$ is the initial process, which is a perfectly stochastic sequence on V_4.

2. $\tilde{\zeta}_\tau$ is a Bernoulli process with strength ϵ_τ in which each $\zeta_{\tau,j}$ denotes the random variable that describes whether or not the type-τ mutation occurs.
3. $\tilde{\ell}^*_\tau$ is a stochastic sequence obeying geometric distribution with the parameter p_τ. Each $\ell^*_{\tau,j}$ then denotes the sequence of the lengths of segments in the jth insertion or deletion.
4. $\tilde{\vartheta}_3$ is a perfectly stochastic sequence on V_4. It is a type-III mutated sequence. That is, it is mutated from the initial sequence by insertion of some "–".

III-2 For a fixed $\tau = 3, 4$, the three stochastic sequences $\tilde{\xi}, \tilde{\zeta}_\tau, \tilde{\ell}^*_\tau$ are independent, and $\tilde{\vartheta}_3$ is independent of $\tilde{\xi}, \tilde{\zeta}_3, \tilde{\ell}^*_3$.
Let $\tilde{v}^*_\tau$ be the renewal process of $\tilde{\zeta}_\tau$ and let $\tilde{j}^*_\tau$ be the corresponding dual renewal process (definitions are given in Sect. 2.2). Let

$$\begin{cases} \mathcal{E}_3 = \{\tilde{a}, \tilde{z}_3, \tilde{\ell}_3, \tilde{u}_3\}, \\ \mathcal{E}_4 = \{\tilde{a}, \tilde{z}_4, \tilde{\ell}_4\} \end{cases} \tag{2.94}$$

be the sample of $\mathcal{E}^*_\tau$. The procedure to determine $\mathcal{E}^*_\tau$ based on $\mathcal{E}_3$ or $\mathcal{E}_4$ is stated as follows:

1. Determine the renewal sequence $\tilde{v}_\tau$ and its corresponding dual renewal process $\tilde{j}_\tau$ based on $\tilde{z}_\tau$. Then
$$1 \le j_{\tau,1} < j_{\tau,2} < j_{\tau,3} < \cdots \tag{2.95}$$
divides the whole set of positive integers $V_+ = (1, 2, 3, \cdots)$ into several segments.
2. The role of the type-IV mutation is to delete the segment
$$a_{(j_{4,k}, j_{4,k}+\ell_{4,k}+1)} = (a_{j_{4,k}+1}, a_{j_{4,k}+2}, \cdots, a_{j_{4,k}+\ell_{4,k}})$$
from the initial sequence $\tilde{a} = (a_1, a_2, a_3, \cdots)$.
3. The role of the type-III mutation is to insert the segment
$$u_{(\psi_{4,k-1}, \psi_{4,k}+1)} = (u_{\psi_{4,k-1}+1}, u_{\psi_{4,k-1}+2}, \cdots, u_{\psi_{4,k}})$$
into $\tilde{a} = (a_1, a_2, a_3, \cdots)$ following $a_{j_{3,k}}$ in which $\psi_{4,k} = \sum_{i=1}^k \ell_{4,j}$ and $u_j \in V_4$ belongs to $\tilde{a}$.

Consequently, we can find the type-III and type-IV mutated sequences from $\tilde{\xi}$ based on the various sequences and generation rules **III-1** and **III-2**.

2.5.2 Estimation of the Errors Caused by Type-III and Type-IV Mutations

We can estimate the local penalty function $w(\tilde{\xi}, \tilde{\eta}; i, j, n)$ of the type-III and type-IV mutated sequence $\tilde{\eta}$ as follows.

Estimation of the Lengths of the Inserted or Deleted Segments in Type-III and Type-IV Mutated Sequences

We can estimate the lengths of the inserted or deleted segments in type-III and type-IV mutated sequences using renewal processes. Let $\psi_{\tau,i,n}$ be the total number of inserted or deleted segments within the region $[i+1, i+n] = \{i+1, i+2, \cdots, i+n\}$, so that the probability of $\psi_{\tau,i,n}$ is independent of i because of the homogeneous nature of stochastic models. Thus, it is enough that we consider

$$\psi_{\tau,n} = \sum_{j=1}^{v^*_{\tau,n}} \ell^*_{\tau,j} . \tag{2.96}$$

Using the limitation property of the compound renewal process, we have

$$E\{\psi_{\tau,n}\} = E\{v^*_{\tau,n}\}E\{\ell^*_{\tau,1}\} = \frac{\epsilon_\tau}{p_\tau} , \quad \tau = 3, 4 . \tag{2.97}$$

Following from the law of large numbers for compound renewal processes, we have

$$\frac{1}{n}\psi_{\tau,n} \sim E\{v^*_{\tau,n}\}E\{\ell^*_{\tau,1}\} = \frac{\epsilon_\tau}{p_\tau} , \quad \tau = 3, 4 . \tag{2.98}$$

$\psi_{3,n}$ is then the combined length of all inserted segments in the region $[1, n]$, and $\psi_{4,n}$ is the combined length of all deleted segments in region $[1, n]$.

Estimation of Local Penalty Function $w(\tilde{\xi}, \tilde{\eta}; i, j, n)$

Without loss of generality, we discuss the estimation of the local penalty function $w(\tilde{\xi}, \tilde{\eta}; i, j, n)$ on type-III mutated sequences as follows:

1. In the case $i = j = 1$, let $\tilde{\zeta}_3$ be the dual renewal process of $\tilde{v}^*_3$. If $v^*_{3,1} = v < n$, then there is no mutation in the region $\delta_1 = [1, v] = (1, 2, \cdots, v)$ but there is a displacement mutation in the region $\delta = [v+1, n] = (v+1, v+2, \cdots, n)$. Consequently, following from the large number law, we have

$$w(\tilde{\xi}, \tilde{\eta}; i, j, n | v^*_{3,1} = v) \sim \frac{(n-v)w}{4n} \tag{2.99}$$

and

$$w(\tilde{\xi}, \tilde{\eta}; i, j, n) = \sum_{v=1}^{n} p_r\{v^*_{3,1} = v\} w(\tilde{\xi}, \tilde{\eta}; i, j, n | v^*_{3,1} = v)$$
$$\sim \sum_{v=1}^{n} \epsilon_3 (1-\epsilon_3)^{v-1} \frac{(n-v)w}{4n} .$$

If n is large enough, we have the relationship

$$w(\tilde{\xi}, \tilde{\eta}; i, j, n) \sim \frac{w}{4}\left(1 - \frac{1}{n\epsilon_3}\right) . \tag{2.100}$$

2. In the case where $i < j$, since $\tilde{\eta}$ is a type-III mutated sequence of $\tilde{\xi}$, we find
$$w(\tilde{\xi}, \tilde{\eta}; i, j, n) \sim \frac{w}{4} . \tag{2.101}$$
3. In the case where $i > j$, let $\tilde{v}_3^*$ be the dual renewal process of $\tilde{\zeta}_3$. If there is a $k > 0$ such that $v_{3,k}^* = i - j$, then following from (2.100), we have
$$w(\tilde{\xi}, \tilde{\eta}; i, j, n | v_{3,k}^* = i - j) \sim \frac{w}{4}\left(1 - \frac{1}{n\epsilon_3}\right) , \tag{2.102}$$
otherwise (2.100) holds. Following from (2.100) and (2.101), we may obtain an estimate of $w(\tilde{\xi}, \tilde{\eta}; i, j, n)$. However, we omit this here because the computation is far too complex.

2.5.3 Stochastic Models of Mixed Mutations

A stochastic model of mixed mutations is the stochastic model that results when we have type-I, type-II, type-III, and type-IV mutations all occurring at the same time.

Definition of Stochastic Model of Mixed Mutations

Stochastic models of mixed mutations can be described by the following relationship:
$$\mathcal{E}^* = \left\{\tilde{\xi}, \tilde{\zeta}_\tau, \tilde{\ell}_\tau^*, \tilde{\vartheta}_{\tau'},\ \tau = 1, 2, 3, 4\,;\ \tau' = 1, 3\right\} \tag{2.103}$$
satisfying the following conditions:

IV-1 Sequences of $\mathcal{E}^*$ are homogeneous and i.i.d. processes, in which:
1. $\tilde{\xi}$ is a perfectly stochastic sequence on V_4 as the initial sequence.
2. $\tilde{\zeta}_\tau$ is a Bernoulli process with strength ϵ_τ, in which, each $\zeta_{\tau,j}$ is a random variable describing whether or not type-τ mutation occurs.
3. $\tilde{\ell}_\tau^*$ is a stochastic sequence obeying a geometric distribution with a parameter p_τ, in which, $\ell_{1,j}^*$, $\ell_{2,j}^*$ are the lengths of the permutated segments if the jth mutation is type-II, while $\ell_{3,j}^*$, $\ell_{4,j}^*$ represent the length of the inserted or deleted segment if the jth mutation is type-III or type-IV, respectively.
4. $\tilde{\vartheta}_1$ is the additional error sequence resulting from type-I mutations, and is defined in **I-3**. $\tilde{\vartheta}_3$ is a perfectly stochastic sequence on V_4, which is the inserted sequence resulting from type-III mutation.

IV-2 All stochastic sequences in $\mathcal{E}^*$ are independent. For any $\tau = 1, 2, 3, 4$, the sequence $\tilde{v}_\tau^*$ is the renewal process of the sequence $\tilde{\zeta}_\tau$, and $\tilde{j}_\tau^*$ is the dual renewal process of $\tilde{\zeta}_\tau$.

IV-3 In the mixed stochastic model $\mathcal{E}^*$, $\{\epsilon_\tau, p_\tau, \tau = 1,2,3,4\}$ is the parameter set, in which, ϵ_τ, $\tau = 1,2,3,4$ represents the strengths of the four types of mutations, p_1, p_2 are parameters of the geometric distribution that describes the type-II mutated sequences, and p_3, p_4 are parameters of the geometric distribution that describes the type-III and type-IV mutated segments, respectively.

Samples of the Stochastic Model of Mixed Mutations

Let

$$\mathcal{E} = \{\tilde{a}, \tilde{z}_\tau, \tilde{\ell}_\tau, \tilde{u}_{\tau'},\ \tau = 1,2,3,4\,;\ \tau' = 1,3\} \tag{2.104}$$

be the sample of $\mathcal{E}^*$, so that we may obtain the mutated sequences through the following steps:

1. Determine the renewal sequence $\tilde{v}_\tau$ and the dual renewal process $\tilde{j}_\tau$, based on $\tilde{z}_\tau$.
2. Determine the type-I mutated sequence b_j according to steps **I-1**–**I-4** in case $z_{1,j} = 1$.
3. Determine the type-II mutated sequence $b_{(j,j+\ell_{2,k})}$, according to steps **II-1** and **II-2** in the case $z_{2,j} = 1$, in which $k = v_{2,j}$.
4. Determine the type-III or type-IV mutated sequences according to steps **III-1** and **III-2** in the cases $z_{3,j} = 1$ or $z_{4,j} = 1$.
5. For the same j, we assume that there are no more than two types such that the state of $z_{\tau,j}, \tau = 1,2,3,4$ is 1, which is based on the following. Since
$$P_r\{z_{1,j} + z_{2,j} + z_{3,j} + z_{4,j} \geq 2\} \leq \sum_{i \neq j \in \{1,2,3,4\}} \epsilon_j \epsilon_j$$
and $\epsilon_1 + \epsilon_2 + \epsilon_3 + \epsilon_4 \ll 1$, we have $P_r\{z_{1,j} + z_{2,j} + z_{3,j} + z_{4,j} \geq 2\} \sim 0$.

Estimation of Crossed Probability of Mixed Mutations

Since type-II, type-III, and type-IV mutations involve only a segment (if the mutation does in fact occur), it implies that different mutation types could happen in the same regions. This is what is meant by the term "crossed mutations". We can estimate the probability of this occurring, which is referred to as the crossed probability of mixed mutations. Typically, the crossed probability of type-III and type-IV is

$$\begin{aligned}&P_r\{\text{in region } N, \text{ type-III and type-IV mutations both happen}\}\\ &= P_r\{\Delta_3^*(n) \cap \delta_4^*(n) \neq \phi\}\ ,\end{aligned} \tag{2.105}$$

in which ϕ is the empty set and $\Delta_\tau^*(n)$, $\tau = 3,4$ are the same as that defined in (2.87), then

$$\Delta_\tau^*(n) = \Delta_\tau^* \cap N\,, \quad \tau = 3,4\ , \tag{2.106}$$

$\Delta_\tau^* = (\delta_{\tau,1}^*, \delta_{\tau,2}^*, \delta_{\tau,3}^*, \cdots)$ and

$$\delta_{\tau,k} = (j_{\tau,k}^* + 1, j_{\tau,k}^* + 2, \cdots, j_{\tau,k}^* + \ell_{\tau,k}^*) \ .$$

We compute the values of (2.104) as follows:

1. Let the value of (2.104) be Γ; we then have

$$\begin{aligned}\Gamma \le & P_r \left\{\text{there is a } j \in \Delta_3^*(n) \text{, such that } \zeta_{4,j} = 1\right\} \\ & + P_r \left\{\text{there is a } j \in \delta_4^*(n) \text{, such that } \zeta_{3,j} = 1\right\} \ . \end{aligned} \tag{2.107}$$

2. Let the two items on the left hand side of (2.107) be Γ_1, Γ_2, then we estimate the values of Γ_1, Γ_2 as follows.
   Using the limit property of compound renewal processes, we have that the length of $\delta_{3,n}^*$ is close to $\frac{n\epsilon_3}{p_3}$. Consequently, the probability that $\zeta_{\delta_{3,n}^*}$ is the 0-vector is $(1-\epsilon_4)^{n\epsilon_3/p_3}$. Then,

$$\begin{aligned}\Gamma_1 &= 1 - (1-\epsilon_4)^{n\epsilon_3/p_3} = [1 - (1-\epsilon_4)^{1/\epsilon_4}]^{(n\epsilon_3\epsilon_4)/p_3} \\ &\sim 1 - \exp\left(-\frac{n\epsilon_3\epsilon_4}{p_3}\right) \ .\end{aligned}$$

   Similarly,

$$\Gamma_2 \sim 1 - \exp\left(-\frac{n\epsilon_3\epsilon_4}{p_4}\right) \ .$$

   Therefore,

$$\Gamma \le 2 - \exp\left(-\frac{n\epsilon_3\epsilon_4}{p_3}\right) - \exp\left(-\frac{n\epsilon_3\epsilon_4}{p_4}\right) \ . \tag{2.108}$$

Based on (2.107), we can say that the combined probability of type-III and type-IV mutations happening in the region $[1, n]$ at the same time is very small if $n \ll \frac{1}{\epsilon_3\epsilon_4}$. We do not give the proof here.

Semistochastic Models of Mutations

We have introduced the stochastic models of mutated sequences, which show many factors affecting the mutations, as well as descriptions of the randomness of these factors. Following from these descriptions, we find some characteristics of the stochastic models of mutated sequences, as well as how to use these stochastic analysis approaches in bioinformatics. Of course, there are many problems still to be solved before these models can become useful in practice.

In the model $\mathcal{E}^*$ of (2.102), we still regard $\tilde{\xi}$ as a stochastic sequence and let

$$T^* = \left\{\tilde{\zeta}_\tau, \tilde{\ell}_\tau^*, \tilde{\vartheta}_{\tau'}, \ \tau = 1, 2, 3, 4, \ \tau' = 1, 3\right\} \tag{2.109}$$

be the stochastic model of $\mathcal{E}^*$. If T is a fixed sample

$$T = \left\{ \tilde{z}_\tau, \tilde{\ell}_\tau, v_{\tau'},\ \tau = 1,2,3,4,\ \tau' = 1,3 \right\} \tag{2.110}$$

then $\mathcal{E}' = \{\tilde{\xi}, T\}$ is a hybrid model of the four types of mutations. We refer to $\mathcal{E}' = \{\tilde{\xi}, T\}$ as the semistochastic model of mutated sequences. We will discuss the structure and analysis of hybrid mutations T in the following chapters.

2.6 Exercises

Exercise 6. Explain the differences between the following terms: i.i.d. sequences, Bernoulli process, Poisson process, geometric distribution sequence, additive sequence, Markov process, and renewal process. Explain the advantages and disadvantages of using each of the above to describe biological sequences.

Exercise 7. Try to extend the application of the Bernoulli process and Poisson process to the case of nonhomogeneous sequences, and use it to describe the model of type-III mutated sequences.

Exercise 8. List some of the important laws of large number and central limit theorems in probability theory, and give examples showing how they apply to biological sequences.

Exercise 9. Prove the following propositions:

1. Properties 1, 2, and 3 of the renewal process given in Sects. 2.2.1–2.2.3
2. Theorems 5 and 7.
3. Formulas (2.82), (2.86) (2.100), (2.101), and (2.102).

Exercise 10. For the stochastic sequences in model (2.103), perform a simulation according to the following cases:

1. The range of the sequence lengths is 1 kbp–1 Mbp (i.e., $n = 1 \times 10^3, 1 \times 10^4, 1 \times 10^5, 5 \times 10^5, 1 \times 10^6$, etc.); the range of ϵ_1 is 0.01–0.4 (i.e., $\epsilon_1 = 0.01, 0.02, 0.05, 0.1, 0.2, 0.3, 0.4$, etc.); the range of $\epsilon_2, \epsilon_3, \epsilon_4$ is 0.01–0.1 (i.e., $0.01, 0.02, \cdots, 0.1$ etc.); and the range of p_1–p_4 is 0.1–0.5 (i.e., 0.1, 0.2, 0.3, 0.4, 0.5, etc.).
2. For the sequence $\tilde{\xi}$ in (2.103), construct the i.i.d. sequence defined on V_4 which obeys uniform distribution.
3. Create the stochastic sequences of model (2.103) according to the parameters given in case 1.

Exercise 11. Based on the simulation results from Exercise 10, align the mutated sequences using the dynamic programming-based algorithm, and compare the time taken by the CPU with the parameters listed as follows:

1. For the parameter ϵ_1–ϵ_4, p_1–p_4, find the relationship between the length of the sequences and the time taken by the CPU, where the length ranges from 1×10^3 to 1×10^6.
2. For a sequence of fixed length (such as 1×10^4), compare the average error of the model (2.103) defined in (2.108) with the time taken by the CPU.

Hint

The simulation of the stochastic sequence and mutated sequences is achieved through the generation of a sequence of random numbers.

3

Modulus Structure Theory

In Chap. 2, we presented the stochastic model of mutated sequences. If samples of a stochastic model are available, then we should be able to estimate its characteristics. To do this, we introduce a new concept called modulus structure, which is a powerful tool for describing the characteristics of these stochastic models. The modulus structure may be determined in many ways, and we will introduce three of these: by expanding or compressing a sequence onto a new sequence; by sequence alignment, and by sequence mutation. The usefulness of modulus structure will become clear after we have discussed these methods. Alternate concepts, such as modulus structure and mode, will also be involved.

3.1 Modulus Structure of Expanded and Compressed Sequences

Expansion and compression are frequent occurrences in mutated sequences and in aligned sequences. They have been defined in Sect. 1.1, and here we give a more detailed description and discussion of their structures.

3.1.1 The Modulus Structures of Expanded Sequences and Compressed Sequences

Definitions of Expansion and Compression

Let A and C be two sequences on V_4 as defined in (1.1), and let $N_a = \{1, 2, \cdots, n_a\}$, $N_c = \{1, 2, \cdots, n_c\}$ be the sets of the positions of sequences A and C, respectively. We begin with the definitions of expansion and compression as follows:

Definition 10. *1. If A is a subsequence of C, then C is an expansion of A. Conversely, A is a compression of C. If A is a real subsequence of C, the* *corresponding expansion and compression are called the real expansion and real compression, respectively.*

2. *Let $\mathcal{A} = \{A_t,\ t = 1, 2, \cdots, m\}$ be a set of sequences defined on V_4. $\mathcal{A}$ then represents the set of homologous expansions of A if all A_t, $t = 1, 2, \cdots, m$ are the expansions of A. A sequence D is called the core of the set $\mathcal{A}$ if D is the common compression of all sequences in $\mathcal{A}$. Similarly, for multiple sequences $\mathcal{A}$, a sequence C is called the envelope of $\mathcal{A}$ if each sequence in $\mathcal{A}$ is a compressed sequence of C.*
3. *D_0 is the maximal core of $\mathcal{A}$ if D_0 is the core of $\mathcal{A}$ and any real larger expansion of D_0 is not the core of $\mathcal{A}$. The longest maximal core is called the maximum core. Similarly, a sequence C_0 is the minimal envelope of $\mathcal{A}$ if it is the envelope of $\mathcal{A}$ and any real compression of C_0 will not be an envelope of $\mathcal{A}$. The smallest such envelope is called the minimum envelope.*

If A is a subsequence of C, then there is a subset $\alpha = \{j_1, j_2, \cdots, j_{n_a}\}$ of N_c such that $1 \le j_1 < j_2 < \cdots < j_{n_a} \le n_c$, and

$$c_\alpha = (c_{j_1}, c_{j_2}, \cdots c_{j_{n_a}}) = (a_1, a_2, \cdots a_{n_a}) = A \ . \tag{3.1}$$

This set α represents the positions if A is embedded into C. Let $\alpha^c = N_c - \alpha$ be the complementary subset of α, then $B = c_{\alpha^c}$ is nothing but the virtual symbol "$-$" to get the expansion of A, in which, $c_{\alpha^c} = (c_j,\ j \in \alpha^c)$. If $\alpha \subset N_c$, and $C = (c_\alpha, c_{\alpha^c}) = (A, B)$, then the binary (A, B) is a decomposition of C, and they are both compressions of C. If sequence C is defined on $V_5 = \{0, 1, 2, 3, 4\}$, then the virtual expanded sequence C based on A is defined in Definition 2.

Modulus Structure of Expanded Sequence and Compressed Sequence

If A is a subsequence of sequence C, then the relationship between A and C can be described by (3.1). However, (3.1) becomes too complex as the lengths of sequences A and C increase. To simplify the description, we introduce some definitions and notations as follows:

$$K_c = (j_0, j_1, j_2, \cdots, j_{2k_c-1}, j_{2k_c}) \ , \tag{3.2}$$

in which $0 = j_0 \le j_1 < j_2 < \cdots < j_{2k_c} \le j_{2k_c+1} = n_c$. Then N_c can be subdivided into many smaller intervals as follows:

$$\delta'_k = [j_k + 1, j_{k+1}] = (j_k + 1, j_k + 2, \cdots, j_{k+1}) \quad k = 0, 1, 2, \cdots, 2k_c \ . \tag{3.3}$$

Based on these small intervals δ'_k, we define

$$\begin{cases} \Delta'_1 = (\delta'_1, \delta'_3, \cdots, \delta'_{2k_c-1}) \ , \\ \Delta'_2 = (\delta'_0, \delta'_2, \cdots, \delta'_{2k_c}) \ . \end{cases} \tag{3.4}$$

Therefore, $N_c = (\Delta_1', \Delta_2')$. The binary (Δ_1', Δ_2') is a decomposition of N_c and is determined by K. It suffices to show that $\Delta_1' \cap \Delta_2' = \phi$ is an empty set and $\Delta_1' \cup \Delta_2' = N_c$.

Definition 11. *1. If K_c is defined by (3.2) and the decomposition of K_c is defined by (3.4), satisfying the following condition:*

$$c_{\Delta_1'} = \left(c_{\delta_1'}, c_{\delta_3'}, \cdots, c_{\delta_{2k_c-1}'}\right) = A \,, \tag{3.5}$$

then K_c is the expanded mode from A to C or the compressed mode from C to A, where $c_{\delta_k'} = (c_{j_k+1}, c_{j_k+2}, \cdots, c_{j_{k+1}})$ is a subvector of C restricted to the interval δ_k'.

2. The pairwise sequence (A, B) in (3.4) is a decomposition of C and the sequence C is an expansion of both A and B. K is a decomposition mode of C related to (A, B), if (3.5) and (3.6) are satisfied.

$$c_{\Delta_2'} = \left(c_{\delta_2'}, c_{\delta_4'}, \cdots, c_{\delta_{2k_c}'}\right) = B \tag{3.6}$$

3. If C is a sequence defined on V_5 and A is a sequence defined on V_4 such that $C = (A, B)$ and $B = (4, 4, \cdots, 4)$ is a sequence of virtual symbols, then C is a virtual expansion of A, and K is the virtually expanded mode for A virtually expanding to C.

Operations for Position-Shifting in Expanded and Compressed Sequences

We add some new notations in (3.2) as follows:

$$\ell_k = j_{2k+1} - j_{2k} \,, \quad L_k = \sum_{k'=1}^{k-1} \ell_{k'} \,, \quad i_k = j_{2k-1} - L_k \,. \tag{3.7}$$

Then ℓ_k is the length of the kth inserted vector and L_k is the shifting function resulting from insertions. If the expanded mode K_c is known, then the relationship between A and C will be determined by K_c, as shown by the following theorem.

Theorem 8. *If C is an expanded sequence of A under the expanded mode K_c, then*

$$A = \left(a_{\delta_1}, a_{\delta_2}, \cdots, a_{\delta_{k_c}}\right) = \left(c_{\delta_1'}, c_{\delta_3'}, \cdots, c_{\delta_{2k_c-1}'}\right) \tag{3.8}$$

in which $a_{\delta_k} = c_{\delta_{2k-1}'}$ and

$$\delta_k = \delta_{2k-1}' - L_k = [i_k + 1, i_{k+1}] = (i_k + 1, i_k + 2, \cdots, i_{k+1}) \,. \tag{3.9}$$

Proof. Since C is an expansion of A, it suffices to prove that the second equation of (3.6) and (3.8) hold. Changing $A = (a_1, a_2, \cdots, a_{n_a})$ into the form of (3.8) gives $a_{\delta_k} = c_{\delta_{2k-1}'}$. Since the distance between the intervals δ_{2k-1}' and δ_{2k+1} is ℓ_k, it follows that (3.9) holds, which concludes the proof.

Equation (3.9) is the position-shifting formula of expanded sequences in the following text.

The Equivalent Representations of Modes

If K_c is given, then the function (i_k, ℓ_k, L_k) is determined by (3.7). Let

$$H_a = \{(i_k, \ell_k),\ k = 1, 2, \cdots, k_c\} \tag{3.10}$$

be the modulus structure expanding A to C. H_a is then the set of all orders describing how to expand A. Typically, the order (i_k, ℓ_k) means that we insert a vector with length ℓ_k following the position i_k. Notice how this is represented in K_c and H_a. Here, K_c targets the positions of C, while H_a is targeting the positions of A. Therefore, we use different subscripts to differentiate them in the following.

Theorem 9. *The expanded modes K_c and the modulus structure H_a are equivalent.*

Proof. Following from (3.7) and (3.10), we know that H_a is determined if K_c is given. Consequently, it is sufficient to show that K_c is also determined by H_a. L_k is then determined by ℓ_k and the second equation of (3.7) if H_a is given. Thus, it follows from (3.10) that j_{2k-1} is determined for all $k = 1, 2, \cdots, k_c$. We can use the first equation of (3.7) to calculate all j_{2k} using the formula $j_{2k} = j_{2k+1} - \ell_k$ for all $k = 1, 2, \cdots, k_c$. Therefore, each of K_c and H_a is determined by the other, concluding the proof.

Let $I_a = \{i_k : k = 1, 2, \cdots, k_c\}$ be the set of all positions after which vectors will be inserted. This is the set of inserting positions to get the expanded sequence C. There is then an alternative expression of H_a as follows:

$$H'_a = (\gamma_0, \gamma_1, \gamma_2, \cdots, \gamma_{n_a}) \tag{3.11}$$

in which $\gamma_i \in V_0 = \{0, 1, 2, \cdots\}$ and

$$\gamma_i = \begin{cases} \ell_k & \text{if } i = i_k \in I_a\,, \\ 0 & \text{otherwise}\,. \end{cases}$$

$\gamma_j = 0$ implies that there will be no virtual symbols inserted after j, and $\gamma_j > 0$ means there is a virtual symbol of length γ_j inserted after j. It is obvious that H_a and H'_a determine each other. Consequently, K_c, H_a and H'_a are all equivalent. The modulus structure of the expanded sequence is demonstrated in Fig. 3.1. Due to their equivalence we do not need to distinguish these names.

In Fig. 3.1, C is the expanded sequence of A. The original part A and the expanded part are represented by different line segments, where i_k, δ_k are positions and intervals of sequence A respectively, $j_{k'}, \delta_{k'}$ are positions and intervals of sequence C, respectively, and where $\ell_k = |\delta'_{2k}|$ is the length of δ'_{2k}.

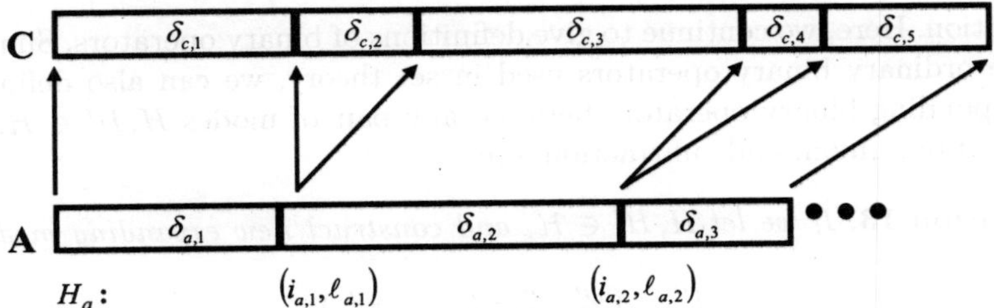

Fig. 3.1. The modulus structure of the expanded sequence

The Modulus Structure of the Compressed Sequences

If C is the expanded sequence of A, then A is the compressed sequence of C. Let

$$H_c = \{(j'_k, \ell'_k),\ k = 1, 2, \cdots, k_c\} , \tag{3.12}$$

in which $j'_k = j_{2k}$, $\ell'_k = j_{2k+1} - j_{2k}$ is determined by K_c of (3.2). It is then referred to as the as compressed mode from C to A. This mode tells us how to delete a vector of length ℓ'_k after position j'_k of C.

3.1.2 The Order Relation and the Binary Operators on the Set of Expanded Modes

Let A be a sequence defined on V_4. There will exist many expanded sequences based on A. These expanded sequences are quite different if the inserted vectors or expanding modes are different. In this section, we illustrate how to compare these modes, and give the binary operators on the set of expanding modes.

The Order Relation and the Binary Operators Defined on the Set of Modes

Let C and C' be two expanded sequences based on A, and let

$$H = (\gamma_0, \gamma_1, \gamma_2, \cdots, \gamma_{n_a}) , \quad H' = (\gamma'_0, \gamma'_1, \gamma'_2, \cdots, \gamma'_{n_a})$$

be the two corresponding expanding modes. We can compare the two modes by defining the "order relation" on the set of all modes as follows.

Definition 12. *Denote a nonempty set of modes by $\mathcal{H}$. We define an order relation on $\mathcal{H}$ as follows: For any pair $H, H' \in \mathcal{H}$, we can say that $H \leq H'$ or $H \subset H'$ if and only if $\gamma_j \leq \gamma'_j$ for all $j \in \{0, 1, 2, \cdots, n_a\}$.*

This order relation is quite natural. If we define this order relation "$\leq$" on the set $\mathcal{H}$ consisting of all modes resulting from a given sequence A, it becomes a semiordered set. We will discuss the properties of order relations in the next subsection. Here, we continue to give definitions of binary operators. Similarly to the ordinary binary operators used in set theory, we can also define the corresponding binary operators between any pair of modes $H, H' \in \mathcal{H}$, e.g., intersection, union, and subtraction, etc.

Definition 13. *If we let $H, H' \in \mathcal{H}$, and construct new expanding modes as below*

$$H'' = (\gamma''_0, \gamma''_1, \gamma''_2, \cdots, \gamma''_{n_a}) ,$$

then we have the following cases:

1. *If $\gamma''_j = \min\{\gamma_j, \gamma'_j\}$ for all j, this mode H'' is called the intersection of H and H', and is denoted by $H'' = H \wedge H'$.*
2. *If $\gamma''_j = \max\{\gamma_j, \gamma'_j\}$ for all j, this mode H'' is called the union of H and H', and denoted by $H'' = H \vee H'$.*
3. *If $\gamma''_j = \gamma'_j - \gamma_j \geq 0$ for all j, then this mode H'' means that H' is subtracted from H, and is denoted by $H'' = H - H'$.*

Based on these definitions, we may find that $H \leq H'$ is the necessary condition for $H'' = H - H'$ to hold. Generally, subtraction for any pair of modes can be defined as follows:

$$H'' = H' \setminus H = H' - (H \wedge H') .$$

Furthermore, we can define the symmetric subtraction of H and H' as follows:

$$H'' = H \Delta H' = H \vee H' - H \wedge H' .$$

The Properties of Order Relation and the Binary Operators

Theorem 10. *If $\mathcal{H}$ is the set of all expanded modes of A, the relationships defined by Definitions 12 and 13, such as the order relations "$\leq$", and the three operations: union, intersection, and subtraction, make $\mathcal{H}$ a Boolean algebra.*

Before proving this theorem, let us first discuss some additional properties. Let $H, H', H'' \in \mathcal{H}$, so that we have the properties for the order relation "$\leq$" as follows:

1. **Transmissivity** If $H \leq H'$ and $H' \leq H''$, then $H \leq H''$.
2. **Reflexivity** If $H \leq H$ and $H \leq H'$, $H' \leq H$, $H = H'$. That is, $\mathcal{H}$ is a semiordered set in this order relation.
3. Furthermore, the three operations: intersection, union, and subtraction, satisfy the properties as follows:
   (a) **Closeness**. For every $H, H' \in \mathcal{H}$, the results of intersection, union and subtraction are still in $\mathcal{H}$. This is, $H \wedge H' \in \mathcal{H}$, $H \vee H' \in \mathcal{H}$ and $H \setminus H' \in \mathcal{H}$.

(b) **Commutative law**. For intersection and union, we have

$$H \wedge H' = H' \wedge H\,, \quad H \vee H' = H' \vee H\,.$$

(c) **Associative law for union and intersection**. For intersection and union, the properties are as follows:

$$(H \wedge H') \wedge H'' = H \wedge (H' \wedge H'')\,, \quad (H \vee H') \vee H'' = H \vee (H' \vee H'')\,.$$

(d) **Distributive law**. When the two operations of intersection and union occur at the same time, we have

$$(H \vee H') \wedge H'' = (H \wedge H'') \vee (H' \wedge H'')\,.$$

All six properties above can be easily validated using Definitions 12 and 13. Theorem 10 is easily proved based on these properties, and consequently satisfies the requirements of a Boolean algebra.

3.1.3 Operators Induced by Modes

The Operator Induced by Compressed Modes

Let C be an expanded sequence of A. Using (3.10) and (3.12), we obtain the corresponding expanded mode H_a (from A to C), and the compressed mode H_c (from C to A). If C and the compressed mode H_c are given, then A is determined by (3.8). Consequently, A can be considered to be the result of having C acted upon by an operator induced by H_c. We denote this induced operator by $\mathbf{H}_c$, and we then state that $A = \mathbf{H}_c(C)$. This operator $\mathbf{H}_c$ induced by the compressed mode is called the compressed operator.

The Operator Induced by Expanded Modes

If the virtual expanded mode of A is given, then the virtual expanded sequence C is unique. Generally, if only A and its expanded mode H_a are known, then we may not be able to completely determine C because (3.8) can only determine one part of C. If sequence B is the inserted part, we can induce an operator $\mathbf{H}_a$ based on H_a as follows:

$$\mathbf{H}_a(A, B) = C = \left(a_{\delta_{1,1}}, b_{\delta_{2,1}}, a_{\delta_{1,2}}, b_{\delta_{2,2}}, \cdots, a_{\delta_{1,h_a}}, b_{\delta_{2,h_a}}\right) \tag{3.13}$$

in which $\delta_{1,k} = \delta_k$ is defined by (3.8) and (3.9), and $b_{\delta_{2,k}}$ is the insertion vector of length $\ell_{2,k}$. Then

$$A = \left(a_{\delta_{1,1}}, a_{\delta_{1,2}}, \cdots, a_{\delta_{1,k_a}}\right)\,, \quad B = \left(b_{\delta_{2,1}}, b_{\delta_{2,2}}, \cdots, b_{\delta_{2,k_b}}\right) \tag{3.14}$$

follows from (3.13). That is, sequence C is determined by A, B, and H_a. The operator $\mathbf{H}_a$ defined by (3.13) is called the expanded operator.

Mutually Converse Operation Between the Expanded Mode and the Compressed Mode

Theorem 11. *If C is the virtual expanded sequence of A, and the expanded mode and compressed mode H_a and H_c are defined in (3.10) and (3.12) satisfying the following properties:*

1. *$k_c = k_a$, and $\ell'_k = \ell_k$ for all $k = 1, 2, \cdots, k_a$.*
2. *$j'_k = i_k + L_k$ for all $k = 1, 2, \cdots, k_a$,*
   then $\mathbf{H}_c[\mathbf{H}_a(A)] = A$ and $\mathbf{H}_a[\mathbf{H}_c(C)] = C$ will always hold.

This theorem can be proved using (3.8), (3.13), and (3.14) under conditions (1–3).

Definition 14. *If H_a, H_c are the modulus structures of A and C defined as in (3.10) and (3.12), and if conditions 1–3 of Theorem 11 are satisfied, then $\mathbf{H}_a$ is the inverse operator of the expanded operator and H_c is the inverse operator of the compressed operator. That is, $\mathbf{H}_a = \mathbf{H}_c^{-1}$, $\mathbf{H}_c = \mathbf{H}_a^{-1}$.*

Operators Induced by Union on Expanded Modes

For any $H, H' \in \mathcal{H}$, let $H'' = H' \vee H$. The relationships among the three operators $\mathbf{H}, \mathbf{H}', \mathbf{H}''$ induced by H, H', H'' are stated as follows: First, we notice that

$$\mathbf{H}''(A, B) = \mathbf{H}'[\mathbf{H}(A, B)] \tag{3.15}$$

does not hold because $\mathbf{H}(A, B)$ is not the same as the sequence A according to (3.13), and the operator $\mathbf{H}'[\mathbf{H}(A, B)]$ is much longer than $\mathbf{H}''(A, B)$ according to the definition of $\mathbf{H}, \mathbf{H}', \mathbf{H}''$. Consequently, we must modify (3.13) to fit this case. Without loss of generality, we may assume that all expanded positions of H, H', and H'' are the same, that is,

$$\begin{cases} H \;\,= \{(i_k, \ell_k)\,,\ k = 1, 2, \cdots, k_c\}\,, \\ H' \;= \{(i_k, \ell'_k)\,,\ k = 1, 2, \cdots, k_c\}\,, \\ H'' = \{(i_k, \ell''_k)\,,\ k = 1, 2, \cdots, k_c\}\,, \end{cases} \tag{3.16}$$

in which $\ell''_k = \max\{\ell_k, \ell'_k\}$ and $0 \le \ell_k, \ell'_k \le \ell''_k$. In addition, we reiterate the following notations:

1. In $C = \mathbf{H}(A, B)$, sequence $B = (B_1, B_2, \cdots, B_{k_c})$ is called the insertion template. Each component B_k is a vector inserted after the component a_{i_k} of sequence A, and its length is ℓ_k.
2. $B' = (B'_1, B'_2, \cdots, B'_{k_c})$ is called the expanded sequence of B if $\ell_k \le \ell'_k$ for all k. Importantly, B'_k is empty if $\ell'_k = \ell_k$. There is then a new sequence formed as follows:

$$B'' = \left[(B_1, B'_1)\,, (B_2, B'_2)\,, \cdots, (B_{k_c}, B'_{k_c})\right]\,.$$

3. Let $H'_c = \{(j_k, \ell''_k - \ell_k), k = 1, 2, \cdots, k_c\}$, in which $j_k = i_k + L_k$ and L_k is defined by (3.7). Then, H'_c is an expanded mode of $C = H(A, B)$.

With the above notations, we formulate the following theorem.

Theorem 12. *If $H, H' \in \mathcal{H}$ and $H'' = H' \vee H$, then the expanded operators induced by H, H', H'' satisfy the following relationship:*

$$\mathbf{H}''(A, B'') = \mathbf{H}'_c[\mathbf{H}(A, B), B'] \ . \tag{3.17}$$

The proof of this theorem can be constructed using the definitions of $\mathbf{H}, \mathbf{H}'_c, \mathbf{H}''$ and notations 1–3.

Definition 15. $\mathbf{H}'_c$ *defined in (3.17) is called the second expanded operator of A, and $j_k = i_k + L_k$ is called the shift of the second expanded operator.*

Theorem 13. *For every pair $H', H \in \mathcal{H}$, $H'' = H' \vee H$ can be decomposed into a second expanded operation. If both C and C' are virtual expanded sequences of A, then the corresponding expanded modes can be simplified so that $B'' = (B, B')$ is composed of virtual symbols q alone.*

Example 5. Comparing the three sequences

$$\begin{cases} A = (010201032213020103211022301)\,, \\ C = (01020[210]103221302[11]010321[03]1022301)\,, \\ C' = (01020[210]103221302[1100]010321[02231]1022301)\,, \end{cases}$$

we can make the following observations:

1. Sequences C and C' are both expanded sequences of A, their lengths are respectively: $n_c = 34$ and $n_{c'} = 39$, and their expanded modes are stated as follows:

$$\begin{cases} H = \{(5,3), (14,2), (20,2)\} = (000003000000002000002000000)\,, \\ H' = \{(5,3), (14,4), (20,5)\} = (000003000000004000005000000)\,. \end{cases}$$

2. $H \le H'$ and

$$H'' = H' - H = \{(14,2), (20,3)\} = (000000000000002000003000000)\,.$$

C' can be seen as a second expanded partition of A as follows:

$$A \overset{H}{\to} C \overset{H''}{\to} C' \,.$$

3. The shifting function of the accumulated length of (A, C) is

$$\bar{L} = (0,0,0,0,3,0,0,0,0,0,0,0,0,5,0,0,0,0,0,7,0,0,0,0,0,0,0)\,.$$

Therefore,

$$H_c = H' - H = (000000[000]00000000[002]00000[0201]0000000)$$

in which all segments in the square brackets connected as 0000020201 comprise the second expanded mode of B.

4. The virtual expanded mode H_c in observation 3 can be simplified as

$$H_c = H' - H = (000000[000]00000000[002]00000[003]0000000)$$

in which 000002003 composed from all segments in the brackets is just the second expanded mode of B. This means that we must insert two virtual symbols after position 5 and insert three virtual symbols after position 7.

3.2 Modulus Structure of Sequence Alignment

In Sect. 3.1, we mentioned that the result of a sequence alignment operation is also a special sequence. For a set of sequences, the results of multiple alignment are also special multiple sequences. We will elaborate on the description of modulus structures for multiple alignment, since multiple alignment is an important tool in analyzing sequence mutations.

3.2.1 Modulus Structures Resulting from Multiple Alignment

Let $\mathcal{A}$ be a set of sequences given in (3.1) and (3.2), defined on V_4, and let $\mathcal{A}'$ be a set of sequences defined on V_5 similar to (1.4). Following the definition of multiple alignment given in Definition 2, we have that A'_s is the virtual expanded sequence of A_s for all $s \in M$, and sequence $a'_{s,j}, s = 1, 2, \cdots, m$ has no two connected virtual symbols for all j. Below, we discuss its structure in detail.

Representation of Modulus Structures Resulting from Multiple Alignment

Based on the representation of the modulus structures resulting from a single expanded sequence, we generate a representation to fit multiple alignment. The two modes of multiple alignment are stated as follows:

$$\mathcal{H} = \{H_s\,,\ s = 1, 2, \cdots, m\}\,, \quad \mathcal{H}' = \{H'_s\,,\ s = 1, 2, \cdots, m\}\,, \tag{3.18}$$

where

$$H_s = \{(i_{s,k}, \ell_{s,k}),\ k = 1, 2, \cdots, k_s\}\,, \quad H'_s = (\gamma_{s,0}, \gamma_{s,1}, \gamma_{s,2}, \cdots, \gamma_{s,n_s})\,. \tag{3.19}$$

The notations in (3.19) are explained as follows:

1. n_s is the length of sequence A_s, and k_s is the number of the inserted vectors so that A_s becomes A'_s.
2. In H_s, $i_{s,k}$ is the starting position of the kth inserted vectors so that A_s becomes A'_s, and $\ell_{s,k}$ is the length of the kth inserted vector. In other words, $\ell_{s,k}$ virtual symbols are inserted after position $i_{s,k}$ in sequence A_s.

3. In H'_s, $\gamma_{s,i} \geq 0$ means that we insert $\gamma_{s,i}$ symbols after the ith position. Typically, $\gamma_{s,i} = 0$ means that there is no symbol to be inserted, while $\gamma_{s,i} > 0$ means that we will insert $\gamma_{s,i}$ virtual symbols after position i.
4. Each of $\mathcal{H}$ and $\mathcal{H}'$ can be determined by the other. Based on H_s, we can define $I_s = \{i_{s,k}\ ,\ k = 1, 2, \cdots, k_s\}$. Then

$$\gamma_{s,j} = \begin{cases} \ell_{s,k}\ , & \text{if } i = i_{s,k} \in I_s\ , \\ 0\ , & \text{otherwise}\ , \end{cases} \tag{3.20}$$

and $\mathcal{H}$ determines $\mathcal{H}'$. Conversely, $\mathcal{H}'$ can determine $\mathcal{H}$ as follows:

$$\begin{cases} I_s = \{i \in N_{s+}\ ,\ \gamma_{s,i} > 0\}\ , \\ \ell_{s,i_k} = \gamma_{s,i}\ , \quad \text{if } i = i_k \in I_s, \end{cases} \tag{3.21}$$

where $N_{s+} = \{0, 1, 2, \cdots, n_s$, and n_s is the length of sequence A_s. Following from (3.20) and (3.21), $\mathcal{H}$ and $\mathcal{H}'$ can be determined by each other. $\mathcal{H}$ and $\mathcal{H}'$ are therefore representations of the modulus structures based on $\mathcal{A}$.

Representation of Modes Resulting from Alignments

Since $\mathcal{A}'$ is the expanded sequence of $\mathcal{A}$, the alignment mode can be represented based on $\mathcal{A}'$. In fact,

$$\delta'_{s,k} = [j_{s,k} + 1, j_{s,k+1}] = (j_{s,k} + 1, j_{s,k} + 2, \cdots, j_{s,k+1})\ , \tag{3.22}$$

$$a'_{\delta'_{s,k}} = \left(a'_{s,j_{s,k}+1}, a'_{s,j_{s,k}+2}, \cdots, a'_{s,j_{s,k+1}}\right)\ . \tag{3.23}$$

If we let

$$\mathcal{K} = \{K_s\ ,\ s \in M\}\ , \quad K_s = \{j_{s,1}, j_{s,2}, \cdots, j_{s,2k_s}\} \tag{3.24}$$

then

1. For every K_s, we choose $j_{s,k} + 1, j_{s,k} + 2, \cdots, j_{s,k+1}$ such that

$$1 = j_{s,0} \leq j_{s,1} < j_{s,2} < \cdots < j_{s,2k_s} \leq j_{s,2k_s+1} = n'_s\ ,$$

where n'_s is the length of sequence A'_s.
2. Since A'_s is the expanded sequence A_s for all s, it means that all definitions and properties and notations in Sect. 3.2 can be used here. For example, the expanded sequence A_s may be represented by

$$A_s = \left(a_{\delta_{s,1}}, a_{\delta_{s,2}}, \cdots, a_{\delta_{s,k_c}}\right) = \left(a'_{\delta'_{s,1}}, a'_{\delta'_{s,3}}, \cdots, a'_{\delta'_{s,2k_c-1}}\right)\ , \tag{3.25}$$

where $a_{\delta_{s,k}} = a'_{\delta'_{s,2k-1}}$ and

$$\delta_{s,k} = [i_{s,k} + 1, i_{s,k+1}] = (i_{s,k} + 1, i_{s,k} + 2, \cdots, i_{s,k+1})$$

are determined by (3.18).

3. Relationships between the parameters of H_s and K_s are stated as follows:

$$\begin{cases} \ell_{s,k} = j_{s,2k+1} - j_{s,2k} \,, \\ L_{s,k} = \sum_{k'=1}^{k-1} \ell_{s,k'} \,, \\ i_{s,k} = j_{s,2k-1} - L_{s,k} \,. \end{cases} \quad (3.26)$$

The shifting formula of each position in sequence A_s and A'_s is stated as follows:

$$\delta_{s,k} = \delta'_{s,2k-1} - L_{s,k} = (j_{s,2k-1}+1-L_{s,k}, j_{s,2k-1}+2-L_{s,k}, \cdots, j_{s,2k}-L_{s,k}) \,. \quad (3.27)$$

4. Vector

$$a'_{s,\delta'_{s,2k}} = (\overbrace{4, 4, \cdots, 4}^{\ell_{s,k}})$$

is a virtual vector with length $\ell_{s,k}$, and $a'_{s,\delta'_{s,2k-1}}$ is a subvector of A_s.

5. Let

$$\begin{cases} \Delta'_{s,1} = \left(\delta'_{s,1}, \delta'_{s,3}, \cdots, \delta'_{s,2h_s-1}\right), \\ \Delta'_{s,2} = \left(\delta'_{s,0}, \delta'_{s,2}, \delta'_{s,4}, \cdots, \delta'_{s,2h_s}\right) \,, \end{cases} \quad (3.28)$$

where $\Delta'_{s,1}, \Delta'_{s,2}$ are the noninserted part and the inserted part of sequence A_s, respectively. Since $a'_{s,j}$, $s = 1, 2, \cdots, m$ has no m connected virtual symbols for every fixed j, we find that

$$\bigcap_{s=1}^{m} \Delta'_{s,2} = \phi \quad (3.29)$$

is an empty set.

Theorem 14. *$\mathcal{H}, \mathcal{H}'$ and $\mathcal{K}$ are three equivalent representations of modes.*

Proof. It follows from (3.4) and (3.5) that $\mathcal{H}$ and $\mathcal{H}'$ can be determined from each other. It also follows from (3.27) that $\mathcal{H}$ and $\mathcal{K}$ can be determined from each other.

3.2.2 Structure Analysis of Pairwise Alignment

In the case $m = 2$, $\mathcal{H}$ and $\mathcal{K}$ are defined by (3.18) and (3.24), respectively, and are referred to as the modes of pairwise alignment. We then represent (A_1, A_2) and (A'_1, A'_2) by (A, B) and (A', B'), respectively. Since (A', B') is the expanded sequence of (A, B), their modes are stated as follows:

$$H = (H_1, H_2) = (H_a, H_b) \,, \quad K = (K_1, K_2) = (K_a, K_b) \,.$$

Following from (3.28) and (3.29), we have that $\Delta'_{1,2} \cap \Delta'_{2,2} = \phi$. Next, we process (A', B') as follows:

Make A' and B' the Same Length

In this process, we make A' and B' have the same length by inserting virtual symbols into the shorter sequence, so that $n_{a'} = n_{b'} = n'$ are the common length of A' and B'. Similarly, let $N_{a'} = N_{b'} = N' = \{1, 2, \cdots, n'\}$ be the regions of positions of A' and B'.

Decomposition of the Region N' Resulting from Alignment

Definition 16. *$N' = (\Delta'_0, \Delta'_1, \Delta'_2)$ is called the decomposition of region N', in which*

$$\Delta'_1 = \Delta'_{1,2}\,, \quad \Delta'_2 = \Delta'_{2,2}\,, \quad \Delta'_0 = N_{a'} - (\Delta'_1 \cup \Delta'_2)\,, \tag{3.30}$$

and $\Delta'_{s,2}$ is defined in (3.28).

Details about the decomposition of $N' = (\Delta'_0, \Delta'_1, \Delta'_2)$ are stated below:

1. Since Δ_τ, $\tau = 0, 1, 2$ are disjoint, we have $\Delta'_0 \cup \Delta'_1 \cup \Delta'_2 = N'$ and $\Delta_\tau \cap \Delta_{\tau'} = \phi$, $\tau \neq \tau' \in \{0, 1, 2\}$.
2. Following the definitions of $\Delta'_{1,2}$ and $\Delta'_{2,2}$, we have that the values of A' on region Δ'_1 and B' on region Δ'_2 are both 4. In the region Δ'_0, there is no virtual symbol in the sequences C and D. Therefore, the decomposition of $N' = (\Delta'_0, \Delta'_1, \Delta'_2)$ actually classifies N' into three parts:
   (a) Part 1 (Δ'_1): In this region, the sequence A' is composed of virtual symbols while B' does not contain virtual symbols.
   (b) Part 2 (Δ'_2): In this region, B' is composed of virtual symbols while A' has no virtual symbols.
   (c) Part 3 (Δ'_0): In this region, both A' and B' have no virtual symbols.

Definition 17. *In the decomposition, Δ'_0 is called the noninserted region for both A' and B', while Δ'_1 and Δ'_2 are referred to as the dual inserting regions of A' and B'. This means that Δ'_1 is the inserting region of A' but not the inserting region of B', and Δ'_2 is the inserting region of B' but not the inserting region of A'.*

Structure Analysis of Decomposition

Each of the three regions Δ'_τ, $\tau = 0, 1, 2$ is composed of many smaller regions. We denote them as follows:

$$\Delta'_\tau = \left\{\delta'_{\tau,1}, \delta'_{\tau',2}, \cdots, \delta'_{\tau,k_\tau}\right\}\,, \quad \tau = 0, 1, 2\,. \tag{3.31}$$

These smaller intervals are disjoint from each other, and their union is N'. Furthermore, we have $\delta'_{\tau,k} \cap \delta'_{\tau',k'} = \phi$ and $\bigcup_{\tau=0}^{2} \bigcup_{k=1}^{k_\tau} \delta'_{\tau,k} = N'$ for all $(\tau', k') \neq (\tau, k)$. Following (3.31), we can alternatively represent all the regions using smaller intervals as follows:

$$\Delta' = \{\Delta'_\tau,\ \tau = 0,1,2\} = \{\delta'_{\tau,k},\ k = 1,2,\cdots,k_\tau,\ \tau = 0,1,2\} . \qquad (3.32)$$

If we arrange them according to their size, they are disjoint with each other and fill all the regions of N'. Thus, we can decompose N' as follows:

$$N' = \left(\delta'_1, \delta'_2, \cdots, \delta'_{k_{a'}}\right) = (j_1, j_2, \cdots, j_{k_{a'}}) , \qquad (3.33)$$

in which $k_{a'} = k_0 + k_1 + k_2$ and

$$j_1 = 0 < j_2 < j_3 < \cdots < j_{k_{a'}} \le j_{k_{a'}+1} = n' ,$$

and

$$\delta'_k = [j_k + 1, j_{k+1}] = \{j_k + 1, j_k + 2, \cdots, j_{k+1}\}$$

Consequently, δ'_j is included in Δ defined in (3.32), and satisfies the following conditions:

1. $\delta'_1 \in \Delta'_0$, and δ'_1 may be empty.
2. For every $j = 1,2,\cdots,k_{a'}$, if $\delta'_j \in \Delta'_1 \cup \Delta'_2$, we must have $\delta'_{j+1} \in \Delta'_0$. Consequently, in the decomposition of (3.33), we have that

$$\delta'_{2k-1} \in \Delta'_0 , \quad \delta'_{2k} \in \Delta'_1 \cup \Delta'_2$$

holds.
3. Let $k_0 = k_1 + k_2$, $k_{a'} = 2k_0$, then δ'_{2k_0} may be an empty set. Then, (3.33) is called the decomposition of region $N_{a'}$ of (A', B'). As well, we may represent the mode of the alignment as follows:

$$K_0 = \{(\delta'_k, \tau_k),\ k = 1,2,\cdots,k_{a'}\} = \{(j_k, \ell_k, \tau_k),\ k = 1,2,\cdots,k_{a'}\} , \qquad (3.34)$$

in which $\ell_k = j_{k+1} - j_k$, j_k, and δ'_k are defined by (3.33), and $\tau_k = 0,1,2$ is determined by the fact that δ'_{2k} belongs to one of $\in \Delta'_0$, Δ'_1 and Δ'_2. The vector $\bar{\tau} = (\tau_1, \tau_2, \cdots, \tau_{k_{a'}})$ is called the structure model of the aligned sequence. Following from (3.34), we have that K_0 has an alternative representation given by:

$$H_0 = \{(i_k, \ell_k),\ k = 1,2,\cdots,k_0\} , \qquad (3.35)$$

in which $j_k = i_{2k} + L_k$, and

$$\ell_k = \begin{cases} ||\delta_{2k}|| , & \text{if } \tau_{2k} = 1 , \\ -||\delta_{2k}|| , & \text{if } \tau_{2k} = 2 . \end{cases}$$

It is easy to prove that H_0 and K_0 as defined in (3.34) and (3.35) are two equivalent modes.

3.2.3 Properties of Pairwise Alignment

If the alignment mode K_0 of the aligned sequence (A', B') is given, then (A', B') can be decomposed as follows:

$$\begin{cases} A' = \left(a'_{\delta'_1}, a'_{\delta'_2}, \cdots, a'_{\delta'_{k_{a'}}}\right), \\ B' = \left(b'_{\delta'_1}, b'_{\delta'_2}, \cdots, b'_{\delta'_{k_{a'}}}\right), \\ \bar{\tau} = \left(\tau_1, \tau_2, \cdots, \tau_{k_{a'}}\right), \end{cases} \tag{3.36}$$

in which $\tau \in \{0, 1, 2\}$ and the vector $\bar{\tau}$ is called the structure type of the aligned sequence. More details are provided below. Let $n = \max\{n_a, n_b\}$, and $H'_s = (\gamma_{s,1}, \gamma_{s,2}, \cdots, \gamma_{s,n})$, $s = a, b$, then the alignment mode of (H_a, H_b) can be rewritten as

$$H_0 = \{(i_h, \ell_h),\ h = 1, 2, \cdots, h_a + h_b\} \quad H'_0 = (\gamma_1, \gamma_2, \cdots, \gamma_{n_0}), \tag{3.37}$$

where

$$\gamma_j = \begin{cases} 0, & \text{if } \gamma_{1,j} = \gamma_{2,j} = 0, \\ \gamma_{1,j}, & \text{if } \gamma_{1,j} > 0,\ \gamma_{2,j} = 0, \\ -\gamma_{2,j}, & \text{if } \gamma_{1,j} = 0,\ \gamma_{2,j} > 0. \end{cases}$$

Since $\gamma_{1,j}$ and $\gamma_{2,j}$ are not greater than 0 simultaneously, it follows that the alignment modes (H_1, H_2) and H_0 can be determined from each other. Let

$$I_0 = \{i \in N_a : \gamma_i \neq 0\}$$

so that $i_h \in H_0$ implies $i_h \in I_0$. Next, let $\ell_h = \gamma_{i_h}$ if $i_h \in I_0$, and let

$$L_s(j) = \sum_{j'=1}^{j-1} \gamma_{s,j'}, \quad j = 1, 2, \cdots, n_s, \quad s = 1, 2. \tag{3.38}$$

These are referred to as the inserted functions of the expanded sequence A', B'. Consequently, we define

$$L_0(j) = L_1(j) - L_2(j), \quad L(j) = L_1(j) + L_2(j), \quad j = 1, 2, \cdots, n_0, \tag{3.39}$$

which are referred to as the relatively inserted length function and the absolutely inserted length function of (A', B'), respectively, in which, $n_0 = \max\{n_1, n_2\}$.

Theorem 15. *If H_0 is the alignment mode defined by (3.38), then we have*

$$L_0(j) = \sum_{j'=1}^{j-1} \gamma_{j'}, \quad L(j) = \sum_{j'=1}^{j-1} |\gamma_{j'}|, \tag{3.40}$$

in which $|\ell|$ is the absolute value of ℓ. The alignment mode H_0 and function $L_0(j)$ can be determined from each other.

Proof. Formula (3.40) follows from the definition of H_0, that is, H_0 determines $L_0(j)$. Conversely, $\gamma_j = L_0(j+1) - L_0(j)$ follows from (3.38). Thus, H_0 is determined. Therefore, each of H_0 and $L_0(j)$ can be determined by the other.

Expression (3.36) decomposes (A', B') into three parts: the noninserting part, the inserting part and the dual-inserting part. Their relationship is demonstrated in Fig. 3.2 using a bold line, a normal line and a broken line.

In this figure, A', B' are the aligned sequences of sequences A and B, respectively, where:

1. The alignment modes are:

$$H_a = \{(i_{a,1}, \ell_{a,1}), (i_{a,2}, \ell_{a,2})\} \quad H_b = \{(i_{b,1}, \ell_{b,1}), (i_{b,2}, \ell_{b,2})\} \ .$$

2. One of the decompositions of A' and B' resulting from alignment is $\Delta' = \{\delta'_1, \delta'_2, \cdots, \delta'_9\}$, in which

$$\delta'_k = [j_k + 1, j_k] = (j_k + 1, j_k + 2, \cdots, j_{k+1}) \ .$$

3. In the decomposition Δ',

$$\Delta'_0 = \{\delta'_1, \delta'_3, \delta'_5, \delta'_7, \delta'_9\} \quad \text{and} \quad \Delta'_3 = \{\delta'_2, \delta'_4, \delta'_6, \delta'_8\}$$

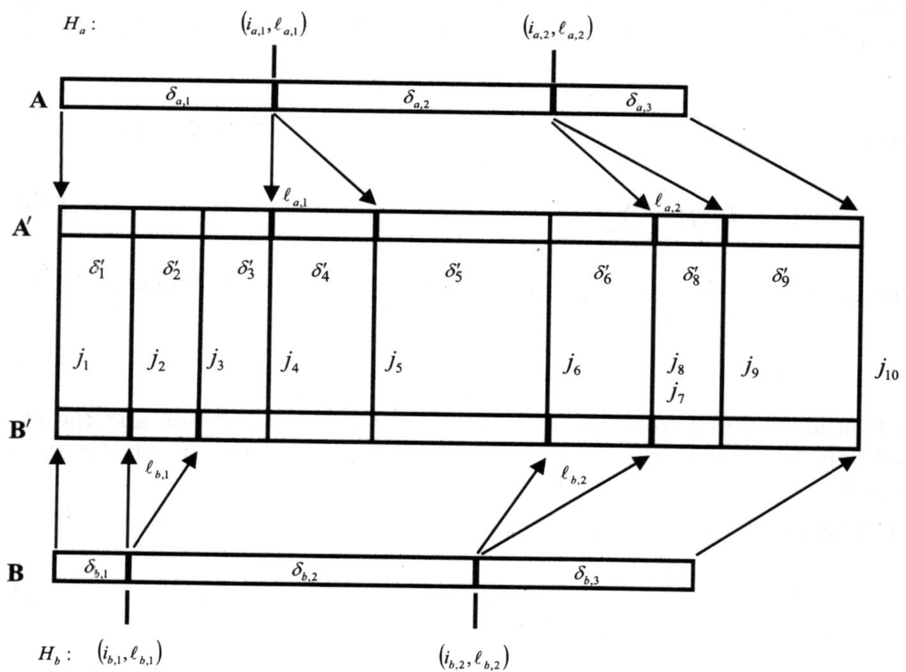

Fig. 3.2. The modulus structure of sequence alignment

are the noninserted part and the inserted part of the alignment, respectively, where δ'_7 is an empty set, and $\Delta'_1 = \{\delta'_4, \delta'_8\}$, $\Delta'_2 = \{\delta'_2, \delta'_6\}$ are the common inserted part of both A' and B'.

By using positive and negative signs, K_0 can be expressed as follows:

$$K_0 = (\theta_1, \theta_2, \cdots, \theta_{n_{a'}}) , \tag{3.41}$$

where $\theta_j = \tau_k \in \{0, 1, 2\}$ if $j \in \delta_k$ and (δ_k, τ_k) is determined by K_0 in (3.35). We then have

$$\Delta_\tau = \{j \in N_{a'},\ \theta_j = \tau\} , \quad \tau = 0, 1, 2 ,$$

where Δ_τ is defined in (3.30).

Example 6. If

$$\begin{cases} A = (11231003221000232103220121120) , \\ B = (1123100323202100023210311200333) , \end{cases}$$

the aligned sequences are

$$\begin{cases} A' = (112310032[444]21000232103[22012]1120[4444]) , \\ B' = (112310032[320]21000232103[44444]1120[0333]) , \end{cases}$$

where the corresponding noninserted part, inserted part and the dual part are indicated by brackets. The part whose value is 4 is the inserted part, the part whose value is not 4 is the dual part, and the rest is the noninserted part. By using the corresponding formula for alignment modes, the decompositions are as stated as follows:

1. The length of sequence A is $n_1 = 29$, and the length of sequence B is $n_0 = n_2 = 31$.
2. Their alignment mode is

$$\begin{cases} H_1 = \{(9,3), (29,4)\} = (000000000300000000000000000004) , \\ H_2 = \{(23,5)\} = (0000000000000000000000500000000) . \end{cases}$$

   The alternative form is $H_0 = (00000000300000000000000(-5)00000400)$.
3. Following from (3.33) as $k_{a'} = 6$, we get the decompositions of K_0 as follows:

$$\begin{aligned} K_0 &= \{(9,0), (3,1), (11,0), (5,2), (4,0), (4,1)\} \\ &= \{(0,0), (9,1), (12,0), (23,2), (28,0), (32,1)\} \end{aligned}$$

   These are the two alternative decompositions about K_0 stated in (3.33).

4. The corresponding decomposition function is

$$\begin{cases} \bar{L}_1 = (00000000333333333333333333333337) , \\ \bar{L}_2 = (0000000000000000000000000444444444) , \\ \bar{L}_0 = (0000000033333333333333333, -1, -1, -1, -1, -1, -1, -1, -1, 3) , \\ \bar{L} = (000000003333333333333333377777777, 11) , \end{cases}$$

in which $\bar{L}_z = (L_z(1), L_z(2), \cdots, L_z(n_0))$. In the vector $\bar{L}_z$, we only use the comma to separate the negative components and the components whose values exceed 10.

5. The decomposition K_0 can be written as:

$$K_0 = (000000000111000000000002222200001111)$$

and

$$\begin{aligned} \Delta_0 &= \{1, 2, 3, 4, 5, 6, 7, 8, 9, 13, 14, 15, 16, 17, 18, 19, 20, \\ &\qquad 21, 22, 23, 29, 30, 31, 32\} , \\ \Delta_1 &= \{10, 11, 12, 33, 34, 35, 36\} , \\ \Delta_2 &= \{24, 25, 26, 27, 28\} . \end{aligned}$$

These are the noninserted and inserted parts of both C and D, respectively.

3.2.4 The Order Relation and the Operator Induced by Modulus Structure

We can expand the order relation and the operator to apply to the modulus structures induced by multiple alignments as follows:

1. Let $H = (H_a, H_b)$ and $H' = (H'_a, H'_b)$ be two modulus structures induced by pairwise alignment. We can say that $H \leq H'$ if both $H_a \leq H'_a$ and $H_b \leq H'_b$ hold.
2. The union operation between two modulus structures H, H' given in condition 1 is defined as $H \vee H' = (H_a \vee H'_a, H_b \vee H'_b)$.
3. Similarly, we can define other binary operations such as intersection and subtraction between two modulus structures H, H', such that the set of all modulus structures obey Boolean algebra with these binary operations.
4. With a similar argument, we may find that corresponding operators can also be induced by the modulus structure. Let (A, B) be a pair of sequences and $H = (H_a, H_b)$ be the corresponding modulus structure pair. There is then a pair of aligned sequences (A', B'), and the corresponding operators $\mathbf{H}$ are induced as

$$(A', B') = \mathbf{H}(A, B) = (\mathbf{H}_a(A), \mathbf{H}_b(B)) . \tag{3.42}$$

3.3 Analysis of Modulus Structures Resulting from Sequence Mutations

We have noted in Chap. 1 that there are four types of biological mutations, and we called them type-I, type-II, type-III, and type-IV mutations. Recall that type-I and type-II mutations do not change the lengths of sequences, and thus they are also called nonshifting mutations. On the other hand, type-III and type-IV mutations necessarily change the lengths of sequences, so they are referred to as shifting mutations. In this chapter, we focus on type III and type IV mutations.

3.3.1 Mixed Sequence Mutations

Following the definitions of type-III and type-IV mutations, their mathematical representations are just a mixture of expanding and compressing operators. That is, the shifting mutations are mixed operations of expanding and compressing operations. Since the cases in which mixed operations occur are more complicated than the operations themselves, we discuss the possible situations in which mixed operations occur. From their definitions, we know that type-I and type-II mutations are different from type-III and type-IV mutations. However, it is possible to consider one type as the combination of other types. This may at first seem counter-intuitive, but it is true. For example:

Example 7. Assume a type-I mutation occurs at the second nucleotide of sequence acg, i.e, it changes acg into aug. This is a type-I mutation, but it can be regarded as the combination of a type-IV mutation that deletes c so that acg becomes ag, followed by a type-III mutation that inserts u into ag, making ag become aug. This demonstrates that type-III and type-IV mutations can be represented by type-I and type-II mutations, and vice versa. Next, we consider the following cases:

1. At the same position, if a type-III mutation (insertion) occurs ℓ times, and a type-IV mutation (deletion) occurs ℓ times, then all insertions and all deletions will counteract each other, and there is no shifting.
2. At the same position, if a type-IV mutation occurs ℓ times and a type-III mutation occurs ℓ times, but the inserted characters and deleted characters are different, it is as if a type-I mutation occurred. Above all, we may conclude that for mixtures of the four types of mutations, type-III and type-IV have no intersection, and intersection regions can be replaced by type-I and type-II mutations.

Example 8. The shifting mutations from sequence A = (acccccuuuuu) to A' = (agggguuuuu) can be divided as follows: deleting five nucleotides c after position 1 and then inserting three nucleotide g after position 1. The mixed mutation can be divided into a type-IV mutation and a type-III mutation acting on the sequence A as follows:

$$\text{acccccuuuuu} \xrightarrow{\text{IV}} \text{auuuuu} \xrightarrow{\text{III}} \text{aggguuuuu} . \tag{3.43}$$

The combined effect of the two mutations is the same as that of a type-I mutation, ccc $\rightarrow$ ggg, and a type-IV mutation that deletes the cc at positions 5 and 6. The process is as follows:

$$\text{acccccuuuuu} \xrightarrow{\text{I}} \text{agggccuuuuu} \xrightarrow{\text{IV}} \text{aggguuuuu} . \tag{3.44}$$

Based on this example, we can conclude that a pair of type-III and type-IV mutations can be decomposed as a nonshifting mutation and a shifting mutation if type-III mutation occurs in the region of the type-IV mutation.

Example 9. In Example 8, if type-III mutation occurs first and type-IV mutation occurs later, then the effect of these two shifting mutations is as follows:

$$\text{acccccuuuuu} \xrightarrow{\text{III}} \text{accccgggcuuuuu} \xrightarrow{\text{IV}} \text{aggcuuuuu} . \tag{3.45}$$

The effect is the same as inserting the nucleotide triplet ggg after position 5 and then deleting nucleotides ccccg at positions 2–6. The final result is the same as: acccccuuuuu $\rightarrow$ aggcuuuuu. It is different from the result: aggguuuu, which is given in Example 8.

Therefore, the results are related to the order in which the type-III mutation and type-IV mutation occur if the regions of the type-III and type-IV mutations intersect. Generally, a different ordering of the type-III and type-IV mutations will lead to different results. On the other hand, the result of the first mutation of (3.44) can be considered as a mixture of a type-I mutation plus a type-IV mutation as follows:

$$\text{acccccuuuuu} \xrightarrow{\text{I}} \text{agccccuuuuu} \xrightarrow{\text{IV}} \text{agccuuuuu} . \tag{3.46}$$

Basic Assumptions About Hybrid Mutation

If B is mutated from sequence A, then the type-III and type-IV mutation regions do not intersect. This means that there is no type-III mutation occurring in the interval $[i+1, i+\ell]$ if a type-IV mutation of length ℓ occurs at position i of sequence A. The inverse is also true.

Theorem 16. *If a mixed mutation satisfies the basic assumptions, then it can be decomposed into several single mutation types and the result is invariant with respect to the mutation order.*

The theorem will be proved later. Assuming that the sequence mentioned here satisfies this basic assumption, the result of Theorem 16 will always hold. Since it is more difficult to deal with a case involving shifting mutation based on the above mentioned sequence mutation theory, we will discuss hybrid mutations without type-I and type-II mutations. This hybrid mutation is called the purely shifting mutation.

Example 10. We begin by discussing the sequence

$$A = (00000\,11111\,11111\,22222\,22222\,22222\,33333\,33333)\ .$$

We perform type-II, type-III, and type-IV mutations defined respectively as follows: (1) type-II mutation – mutually exchanging the data within [11, 15] to that within [16, 20], and mutually exchanging the data within [26, 30] to that within [31, 35]; (2) type-III mutation – inserting the segment 00112233 after position 13; (3) type-IV mutation – deleting a segment with length 3 after position 27. We then have the following output for each stage:

1. On one hand, the type-II mutation mutates A to B,

   $$B = (00000\,11111\,22222\,11111\,22222\,33333\,22222\,33333)\ ,$$

   and then type-III mutation and type-IV mutation continuously mutate B to C,

   $$C = (00000\,11111\,222[00\,11223\,3]2211\,11122\,22233\,22222\,33333)\ .$$

2. On the other hand, the type-III and type-IV mutations mutate A to D,

   $$D = (00000\,11111\,111[00\,11223\,3]1122\,22222\,22222\,33333\,33333)\ ,$$

   and then the type-II mutation continuously mutates D to E,

   $$E = (00000\,11111\,112[23\,11100\,3]1122\,22222\,33333\,22222\,33333)\ .$$

Then C and E are different. As a result, we conclude that the action order between type-II mutations and type-III and type-IV mutations do not generally permit an exchange. However, an exchange is permitted if there is no type-II mutation in any region of insertion.

3.3.2 Structure Analysis of Purely Shifting Mutations

Decomposition of Shifting Mutations

If sequence B is mutated from A through purely shifting mutations, we then denote the intermediate result as $C = (c_1, c_2, \cdots, c_{n_{a'}})$ mutated from A through type-III mutation, and then to B through type-IV mutation. The procedure is represented as follows:

$$A = (a_1, a_2, \cdots, a_{n_a}) \xrightarrow{\text{III}} C = (c_1, c_2, \cdots, c_{n_c}) \xrightarrow{\text{IV}} B = (b_1, b_2, \cdots, b_{n_b})\ . \tag{3.47}$$

The sequence C is an envelope of (A, B), and we may decompose C as follows:

1. Let Δ'_1 and Δ'_2 be the expanded part of sequence C related to sequences A and B, respectively. Then, $c_{N_c-\Delta'_1} = A$, $c_{N_c-\Delta'_2} = B$. Following from the basic assumption on shifting mutations, we have that Δ'_1 is disjoint from Δ'_2, i.e., $\Delta'_1 \cap \Delta'_2 = \emptyset$.
2. Let $\Delta'_0 = \{N_c - \Delta'_1\} \cap \{N_c - \Delta'_2\} = N_c - \Delta'_1 - \Delta'_2$, then Δ'_0 and Δ'_1, Δ'_2 are disjoint and $\Delta'_0 \cup \Delta'_1 \cup \Delta'_2 = N_c$. We can then obtain the following expressions:
$$(c_{\Delta'_0}, c_{\Delta'_1}) = A\,, \quad (c_{\Delta'_0}, c_{\Delta'_2}) = B\,. \tag{3.48}$$
We call $(\Delta'_0, \Delta'_1, \Delta'_2)$ the decomposition of the shifting mutation procedure $A \to C \to B$, in which, Δ'_0 is the nonshifting mutation region, Δ'_1 is the inserting region and Δ'_2 is the deleting region.
3. Regions Δ'_0, Δ'_1 and Δ'_2 can be decomposed into small intervals as follows:
$$\Delta'_\tau = \{\delta_{\tau,k},\ k = 1, 2, \cdots, k_\tau\}\,, \quad \tau = 0, 1, 2\,. \tag{3.49}$$
They are disjoint from each other. If we arrange these small intervals according to their lengths, we then have
$$\Delta' = \{\delta'_0, \delta'_1, \delta'_2, \cdots, \delta'_{k_{a'}}\}\,, \tag{3.50}$$
where $k_{a'} = k_0 + k_1 + k_2$. Since the nonshifting mutation and shifting mutation occur in the small intervals Δ given in (3.49), it follows that
$$\begin{cases} \Delta'_0 = (\delta'_1, \delta'_3, \cdots, \delta'_{2k_0-3}, \delta'_{2k_0-1})\,, \\ \Delta'_1 \cup \Delta'_2 = (\delta'_0, \delta'_2, \cdots, \delta'_{2k_0-2}, \delta'_{2k_0})\,. \end{cases} \tag{3.51}$$
So, $k_{a'} = 2k_0$, and δ'_0 and $\delta_{k_{a'}}$ may be empty sets. Similarly, we call (3.50) and (3.51) the decomposition of the shifting mutations, so the sequence C given in (3.47) satisfies $c_{N_c-\Delta'_1} = A$, $c_{N_c-\Delta'_2} = B$.

Modulus Structure of Shifting Mutation

If sequence B is mutated from A through shifting mutations, then we may express (3.47) using a shifting mutation T, and denote this by $A \xrightarrow{T} B$, with the associated mutation mode denoted as follows:
$$\begin{cases} T = \{(i_k, \ell_k),\ k = 1, 2, \cdots, k_a\}\,, \quad \ell_k \neq 0\,, \\ T' = (\gamma_1, \gamma_2, \cdots, \gamma_{n_a})\,, \quad \gamma_i \in \mathbf{Z}\,, \end{cases} \tag{3.52}$$
where $\mathbf{Z} = \{\cdots, -2, -1, 0, 1, 2, \cdots\}$ is the set of all integers. More details about (3.52) follow:

1. The k_a in T is the total number of shifting mutations in sequence A. The notation (i_k, ℓ_k) represents the position and length of the kth shifting mutation, where i_k is the position of shifting mutation, and ℓ_k includes the information of both the type and length of the kth shifting mutation. Typically, it is a type-III mutation if $\ell_k > 0$, and a type-IV mutation if $\ell_k < 0$. $|\ell_k|$ is the length of the inserted or deleted segment.

2. Similarly, γ_i in T' includes information about the length and type of the shifting mutation at position i. Typically, there is no shifting mutation after a_i if $\gamma_i = 0$. There is a type-III mutation occurring after a_i if $\gamma_i > 0$, i.e., a fragment with γ_i length is inserted after a_i if $\gamma_i < 0$. There is a type-IV mutation occurring after a_i, i.e., a vector with $|\gamma_i|$ length is deleted after a_i.
3. Let $I_T = \{i_k,\ k = 1, 2, \cdots, k_a\}$ denote the set of all the positions of the shifting mutations. We arrange them according to their values in increasing order as follows:

$$i_0 = 0 \le i_1 < i_2 < \cdots < i_{k_a} \le i_{k_a+1} = n_a \ . \tag{3.53}$$

Since the regions of type-III mutation and type-IV mutation may not overlap, we assume that

$$i_k + |\ell_k| < i_{k+1} \ , \quad i = 1, 2, \cdots, k_a \ . \tag{3.54}$$

4. Similarly, for the two mutation modes T, T' given in (3.54), one determines the other. The corresponding expression is found as follows: If T is given, then

$$\gamma_i = \begin{cases} \ell_{i_k} \ , & \text{if } i = i_k \in I_T \ , \\ 0 \ , & \text{otherwise} \ . \end{cases} \tag{3.55}$$

On the other hand, if T' is given, the variables in T are rewritten as follows:

$$\begin{cases} I_T = \{i \in N_a \colon\ |\gamma_i| > 0\} \ , \\ \ell_k = \gamma_{i_k} \ , \quad \text{if } i = i_k \in I_T \ , \end{cases} \tag{3.56}$$

where set I_T arranged in the order of increasing values. The expression showing how T and T' determine each other is given in (3.55) and in (3.56), with T and T' as the two equivalent mutation modes, with both mutation modes relying on A.

Definition 18. *With T and T' defined as in (3.52) and the mutations determined by the values of γ_i, i_k, ℓ_k, we have T and T' as two equivalent modes of shifting mutation. Because T and T' determine each other, we will use T and T' interchangeably in the future. In addition, the final result B is not unique if the mutation mode (A, T) is given. This is because there is no restriction on the insertion of nucleotides following from (A, T), so we must add rules to restrict the insertion data.*

Definition 19.
1. *Let T and T' be two shifting mutation modes of a sequence based on A; sequence A is then called the initial data (or initial template) of the corresponding type-III mutation.*
2. *For a type-III mutation with initial template A, if the inserted data is selected from a fixed sequence $A' = (a'_1, a'_2, \cdots)$ in order, then A' is called the inserted data template.*

Example 11. In Example 6, let B be mutated from sequence A, so the mutation result B is uniquely determined if T along with the original and inserted data templates are both known. By inserting 320 and 0333 and deleting 22012 in sequence A, we obtain sequence B. This mutation mode is described as follows: $T = \{(9, +3), (20, -5), (29, 4)\}$,

$$T' = (000000000300000000000, -5, 0000000004) .$$

Here the inserted data template is $A' = (320333)$.

The Decomposition of the Modulus Structure of Shifting Mutation

The mutation modes T and T' defined in (3.52) can be decomposed further according to the signs of the parameters γ_i:

$$T_\tau = \{(i_{\tau,k}, \ell_{\tau,k}),\ k = 1, 2, \cdots, k_\tau\}\ , \quad T'_\tau = (\gamma_{\tau,1}, \gamma_{\tau,2}, \cdots, \gamma_{\tau,n_a})\ , \tag{3.57}$$

where $\tau = +, -$, and

$$\gamma_i = \begin{cases} \gamma_i\,, & \text{if } \gamma_i > 0\,, \\ 0\,, & \text{otherwise}\,, \end{cases} \qquad \gamma_{-,i} = \begin{cases} -\gamma_i\,, & \text{if } \gamma_i < 0\,, \\ 0\,, & \text{otherwise}\,. \end{cases} \tag{3.58}$$

If

$$\begin{cases} \theta_{\tau,i} = 0\,, & \text{if } \gamma_{\tau,i} = 0\ , \\ 1\,, & \text{if } \gamma_{\tau,i} > 0\ , \end{cases}$$

we can obtain the value of $(i_{\tau,k}, \ell_{\tau,k})$ restricted by

$$i_{\tau,k} = \min\left\{i \colon \sum_{i'=1}^{i} \theta_{\tau,i'} \geq k\right\}\ , \qquad \ell_{\tau,k} = \gamma_{\tau,k}\ . \tag{3.59}$$

Therefore, $\gamma_{\tau,i}$ and $(i_{\tau,k}, \ell_{\tau,k})$ in (3.57) can determine each other according to (3.58) or (3.59). The following relationships are also satisfied:

1. $k_+ + k_- = k_a$, where k_a, k_+, k_- are the number of mutations described in mutation modes T, T_+, T_-.
2. In (3.57), $(i_{+,k}, \ell_{+,k})$ is the binary of the position and length of the type-III mutation, while $(i_{-,k}, \ell_{-,k})$ is the position and length of the type-IV mutation, satisfying the following relationship:
$$1 = i_{\tau,0} \leq i_{\tau,1} < i_{\tau,2} < \cdots < i_{\tau,k_\tau} \leq i_{\tau,k_\tau+1} = n_a\ . \tag{3.60}$$
3. T_τ and T'_τ determine each other and the corresponding expressions are similar to (3.55) and (3.56).

Definition 20. *The pairwise (T_+, T_-) defined in (3.57) is called the decomposition of mutation mode T. Conversely, T is called the combination of (T_-, T_+). T_+, T_-, T are respectively called the mutation mode of the type-III mutation (insertion), type-IV mutation (deletion), and the shifting mutation, alternatively to T', T'_+, T'_-.*

The Decomposition of Modulus Structure of Purely Shifting Mutation

The purely shifting mutation defined in Sect. 3.3.1 can be decomposed as follows:

$$A = \left(c_{\Delta'_0}, c_{\Delta'_1}\right) \overset{I(c_{\Delta_2})}{\longrightarrow} C = \left(c_{\Delta'_0}, c_{\Delta'_1}, c_{\Delta'_2}\right) \overset{D(c_{\Delta_1})}{\longrightarrow} B = \left(c_{\Delta'_0}, c_{\Delta'_2}\right) \ , \quad (3.61)$$

where $I(c_{\Delta_2})$ is insertion sequence c_{Δ_2}, $D(c_{\Delta_1})$ is deletion c_{Δ_1}. Next, we discuss the relationship between this decomposition and the modulus structure $T = (T_+, T_-)$:

1. The region demonstrated in (3.50) can be decomposed as follows:

$$\Delta' = \{j_0, j_1, j_2, \cdots, j_{k_{a'}}, j_{k_{a'}+1}\} \ , \quad (3.62)$$

where Δ' is defined by (3.50) and

$$\begin{aligned} j_0 &= 0 \le j_1 < j_2 < \cdots < j_{k_{a'}} \le j_{k_{a'}+1} = n_{a'} \ , \\ \delta'_k &= [j_k + 1, j_{k+1}] = (j_k + 1, j_k + 2, \cdots, j_{k+1}) \ . \end{aligned} \quad (3.63)$$

Since C is the expansion of $D = c_{\Delta'_0}$, then the expanded mode from C to D can be determined by (3.51) and is written as

$$H_0 = \{(i_{0,k}, \ell_{0,k}) \ , \ k = 1, 2, \cdots, k_a\} \ , \quad (3.64)$$

where k_a is the number of mutations in sequence A and $2k_a = k_{a'} + 1$. We then have

$$\begin{cases} \ell_{0,k} = j_{2k+1} - j_{2k} \ , \\ i_{0,k} = j_{2k} - L_k \ , \end{cases} \quad (3.65)$$

where $L_k = \sum_{k'=1}^{k-1} \ell_{d,k'}$. In view of the theory of expanded sequences, we infer that (3.50), (3.62), and H_0 determine each other.

2. D is the compressed sequence from A; its compressed mode is the same as the mode caused by type-IV mutation on sequence A. Let the expanded mode from D to A be H_1, then

$$H_1 = (T_-)^{-1} = \{(i_{1,k}, \ell_{1,k}), \ k = 1, 2, \cdots, k_-\} \ , \quad (3.66)$$

where $(T_-)^{-1}$ is the inverse of T_-. According to the definition of (3.7), we have the formula for shifting mutation as follows:

$$\ell_{1,k} = \ell_{-,k} \ , \quad i_{1,k} = i_{-,k} - L_{-,k} \ , \quad (3.67)$$

where $L_{-,k} = \sum_{k'=1}^{k-1} \ell_{-,k'}$. The symbols in (3.63) and (3.67) are defined as follows.

3. C is the expanded sequence of A and its expanded mode is just the mutation mode of the type-III mutation on A. Therefore, the expanded mode from A to C is $H_2 = T_+$, and we can obtain the transformation relationship of sequence A in C as follows: $a_i = c_j$ if
$$j = i + L_+(i)\,, \quad L_+(i) = \sum_{k:i_{+,k}<i} \ell_{+,k}\,. \tag{3.68}$$
4. B is the compressed sequence of the intermediate sequence C. In fact, there is a type-III mutation which mutates A to C, and a type-IV mutation which mutates C to B. Let H_3 be the compressed mode from C to B, so we have
$$H_3 = \{(i_{3,k}, \ell_{3,k}),\ k = 1, 2, \cdots, k_-\}\,, \tag{3.69}$$
where
$$\ell_{1,k} = \ell_{-,k}\,, \quad i_{3,k} = i_{-,k} + L_{+,k} \tag{3.70}$$
and $L_{+,k} = \sum_{k'=1}^{k-1} \ell_{+,k'}$. As a result, the decomposition of (3.50) and (3.51) and mutation mode $T = (T_+, T_-)$ determine each other.

Their transformation is described in Steps 1–4. Keeping in mind the above notation, we note the difference between $L_{\tau,k}$ and $L_\tau(i)$, in which, $L_{\tau,k}$ is the total length of the first $k-1$ occurrences of type-τ mutations, but $L_\tau(i)$ is the total length of all the mutation types τ that happened before position i. The modulus structure is demonstrated in Fig. 3.3.

Sequence B is mutated from A, and the mutation mode is
$$T = \{(i_1, \ell_1),\ (i_2, \ell_2)\}\,,$$
in which i_1, i_2 are the mutated positions. The mutation is a type-III mutation if $\ell_1 > 0$ and it is a type-IV mutation if $\ell_2 < 0$. In light of Fig. 3.3, the

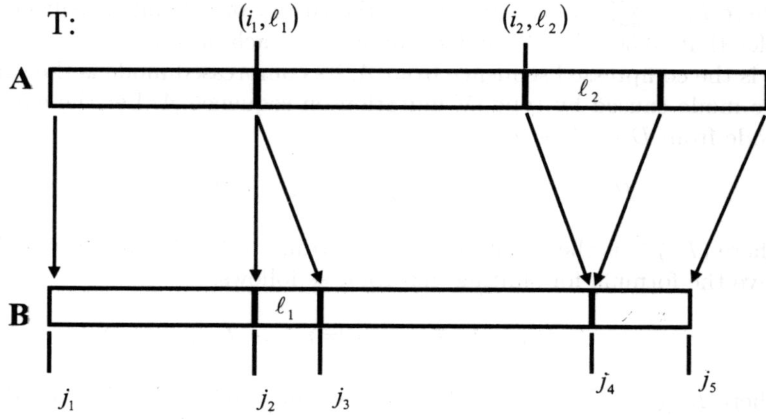

Fig. 3.3. The modulus structure of sequence mutation

relationships are as follows: If we insert segment $b_{\delta_{b,2}}$ into A after position i_1, we obtain sequence C, which is just the envelope of sequences A and B. If we delete the segment $c_{\delta_{c,4}}$ from C after position $i_2 + l_1$, we obtain B, which is just is the output mutated from A directly. If we delete segment $a_{\delta_{a,3}}$ from A after position i_2, we get D, which is the core of sequences A and B.

3.3.3 Structural Representation of Mixed Mutation

If B is mutated from A by type-I, type-II, type-III, and type-IV mutations in a hybrid, we must describe the procedure leading to the mutations. From the basic assumption of shifting mutations, we begin to determine the modulus structure of the shifting mutation, $T = (T_+, T_-)$. Then,

$$A \xrightarrow{T_+} C \xrightarrow{T_-} D \xrightarrow{T_0} B \ , \tag{3.71}$$

in which $T = (T_+, T_-)$ is the shifting mutation, and T_0 is the nonshifting mutation. According to the decomposition expression for the modulus structure of shifting mutations, we decompose sequence C into three parts as follows. $C = (c_{\Delta'_0}, c_{\Delta'_1}, c_{\Delta'_2})$, in which $c_{\Delta'_2}$ is directly drawn from B. Since $c_{\Delta'_1}$ is deleted by the operation T_-, it follows that mutations of type-I and type-II only occur in $c_{\Delta'_0}$. Therefore, the effect of a hybrid of four types of mutation is equivalent to a hybrid of a purely shifting mutation and a nonshifting mutation, and the type-I and type-II mutations only occur in Δ'_0. This illustrates the data structural relationship of hybrid mutations.

3.4 The Binary Operations of Sequence Mutations

Operations of sequence mutations include all operators caused by the modulus structure of the mutations and the transformations between the mutation modes.

3.4.1 The Order Relationship Among the Modes of Shifting Mutations

In Sect. 3.1, we mentioned the order relationship and binary operations for expanded modes. Here, we expand it to the modes of the shifting mutations.

The Order Relationship and Binary Operations for Shifting Mutations

Let us begin with the definitions of relationships between and operations on the modes caused by sequence mutations. Let T, T', T'' be three different mutation modes; following from (3.52) their expressions are as follows:

$$\begin{cases} T = (\gamma_1, \gamma_2, \cdots, \gamma_{n_a}) = \{(i_k, \ell_k) \ , \ k = 1, 2, \cdots, k_a\} \ , \\ T' = (\gamma'_1, \gamma'_2, \cdots, \gamma'_{n_a}) = \{(i'_k, \ell'_k) \ , \ k = 1, 2, \cdots, k'_a\} \ , \\ T'' = (\gamma''_1, \gamma''_2, \cdots, \gamma''_{n_a}) = \{(i''_k, \ell''_k) \ , \ k = 1, 2, \cdots, k''_a\} \ . \end{cases} \tag{3.72}$$

The notations $\gamma'_k, \gamma''_k, i'_k, i''_k$ in T', T'' are similar to ℓ'_k, ℓ''_k in T. Then T, T', T'' satisfy the condition in (3.54) since they are shifting mutations. We can then define the binary operations and order relationship.

Definition 21. *Let T, T' be two shifting mutation modes based on the initial template A. Then:*

1. *T and T' are consistent if for each $i = 1, 2, \cdots, n_a$, $\gamma_i > 0$ implies $\gamma'_i \geq 0$, and $\gamma_i < 0$ implies $\gamma'_i \leq 0$. The converse is also true.*
2. *We say that T is included in T' if T, T' are consistent and $|\gamma_i| \leq |\gamma'_i|$ for each $i = 1, 2, \cdots, n_a$. In this case, we say that T is less than T', and denote it by $T \leq T'$ or $T \subset T'$.*

Definition 22. *Let T and T' be two consistent modes of shifting mutation. We then define the binary operations as follows:*

1. ***Union**: The union operation is denoted by $T'' = T \vee T'$, and each term of T'' is a new mode defined by*

$$\gamma''_i = \begin{cases} \max\{\gamma_i, \gamma'_i\} \, , & \text{if } \gamma_i, \gamma'_i \geq 0 \, , \\ \min\{\gamma_i, \gamma'_i\} \, , & \text{if } \gamma_i, \gamma'_i \leq 0 \, . \end{cases} \tag{3.73}$$

2. ***Intersection**: The intersection operation is denoted by $T'' = T \wedge T'$, and each term of T'' is defined by*

$$\gamma''_i = \begin{cases} \min\{\gamma_i, \gamma'_i\} \, , & \text{if } \gamma_i, \gamma'_i \geq 0 \, , \\ \max\{\gamma_i, \gamma'_i\} \, , & \text{if } \gamma_i, \gamma'_i \leq 0 \, . \end{cases} \tag{3.74}$$

3. ***Subtraction**: Subtraction is denoted by $T'' = T' \ominus T$, and each term of T'' is defined by*

$$\gamma''_i = \begin{cases} \max\{\gamma'_i - \gamma_i, 0\} \, , & \text{if } \gamma_i, \gamma'_i \geq 0 \, , \\ \min\{\gamma'_i - \gamma_i, 0\} \, , & \text{if } \gamma_i, \gamma'_i \leq 0 \, . \end{cases} \tag{3.75}$$

   Typically, if $T \leq T'$, then the expression $\gamma''_i = \gamma'_i - \gamma_i$ holds.
4. ***Symmetric subtraction**: Symmetric subtraction is denoted by $T'' = T' \Delta T$, and each term of T'' is defined by*

$$\gamma''_i = \begin{cases} \max\{\gamma_i, \gamma'_i\} - \min\{\gamma_i, \gamma'_i\} \, , & \text{if } \gamma_i, \gamma'_i \geq 0 \, , \\ \min\{\gamma_i, \gamma'_i\} - \max\{\gamma_i, \gamma'_i\} \, , & \text{if } \gamma_i, \gamma'_i \leq 0 \, . \end{cases} \tag{3.76}$$

Remark 1. Binary operations on the modes for shifting mutations are different from those on modes for expanding sequences. Here, the binary operations are only defined on the set of all consistent modes for shifting mutations. It would be more complicated if we removed this restriction.

Properties of the Binary Operations and Order Relationships on the Set of Modes

1. The order relationship satisfies transferability and self-reflectivity properties described below. **Transferability**: If $T \leq T'$ and $T' \leq T''$, then we have $T \leq T''$. **Self-reflectivity**: If $T \leq T'$ and $T' \leq T$, then we have $T = T'$.
2. The operations of intersection and subtraction defined in Definition 22 are closed. In other words, for any pair of consistent modes T and T', intersection and subtraction are still consistent modes.
3. The union operation defined in Definition 22 is not closed. For example, let
$$T = \{(3,4),(10,-4)\} \quad T' = \{(3,2),(7,-5)\} .$$
These satisfy the consistency condition of Definition 21, but their union
$$T'' = T \vee T' = \{(3,4),(7,-5),(10,-4)\}$$
does not satisfy the basic assumption of shifting mutation because
$$i''_2 + |\ell''_2| = 12 > i''_3 = 10 .$$
4. If the modes T and T' of a shifting mutation are consistent and their intersection, union and subtraction are closed, the following properties are satisfied:
Commutative law of intersection and union:
$$T \wedge T' = T' \wedge T , \quad T \vee T' = T' \vee T .$$
Associative law of intersection and union:
$$(T \wedge T') \wedge T'' = T \wedge (T' \wedge T'') , \quad (T \vee T') \vee T'' = T \vee (T' \vee T'') .$$
Distributive law:
$$(T \vee T') \wedge T'' = (T \wedge T'') \vee (T' \wedge T'') .$$
These properties can be verified using Definition 22; we omit this discussion here. As a result, we conclude that the set of all consistent modes endowed with the operations of intersection, union and subtraction does not form a Boolean algebra.

3.4.2 Operators Induced by Modes of Shifting Mutations

In expression (3.71), we showed how A becomes B, and the corresponding operators $\mathbf{T}, \mathbf{T}_+, \mathbf{T}_-, \mathbf{T}_0$ induced by the modes $T = (T_+, T_-, T_0)$, respectively. If B is mutated from A by a purely shifting mutation, then $\mathbf{T}_0$ is an identical operation. We discuss the data structural characteristics for these four operators.

The Representation of Operators of Shifting Mutations

Since C is the expanded sequence of both A and B determined by shifting mutation T, its decomposition is $C = (c_{\Delta'_0}, c_{\Delta'_1}, c_{\Delta'_2})$ in Sect. 3.3.2, in which, Δ'_τ is composed of several small intervals $\delta_{\tau,k}$. Then,

$$A = \left(a_{\Delta_{a,0}}, a_{\Delta_{a,1}}\right) = \left(c_{\Delta'_0}, c_{\Delta'_1}\right) , \quad B = \left(b_{\Delta_{b,0}}, b_{\Delta_{b,1}}\right) = \left(c_{\Delta'_0}, c_{\Delta'_2}\right) . \tag{3.77}$$

This relationship is described in expressions (3.61)–(3.70). The corresponding operators are

$$\begin{cases} C = \mathbf{T}_+(A, B) = \left(c_{\Delta'_0}, c_{\Delta'_1}, c_{\Delta'_2}\right) = \left(a_{\Delta_{a,0}}, a_{\Delta_{a,1}}, b_{\Delta_{b,2}}\right) , \\ B = \mathbf{T}_-(C) = \left(c_{\Delta'_0}, c_{\Delta'_2}\right) = \left(b_{\Delta_{b,0}}, b_{\Delta_{b,1}}\right) , \end{cases} \tag{3.78}$$

where $a_{\Delta_{a,0}} = b_{\Delta_{b,0}}$.

The Commutative Property of Operators of Shifting Mutations

In (3.71), the procedure that mutates A to B is chosen as follows: first apply T_+ and then T_-. An alternative procedure would be first apply T_- and then T_+. The commutative property is that the result of B is independent of the order. While discussing shifting mutations, we always assume that the modulus structure T in (3.72) satisfies

$$i_{k+1} \geq i_k + |\ell_k| , \quad k = 1, 2, \cdots , k_a , \tag{3.79}$$

where $i_{k_a+1} = n_a$.

Theorem 17. *1. Let $\mathbf{T}_+$, $\mathbf{T}_-$ be the operators induced by type-III and type-IV mutation based on the initial template A, and let T be the modulus structure determined by (3.72). If T satisfies the condition of (3.79), the operators are commutative.*

2. The theorem can be expanded to address hybrid mutation. If shifting mutations $\mathbf{T}_+$, $\mathbf{T}_-$ of a hybrid mutation satisfy the conditions in expression (3.72), then the operations of operators $\mathbf{T}_+$, $\mathbf{T}_-$, $\mathbf{T}_0$ are all commutative, where $\mathbf{T}_0$ are operators induced by type-I or type-II mutations.

Proof. Omitting the details of the proof, we use the following examples to explain the theorem. Let $k_1 = k_2 = 2$, $n_a = 20$, and

$$T_+ = \{(4,2),(12,4)\}\,, \quad T_- = \{(7,4),(16,3)\}\;.$$

Then, $T = \{(4,2),(7,-4),(12,4),(16,-3)\}$. It follows that T is a purely shifting mutation from $12 > 7+4 = 11$. The shifting mutation is decomposed as follows:

$$\begin{aligned}
A &= (a_1, a_2, \cdots, a_{20}) \\
\xrightarrow{T_+} C_1 &= (a_1, a_2, a_3, a_4, a_1', a_2', a_5, a_6, \cdots, a_{12}, a_3', a_4', a_5', a_6', \\
&\quad a_{13}, a_{14}, \cdots, a_{20}) \\
\xrightarrow{T_-} B_1 &= (a_1, a_2, a_3, a_4, a_1', a_2', a_5, a_6, a_7, a_{12}, a_3', a_4', a_5', a_6', \\
&\quad a_{13}, a_{14}, a_{15}, a_{16}, a_{20})\;,
\end{aligned}$$

where $(a_1', a_2', a_3', a_4', a_5', a_6')$ are the inserted data. If we exchange the operation order of T_+ and T_-, we find

$$\begin{aligned}
A &= (a_1, a_2, \cdots, a_{20}) \\
\xrightarrow{T_-} C_2 &= (a_1, a_2, a_3, a_4, a_5, a_6, a_7, a_{12}, a_{13}, a_{14}, a_{15}, a_{16}, a_{20}) \\
\xrightarrow{T_+} B_2 &= (a_1, a_2, a_3, a_4, a_1', a_2', a_5, a_6, a_7, a_{12}, a_3', a_4', a_5', a_6', \\
&\quad a_{13}, a_{14}, a_{15}, a_{16}, a_{20})\;.
\end{aligned}$$

Thus, $B_1 = B_2$. That is, the mutation operations T_+ and T_- are commutative. In addition, the operator $\mathbf{T}_0$ only acts on the region of Δ_0', since it cannot act on the region of Δ_1' and is useless if it happens in the region of Δ_2' because its action will be deleted. Therefore, it is commutative with the operators of $\mathbf{T}_1, \mathbf{T}_2$. The theorem is therefore proved.

As a consequence of this theorem, we find that the mode T of a shifting mutation may be decomposed into several locally commutative shifting mutations if T satisfies condition (3.79).

The Inverse of Shifting Mutation Operator T

If B is mutated from A by the shifting mutation operator T_a, then the operator induced by the mode of the shifting mutation that mutates B to A is the inverse of T_a. We denote this by T_b giving $T_b = T_a^{-1}$.

If the shifting mutation operator T is defined by (3.72), then we have

$$T_a = (\gamma_1, \gamma_2, \cdots, \gamma_{n_a}) = \{(i_k, \ell_k)\,,\ k = 1, 2, \cdots, k_a\}\;,$$

where $\gamma_i \in \mathbf{Z}$, $\ell_k \neq 0$ and i_k is the mutation position of sequence A.

Theorem 18. *If the shifting mutation operator T defined by (3.72) satisfies the basic assumption (3.79), then the operator $T_b = T_a^{-1}$ mutating B to A is also a shifting mutation operator with the following properties:*

1. *The number of the shifting mutations from B to A are equal to the number of the shifting mutations caused by T_a; hence, $k_b = k_a$.*
2. *Mutation positions in the sequence B are*

$$i'_k = i_k + L_k \ , \quad k = 1, 2, \cdots, k_a \ , \tag{3.80}$$

   where $L_k = \sum_{k'=1}^{k-1} \ell_{k'}$ is the shifting function induced by the shifting mutation operator T_a.
3. *The i'_k defined by (3.80) is strictly monotonic increasing, that is,*

$$1 = i'_0 \leq i'_1 < i'_2 < \cdots < i'_{k_a} \leq i'_{k_a+1} = n_b \ ,$$

   where $n_b = n_a + L_{0,k_a+1}$.
4. *The insertion operation of $T_b = T_a^{-1}$ simply corresponds to the deletion operation of T_a. In other words, the insertion of T is the deletion of T^{-1}, and the deletion of T is the insertion of T^{-1}. It can also be found that*

$$T_a^{-1} = \{(i'_k, -\ell_k) \ , \ k = 1, 2, \cdots, k_a\} \ . \tag{3.81}$$

Proof. The proofs of propositions 1, 2, and 4 in Theorem 18 only involve the decomposition expression (3.61). In fact, the decomposition expression $(\Delta'_0, \Delta'_1, \Delta'_2)$ of N_c is determined if T_a is known. We then have $(c_{\Delta'_0}, c_{\Delta'_2}) = B$, and obtain C by inserting $c_{\Delta'_1}$ into B. Obviously, this data fragment $c_{\Delta'_1}$ is just the part deleted from A. On the other hand, to obtain A, we need only delete the data fragment $c_{\Delta'_2}$ from sequence C. This fragment $c_{\Delta'_2}$ is inserted into A to find N_c. Therefore, the mutation from B to A is just the inverse of the mutation from A to B. Since $c_{\Delta'_1}$ and $c_{\Delta'_2}$ in C are disjoint, it follows that T^{-1} is a purely shifting mutation, and expressions (3.79) and (3.80) hold. Thus, the propositions 1, 2, and 4 of Theorem 18 are proved. To prove proposition 3, we involve the conditions of shifting mutations and expression (3.79). We then have

$$i'_{k+1} - i'_k = (i_{k+1} + L_{0,k+1}) - (i_k + L_{0,k}) = i_{k+1} - i_k + \ell_k > 0 \ .$$

Following from the conditions of purely shifting mutations, we have $i_{k+1} - i_k + |\ell_k| > 0$ for every $k = 1, 2, \cdots, k_a$. Thus, proposition 3 holds, and the theorem is proved.

Example 12. In Example 6, the shifting function is

$$\bar{L} = (L_{0,1}, L_{0,2}, L_{0,3}, L_{0,4}) = (0, 4, 1, -3, -1)$$

and the mutation mode from B to A is

$$T^{-1} = \{(7,-4),(17,3),(21,4),(26,-2)\} .$$

Next, we let $\mathbf{T}_b = \mathbf{T}_a^{-1}$ be the inverse of $\mathbf{T}_a$. Thus, B is completely determined by $B = \mathbf{T}_a(A)$, if $A' = c_{\Delta'_1}$ is given. However, A is determined by $A = \mathbf{T}^{-1}(B)$, if $B' = c_{\Delta'_2} = a_{\Delta_2}$ is known.

Addition of Operators of Consistent Mutations

Let T^1, T^2 be two mutation modes based on the initial template A and satisfying the consistent condition, and let $T^3 = T^2 \vee T^1$. The operators $\mathbf{T}^1$, $\mathbf{T}^2$, $\mathbf{T}^3$ induced by T^1, T^2, T^3 are then shifting operators if T^3 satisfies the basic assumption of shifting mutations. Their mutual relationships are described as follows:

1. Following from the consistent conditions and the definition of $T^3 = T^2 \vee T^1$, we have that $T^1, T^2 \le T^3$. Therefore, all mutation positions of A can be uniformly denoted by $I^3 = \{i_1, i_2, \cdots, i_{k_a}\}$, and their modulus structures can be also uniformly represented by
$$T^\theta = \{(i_k, \ell_k^\tau) \ , \ k = 1,2,\cdots,k_a\} \ , \quad \theta = 1,2,3 \ , \tag{3.82}$$
where
$$\ell_k^3 = \begin{cases} \max\left\{\ell_k^1, \ell_k^2\right\} \ , & \text{when } \ell_k^1, \ell^2 > 0 \ , \\ \min\left\{\ell_k^1, \ell_k^2\right\} \ , & \text{when } \ell_k^1, \ell^2 < 0 \ . \end{cases}$$
2. T^1, T^2, T^3 can be decomposed as follows:
$$T^1 = \left(T_+^1, T_-^1\right) , \quad T^2 = \left(T_+^2, T_-^2\right) , \quad T^3 = \left(T_+^3, T_-^3\right) ,$$
in which T_+^1, T_+^2, T_+^3 are the modulus structures induced by type-III mutation, and T_-^1, T_-^2, T_-^3 are the modulus structures induced by type-IV mutation. The uniform expression is given as follows:
$$T_\tau^\theta = \left\{\left(i_{\tau,k}^\theta, \ell_{\tau,k}^\theta\right) \ , \ k = 1,2,\cdots,k_\tau^\theta\right\} \ , \quad \theta = 1,2,3 \ , \quad \tau = +,- \ . \tag{3.83}$$
3. The operator induced by T^θ is denoted by $\mathbf{T}^\theta$ and defined by
$$\mathbf{T}^\theta(A, E^\theta) = B^\theta \ , \quad \theta = 1,2,3 \ , \tag{3.84}$$
in which $E^\theta = (E_1^\theta, E_2^\theta, \cdots, E_{k_+^\theta}^\theta)$, $\theta = 1,2,3$ are templates of inserted data. Since $||E_k^\theta|| = \ell_+^\theta$, we have $||E_+^1||$, $||E_+^2|| \le ||E_+^3||$. As a result, we have the following theorem.

Theorem 19. *If T^1 and T^2 are two consistent modes based on the template A, and $T^3 = T^2 \vee T^1$ satisfies the basic assumption, then the three operators $\mathbf{T}^1, \mathbf{T}^2, \mathbf{T}^3$ induced by T^1, T^2, T^3 are shifting operators and satisfy the properties as follows:*

1. *Let* $E^4 = (E_1^4, E_2^4, \cdots, E_{k_a}^4)$, *in which* $E_k^4 = E_k^3 - E_k^1$ *for each* k *and let* $B^1 = \mathbf{T}^1(A, E^1)$. *If* E_k^1 *is a subvector of* E_k^3 *for each* k, *then we have*

$$B^3 = \mathbf{T}^3(A, E^3) = \mathbf{T}_1^2(B^1, E^4) = \mathbf{T}_1^2[\mathbf{T}^1(A, E^1), E^4] \ , \qquad (3.85)$$

in which $\mathbf{T}_1^2$ *is a shifting operator on* B^1 *and its modulus structure is defined as follows:*

$$T_1^2 = \left\{ \left(j_k, \ell_k^2\right) \ , \ k = 1, 2, \cdots, k_a \right\} \ , \qquad (3.86)$$

where ℓ_k^2 *is defined in (3.82), and* $j_k = i_k^2 + L_k^1$, $L_k^1 = \sum_{k'=1}^{k-1} \ell_{k'}^1$ *is the shifting function of* T^1.

2. *If* $E_k^1 \cup E_k^2 = E_k^3$ *for each* k, *then let* $E_k^5 = E_k^3 - E_k^2$, $E^5 = (E_1^5, E_2^5, \cdots, E_{k_a}^5)$ *and* $B^2 = \mathbf{T}^2(A, E^2)$, *we have*

$$B^3 = \mathbf{T}^3(A, E^3) = \mathbf{T}_2^1(B^2, E^5) = \mathbf{T}_1^2[B^1, E^4] \ , \qquad (3.87)$$

in which $\mathbf{T}_2^1$ *is the shifting operator on the sequence* B^2. *Its modulus structure is defined by:*

$$T_2^1 = \left\{ \left(j_k', \ell_k^1\right) \ , \ k = 1, 2, \cdots, k_a \right\} \ , \qquad (3.88)$$

where ℓ_k^1 *is defined by (3.82),* $j_k' = i_k^1 + L_k^2$ *and* $L_k^2 = \sum_{k'=1}^{k-1} \ell_{k'}^2$ *is the shifting function of* T^2. *It follows from (3.88) that the multiplication of shifting operators* $\mathbf{T}^1$ *and* $\mathbf{T}^2$ *is commutative.*

The proof of this theorem follows from properties 1 and 2 above.

3.5 Error Analysis for Pairwise Alignment

If B is mutated from A, then (A', B') is the aligned sequence of (A, B). In this section, we analyze the error problem. These four sequences A, B, A', B' transform into the uniform form as follows:

$$U = (u_1, u_2, \cdots, u_{n_u}) \ , \quad U = A, B, A', B' \ , \quad u = a, b, a', b' \ ,$$

where $n_{a'} = n_{b'}$ and $a_i, b_i \in V_q$, $a_j', b_j' \in V_{q+1}$, and $q \in V_{q+1}$ is a virtual symbol. For simplicity, we still assume $q = 4$.

3.5.1 Uniform Alignment of Mutation Sequences

Definition of the Uniform Alignment of Mutation Sequences

In Sects. 3.2 and 3.3, we mentioned mutation modes for both the sequence alignment and mutation. If the mutation positions are just the inserting positions raised by the alignment, then the alignment is a uniform alignment. The definition is as follows:

Definition 23. *Let* B *be the sequence mutated from* A, *and* (A', B') *is the alignment of* (A, B). *If the insertion part of* A' *is caused by type-III mutation, and the insertion part of* B' *is caused by type-IV mutation acting on* A, *then* (A', B') *is the uniform alignment of* (A, B).

Properties of Uniform Alignment

By Definition 23 we can determine the relationship between the modulus structure induced by uniform alignment and the modulus structure induced by mutations:

1. If B is mutated from A, and its mutation mode T_A is defined by (3.53) or (3.54), then the mutation mode from B to A is $T_b = T_a^{-1}$. Modes T_a and T_b can be decomposed as follows:
$$T_a = (T_{a,+}, T_{a,-}) , \quad T_b = (T_{b,+}, T_{b,-}) ,$$
where $T_{a,+}, T_{b,+}$ are the modes induced by type-III mutation based on the initial templates A and B, respectively, and $T_{a,-}, T_{b,-}$ are modes resulting from type-IV mutation based on the initial templates A and B, respectively.
2. If (A', B') is the uniform alignment of sequence (A, B), the expanded mode from A to A' is $H_a = T_{a,+}$, and the expanded mode from B to B' is $H_b = T_{b,+}$.
3. If B is mutated from A purely by shifting mutations, then the uniform alignment can be represented using the core D and envelope C, where $C = (c_{\Delta'_0}, c_{\Delta'_1}, c_{\Delta'_2})$, $D = c_{\Delta'_0}$, $A = (c_{\Delta'_0}, c_{\Delta'_1})$, and $B = (c_{\Delta'_0}, c_{\Delta'_2})$.
4. If (A', B') is the alignment of (A, B) and the following conditions are satisfied:
$$a'_j = \begin{cases} c_j , & \text{if } \ j \in \Delta'_0 \cup \Delta'_1 , \\ 4 , & \text{if } j \in \Delta'_2 , \end{cases} \qquad b'_j = \begin{cases} c_j , & \text{if } j \in \Delta'_0 \cup \Delta'_2 , \\ 4 , & \text{if } j \in \Delta'_1 . \end{cases} \tag{3.89}$$
Then (A', B') is the uniform alignment of (A, B) and $n_{a'} = n_{b'} = n_c$.

Example 13. In Example 12, the modulus structures of A and B are given as follows:
$$T_A = \{(7,4), (13,-3), (20,-4), (29,2)\} ,$$
$$T_B = \{(7,-4), (17,3), (21,4), (26,-2)\} .$$

If (A', B') is the uniform alignment of (A, B), it follows that the modulus structures from A to A' and from B to B' are:
$$H_a = T_{a,+} = \{(7,4), (29,2)\} , \quad H_b = T_{b,+} = \{(17,3), (21,4)\} .$$

Therefore, we find the representation of the uniform alignment (A', B'),
$$\begin{cases} A' = (0000000444411111133300003333111111440000) , \\ B' = (0000000222211111144400004444111111220000) , \end{cases}$$

and the core D and envelope C are
$$\begin{cases} D = (00000001111113330000333311111110000) , \\ C = (0000000222211111133300004333111111220000) . \end{cases}$$

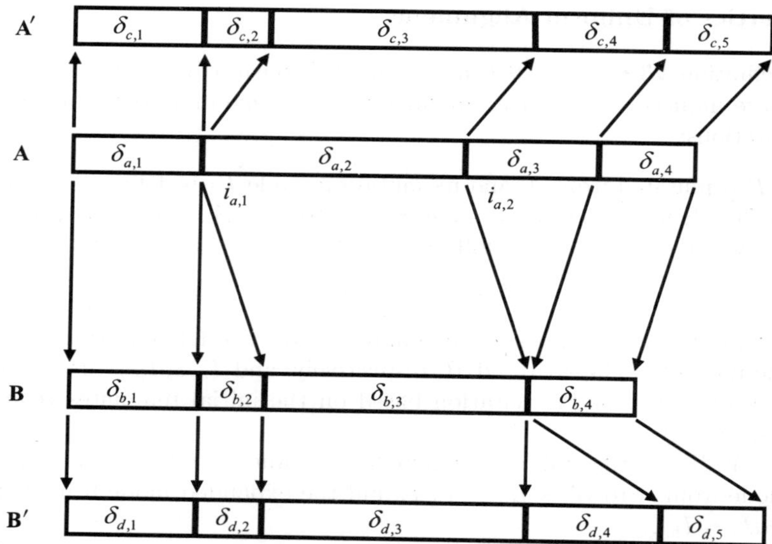

Fig. 3.4. The modulus structure of uniform alignment

The relationship between sequence mutation and uniform alignment is illustrated in Fig. 3.4.

In Fig. 3.4, (A', B') is the alignment of (A, B), and $T = (T_+, T_-)$ is the modulus structure, where

$$T_+ = \{i_1, \ell_1\}, \quad T_- = \{i_2, \ell_2\} .$$

Thus, the modulus structure of uniform alignment is

$$H_a = \{i_1, \ell_1\}, \quad H_b = \{i_2, \ell_2\} ,$$

with

$$\ell_1 = ||\delta_{b,2}|| = ||\delta_{c,2}||, \quad \ell_2 = ||\delta_{a,3}|| = ||\delta_{d,4}|| .$$

3.5.2 Optimal Alignment and Uniform Alignment

The definition of optimal alignment was presented in Chap. 1. In this section, we study the relationship between optimal alignment and uniform alignment. A uniform alignment is optimal if $\epsilon_1 \ll 1/2$. However, this is not generally the case.

Example 14. We delete segment (23210) from the initial template $A = (11231003221000[23210]3221\{\}02112003)$, and insert a segment (11230) into the large bracket of A between segments (3221) and (0211), and perform type-I and type-II mutations outside the insertion and deletion region. Let the final output mutated from A be

$$B = (112310032210003220[11230]12112003)\ .$$

Then the uniform alignments of A, C are given as follows:

$$\begin{cases} A'_1 = (11231(0)032210(0)0[23210]322(1)[44444](0)2112003)\ , \\ B'_1 = (11231(2)032210(1)0[44444]322(0)[11230](1)2112003)\ , \end{cases}$$

where the components in parentheses are caused by type-I and type-II mutations, and components in square brackets are from type-III and type-IV mutations. The Hamming matrix between them is $d_{\mathrm{H}}(A', B') = 14$ and the optimal alignment is

$$\begin{cases} A'_0 = (11231(0)032210(0)0[23210]322[4]1[444]0[4]2112003)\ , \\ B'_0 = (11231(2)032210(1)0[44444]322[0]1[123]0[1]2112003)\ , \end{cases}$$

where $d_{\mathrm{H}}(C', D') = 12 < d_{\mathrm{H}}(A', B')$. Therefore, (A', B') is the uniform alignment of (A, B) but is not an optimal alignment. Making use of Example 8, we find that a uniform alignment can be represented by a minimum penalty alignment through a local modification. The local modification permutes the insertion symbols with their nearest nucleotides. For example, (A'_0, B'_0) is a local modification of (A'_1, B'_1). The details of local modification will be discussed later.

It is worth noting that a uniform alignment may not be a minimal penalty alignment even if $\epsilon_1 = \epsilon_2 = 0$; this corresponds to the cases where type-I and type-II mutations do not work.

Example 15. If we delete the segment (0000) from the initial template $A =$ (121200001212121212001212) and insert the segment (121200) into A between (01212) and (121200), then A and the output B mutated from A are compared as follows:

$$\begin{cases} A = (1212[0000]12121212001212)\ , \\ B = (12121212[121200]1212001212)\ . \end{cases}$$

The uniform alignment is given by:

$$\begin{cases} A'_1 = (1212[0000]1212[444444]1212001212)\ , \\ B'_1 = (1212[4444]1212[121200]1212001212)\ . \end{cases}$$

Therefore $d_{\mathrm{H}}(A'_1, B'_1) = 10$, and we can construct a new alignment as follows:

$$\begin{cases} A'_0 = (1212[00]001212[4444]1212001212)\ , \\ B'_0 = (1212[44]121212[1200]1212001212)\ . \end{cases}$$

Hence, $d_{\mathrm{H}}(A'_0, B'_0) = 8 < d_{\mathrm{H}}(A'_1, B'_1)$. It follows that an optimal alignment (A', B') may not be the minimal penalty alignment.

3.5.3 Error Analysis of Uniform Alignment

Let B^* be a stochastic sequence mutated from the stochastic sequence A^*, which is determined by expression (2.93). Let (C^*, D^*) be the uniform alignment of (A^*, B^*). The error analysis of (C^*, D^*) is then an estimate of the absolute error

$$w(C^*, D^*) = \sum_{j=1}^{n_{c^*}} w\left(c_j^*, d_j^*\right) . \tag{3.90}$$

For simplicity, we assume the penalty matrix is the Hamming matrix; i.e., $w(a, b) = 0$ or 1, for $a = b$ or $a \neq b$:

1. We can decompose the expression (3.90) as follows:

$$w(C^*, D^*) = \sum_{j\in\Delta_0^*} w\left(c_j^*, d_j^*\right) + \sum_{j\in\Delta_3^*} w\left(c_j^*, d_j^*\right) + \sum_{j\in\Delta_2^*} w\left(c_j^*, d_j^*\right) . \tag{3.91}$$

   Since $c_j^* = 4, d_j^* \neq 4$ in the second part, and $d_j^* = 4, c_j^* \neq 4$ in the third part, it follows that

$$\sum_{j\in\Delta_3^*} w\left(c_j^*, d_j^*\right) = n_3^* , \quad \sum_{j\in\Delta_2^*} w\left(c_j^*, d_j^*\right) = n_2^* .$$

   Therefore, expression (3.91) becomes

$$w(C^*, D^*) = \sum_{j\in\Delta_0^*} w\left(c_j^*, d_j^*\right) + n_1^* + n_2^* , \tag{3.92}$$

   where n_1^* and n_2^* are the total lengths of the type-III and type-IV mutations respectively. Both are random numbers.
2. The estimate of n_1^* and n_2^* in the expression (3.92). Since $T^* = \{(i_k^*, \ell_k^*), k = 1, 2, \cdots, k_a^*\}$ is a random modulus structure of random mutation, and based on the assumptions of type-III and type-IV mutations and the large number law, we estimate n_1^*, n_2^* as follows:

$$\begin{cases} n_1^* = \sum_{k:\ell_k^*>0} \ell_k^* \sim n_a \left(1 + \frac{\epsilon_3}{p_3}\right) , \\ n_2^* = \sum_{k:\ell_k^*<0} \ell_k^* \sim n_a \left(1 + \frac{\epsilon_4}{p_4}\right) . \end{cases} \tag{3.93}$$

3. To estimate the first term in expression (3.92), we begin with the following notations:
   (a) $(c_j^*, d_j^*) = (a_j^*, a_j^* + \zeta_j)$ for all $j \in \Delta_0^*$ is an i.i.d. sequence. a_j^* is a uniform distribution on V_4, ζ_j is defined by (3.96), and $\tilde{\zeta}$ and A^* are

independent. Therefore,

$$\begin{cases} E\left\{w(c_j^*, d_j^*)\right\} = \sum_{a,b \in V_4} p(a,b) w(a,b) = \epsilon \,, \\ D\left\{w(c_j^*, d_j^*)\right\} = \sum_{a,b \in V_4} p(a,b) [w(a,b) - \epsilon]^2 \\ \qquad\qquad\qquad\quad = \sum_{a,b \in V_4} p(a,b) w^2(a,b) - \epsilon^2 = \epsilon(1-\epsilon) \,. \end{cases} \tag{3.94}$$

(b) For any pair $j \neq j' \in \Delta_0'$, the two components of $(c_j^*, d_{j'}^*) = (a_j^*, a_{j'}^* + \zeta_{j'})$ are independent and both are uniform distributions on V_4, therefore

$$\begin{cases} E\left\{w(c_j^*, d_j^*)\right\} = \sum_{a,b \in V_4} p(a) p(b) w(a,b) = 3/4 \,, \\ D\left\{w(c_j^*, d_j^*)\right\} = \sum_{a,b \in V_4} p(a,b) [w(a,b) - 3/4]^2 \\ \qquad\qquad\qquad\quad = \sum_{a,b \in V_4} p(a,b) w^2(a,b) - (3/4)^2 = 3/16 \,. \end{cases} \tag{3.95}$$

(c) Based on the definition of Δ_0^*, we have that Δ_0^* is composed of several small intervals as illustrated in (3.59). Let

$$n_{0,h}^* = ||\delta_{0,h}^*||$$

be the length of the small interval $\delta_{0,h}^*$. In the event that $n_{0,h}^*$ is large enough, we may estimate using the law of large numbers:

$$\frac{1}{n_{0,h}} d_w \left(c_{\delta'_{0,h}}^*, d_{\delta_{0,h}^* + L}^* \right) \sim \begin{cases} \epsilon \,, & \text{if } L = 0 \,, \\ \dfrac{3}{4} \,, & \text{if } L \neq 0 \,. \end{cases} \tag{3.96}$$

The central limit theorem yields

$$\frac{1}{n_{0,h}^*} d_w \left(c_{\delta'_{0,h}}^*, d_{\delta'_{0,h} + L}^* \right) \sim \begin{cases} N\left(\epsilon, \dfrac{\epsilon(1-\epsilon)}{n_{0,h}^*}\right) \,, & \text{if } L = 0 \,, \\ N\left(\dfrac{3}{4}, \dfrac{3}{16 n_{0,h}^*}\right) \,, & \text{if } L \neq 0 \,, \end{cases} \tag{3.97}$$

where $N(\gamma, S)$ is a normal distribution with the expectation of γ and variance of S.

4. Using (3.93) and (3.96) and the properties of composite renewal processes, we find that for the hybrid mutations, the error estimate of the uniform alignment of the stochastic sequences is given by

$$\frac{1}{n_a} w(C^*, D^*) \sim \epsilon_1 + \epsilon_2 \left(\frac{1}{p_1} + \frac{2}{p_2} \right) (1 - \epsilon_1) + \frac{\epsilon_3}{p_3} + \frac{\epsilon_4}{p_4} \,, \tag{3.98}$$

where ϵ_τ, $\tau = 1,2,3,4$ are the strengths of the τ mutation stream, and that p_τ, $\tau = 1,2,3,4$ are the lengths of the type-II, type-III and type-IV mutations, respectively. The reader is referred to Chap. 2 for more detail.

3.5.4 Local Modification of Sequence Alignment

In Examples 13 and 14, and in the proof of Theorem 16, we may find that uniform alignment and the minimal penalty alignment are not perfectly compatible. There is a minor difference that can be addressed with a local modification. The basic purpose of sequence alignment is to find the mutation relationship between different sequences, that is, to determine their uniform alignments. However, we do not know whether the uniform alignment of two sequences has been realized. Therefore, we instead determine the minimal penalty alignment to approximate the uniform alignment because the minimal penalty alignment may be accurately computed. A local modification allows us to estimate the uniform alignment since the minimal penalty alignment approximately determines the boundary of uniform alignment. Local modification is a re-computation based on the output from sequence alignment in order to reduce the penalty. If (A', B') is the alignment output, the local modification is stated as follows:

W-1 Permute the virtual symbols in A' and B' with the nearest nonvirtual symbols.

W-2 Insert or delete one or more virtual symbols in both A' and B' simultaneously.

W-3 At the tails of the aligned sequences, we may insert or delete a number of virtual symbols, in the hopes that this process reduces the magnitude of the total error.

Example 16. The following shows three local modifications:

1. If

$$\begin{cases} C = (00231(444)312(0021)3200231) , \\ D = (00231(221)130(4444)3200231) , \end{cases}$$

then the alignment error is $d_{\mathrm{H}}(A', B') = 10$. If we perform local modification **W-2** such that the output (A', B') is

$$\begin{cases} C' = (00231312002(1)3200231) , \\ D' = (00231221130(4)3200231) , \end{cases}$$

then $d_{\mathrm{H}}(C', D') = 7 < d_{\mathrm{H}}(A', B')$.

2. If

$$\begin{cases} C = (00231(44444)3123200231) , \\ D = (00231(22312)1303200231) , \end{cases}$$

the alignment error is $d_{\mathrm{H}}(A', B') = 8$. If we perform local modification **W-1** such that (A', B') is

$$\begin{cases} C' = (00231(44)312(444)3200231) , \\ D' = (00231(22)312(130)3200231) , \end{cases}$$

then $d_{\mathrm{H}}(C', D') = 5 < d_{\mathrm{H}}(A', B')$.

3. If

$$\begin{cases} C = (00231312002(1)32000231)\,, \\ D = (00231221130(4)32002310)\,, \end{cases}$$

the alignment error is $d_{\mathrm{H}}(A', B') = 11$. If we perform local modification **W-3** such that (A', B') is

$$\begin{cases} C' = (00231312002(1)3200(0)231(4))\,, \\ D' = (00231221130(4)3200(4)231(0))\,, \end{cases}$$

then $d_{\mathrm{H}}(C', D') = 9 < d_{\mathrm{H}}(A', B')$.

Based on these examples, we find that all of the local modification operations can reduce alignment error. Through a modification, we can approximate a uniform alignment by a minimal penalty alignment, and obtain the modulus structure of type-III and type-IV mutations.

3.6 Exercises

Exercise 12. Describe the relationship between modulus structure theory and random mutations. Give the modulus structure of the sequence obtained in Exercise 11.

Exercise 13. Prove Theorems 9, 10, 11, 12, and 15.

Exercise 14. For the given sequences

$$\begin{cases} A = 1101020103221302010321102230 1\,, \\ B = 11010202101032200130211010321031102230 1\,, \\ C = 11010333202101032200130211000103210223110223 01\,, \end{cases}$$

complete the following tasks:

1. Show that B is the expanded sequence of A, and C is the expanded sequence of B, and therefore, that C is the expanded sequence of A.
2. Give the three expanded modes by which B mutated from A, C mutated from B, and C mutated from A. Also give the corresponding shifting functions.
3. Give the three different compressed modes by which B compressed into A, C compressed into B, and C compressed into A.
4. Give the operators of the expanded modes by which B mutated from A, C mutated from B, and C mutated from A, showing the computation procedure.
5. Compute the minimum penalty alignments based on the Hamming penalty matrix and the SP-function condition, and compute the value of the SP-penalty.

Exercise 15. For the pair of sequences

$$\begin{cases} A = 1101020210103220111013021101032103110232323 01\,, \\ B = 1101033320210103220013021100010321022311022301\,, \end{cases}$$

answer the following questions:

1. If B is considered to be the sequence mutated from A, compute the three different modes of type-II, type-III, and type-IV mutations, and write down the corresponding shifting functions.
2. Find the optimal alignment (A', B') of (A, B) using the dynamic programming-based algorithm.
3. Give the alignment modulus structures by which (A, B) mutates to (A', B') and the corresponding function.
4. Analyze the relationship between the mutation mode and the alignment mode of (A, B).

4

Super Pairwise Alignment

In Chap. 1, we introduced dynamic programming-based algorithms that are used comprehensively in many fields. Next, we propose several modulus structure-based and statistical decision-based algorithms. The key concept giving rise to these algorithms was published in [90], and is referred to as super pairwise alignment (SPA).

4.1 Principle of Statistical Decision-Based Algorithms for Pairwise Sequences

In this section, we introduce pairwise alignment and the principle of SPA.

4.1.1 Uniform Alignment and Parameter Estimation for Pairwise Sequences

The description of pairwise alignment was given in Chap. 1. The definitions of minimum penalty alignment and uniform alignment were mentioned in Sect. 1.2. The uniform alignment has been mentioned in Definition 23. In this chapter, we focus on the solution for uniform alignment of pairwise sequences.

We continue to use the symbols given in (1.1). If sequence B is mutated from A, and the mutation mode T is defined as in (3.55), then the solution for uniform alignment of the pairwise sequence estimates the parameters $\{(i_k, \ell_k), k = 1, 2, \cdots, k_a\}$ and k_a.

If we estimate the parameter set T, which denotes the positions and lengths of the mutations based on sequence (A, B), then sequence (A', B') constructed by Definition 23 is an estimate for the uniform alignment of sequence (A, B). Therefore, the key to solving the uniform alignment of the pairwise sequence is knowing how to estimate the parameters in the mutation mode T based on sequence (A, B). Then T in expression (3.55) is a group of statistical parameters and

$$\hat{T} = \left\{ \left(\hat{i}_k, \hat{\ell}_k\right) , \; k = 1, 2, \cdots , \hat{k}_a \right\} \tag{4.1}$$

is a set of statistics determined by (A, B), and an estimate of the parameter set T. The vital problem of uniform alignment of pairwise sequences is the estimate of the parameters in T. The approach to solving this problem is outlined as follows:

Sequential Estimate of Parameter Set T

The so-called sequential estimate for the parameter set T is briefly described below:

1. To estimate the parameters in T alternately, we estimate (i_k, ℓ_k), $k = 1, 2, \cdots , k_a$ one after the other, that is, we estimate $(i_{k'}, \ell_{k'})$ based on (i_k, ℓ_k), $k = 1, 2, \cdots , k' - 1$.
2. To estimate each (i_k, ℓ_k), we need not have the entire data of sequence (A, B), but depend on only part of the data. Therefore, choosing the data to use becomes one of the most important aspects of the statistical decision algorithm.
3. The estimate of the parameter set T includes an estimate of the parameter k_a.

Since this estimation procedure is identical to sequential estimation in statistics, we call it the sequential estimation of the parameter set T.

The Locally Uniform Alignment of Sequence (A, B)

From the definition of uniform alignment in Definition 23, we generalize it to formulate the definition of locally uniform alignment as follows:

Definition 24.

1. *Let T be the shifting mutation mode given in expression (3.55), then*

$$T_{k'} = \{(i_k, \ell_k), \; k = 1, 2, \cdots , k'\} , \quad k' \leq k_a \; , \tag{4.2}$$

   is called the local shifting mutation of T.
2. *Let B be the sequence mutated from A through shifting mutation mode T and type-I and type-II mutations. If sequence $B_{k'}$ is mutated from A through shifting mutation $T_{k'}$ and type-I and type-II mutations, then $B_{k'}$ is called the local shifting mutation of A, and $T_{k'}$ is corresponding local shifting mutation mode.*
3. *If (A', B'), $(C_{k'}, D_{k'})$ are two uniform alignmemt sequences of (A, B) and $(A, B_{k'})$ respectively, then $(C_{k'}, D_{k'})$ is the locally uniform alignment sequence of (A', B').*

The Outline of the Sequential Estimation for Parameter Set T

To estimate the parameter set T, we actually estimate (i_k, ℓ_k) and the locally uniform alignment alternately. If the sequence pair (A, B) is given, then the sequential estimation for parameter set T is summarized as follows:

1. Letting $\bar{a}_0 = (a_1, a_2, \cdots, a_n)$, $\bar{b}_0 = (b_1, b_2, \cdots, b_n)$, we begin to estimate (i_1, ℓ_1), and let $(\hat{i}_1, \hat{\ell}_1)$ denote the estimation. The methods to select n and to compute $(\hat{i}_1, \hat{\ell}_1)$ will be detailed later.
2. After obtaining the estimation of the local shifting mutation $T_{k'}$, we denote it by $\hat{T}_{k'}$ similarly to (4.1), and then obtain the local alignment sequence $(\hat{C}_{k'}, \hat{D}_{k'})$ based on $\hat{T}_{k'}$ and (A, B). Computation of the local alignment sequence $(\hat{C}_{k'}, \hat{D}_{k'})$ will be described later.
3. Select vectors $\bar{c}_t = (c_{t+1}, c_{t+2}, \cdots, c_{t+n})$ and $\bar{d}_t = (d_{t+1}, d_{t+2}, \cdots, d_{t+n})$ from the local alignment sequence $(\hat{C}_{k'}, \hat{D}_{k'})$, so we can estimate $(\hat{i}_{k'+1}, \hat{\ell}_{k'+1})$ of $(i_{k'+1}, \ell_{k'+1})$. The method to select $\bar{c}_t$ and $\bar{d}_t$ will be explained later in the block.

4.1.2 The Locally Uniform Alignment Resulting from Local Mutation

Now, we detail the key steps 1–3 given in Sects. 4.1.1–4.1.3. We begin by discussing the locally uniform alignment induced by a local mutation.

Let $T_{k'}$ be a known local shifting mutation mode, which is defined by expression (4.2). We can then obtain its local alignment sequence following from the local shifting mutation mode $T_{k'}$ discussed in Chap. 3. We recall this process briefly, and comment on the specifics. For simplicity, we omit the subscript k' of $T_{k'}$.

Some Symbols

The expansion or decomposition induced by a shifting mutation mode T are addressed in Sect. 3.3.2. Here:

1. Let the decomposition of (C', D') expanded from (A, B) based on mode T be

$$\Delta' = \{\Delta'_\tau\,,\ \tau = 0, 1, 2\} = \left\{\delta'_{\tau,k}\,,\ k = 1, 2, \cdots, \hat{k}_\tau\,,\ \tau = 0, 1, 2\right\}\,. \quad (4.3)$$

If we arrange the areas in Δ' in order, then we obtain $\Delta' = N' = \{1, 2, \cdots, n'\}$, and the decomposition expression is:

$$\Delta' = \{\Delta'_0, \Delta'_1, \Delta'_2\} = (\delta'_1, \delta'_2, \cdots, \delta'_{2k'})\,, \quad (4.4)$$

where k' is the number of times the shifting mutation occurred in T, $n' = n_a + L_+$, $L_+ = \sum_{k:\ell_k > 0} \ell_k$. Then,

$$\delta'_{2k-1} \in \Delta'_0\,, \quad \delta'_{2k} \in \Delta'_1 \cup \Delta'_2\,.$$

2. The decompositions of N_a and N_b are determined by the decomposition of Δ':

$$\Delta_\tau = \{\Delta_{\tau,0}, \Delta_{\tau,1}\} = (\delta_{\tau,1}, \delta_{\tau,2}, \cdots, \delta_{\tau,2k'_\tau}), \quad \tau = a, b , \tag{4.5}$$

where k'_a, k'_b are the numbers of times the type-IV and type-III mutations occurred in the shifting mutation mode T. Thus, $k'_a + k'_b = k'$.
The intervals in (4.5) form a set produced by two steps: deletion Δ'_1, Δ'_2 from Δ', and then rearrangement of the rest of the intervals. If we denote this set by

$$\Delta_\tau = \{i_{\tau,1}, i_{\tau,2}, \cdots, i_{\tau,2k'_\tau}, i_{\tau,2k'_\tau+1}\}, \quad \tau = a, b , \tag{4.6}$$

in which

$$i_{\tau,1} = 0 \le i_{\tau,2} < i_{\tau,3} < \cdots < i_{\tau,2k'_\tau} \le i_{\tau,2k'_\tau+1} = n_\tau$$

and

$$\delta_{\tau,k} = [i_{\tau,k}+1, i_{\tau,k+1}] = (i_{\tau,k}+1, i_{\tau,k}+2, \cdots, i_{\tau,k+1}) ,$$

then by (3.78) and (3.82), we obtain the interrelationship between Δ_a, Δ_b, and Δ' as follows:

$$\begin{cases} i_{a,2k-1} = j_{0,2k-1} - L_{1,k} , \\ i_{a,2k} = j_{0,2k} - j_{0,2k-1} , \end{cases} \qquad \begin{cases} i_{b,2k-1} = j_{1,2k-1} - L_{2,k} , \\ i_{b,2k} = j_{1,2k} - j_{1,2k-1} , \end{cases} \tag{4.7}$$

where

$$L_{\tau,k} = \sum_{k'':\delta'_{k''} \in \Delta'_\tau, j_{1,2k''-1} < j_{0,2k-1}} | \delta'_{k''} | , \quad \tau = 1, 2 . \tag{4.8}$$

3. Moreover, we have

$$C' = \{a_{\Delta_{a,0}}, a_{\Delta_{a,1}}, b_{\Delta_{b,1}}\}, \quad D' = \{b_{\Delta_{b,0}}, b_{\Delta_{b,1}}, a_{\Delta_{a,1}}\} , \tag{4.9}$$

where the local vectors are arranged according to the order in Δ', and then

$$\begin{cases} \Delta_{\tau,0} = \{\delta_{\tau,1}, \delta_{\tau,3}, \cdots, \delta_{\tau,2k'-1}\} , \\ \Delta_{\tau,1} = \{\delta_{\tau,2}, \delta_{\tau,4}, \cdots, \delta_{\tau,2k'}\} , \end{cases}$$

where $\tau = 1, 2$ and $\delta_{\tau,k} = [i_{\tau,k}+1, i_{\tau,k+1}]$.

Uniform Alignment Induced by the Shifting Mutation Mode T

From the shifting mutation mode T and sequence (A, B), we find the expanded sequence (C', D'), and we then obtain the alignment (A', B') of (A, B) through the following steps:

1. Modify the sequence C' to become sequence A', by replacing each component in the subvector $a_{\Delta_{a,1}}$ with virtual symbols.

2. Modify the sequence D' to become sequence B', by replacing each component in the subvector $b_{\Delta_{b,1}}$ with virtual symbols.

Therefore, (A', B') is the uniform alignment if $T = \hat{T}_{k'}$ is the uniform estimation for the local mutation mode $T_{k'}$, and then $(\hat{C}_{k'}, \hat{D}_{k'})$, created by the above process, is the uniform estimation statistic of (A, B) according to the local mutation $T_{k'}$. We can estimate the next mutation position of $(C_{k'}, D_{k'})$ based on $(\hat{C}_{\delta_{2k'+1}}, \hat{D}_{\delta_{2k'+1}})$.

4.1.3 The Estimations of Mutation Position and Length

We propose the statistical approach for estimating the mutation position and length (i_k, ℓ_k) in this subsection.

Sliding Window Function of Sequence

The sliding window function of sequence A, B is defined by

$$w(A, B; i, j, n) = w\left(a_{[i+1,i+n]}, b_{[j+1,j+n]}\right) = \frac{1}{n}\sum_{k=1}^{n} w\left(a_{i+k}, b_{j+k}\right) \,, \quad (4.10)$$

where $w(a, b)$ is the Hamming matrix. Similarly, the sliding window function of the stochastic sequence A^*, B^* is defined as

$$w(A^*, B^*; i, j, n) = w\left(a^*_{[i+1,i+n]}, b^*_{[j+1,j+n]}\right) = \frac{1}{n}\sum_{k=1}^{n} w\left(a^*_{i+k}, b^*_{j+k}\right) . \quad (4.11)$$

We will see later that the sliding window function is the same as the local penalty function.

If B^* is mutated from A^*, we can use $w(A^*, B^*; i, j, n)$ to estimate the first mutation position of (A^*, B^*). Therefore, we can separate the random variable pair (a^*_{i+k}, b^*_{j+k}) in expression (4.11) into two segments as follows: The first segment: (a^*_{i+k}, b^*_{j+k}), $k = 1, 2, \cdots, n_1$ is the segment without shifting mutations, here b^*_{j+k} is the random variable mutated from a^*_{i+k} after type-I and type-II mutations. The joint probability distribution is

$$P_r\left\{a^*_{i+k} = b^*_{j+k}\right\} = 1 - \epsilon \,, \quad P_r\left\{a^*_{i+k} \neq b^*_{j+k}\right\} = \epsilon \,.$$

The second segment: (a^*_{i+k}, b^*_{j+k}), $k = n_1 + 1, n_1 + 2, \cdots, n$ is the segment with shifting mutation, here mutations III and IV occur in front of b^*_{j+k} and a^*_{i+k}. The joint probability distribution is

$$P_r\left\{a^*_{i+k} = b^*_{j+k}\right\} = \frac{1}{4} \,, \quad P_r\left\{a^*_{i+k} \neq b^*_{j+k}\right\} = \frac{3}{4} \,.$$

Let $\delta_1 = [1, n_1], \delta_2 = [n_1 + 1, n]$. Then we have

$$E\{w(a_{i+k}, b_{j+k})\} = \begin{cases} \epsilon \,, & \text{if } (a_{i+k}, b_{j+k}) \in \delta_1 \,, \\ \dfrac{3}{4} \,, & \text{if } (a_{i+k}, b_{j+k}) \in \delta_2 \,. \end{cases} \quad (4.12)$$

If (A^*, B^*) are two independent stochastic sequences, then applying the law of large numbers we have

$$w(A^*, B^*; i, j, n) \sim \frac{\epsilon n_1}{n} + \frac{3n_2}{4n} , \tag{4.13}$$

where $n_\tau = ||\delta_\tau||$ is the length of the interval δ_τ.

Statistics for the Sliding Window Function

From the computation formula for the sliding window function (4.13), we can develop statistical estimates of (n_1, n_2). If the parameters ϵ, n are known, following from (4.13) and $n_1 + n_2 = n$ we can estimate (n_1, n_2) as:

$$\begin{cases} \hat{n}_1 = n\dfrac{3-4w}{3-4\epsilon} = n\left(1 - \dfrac{w-\epsilon}{0.75-\epsilon}\right) , \\ \hat{n}_2 = n\dfrac{w-\epsilon}{0.75-\epsilon} , \end{cases} \tag{4.14}$$

in which $w = w(A^*, B^*; i, j, n)$ is calculated directly using (4.11) and (A^*, B^*).

Next, we discuss the estimate of (n_1, n_2) when the parameter n is known while ϵ is still to be determined. Let (A, B) be a fixed sample. We then have:

1. We assume that δ_1 and δ_2 are two fixed intervals, say,
$$\delta_1 = [1, 2, \cdots, n_1] , \quad \delta_2 = [n_1 + 1, n_1 + 2, \cdots, n] ,$$
in which $n_2 = n - n_1$. Then, $a_{i+\delta_1}$ and $b_{j+\delta_1}$ are segments without the shifting mutations, and $a_{i+\delta_2}$ and $b_{j+\delta_2}$ are segments with shifting mutations. Let
$$\delta = (\delta_1, \delta_2) = (1, 2, \cdots, n) .$$
Then, (4.13) is also true and
$$w = w(A, B; i, j, n) = w(a_{i+\delta}, b_{j+\delta}) .$$
2. If we add a shifting parameter h to the penalty function, that is, if we calculate
$$w' = w(A, B; i+h, j+h, n) = w(a_{i+h+\delta}, b_{j+h+\delta}) ,$$
in which $0 < h < n$ is a positive integer but not too large, then the interval δ can be decomposed as:
$$\delta = (\delta_1', \delta_2') = ((1, 2, \cdots, n_1 - h), (n_1 - h + 1, n_1 - h + 2, \cdots, n)) .$$
If $a_{i+h+\delta_1'}$ and $b_{j+h+\delta_1'}$ are segments without shifting mutations, and $a_{i+h+\delta_2'}$ and $b_{j+h+\delta_2'}$ are segments with shifting mutations, the formula for w' is found to be:
$$w' = w(A, B; i+h, j+h, n) \sim \frac{\epsilon(n_1 - h)}{n} + \frac{3(n_2 + h)}{4n} . \tag{4.15}$$

Therefore, following from (4.13) and (4.15) we obtain the following coupled algebraic equations:

$$\begin{cases} \dfrac{\epsilon n_1}{n} + \dfrac{3n_2}{4n} = w \ , \\ \dfrac{\epsilon(n_1 - h)}{n} + \dfrac{3(n_2 + h)}{4n} = w' \ , \end{cases} \tag{4.16}$$

in which $n_1 + n_2 = n$, and w, w' are constants that can be directly calculated based on A, B, i, j, n, h. Therefore, the value of n_1, ϵ can be solved using (4.16).

The simultaneous equations of (4.16) can be reduced to the following form:

$$\begin{cases} 4n_1\epsilon + 3(n - n_1) = 4nw \ , \\ 4(n_1 - h)\epsilon + 3(n - n_1 + h) = 4nw' \ . \end{cases}$$

Thus, we have

$$\begin{cases} \epsilon = \dfrac{3}{4} + \dfrac{n}{h}(w - w') \ , \\ n_1 = \dfrac{h}{w' - w}\left(\dfrac{3}{4} - w\right) \ . \end{cases} \tag{4.17}$$

If we specify w, w' in (4.17) by $w(A^*, B^*; i, j, n)$ and $w(A^*, B^*; i + h, j + h, n)$, respectively, then the results of (4.17) are denoted by $\hat{\epsilon}, \hat{n}_1$, which is the estimate of ϵ, n_1.

4.2 Operation Steps of the SPA and Its Improvement

Continuing from the above section, we introduce the SPA for pairwise alignment. Let us begin by introducing the operation steps of SPA.

4.2.1 Operation Steps of the SPA

Let (A, B) be two fixed sequences. Every algorithm has the goal of estimating all the parameters in the mutation mode T. The SPA is no exception. Specifically, we first select the important parameters $n, h, \theta, \theta', \tau$. Here, n is selected according to the convergence of the law of large numbers or the central limit theorem. Typically, we choose $n = 20, 50, 100, 150$, etc. θ, θ' are selected based on the error rate of the type-I and type-II mutations and the error rate of two independently random variables. Thus, we choose $0 < \theta < \theta' < 0.75$. For the parameters h, τ as two local modifications, we choose them to be proportional to n; typically, $\tau = \alpha n$, $h = \beta n$, $0 < \alpha$, $\beta < 0.5$, etc. The SPA is described below:

Step 4.2.1 Estimate the first mutation position i_1 in T:

1. Initialize $i = j = 0$ and calculate $w(A, B; i, j, n)$. If
$$w(A, B; i, j, n) = w \geq \theta' ,$$
then let $\hat{i}_1 = 0$. This means the shifting mutation occurs at the beginning of $[1, n]$. Otherwise, go to step 4.2.1-(2).
2. In Step 4.2.1, procedure 1, if $w \leq \theta$, meaning no shifting mutation occurs in $[1, n]$, we put the starting point forward and consider $i = j = n - \tau$. Next, we calculate the corresponding $w(A, B; i, j, n)$. If
$$w(A, B; i, j, n) = w \leq \theta ,$$
then let $i = j = 2(n - \tau)$ and repeat Step 4.2.1, procedure 2 until $w(A, B; i, j, n) > \theta$. Let k_1 be the integer satisfying the following requirements:
$$w(A, B; i, j, n) = w \leq \theta$$
if $i = j = k_1(n - \tau)$, and $w(A, B; i, j, n) = w > \theta$ if $i = j = (k_1 + 1)(n - \tau)$. Proceed to Step 4.2.1, procedure 3 or procedure 4.
3. For $i = j = (k_1 + 1)(n - \tau)$, if $w(A, B; i, j, n) = w \geq \theta'$, then set $\hat{i}_1 = (k_1 + 1)(n - \tau)$. Otherwise, go to Step 4.2.1, procedure 4.
4. Following Step 4.2.1, procedures 1–3, we have $\theta < w < \theta'$ if $i = j = (k_1 + 1)(n - \tau)$. Therefore, for the same n, compute $w' = w(A, B; i + h, j + h, n)$. If $w' > w$, calculate $\hat{i}_1$ according to (4.17). Otherwise, repeat Step 4.2.1, procedures 1–4 for the larger h and n, until $w' > w$.

Therefore, through the use of Step 4.2.1, we may estimate $\hat{i}_1$ of i_1.

Step 4.2.2 Estimate ℓ_1 based on the estimation $\hat{i}_1$ of the first mutation position in T. Typically,
$$w\left(A, B; \hat{i}_1 + \ell, \hat{i}_1, n\right) , \quad w\left(A, B; \hat{i}_1, \hat{i}_1 + \ell, n\right) , \quad \ell = 1, 2, 3, \cdots .$$

If pair $(\hat{i}_1 + \ell, \hat{i}_1)$ or pair $(\hat{i}_1, \hat{i}_1 + \ell)$ satisfies $w \leq \theta = 0.3$ or 0.4, where w is its corresponding sliding window function, then this ℓ is the length of the shifting mutation. Specifically:

1. If $w(A, B; \hat{i}_1 + \ell, \hat{i}_1, n) \leq \theta$, we note that $\hat{\ell}_1 = -\ell$ and we insert ℓ virtual symbols into sequence B following the position $\hat{i}_1$, while keeping sequence A invariant.
2. If $w(A, B; \hat{i}_1, \hat{i}_1 + \ell, n) \leq \theta$, we note that $\hat{\ell}_1 = \ell$ and we insert ℓ virtual symbols into sequence A following the position $\hat{i}_1$, while keeping sequence B invariant.

Through the use of these two steps, we may estimate the local mutation mode $T_1 = \{(i_1, \ell_1)\}$, and its corresponding locally uniform alignment (C_1, D_1). It is decomposed as follows:
$$C_1 = (C_{1,1}, A_{2,1}) , \quad D_1 = (D_{1,1}, B_{2,1}) .$$

Denote the length of vector $C_{1,1}$ and $D_{1,1}$ by $\hat{i}_1 + |\hat{\ell}_1|$. Since there is no shifting mutation occurring in the first n positions of $A_{2,1}, B_{2,1}$, we let $L_1 = \hat{i}_1 + |\hat{\ell}_1| + n$ be the starting point for next alignment.

Step 4.2.3 After obtaining the estimation $(\hat{i}_1, \hat{\ell}_1)$, we continue to estimate i_2 based on (C_1, D_1). We initialize $i = j = L_1$ and calculate $w(A, B; i, j, n)$ by repeating Step 4.2.1, procedures 1–4 to obtain the estimation $\hat{i}_2$ for i_2.

Step 4.2.4 Estimate ℓ_2 based on the estimations $\hat{i}_1, \hat{\ell}_1, \hat{i}_2$. Here, we calculate

$$w\left(C_1, D_1; \hat{i}_2 + \ell, \hat{i}_2, n\right), \quad w\left(C_1, D_1; \hat{i}_2, \hat{i}_2 + \ell, n\right), \quad \ell = 1, 2, 3, \cdots .$$

We repeat Step 4.2.2 to get $\hat{\ell}_2$ and the local alignment (C_2, D_2).

Step 4.2.5 Continuing the above process, we find the sequence $(\hat{i}_k, \hat{\ell}_k)$ and the corresponding sequence (C_k, D_k) for all $k = 1, 2, 3, \cdots$. The process will terminate at some k_0 such that $C_{k_0} = (C_{1,k_0}, A_{2,k_0})$ and $D_{k_0} = (D_{1,k_0}, B_{2,k_0})$ have shifting mutations occurring in (A_{2,k_0}, B_{2,k_0}). Let L_{k_0} denote the length of sequence C_{1,k_0}, D_{1,k_0} and $i = j = L_{k_0}$. The corresponding ℓ is the length of the shifting mutation if pair $(\hat{i}_{k_0} + \ell, \hat{i}_{k_0})$ or pair $(\hat{i}_{k_0}, \hat{i}_{k_0} + \ell)$ satisfies $w \le \theta$, and then $w(C_{k_0}, D_{k_0}; i, j, n') \le \theta$ in which n' is the shorter of the lengths of A_{2,k_0} and B_{2,k_0}.

Finally, we equalize the lengths of A_{2,k_0} and B_{2,k_0}. In other words, if the length of A_{2,k_0} is shorter than that of B_{2,k_0}, we insert several virtual symbols at the end of A_{2,k_0} so that its length is same as that of B_{2,k_0}.

Example 17. The following two RNA sequences are drawn from 195 sRNAs in [77]. We show how to align them using SPA:

$$\begin{cases} E.co': & \text{ugccuggcgg ccguagcgcg guggucccac cugaccccau gccgaacuca gaagugaaa} \\ B.st': & \text{ccuagugaca auagcggaga ggaaacaccc gucccauccc gaacacggaa guuaag} \end{cases}$$

Here, $n_a = 59, n_b = 56$. These two sequences seem disorderly and unsystematic, but we can adapt them by inserting the virtual symbol "–" several times, which is restricted so that the penalty functions of the two sequences are at a minimum. By performing Steps 4.2.1–4.2.4, we obtain:

$$\begin{cases} E.co': & \text{ugccuggcgg ccguagcgcg guggucccac cugaccccau gccgaacuca gaagugaaa} \\ B.st': & \text{—ccuaguga caauagcgga gaggaaacac ccgucc-cau cccgaacacg gaaguuaag} \end{cases}$$

The specific computational procedure is detailed as follows:

1. Let $i = j = 0$, $n = 15$ and calculate the sliding window function of B.st and E.co: $w(A, B; i, j, n) = \frac{14}{15} = 0.933 > \theta' = 0.6$. Thus, $\hat{i}_1 = 0$.
2. For fixed $\hat{i}_1 = 0$, let $n = 20$ and calculate the sliding window functions

$$w(A, B; 0, 1, n) = \frac{13}{20} = 0.65\,, \quad w(A, B; 0, 2, n) = \frac{18}{20} = 0.9\;,$$
$$w(A, B; 0, 3, n) = \frac{15}{20} = 0.75\,, \quad w(A, B; 0, 4, n) = \frac{16}{20} = 0.8\;,$$
$$w(A, B; 1, 0, n) = \frac{16}{20} = 0.8\,, \quad w(A, B; 2, 0, n) = \frac{7}{20} = 0.35\;,$$

in which $w(A,B;2,0,n) = \frac{7}{20} = 0.35 < 0.4$ and the values of the other functions are all greater than 0.6. Therefore, following from Step 4.2.2, we have $\hat{\ell}_1 = -2$. We then insert two virtual symbols at the beginning of sequence B, and we find the local aligned sequences (C_1, D_1) as follows:

$$\begin{cases} C_1: & \text{ugccuggcgg ccguagcgcg gugguucccac cugaccccau gccgaacuca gaagugaaa} \\ D_1: & \text{—ccuaguga caauagcgga gaggaaacac ccgucccauc ccgaacacgg aaguuaag} \end{cases}$$

3. Aligning (C_1, D_1). If we put $i = j = 22$, $n = 25$ and calculate the sliding window function, we get $w = w(A,B;i,j,n) = \frac{12}{25}$. Putting $h = 10$, we get $w' = w(A,B;i+h,j+h,n) = \frac{16}{25}$. We input $w = \frac{16}{25}$, $w' = \frac{12}{25}$, $n = 25$, $h = 10$ into (4.17), giving

$$\hat{i}_2 = L_1 + \frac{h}{w' - w}\left(\frac{3}{4} - w\right) = 20 + \frac{250}{4}\left(\frac{3}{4} - \frac{12}{25}\right) \sim 20 + 17 = 37 .$$

4. Letting $i = j = \hat{i}_2$, $n = 15$, we calculate $w(A,B;i+\ell,j,n)$ and $w(A,B;i,j+\ell,n)$. The results are

$$w(A,B;i,j+1,n) = \frac{12}{15} = 0.8 > 0.6 ,$$
$$w(A,B;i+1,j,n) = \frac{3}{15} = 0.2 < 0.3 .$$

Thus, $\hat{\ell}_2 = -1$, and the local aligned sequences (C_2, D_2) are given as:

$$\begin{cases} C_2: & \text{ugccuggcgg ccguagcgcg gugguucccac cugaccccau gccgaacuca gaagugaaa} \\ D_2: & \text{—ccuaguga caauagcgga gaggaaacac ccgucc-cau cccgaacacg gaaguuaag} \end{cases}$$

5. Let $i = j = \hat{i}_2$, $n = 21$, and calculate $w(A,B;i+1,j,n) = \frac{5}{21} = 0.24 < 0.3$. We have aligned the entire sequences (C_2, D_2). Therefore, (C_2, D_2) is the uniform alignment of sequences (A, B).

The implications of Example 17: We have performed the alignment of the pair of E.co and B.st using the SPA, and obtained (C_2, D_2), in which both sequences are of the same length 59. The total penalty is $w(C_2, D_2) = \frac{21}{59} \sim 0.356$. This value is much less than 0.75, so we can declare that sequences E.co and B.st are homologous.

4.2.2 Some Unsolved Problems and Discussions of SPA

Some Unsolved Problems

In the last subsection, we introduced the operation steps of the SPA. Because of the complexity of biological data, some problems arise while running the SPA. Therefore, the SPA does not represent the final word, as there are still unsolved problems in the sense of both theoretical analysis and the design of the program. For example:

1. Selection of parameters. The parameters involved in the SPA are chosen as: $n, h, \tau, \theta, \theta'$, etc. This is a specific case. However, how are the parameters selected, in general, from the parameter set? How may we adjust the selected parameters automatically in the alignment? Which principle and algorithm should be used to adjust these parameters?
2. What can be said about the validity of the statistic $\hat{T} = \{(\hat{i}_k, \hat{\ell}_k), k = 1, 2, \cdots, \hat{k}_a\}$? What can be concluded about the stability of the algorithm, and how could the algorithm be adjusted or modified if errors occur?
3. How might one estimate the computational complexity of the SPA? How might one design a simpler, comprehensive algorithm?
4. How might one implement the minimum penalty alignment based on the uniform alignment of sequences?

All the abovementioned problems (and many more) are important for the improvement of the SPA. Many of these issues involve probability and statistics. In this section, we focus on these problems, which are essential for this kind of algorithm.

Discussion of the Selection of Parameters

The principle of the selection of parameters n, θ, θ' is the law of large numbers or the central limit theorem. Through statistical analysis based on the benchmark dataset, the GenBank Database, we see the randomness of DNA (or RNA) sequences. Therefore, for the cases in which those segments are mutated by shifting mutations, we note that a^*_{i+t} and b^*_{j+t} are independent, obey uniform distribution on V_4, and have changes which are easily controlled. These special properties play an important role in the following formula:

$$w\left(a^*_{[i+1,i+n]}, b^*_{[j+1,j+n]}\right) = \frac{1}{n}\sum_{t=1}^{n} w\left(a^*_{i+t}, b^*_{j+t}\right) .$$

Then

$$P_r\left\{w\left(a^*_{[i+1,i+n]}, b^*_{[j+1,j+n]}\right) < \frac{3}{4} - \kappa\sqrt{\frac{3}{16n}}\right\} \sim \phi(-\kappa) ,$$

in which $\phi(-\kappa)$ is the density function of the standard normal distribution. Thus, following from the normal distribution table, we have

$$\phi(-1.5) = 0.0668 , \quad \phi(-2) = 0.0228 , \quad \phi(-2.5) = 0.0062 ,$$
$$\phi(-3.0) = 0.0013 , \quad \phi(-3.5) = 0.0002 , \quad \phi(-4.0) < 0.0001 .$$

Hence, $1 - \phi(\kappa)$ is considered the degree of confidence. Therefore, we can use different κ's to control the parameter

$$\theta' = \frac{3}{4} - \kappa\sqrt{\frac{3}{16n}} .$$

Table 4.1. The relationship table of the selection of parameters n, κ, ϱ

κ	$\varrho \vert n$	15	20	25	30	40	50	75	100
1.0	0.8413	0.45505	0.43874	0.42761	0.41940	0.40787	0.40000	0.38777	0.38047
1.5	0.9332	0.58229	0.60476	0.62010	0.63141	0.64730	0.65814	0.67500	0.68505
2.0	0.9772	0.52639	0.55635	0.57679	0.59189	0.61307	0.62753	0.65000	0.66340
2.5	0.9938	0.47049	0.50794	0.53349	0.55236	0.57884	0.59691	0.62500	0.64175
3.0	0.9987	0.41459	0.45953	0.49019	0.51283	0.54460	0.56629	0.60000	0.62010
3.5	0.9998	0.35869	0.41111	0.44689	0.47330	0.51037	0.53567	0.57500	0.59845
4.0	>0.9999	0.30279	0.36270	0.40359	0.43377	0.47614	0.50505	0.55000	0.57679

The values of κ, ϱ, and n are listed in Table 4.1, where $\varrho = 1 - \phi(\kappa)$ is the confidence level.
Based on this table we can draw the following conclusions:

1. If $w(a^*_{[i+1,i+n]}, b^*_{[j+1,j+n]}) < \theta'(\kappa, n)$, then a^*_{i+k} and b^*_{j+k} are dependent and $b^*_{[j+1,j+n]}$ is mutated from $a^*_{[i+1,i+n]}$ through type-I mutation with a confidence level ϱ, where pairs (a^*_{i+k}, b^*_{j+k}) correspond to the vectors $a^*_{[i+1,i+n]}$ and $b^*_{[j+1,j+n]}$ for all $k = 1, 2, \cdots, n$.
2. For the sequences without shifting mutations, we assume that

$$P_r\left\{a^*_{i+t} \neq b^*_{j+t}\right\} < \frac{1}{3} .$$

We then have

$$P_r\left\{w\left(a^*_{[i+1,i+n]}, b^*_{[j+1,j+n]}\right) > \frac{1}{3} + \kappa\sqrt{\frac{2}{9n}}\right\} \sim \phi(-\kappa) .$$

Therefore, we can use different κ to control the parameter

$$\theta = \frac{1}{3} + \kappa\sqrt{\frac{2}{9n}} .$$

The values κ, ϱ, and n are listed in Table 4.2.

Here, if $w(a^*_{[i+1,i+n]}, b^*_{[j+1,j+n]}) > \theta(\kappa, n)$, we may conclude that type-III or type-IV mutations occur in vector $a^*_{[i+1,i+n]}$ and $b^*_{[j+1,j+n]}$ with a confidence level ϱ.

Estimation of Parameter T

In the data of DNA sequences, the occurrence of type-I and type-II mutations actually fluctuates, affecting the validity of the estimation of $\hat{i}_k, \hat{\ell}_k$, especially the computation of (4.17). Consequently, there are still many problems to be discussed when estimating $\hat{i}_k, \hat{\ell}_k$.

If the fluctuation of the error ϵ resulting from type-I or type-II mutations is large, we may improve the val idity of the estimation of $\hat{i}_k, \hat{\ell}_k$ in several ways:

Table 4.2. Relationship table of the selection of parameters n, κ, ϱ

κ	$\varrho\|n$	15	20	25	30	40	50	75	100
1.0	0.8413	0.62090	0.63820	0.65000	0.65871	0.67094	0.67929	0.69226	0.70000
1.5	0.9332	0.51591	0.49145	0.47475	0.46243	0.44514	0.43333	0.41498	0.40404
2.0	0.9772	0.57677	0.54415	0.52190	0.50547	0.48240	0.46667	0.44220	0.42761
2.5	0.9938	0.63762	0.59686	0.56904	0.54850	0.51967	0.50000	0.46942	0.45118
3.0	0.9987	0.69848	0.64956	0.61618	0.59153	0.55694	0.53333	0.49663	0.47475
3.5	0.9998	0.75934	0.70227	0.66332	0.63457	0.59421	0.56667	0.52385	0.49832
4.0	>0.9999	0.82020	0.75497	0.71046	0.67760	0.63148	0.60000	0.55107	0.52190

1. Enlarge n. As a result of our actual computational experience, we may choose n: $n = 20$ if the fluctuation of the mutation error ϵ is not larger than that seen in Example 17. We choose $n = 40$ if the fluctuation of the mutation error ϵ is larger. If the fluctuation of the mutation error ϵ is much larger, n may be chosen much greater than 40.
2. Improve the decision-making process in (4.17). Enlarging the parameter n as mentioned in step 1 does improve the validity of the estimation, but the effect is limited since the fluctuation of the statistic $\hat{i}_k$ will increase as n increases. Therefore, a more important improvement would be to refine the discrimination in equations (4.17). In the next subsection, we introduce the fluctuation range of the statistic of $\hat{i}_k$, and use other mathematical methods, such as regression analysis, wavelet analysis, etc.

4.2.3 Local Modifications for Sequence Alignment

We have mentioned the procedure for the local modification for sequence alignment in Sect. 3.5.4. Its purpose is to modify the output sequences A' and B' by locally permutating the inserted virtual symbols so that the penalty score of the alignment becomes smaller. The local modifications can be classified as follows:

1. Permutation occurring in the neighborhood of the inserted region. For example, if we obtain the locally aligned sequences:

$$\begin{pmatrix} \text{ac———cg} \\ \text{guuuccaau} \end{pmatrix},$$

the local modification becomes

$$\begin{pmatrix} \text{a—cc—g} \\ \text{guuuccaau} \end{pmatrix},$$

while keeping the other segments unchanged. The penalty score of the new alignment is then reduced.

2. If there are almost the same number of inserted symbols in both sequences, we may reduce the penalty by deleting the same number of virtual symbols from both sequences simultaneously. For example, if the local alignment sequence is:
$$\begin{pmatrix} \text{ac—cguua} \\ \text{aguuuc—a} \end{pmatrix},$$
we delete two "–" from both sequences to find
$$\begin{pmatrix} \text{accguua} \\ \text{aguuuca} \end{pmatrix},$$
while keeping the other segments unchanged. The penalty score then falls from 7 to 4, and the rate of the penalty score is reduced.
3. It is possible to reduce the penalty score by adding or deleting several "–" symbols at the tail ends of the sequences. For example, if the initial alignment is:
$$\begin{pmatrix} \text{aaccacg} \\ \text{acacgga} \end{pmatrix},$$
which is the tail of some pair, we insert the "–" symbols as follows:
$$\begin{pmatrix} \text{aaccacg—} \\ \text{a–c–acgga} \end{pmatrix},$$
while keeping the other segments unchanged. The penalty score of this local segment falls from 5 to 4. The insertions occur at the end of the sequence, so they do not affect the alignment, while reducing the penalty score.

The problem of how to standardize the local modifications is still unsolved. We wish to use computational complexity as the measure of whether this goal has been achieved, but it is also a valuable problem in its own right to be studied further.

4.3 Index Analysis of SPA

For the most functional indices involved in the SPA, we analyze them theoretically or simulate them computationally. We detail this in the discussion below.

4.3.1 The Statistical Features of Estimations

In (4.14) and (4.17), we presented the formulas for estimating the mutation position s and mutation probability ϵ. Here, we discuss the statistical features of these estimations.

Statistical Model for the Estimation of Mutation Position s

In Chap. 2, we presented the stochastic and semistochastic models of the random mutation induced by stochastic sequences. To estimate the mutation position s is to make a statistical estimation based on its local sequences. Therefore, this problem can be solved by the following model. Let

$$\mathcal{E}^*(s) = \{\bar{a}^*, \bar{b}^*, s\} , \tag{4.18}$$

where

$$\bar{a}^* = \left(a_1^*, a_2^*, \cdots, a_{n+h}^*\right) , \quad \bar{b}^* = \left(b_1^*, b_2^*, \cdots, b_{n+h}^*\right)$$

is a stochastic model with one shifting mutation. The mutation position is s. We call this the single mutation stochastic model. Based on the discussion in Chap. 2, we translate the stochastic models with shifting mutation into stochastic sequences satisfying the following requirements:

1. If $w(a, b)$ is the Hamming matrix, then the stochastic sequence
$$w\left(a_j^*, b_j^*\right) , \quad j = 1, 2, \cdots, n+h$$
is an independent random vector sequence.
2. Each $w(a_j^*, b_j^*)$ is a Bernoulli test, and there exists a positive integer $s < n+h$ such that:
$$w\left(a_j^*, b_j^*\right) \sim \begin{pmatrix} 0 & 1 \\ 1-\epsilon & \epsilon \end{pmatrix} \quad \text{if } j \leq s ,$$
and
$$w(a_j^*, b_j^*) \sim \begin{pmatrix} 0 & 1 \\ 1/4 & 3/4 \end{pmatrix} \quad \text{if } j > s .$$

This model is in fact a hybrid of the four types of mutations. The background information and relevant definitions are discussed in Chap. 2. Here, the vectors $\bar{a}^*$, $\bar{b}^*$ are deemed to be the subvectors of the mutated sequences (A^*, B^*) respectively, and the starting positions have no shifting mutations. Conditions 1 and 2 are therefore satisfied.

The Characteristic Property of the Estimations $\hat{s}$, $\hat{\epsilon}$

Let $\bar{a}^*$, $\bar{b}^*$ be two random vectors defined by (4.18) and satisfying conditions 1 and 2 above; then the values of the parameter s are called mutation positions and their penalty function is defined as follows:

$$\begin{cases} w^* = w\left[(a_1^*, a_2^*, \cdots, a_n^*), (b_1^*, b_2^*, \cdots, b_n^*)\right] = \sum_{j=1}^{n} w\left(a_j^*, b_j^*\right) , \\ (w^*)' = w\left[\left(a_{h+1}^*, a_{h+2}^*, \cdots, a_{h+n}^*\right), \left(b_{h+1}^*, b_{h+2}^*, \cdots, b_{h+n}^*\right)\right] \\ \quad = \sum_{j=1}^{n} w\left(a_{j+h}^*, b_{j+h}^*\right) . \end{cases} \tag{4.19}$$

Using the above notations, we estimate the mutation position s as follows:

1. If the parameter ϵ is known, then

$$\hat{s}^* = n\frac{3-4w^*}{3-4\epsilon} \ . \tag{4.20}$$

2. If the parameter ϵ is unknown, then

$$\begin{cases} \hat{\epsilon}^* = \dfrac{3}{4} + \dfrac{n}{h}\left(w^* - (w^*)'\right) \ , \\ \hat{s}^* = \dfrac{h}{(w^*)' - w^*}\left(\dfrac{3}{4} - w^*\right) \ , \end{cases} \tag{4.21}$$

where h is given in Sect. 4.1.3.

Theorem 20. *If the random vectors $\bar{a}^*$, $\bar{b}^*$ satisfy conditions 1 and 2 above, then the following properties are satisfied:*

1. *If the parameter ϵ is known, then the estimation $\hat{s}^*$ in expression (4.20) is a uniformly unbiased estimation. Its lack of bias means that $E\{\hat{s}^*\} = s$ and its uniformity means that*

$$\lim_{n\to\infty} \frac{\hat{s}^*}{n} = \lambda \quad a.e. \tag{4.22}$$

holds for any fixed $\lambda = \frac{s}{n}$.

2. *If the parameter ϵ is unknown, then $\hat{\epsilon}^*$ in expression (4.21) is a uniformly unbiased estimation, and $\hat{s}^*$ is a uniform estimation.*

Proof.

1. We attempt to prove conclusion 1. If ϵ is a known fixed constant, then

$$\begin{aligned} E\{\hat{s}^*\} &= E\left\{\frac{n}{3-4\epsilon}(3-4w^*)\right\} = \frac{n}{3-4\epsilon}E\{3-4w^*\} \\ &= \frac{n}{3-4\epsilon}(3-4E\{w^*\}) = \frac{n}{3-4\epsilon}\left\{3 - \frac{4}{n}\left[s\epsilon + \frac{3}{4}(n-s)\right]\right\} \\ &= \frac{1}{3-4\epsilon}[3n - 4s\epsilon - 3(n-s)] = \frac{1}{3-4\epsilon}(3s - 4s\epsilon) = s \ . \end{aligned}$$

It means that $\hat{s}^*$ is an unbiased estimation. Next, we prove it is also a uniform estimation. Since

$$\lim_{n\to\infty}\frac{\hat{s}^*}{n} = \lim_{n\to\infty}\frac{\frac{n}{3-4\epsilon}(3-4w^*)}{n} = \lim_{n\to\infty}\frac{1}{3-4\epsilon}(3-4w^*) \ ,$$

then using the law of large numbers, we have

$$\begin{aligned} \lim_{n\to\infty} w^* &= \lim_{n\to\infty}\frac{1}{n}\left[\sum_{j=1}^{s} w\left(a_j^*, b_j^*\right) + \sum_{j=1}^{n-s} w\left(a_{s+j}^*, b_{s+j}^*\right)\right] \\ &= \lim_{n\to\infty}\left[\frac{\lambda}{s}\sum_{j=1}^{s} w\left(a_j^*, b_j^*\right) + \frac{1-\lambda}{n-s}\sum_{j=1}^{n-s} w\left(a_{s+j}^*, b_{s+j}^*\right)\right] \\ &= \lambda\epsilon + \frac{3}{4}(1-\lambda) \quad \text{a.e.} \end{aligned}$$

Therefore,

$$\lim_{n\to\infty}\frac{\hat{s}^*}{n} = \frac{1}{3-4\epsilon}\left\{3-4\left[\lambda\epsilon+\frac{3}{4}(1-\lambda)\right]\right\}$$
$$= \frac{1}{3-4\epsilon}\left[3-4\lambda\epsilon-3(1-\lambda)\right] = \frac{1}{3-4\epsilon}(3\lambda-4\lambda\epsilon) = \lambda\,. \quad \text{a.e.}$$

This concludes the proof of conclusion 1.

2. Secondly, we prove that the estimation $\hat{\epsilon}^*$ in expression (4.21) is uniformly unbiased. Since

$$E\{\hat{\epsilon}^*\} = \frac{3}{4}+\frac{n}{h}E\{w^*-(w^*)'\}$$
$$= \frac{3}{4}+\frac{1}{h}\left\{\left[s\epsilon+\frac{3}{4}(n-s)\right]-\left[(s-h)\epsilon+\frac{3}{4}(n-s+h)\right]\right\}$$
$$= \frac{3}{4}+\frac{1}{h}\left(h\epsilon-\frac{3h}{4}\right) = \frac{3}{4}+\epsilon-\frac{3}{4} = \epsilon\,,$$

it implies that $\hat{\epsilon}^*$ is an unbiased estimation of ϵ. To prove that $\hat{\epsilon}^*$ is the uniform estimation of ϵ, we still must assume that $\lambda = \frac{s}{n}$, $\delta = \frac{h}{n}$ are fixed as constants, and $0<\delta<\lambda<1$. Then

$$\lim_{n\to\infty}\hat{\epsilon}^* = \frac{3}{4}+\lim_{n\to\infty}\frac{n}{h}[w^*-(w^*)']\,.$$

Drawing agian on the law of large numbers, we have

$$\begin{cases}\lim\limits_{n\to\infty} w^* = \lambda\epsilon+\dfrac{3}{4}(1-\lambda)\,,\\ \lim\limits_{n\to\infty}(w^*)' = (\lambda-\delta)\epsilon+\dfrac{3}{4}(1-\lambda+\delta)\,.\end{cases} \tag{4.23}$$

Therefore,

$$\lim_{n\to\infty}[w^*-(w^*)'] = \lambda\epsilon+\frac{3}{4}(1-\lambda)-\left[(\lambda-\delta)\epsilon+\frac{3}{4}(1-\lambda+\delta)\right]$$
$$= \delta\epsilon-\frac{3}{4}\delta\,. \tag{4.24}$$

Since $\frac{n}{h}=1/\delta$, we find

$$\lim_{n\to\infty}\hat{\epsilon}^* = \frac{3}{4}+\frac{1}{\delta}\left(\delta\epsilon-\frac{3}{4}\delta\right) = \epsilon\,;$$

hence, the estimation $\hat{\epsilon}^*$ is uniform.

3. Finally, we prove that $\hat{s}^*$ in (4.21) is uniform. We let $\lambda=\frac{s}{n}$, $\delta=\frac{h}{n}$, and $0<\delta<\lambda<1$, then

$$\lim_{n\to\infty}\frac{\hat{s}^*}{n} = \lim_{n\to\infty}\frac{h}{n[(w^*)'-w^*]}\left(\frac{3}{4}-w^*\right)\,.$$

We substitute the result of (4.23) and (4.24) for the above limit formula, to arrive at

$$\lim_{n\to\infty} \frac{\hat{s}^*}{n} = \frac{1}{\frac{3}{4}-\epsilon}\left[\frac{3}{4} - \lambda\epsilon - \frac{3}{4}(1-\lambda)\right]$$
$$= \frac{1}{\frac{3}{4}-\epsilon}\left(\frac{3\lambda}{4} - \lambda\epsilon\right) = \lambda\ .$$

Hence, $\hat{s}^*$ is the uniform estimation for s, and the theorem is proved.

Distributed Property of the Estimation $\hat{s}^*$

We discuss the case where ϵ is known. Using expression (4.20) and the central limit theorem, we know that $\frac{\hat{s}^*-s}{\sqrt{n}} \sim N(0, \sigma_{n,s}^2)$ is a normal distribution. Thus, we need only estimate $\sigma_{n,s}^2$, which is evaluated below.

$$\begin{aligned}
\sigma_{n,s}^2 &= D\{\hat{s}^*\} = E\{(\hat{s}^*-s)^2\} = E\{(\hat{s}^*)^2\} - s^2 \\
&= E\left\{\left[\frac{n}{\frac{3}{4}-\epsilon}\left(\frac{3}{4}-w^*\right)\right]^2\right\} - s^2 = \frac{n^2}{(\frac{3}{4}-\epsilon)^2}E\left\{\left(\frac{3}{4}-w^*\right)^2\right\} - s^2 \\
&= \frac{n^2}{\left(\frac{3}{4}-\epsilon\right)^2}\left[\frac{9}{16} - \frac{3}{2}E\{w^*\} + E\{(w^*)^2\}\right] - s^2
\end{aligned} \tag{4.25}$$

with the help of

$$\begin{cases}
E\{w^*\} = \frac{1}{n}\left[s\epsilon + \frac{3}{4}(n-s)\right]\ , \\
E\{(w^*)^2\} = (E\{w^*\})^2 + \frac{1}{n^2}\left[s\epsilon(1-\epsilon) + \frac{3}{16}(n-s)\right]\ , \\
\qquad = \frac{1}{n^2}\left\{s^2\epsilon^2 + \frac{3}{2}s\epsilon(n-s) + \frac{9}{16}(n-s)^2 \right. \\
\qquad\qquad \left. + \left[s\epsilon(1-\epsilon) + \frac{3}{16}(n-s)\right]\right\}
\end{cases}$$

(4.25) leads to

$$\sigma_{n,s}^2 = \frac{1}{\left(\frac{3}{4}-\epsilon\right)^2}\left[s\epsilon(1-\epsilon) + \frac{3}{16}(n-s)\right]\ . \tag{4.26}$$

It is known that the range of the estimation $\hat{s}$ is controlled by the size of $\sigma_{n,s}^2$ and that $\sigma_{n,s}^2$ is a function of ϵ, n and s. For example, let $\epsilon = 1/4$, $n = 36$, $s = 18$, then

$$\sigma_{n,s} = \sqrt{\frac{1}{\left(\frac{3}{4}-\epsilon\right)^2}\left[s\epsilon(1-\epsilon) + \frac{3}{16}(n-s)\right]} = 3\sqrt{3} \sim 5.19615\ .$$

This means that the uncertainty in the estimation of $\hat{s}^*$ is comparatively large. Therefore, we must improve the local modification operations or the algorithm to reduce uncertainty in $\hat{s}^*$.

Discussion of the Models of Stochastic Mutation

In (4.18), we talked about the single mutation stochastic model $\mathcal{E}^*(s) = \{\bar{a}^*, \bar{b}^*, s\}$, in which the mutation position s is a fixed value. Therefore, $\mathcal{E}^*$ is a semistochastic mutation model. If the mutation position s^* is a random variable, then we let

$$\mathcal{E}^*(s^*) = \{\bar{a}^*, \bar{b}^*, s^*\} . \tag{4.27}$$

This $\mathcal{E}^*(s^*)$ satisfies the following additional conditions:

1. $\mathcal{E}^*(s)$ is a single mutation stochastic model if $s^* = s$ and the corresponding probability distribution satisfies conditions 1 and 2 as a single mutation statistic model.
2. s^* obeys the geometric distribution: $P_r\{s^* = s\} = \delta(1-\delta)^{s-1}$, $s = 1, 2, 3, \cdots$. Here, we also assume that the length of $(\bar{a}^*, \bar{b}^*)$ can be extended arbitrarily. We still discuss the properties of the estimation $\hat{s}^*$ of s^* as follows:
   (a) In (4.14) and (4.17), $\hat{s}^*$ and $\hat{\epsilon}$ do not depend on s^*. Therefore, (4.14) and (4.17) can be also used to estimate $\hat{s}^*$, $\hat{\epsilon}$.
   (b) If ϵ is known, the formula for the mean value of the estimation $\hat{s}^*$ of s^* is

$$E\{\hat{s}^*\} = E\{E[\hat{s}^*|s^*]\} = E\{s^*\} = \sum_{s=1}^{\infty} s\delta(1-\delta)^{s-1} = \frac{1}{\delta} , \tag{4.28}$$

   where $E[\hat{s}^*|s^*]$ is the conditional expectation, which is a function of s^*. The first equation in (4.28) can be acquired from the property of conditional expectation.
   We can also calculate the variance as follows:

$$\begin{aligned} D\{\hat{s}^*\} &= E\{[\hat{s}^* - E(\hat{s}^*)]^2\} = E\{E[(\hat{s}^* - E\hat{s}^*)^2|s^*]\} \\ &= E\left\{\frac{s^*\epsilon(1-\epsilon) + \frac{3}{16}(n - s^*)}{\left(\frac{3}{4} - \epsilon\right)^2}\right\} \\ &\sim \frac{\epsilon(1-\epsilon) + \frac{3}{16}(n\delta - 1)}{\delta\left(\frac{3}{4} - \epsilon\right)^2} , \end{aligned} \tag{4.29}$$

   in which n is large enough so that

$$\sum_{s=n+1}^{\infty} s^2\delta(1-\delta)^s = \left(n + \frac{1}{\delta}\right)^2 (1-\delta)^{n+1} \sim 0$$

   holds.

4.3.2 Improvement of the Algorithm to Estimate $\hat{s}^*$

In the SPA, we demonstrated that $\hat{s}^*$ is estimated based on the solution of (4.14). This algorithm has two problems, namely:

Problem 1 The uncertainty in the estimation $\hat{s}^*$ based on the set of (4.14) is too large as the parameter n increases.

Problem 2 The mutation position s^* is a random variable, and the distance between two adjacent mutation positions i_k and i_{k+1} is also a random variable. For example, in the Example 18, the distances between three adjacent pairs are shown as follows:

$$i_5 - i_4 = 68 , \quad i_{10} - i_9 = 20 , \quad i_{26} - i_{25} = 402 .$$

The operations on (4.17) have no self-adaptive property. In other words, it cannot automatically search for the mutation positions with different separations.

To solve these two problems, we propose the regression analysis discriminative algorithm as follows.

The Outline of Self-Adaptive Regression Analysis Discriminative Algorithm

The statistical decision model we will discuss is still in the form $\mathcal{E}(s^*) = \{\bar{a}^*, \bar{b}^*, s^*\}$, which satisfies the conditions of a single mutation stochastic model. Here, the length n of vector $\bar{a}^*, \bar{b}^*$ may be quite large, and the distribution of s^* may be widely separated. The outline of the self-adaptive regression analysis discriminative algorithm is presented next:

1. For a fixed sequence $(\bar{a}, \bar{b})$, construct the self-adaptive window function

$$w(k, n_0) = w\left(a_{[k+1,k+n_0]}, b_{[k+1,k+n_0]}\right) = \sum_{j=1}^{n_0} w(a_{k+j}, b_{k+j}) , \qquad (4.30)$$

   where n_0 is a fixed positive integer, for example, $n_0 = 15, 20$, etc.
2. For the self-adaptive window function $w(k, n_0)$, $k = 1, 2, 3, \cdots$, we analyze the tendency. In other words, we perform trend analysis based on one part of the data

$$w_{[k_0+1,k_1]} = \{w(k, n_0),\ k_0 < k \le k_1\} , \qquad (4.31)$$

   and determine linear regression $\Gamma_{[k_0+1,k_1]}$. Here we have several types of linear regression $\Gamma_{[k_0+1,k_1]}$ such as:
   (a) Type 1, where $\Gamma_{[k_0+1,k_1]}$ is a horizontal line of height $y = \rho < 3/4$. We denote it by Γ_1.
   (b) Type 2, where $\Gamma_{[k_0+1,k_1]}$ is a horizontal line of height $y = \rho' \ge 3/4$. We denote it by Γ_3.
   (c) Type 3, where $\Gamma_{[k_0+1,k_1]}$ is a monotonically increasing straight line. We denote this straight line by Γ_2.

3. We choose the intersection point of the straight lines Γ_2 and Γ_3 as the estimation of the mutation position s.
4. The value of (k_0, k_1) may be different. Therefore, straight-line regression may also be different. For the selection of the straight line Γ_2, we should select the straight line with the largest slope as the solution in order to minimize the penalty score.

The Outline of Self-Adaptive Regression Analysis Discriminative Algorithm

We present the steps of the algorithm according to the outline of the self-adaptive regression analysis discriminative algorithm. Steps 4.2.1–4.2.5 of the SPA can continue to be used and we need only modify the corresponding operations in Steps 4.2.1 and 4.2.3. We then have:

Step 4.3.1 We use the function $w_k = \frac{1}{n_0} w(k, n_0)$ to estimate the first mutation position i_1 in T:

1. Initialize $k = 0$ and calculate $w(k, n_0)$. If $w_k \geq \theta'(\theta' \in (0.6, 0.8))$, set $\hat{i}_1 = 0$. Otherwise, go to the next step.
2. If $w_k \leq \theta(\theta \in (0.3, 0.5))$, continue to calculate w_{k+1} for all $k = 0, 1, 2, \cdots$. If there are several connected k such that

$$w_k \leq \theta , \quad w_{k+1} < \theta ,$$

   say, for $k = 0, 1, \cdots, k_1$, then perform regression analysis on these points. The corresponding straight line of the regression analysis is a horizontal line and $\Gamma_1 : y = \rho_1$, where the value of ρ_1 is the solution of equation

$$\sum_{k=0}^{k_1} (w_k - \rho_1)^2 = \min \left\{ \sum_{k=0}^{k_1} (w_k - \rho)^2 , \ \rho > 0 \right\} . \tag{4.32}$$

   Then

$$\sigma_1^2 = \frac{1}{k_1 + 1} \sum_{k=0}^{k_1} (w_k - \rho_1)^2 \tag{4.33}$$

   is the regression error.
3. After the straight line Γ_1 is determined, we continue to calculate w_k, $k = k_1 + 1, k_1 + 2, k_1 + 3, \cdots$, if there exist points k_2, k_3 such that

$$\begin{cases} \theta < w_k < \theta' & \text{for any } k_2 < k \leq k_3 , \\ \theta' < w_k & \text{is true for any } k_3 < k . \end{cases}$$

   We then perform regression analysis based on the data:

$$w_k , \quad k = k_2 + 1, k_2 + 2, \cdots, k_3 , \quad k = k_3 + 1, k_3 + 2, k_3 + 3 \cdots .$$

The corresponding straight lines in the regression analysis are

$$\begin{cases} \Gamma_2: \quad y = \rho_2 x + \rho_2' , \\ \Gamma_3: \quad y = \rho_3 , \end{cases}$$

respectively, which satisfy the following conditions:

$$\begin{cases} \sum_{k=k_2}^{k_3} (w_k - \rho_2 k - \rho_2')^2 = \min \left\{ \sum_{k=k_2}^{k_3} (w_k - \rho k - \rho')^2 ,\ \rho, \rho' > 0 \right\} , \\ \sum_{k=1}^{n'} (w_{k_3+k} - \rho_3)^2 = \min \left\{ \sum_{k=1}^{n'} (w_{k_3+k} - \rho)^2 ,\ \rho > 0 \right\} , \end{cases} \tag{4.34}$$

where $n_0 \leq n' < n_a - k_3$. Both equations in (4.34) can be solved using the least-squares method.

4. The intersection point of the straight lines Γ_2 and Γ_3 is the $\hat{s}$ we need. Replace Step 4.2.1 of the SPA by the new Step 4.3.1 we obtain for the improved SPA, which is a self-adaptive regression analysis discriminative algorithm.

Example 18. For the two RNA sequences E.co and B.st given in Example 17, we use the improved SPA to recalculate the corresponding results:

1. If we let $n_0 = 15$, we have $w_0 = \frac{14}{15} = 0.933 > \theta' = 0.6$ given in Example 18. Therefore, $\hat{i}_1 = 0$.
2. As mentioned in Example 17, $\hat{\ell} = -2$ and (C_1, D_1) is

$$\begin{cases} C_1: & \text{ugccuggcgg ccguagcgcg guggucccac cugaccccau gccgaacuca gaagugaaa} \\ D_1: & \text{—ccuaguga caauagcgga gaggaaacac ccgucccauc ccgaacacgg aaguuaag} \end{cases}$$

3. Based on (C_1, D_1), we calculate the function w_k, $k = 2, 3, 4, \cdots$ as follows:

k	2	3	4	5	6	7	8	9	10	11	12
w_k	0.333	0.333	0.400	0.467	0.400	0.467	0.400	0.400	0.400	0.467	0.467
k	13	14	15	16	17	18	19	20	21	22	23
w_k	0.400	0.400	0.400	0.400	0.467	0.467	0.467	0.400	0.400	0.333	0.400
k	24	25	26	27	28	29	30	31	32	33	34
w_k	0.467	0.467	0.467	0.400	0.467	0.533	0.533	0.600	0.677	0.677	0.667
k	35	36	37	38	39	40	41	42	43		
w_k	0.733	0.800	0.800	0.800	0.800	0.733	0.733	0.733	0.733		

4. Performing regression analysis based on these data, we get three straight lines of regression as follows:

$$\begin{cases} \Gamma_1: & y = \rho_1 = 0.433\,, \quad k = 2, 3, \cdots, 28\,, \\ \Gamma_2: & y = 0.019 \times k + 0.087\,, \quad (\rho_2 = 0.019, \rho_2' = 0.087)\,, \\ & \qquad\qquad k = 27, 28, \cdots, 35\,, \\ \Gamma_3: & y = \rho_3 = 0.764\,, \quad k = 35, 36, \cdots, 43\,. \end{cases}$$

Then we solve equation $0.764 = 0.019 \times k + 0.087$, obtaining

$$\hat{i}_2 = \frac{0.764 - 0.087}{0.019} = 35.63 \sim 36\,.$$

This is the estimation of the second mutation position.

We compare it with Example 17, in which the second mutation position may be chosen as $\hat{i}_2 = 35, 36$, or 37, while the estimation in Example 18 is $35.63 \sim 36$. Thus, their alignments are in fact the same, as both are the alignments of sequences (A, B) with a minimum penalty score.

4.3.3 The Computational Complexity of the SPA

Let $\mathcal{E}^*(T^*) = \{A^*, B^*, T^*\}$ be the stochastic mutation model defined in Sect. 3.3.3, where T^* is a composite renewal process with intensity $\{(\epsilon_\tau, p_\tau),\ \tau = 1, 2, 3, 4\}$. Now we discuss the computational complexity involved to align $\mathcal{E}^*(T^*)$ using the SPA.

Theorem 21. *If the stochastic mutation model $\mathcal{E}^*(T^*)$ satisfies similarity conditions (see Definition 19), then the computational complexity of the SPA is proportional to n_a.*

Proof. Following from the definition of the SPA, the computational complexity of the decision for each (i_k, ℓ_k) is

$$O\left(i_{k+1} - i_k\right) + O\left(n_0^2 \ell_k\right)\,,$$

where n_0, ℓ_k are fixed constants which are not too large. The second term of the complexity does not depend on the length of sequence, and we may assume that it is constant. Therefore, the total computational cost of the SPA can be controlled by

$$\sum_{k=1}^{k_a} \left[O\left(i_{k+1} - i_k\right) + O\left(n_0^2 \ell_k\right)\right] = O(n_a)\,.$$

Thus ends the proof.

Besides theoretical analysis, simulation analysis and comparison of algorithms are also significant analysis methods to study the computational complexity of the SPA. The relationship between computation time and sequence length is shown in Fig. 4.1. We compare the result of the SPA with several popular algorithms in Tables 4.3 and 4.4 and Fig. 4.1, which were obtained in 2002.

Table 4.3. Comparison of alignment speed (CPU time, seconds) and similarity

Initial sequences	Alignment speed 1,2	3,4	5,6	7,8	9,10	Similarity 1,2	3,4	5,6	7,8	9,10
SPA	0.07	0.03	0.07	0.01	0.05	0.8060	0.7624	0.7647	0.9945	0.7014
Band	0.96	0.35	1.63	0.25	0.18	0.8352	0.7868	0.8423	0.9945	0.7223
GlobalS	1.40	1.77	3.21	NA	1.49	0.8251	0.8078	0.8411	NR	0.7379
CDA	6.13	4.75	5.67	NA	5.05	0.8300	0.7600	0.7900	NR	0.7300

1 Toxoplasma gondii strain RH heat shock protein 70 (HSP70) gene, complete cds (ID U85648.1);
2 Toxoplasma gondii heat shock protein 70 mRNA, complete cds (ID U82281.1);
3 Human CCAAT-box-binding factor (CBF) mRNA, complete cds (ID M37197.1);
4 Mus musculus putative CCAAT binding factor 1 (mCBF) mRNA, alternatively spliced transcript mCBF1, complete cds (ID U19891.1);
5 Rattus norvegicus peripheral plasma membrane protein CASK mRNA, complete cds (ID U47110.1);
6 Mus musculus mRNA for mCASK-B (ID Y17138.1);
7 Ca^{2+}/calmodulin-dependent protein kinase IV kinase isoform (rats, brain, mRNA, 3429 NT) (ID S83194.1);
8 Mus musculus putative CCAAT binding factor 1 (mCBF) mRNA, alternatively spliced transcript mCBF1, complete cds (ID U19891.1);
9 Mc.vanniel .757 1385 Methanococcus vannielii strain EY33. (ID M36507);
10 Mb.tautotr, RNA, Methanobacterium thermoautotrophicum strain Marburg. (ID RDP-II);
NR No result

Table 4.4. Comparison table of alignment speed and similarity

Initial sequences	Speed ratio 1,2	3,4	5,6	7,8	9,10	Average speed ratio
Band/SPA	13.7	11.7	23.3	25.0	3.6	15.5
GlobalS/SPA	20.0	59.0	33.0	NA	29.8	35.5
CDA/SPA	87.6	158.0	81.0	NA	101.0	106.9
Aligned sequences	**Similarity difference 1,2**	**3,4**	**5,6**	**7,8**	**9,10**	**Average similarity difference**
Band-SPA	0.0292	0.0776	0.00	0.0209	3.6	0.0304
GlobalS-SPA	0.0191	0.0454	33.0	NA	0.0385	0.0449
CDA-SPA	0.0200	0.0000	81.0	NA	0.0300	0.0200

See Table 4.3 for an explanation of numbers *1–10*.

4.3.4 Estimation for the Error of Uniform Alignment Induced by a Hybrid Stochastic Mutation Sequence

We aim to estimate the error of the uniform alignment, but we begin by outlining the simplest case of a hybrid stochastic mutation.

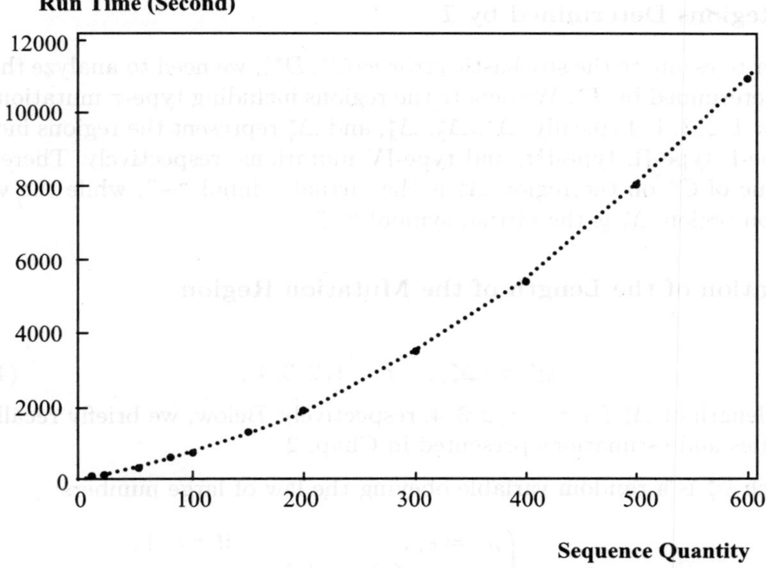

Fig. 4.1. Relationship table of computation time of alignment of sequences with different length

Estimate the Errors Resulting from a Uniform Alignment or Hybrid Stochastic Mutation

The stochastic model of sequence mutation was presented in (2.99). We may decompose $\mathcal{E}^*$ into $\mathcal{E}^* = \{A^*, T^*\}$, where $A^* = (a_1^*, a_2^*, \cdots, a_{n_a}^*)$ is the initial sequence and $a_1^*, a_2^*, \cdots, a_{n_a}^*$ is independently and identically distributed. Moreover, the common distribution is a uniform distribution on V_4. The complete form of T^* is:

$$T^* = \left\{\tilde{\zeta}_\tau, \tilde{\ell}_\tau^*, \tilde{\varrho}_{\tau'}\,,\ \tau = 1,2,3,4\,,\ \tau' = 1,2\right\}\,, \tag{4.35}$$

in which $\tilde{\zeta}_\tau$ is the set of mutation positions of mutation type-τ, and is also a multiple Bernoulli test or Poisson flow for each fixed τ. $\tilde{\ell}_\tau^*$ is the interval length of type-II, type-III, and type-IV mutations, respectively. It obeys a geometric or an exponential distribution. $\tilde{\varrho}_1$, $\tilde{\varrho}_3$ are superpositions of type-I mutations and inserted stochastic sequences of type-III mutations, respectively. Here, the original sequence A^* can be considered as a fixed or stochastic sequence. If $\mathcal{E}^* = \{A^*, T^*\}$ is known, we find B^* mutated from A^* based on the mode T^*. We let (C^*, D^*) be the uniform alignment of (A^*, B^*), and estimating the errors resulting from the uniform alignment or hybrid mutation is the alternative to estimating $w(C^*, D^*)$ which is the error of (C^*, D^*) as A^*, T^* are known.

The Regions Determined by T^*

In order to estimate the stochastic error $w(C^*, D^*)$, we need to analyze the regions determined by T^*. We denote the regions including type-τ mutations by Δ_τ^*, $\tau = 1, 2, 3, 4$. Typically, Δ_1^*, Δ_2^*, Δ_3^*, and Δ_4^* represent the regions including type-I, type-II, type-III, and type-IV mutations, respectively. Therefore, the value of C^* on the region Δ_3^* is the virtual symbol "−", while the value of D^* on region Δ_4^* is the virtual symbol "−".

Estimation of the Length of the Mutation Region

Let

$$\psi_\tau^* = |\Delta_\tau^*|, \quad \tau = 1, 2, 3, 4 , \tag{4.36}$$

be the length of Δ_τ^* for $\tau = 1, 2, 3, 4$, respectively. Below, we briefly recall the properties and estimations presented in Chap. 2.

1. Each ψ_τ^* is a random variable obeying the law of large numbers:

$$\frac{1}{n_a}\psi_\tau^* \stackrel{\text{(a.e.)}}{\longrightarrow} \begin{cases} \mu_1 = \epsilon_1 , & \text{if } \tau = 1 , \\ \mu_2 = \epsilon_2 \left(\dfrac{1}{p_1} + \dfrac{1}{p_2}\right) , & \text{if } \tau = 2 , \\ \mu_\tau = \dfrac{\epsilon_\tau}{p_\tau} , & \text{if } \tau = 3, 4 , \end{cases} \tag{4.37}$$

   where $\stackrel{\text{(a.e.)}}{\longrightarrow}$ shows convergence almost everywhere, ϵ_τ is the intensity of the mutation flow $\tilde{\zeta}_\tau$, and p_τ is the intensity of the geometric distribution.
2. The random variable $\frac{1}{\sqrt{n_a}}\psi_\tau^* - n_a\mu_\tau$ obeys the normal distribution $N(0, \sigma_\tau^2)$, in which

$$\sigma_\tau^2 = \begin{cases} \epsilon_1(1 - ep_1) , & \text{if } \tau = 1 , \\ \epsilon_2 \left[2(1 - \epsilon_2)\left(\dfrac{1}{p_1^2} + \dfrac{1}{p_2^2}\right) - \left(\dfrac{1}{p_1} + \dfrac{1}{p_2} + \dfrac{2\epsilon_2}{p_1 p_2}\right)\right] , & \text{if } \tau = 2 , \\ \dfrac{\epsilon_\tau}{p_\tau^2}(2 - p_\tau - \epsilon_\tau) , & \text{if } \tau = 3, 4 . \end{cases} \tag{4.38}$$

Estimations of the Errors Resulting from Different Type of Mutations

If we take the Hamming matrix as the penalty function of the alignment, then different types of mutation will occur in different regions. Let w_τ^* for $\tau = 1, 2, 3, 4$ be the errors induced by type-τ mutation in region Δ_τ^*. The estimations of these errors are computed as:

$$\begin{cases} w_\tau^* = n_a\psi_\tau^* , & \tau = 1, 3, 4 , \\ w_2^* = \dfrac{3n_a}{4}\psi_2^* . \end{cases} \tag{4.39}$$

This is because these errors are proportional to the lengths of the mutation regions for type-I, type-III, and type-IV mutations occuring in uniform alignment, and the elements in the exchanged segments may coincide with probability 1/4 for the case where type-II mutation occurs.

4.4 Applications of Sequence Alignment and Examples

4.4.1 Several Applications of Sequence Alignment

The motivation for developing alignment is to understand the homology of the genome, and to discuss the evolutionary relationships among genes in different organisms. However, as the field of investigations progresses, researchers find that it is a very complicated problem involving many applied fields. Here, we continue to discuss sequence alignment in a more advanced way than that which we introduced in Chap. 1.

Gene Positioning

Gene positioning or gene localizing is what determines the position of a gene. It includes two aspects; one determines the position of a short sequence within a long sequence, and the other determines the mutual positions for two long sequences.

The position of a short sequence within a long sequence means the region in a long sequence B (i.e., a couple of million or tens or hundreds of million base pairs) where the shorter segment A (i.e., a couple of thousand base pairs) is the same as the short sequence. Its biological meaning is to search for the segment in other organisms such that it is most similar to a known gene in the specific organism. While deciding the mutual positions for a pair of long sequences A, B that may come from different organisms (or species, or organs, or from the same organism but different growth periods), we frequently compare the order and condition of mutation of these segments in sequences A, B because each sequence has many peculiar segments (such as gene, transcription factor, etc.).

GenBank Searching

GenBank searches are actually an extension of gene positioning. In fact, it extends the benchmark set as a database rather than a fixed long sequence. If A is a fixed sequence, then we want to understand whether or not it is in the GenBank. If A belongs to a specific GenBank, then we want to know the location, similarity and mutation mode, etc. The most popular databases are GenBank, cDNA, dbEST; specific biological databases, especially, the human genome, etc. The common functional indices for searching a gene bank are given below:

1. Similarity: This index generally involves a threshold in a specific case. For example, we may search for those sequences in a database Ω with a similarity based on a comparison with sequence A that is greater than 70%. The value of 70% is considered a threshold.
2. Computational speed: The time to complete searching the database.
3. Sensitivity: The rate $\frac{TP}{TP+NF}$, in which TP is the total number of these sequences that are very similar to a sequence A and the given algorithm also declares them similar to A under a given similarity; $TP + NF$ is the total number of these sequences that are similar to A. A is a sequence contained in the database Ω.

Computational speed and sensitivity are two critical indices for all alignment procedures. With the development of sequencing technology, huge amounts of data enter the databases all the time. For example, the size of GenBank is doubling every 2–3 years. At the same time, the requirements, such as the lengths of the query sequences, are also increasing. Therefore, the demand to develop new alignment or to improve existing alignment is unending.

Gene positioning and gene searching are the basis of many types of large-scale software packages. The quality of gene positioning and gene searching required determines the function of the software. For example, BLAST and FASTA software packages have both of these functions. SPA is an ultrafast algorithm, and therefore it is especially suited to gene positioning and gene searching. The corresponding software package based on the SPA is called FLAG. Several functional indices of FLAG exceed those of BLAST or FASTA.

Repetition Searches and Gene Assembly

Repetition searches and gene assembly were introduced in Chap. 1. They are essential techniques of sequencing technology. Improving the accuracy of repetition searches and gene assembly will directly improve the quality of sequencing technology. Of all the methods, sequence alignment is an important tool for solving these two problems simultaneously. For example, the human genome project is a good example showing that sequence alignment is an important tool for solving these problems.

The General Method for Applying Sequence Alignment

In the overview of gene positioning, gene searching, gene assembly and repetition searches mentioned above, the common procedure can be divided into four steps to be processed as outlined below:

1. Choose a "seed" vector to search in the target sequence. That is, if we know that sequence B is the target of A but do not know the position of A, we frequently select one (or several) short vector(s) $\bar{a} = (a_1, a_2, \cdots, a_n)$ from A, called the "seed" and whose length is often assigned as 8–13.

2. Search for the same segments as in the "seed" $\bar{a}$ in the target sequence B thoroughly. In other words, we record all subvectors of B which are the same as $\bar{a}$. Let

$$\bar{b}_k = (b_{j_k+1}, b_{j_k+2}, \cdots, b_{j_k+n}) , \quad k = 1, 2, \cdots, k_0 ,$$

   in which $\bar{b}_k = \bar{a}$ for all $k = 1, 2, \cdots, k_0$. These are the labels of all candidate positions of A in B.
3. For each fixed k, start from the position of $\bar{b}_k$ in B to extend $\bar{b}_k$ to a new sequence B_k such that its length is the same as or close to that of A.
4. Aligning (A, B_k). If B_k has the required similarity compared with A, then that region of B_k is the position of A in B, otherwise we ignore it.

This is the basic procedure of gene positioning. The same procedure is also suited to gene searching, gene assembly, and repetition searching with some changes in skill and method for different software packages. Here, we omit these minor differences.

4.4.2 Examples of Pairwise Alignment

In order to understand pairwise alignment better, we perform pairwise alignment based on two RNA sequences: Mc.vanniel and Mb.tautotr.

RNA Sequences: C.vanniel and Mb.tautotr

The data for RNA sequences Mc.vanniel and Mb.tautotr are downloaded from [77]. Their lengths are 2977 bp and 3018 bp, respectively. For simplicity, we denote the two sequences by A and B, respectively. For convenience, we list these data in Table 4.5.

The output obtained by the SPA is shown in Table 4.6.

Remark 2. In the alignment computation, we choose the parameters $h = 5$, $n = 50$, $\theta = 0.45$. For the output in Table 4.6, the upper row is A', the lower row is B'. A' and B' are the extensions of A and B, respectively, in which "–" represent virtual symbols.

The computation of this algorithm is summarized as follows:

1. Computation takes 26 steps, the length of the aligned sequences is 3061 bp.
2. The penalty function of the original sequences used here is

$$w(A, B; i, j, n) \sim 0.75 , \quad \text{for each } i, j, \tau = |i - j| \geq 50 ,$$

   where $n = n_a - \max\{i, j\}$ if $w(A, B; 1, 1, n_a) = 0.739$, etc.
3. Penalty of the aligned sequences is: $w(A', B'; 1, 1, n_{a'}) = 0.331$.

Table 4.5. Data for two RNA sequences: Mc.vanniel and Mb.tautotr

```
   1 uaucuauuac ccuacccugg ggaauggcuu ggcuugaaac gccgaugaag gacgugguaa gcugcgauaa gccuaggcga
     ggcgcaacag ccuuugaacc uaggauuucc gaaugggacu uccuacuuuu guaauccgua aggauuggua acgcggggga
 161 uugaagcauc uuaguacccg caggaaaaga aaucaacuga gauuccguua guagaggcga uugaacacgg aucagggcaa
     acugaauccc uucggggaga ugugguguua uagggccuuc uuuucgccug uugagaaaag cugaaguuga cuggaacguc
 321 acacuauaga gggugaaagu cccguaagcg caaucgauuc agguuugaag ugucccugag uaccgugcgu uggauaucgc
     gcgggaauuu gggaggcauc aacuuccaac ucuaaauacg uuucaagacc gauagcguac uaguaccgcg agggaaagcu
 481 gaaaagcacc cuuaacaggg uggugaaaag agccugaaac ccagguaggu auggaauggc guggccccaa aggcaacugu
     ucugaaggaa accgucgcaa ggcggcugua cgaagaacag agccaggguu gcguccuccg uuucgaaaaa cgggccgggg
 641 aguguauugu uguggcgagc uuaagaucuu cacgaucgaa ggcguaggga aaccaacaag uccgcagaau cuuuagggac
     ggggucuuaa gggcccggag ucacagcaau acgacccgaa accgggcgau cuaggccggg gcaaggugaa gucccucaau
 801 ugagggaugg aggccugcag aguuguugcc guucgaagca cucuucugac cucggucuag gggugaaagg ccaaucgagc
     ccggagauag cugguucccc ucgaagugac ucucagguca gccagaguuc agguagucgg cagguagag cacugauaag
 961 augguuaggg gaagaaauuc cucgcuguuu ugucaaacuc cgaaccuguc gucgccguag gcucugagug agggcauacg
     ggguaagcug uauguccgag acgggaauag ccgagacuug gguuaaggcc ccuaaaugcc gauuaagugu gaacacgaag
1121 ggcguccuug gucuaagaca gcagggaggu uggcuuagaa gcagccaccc uuuaaagagu gcguaacagc ucaccugucg
     agaucaaggg ccccgaaaau ggacggggcu aaaucggcug ccgagaccca aagggcaccg caaggugauc cccguagggg
1281 ggcguucugc gagggcagaa guucggcugu gaagucgagu ggaccucgua gaaaugaaga ucccgguagu aguaacagca
     uaagugggu gagaaucccc accgccgaag gggcaagggu uccacagcaa uguuugucag cugugguaa gccgguccua
1441 acucucgagg uaacuccuuu gagaggaaag ggaaacaggu uaauauuccu gugccaucua gauacgcgug gcaacacaag
     guuaguuucc aacgcuucug gguaggcuga guguucuugu cuggacauuc aagcuuauaa guccggggag aguuguaaua
1601 acgagaaccg gaugaaagag ugaugagcuc uccguuagga gaguucggcc gaucucugga gcccgugaaa agggaacuag
     caaggauucu agauguccgu acccagaacc gacacuggug ccccuaggug aguauccuaa ggcguagcgg gaugaaucua
1761 gucgagggaa gucggcaaau ugguuccgua acuucgggag aaggagugcc agugaucuug uuuaaauaug ggaucgcugg
     ucgcagugac cagggagguc cgacuguuua auacaaacau aggucuuagc gagccugaaa aggugguac uaaggccgac
1921 gccugcccag ugcugguacg ugaaccccgg uuccaaccgg gcgaagcgcc aguaaacggc gggguaacu auaacccucu
     uaagguagcg aaauuccuug ucgggcaagu uccgaccugc augaauggcg uaacgagacc uccacugucc ccgacuagaa
2081 uccggugaac cuaccauucc ggcgcaaagg ccggagacuu ccagugggaa gcgaagaccc cguggagcuu uacugcagcc
     ugucguuggg gcaugguugu gaguguacag uguaggguggg agccaucgaa accuuuucgc caggaaaggu ggaggcgauc
2241 cugggacacc acccucucau gaccauguuc cucacccuuu uaggggacac cgguaggugg gcaguuuggc uggggcggua
     cccuccuaaa aaugcaucag gagggcccca aagguuggcu caagcggguc aggacuccgc uguugagugu aagggcaaaa
2401 gccagccuga cuuuguugcc aacaaaacgc aacgaagagg cgaaagccgg gccuaacgaa ccccugugcc ucacugaugg
     gggccaggga ugacaaaaa gcuaccccgg ggauaacaga guugucgcgg gcaagagccc auaucgaccc cgcggcuugc
2561 uaccucgaug ucgguuuuuc ccauccuggg ucugcagcag gacccaaggg uggggcuguu cgcccauuaa aggggaucau
     gagcuggguu uagaccgucg ugagacaggu ugguugcuau cugcuggaug uguuaggcug ucugagggaa agguggcucu
2721 aguacgagag gaacgggccg ucggcgccuc uagucgaucg guugucugac aaggcacugc cgagcagcca cgcgccaaga
     gccguuuccu ucgggaacga gaacucccgu agaagacggg uuugauaggc uaggggugua cgcaucaagg uucuuccgag
2881 auguucagcc cgcuaguacu aacaguucga gagauaauuu aggcauc
```

4. Based on the above output, we determine that the sequences Mc.vanniel and Mb.tautotr are homologous. Following from the data of (A', B'), we know that this alignment is not the minimum penalty alignment, and we use local modification operations to reduce the penalty.
5. In 1999, we performed this alignment using a Pentium 586 computer, which took 0.679 s.

The Modulus Structure of this Sequence Alignment

Based on the initial data (A, B) in Table 4.5 and the aligned sequence (A', B') in Table 4.6 using the SPA, we obtain the modulus structure from (A, B) to (A', B') as follows:

1. The numbers of shifting mutations are $k_{a'} = 19, k_{b'} = 9$, and the length of the aligned sequence is $n_{a'} = n_{b'} = 3061$. The total error of the aligned sequences is $w(A', B') = 1013$. The similarity of the aligned sequences is

$$r(A', B') = 1 - \frac{1}{n_{a'}} w(A', B') = 1 - \frac{1013}{3061} = 0.6690 \ .$$

Table 4.5. (continued)

```
   1 cuuuuuuaug ccgucugggg gauggcuugg cuugagucgc ugaugaaggc cguggcaagc ugcgauaagc ccaggggagg
     agcagcaucc uuggauccug ggauugccga augggacuuc ccagccaacc cuucgggguu gugcuacucc cuguuauggg
 161 gagggggaac ccgccgaacu gaaacaucuu aguaggcgga ggaagagaaa gcaaauugcg acugccguga guaauggcga
     augaaagcgg ugcaggacaa acugaacccc uucgcaguga uguguugggg gauguggugu ugucgaucgg ugcguauggg
 321 ggugccgggu guguggugu gaacuugggc uggaaugccc gggccguaga ggguuaaagc cccguagaug cccaugcuug
     gcucccugca ccuuuccuga guagcgucca uuggauauug ggcgugaagc ugggaggcau cgacuccuaa uccuaaacac
 481 gucucaaguc cgauagcgaa cuaguaccgu gagggaaagc ugaaaaguac cccugauagg ggugugaaaa gugccugaaa
     ccaggcggug acagcccggc acggcaugga aggaaugugg cugccccugu aagaaaccau gguaacaugg gaguaugugu
 641 gggugguuga acagugucgu gucguccguc uugaaacacg ggccagggag uuuagugguu guggcgaggc uaagaagugu
     gucgcuuugu agucguaggg aaaccgacag guccgcagca gccuuugugc ugugagggac ggggucuuaa uaggccugg
 801 agucacagcu cuaaaacccg aagccggucg aucuagcccu ggguagggug aagucgcucu uacgagugau ggaggcccgc
     aggggguguug ucgugcgaaa cauuccucua accuggguu aguggugaaa ggccaaucaa ggccggugac agcugguucc
 961 acccgaaaug gcucguaggc cagccugacu ggagauaggu ggcggguag agcacuuauu ggguguuuag ggggagagau
     cccucggcau ccuguaaaac uccgaacucg ucaccgucgu ugaagguugg agucaggggc gcggguaag ccugugucc
1121 gagagaggaa caacucagac uggguuaag gucccuaaau gccggcuaag ucuaaggggg ucuuuggccc uagacaaugg
     gaaggugggc uuagaagcag ccauccuuua aagaguucgu aacagaucac ccaucgaggu caaaggcacc gaaaauggag
1281 gggaauuaag ccggcuaccg auaccucaga gcaccacugg uguggugguc uuguagggug gcguccgguu gggguugaag
     ugggggcgug agcuccugug gacccggcug gaaugaggau ccugguagua guagcagcga agugaggugu gaauccuuac
1441 cgccggaggg gcuagguuc cuuggcaaug uucgucagcc aaggguuagu cgguccuaag gccgugggua auguccauuu
     uggucgaaag gguaacgggu uaauauuccu guacggucca gguacuugcg gugacgcugg guugggcuuc ugacgcuuug
1601 ggguaggcug agcgggauuu ucguccuguu uaaggguuga agccuggga gagccguaau ggcgagaacc auggugaagg
     ccugaauagc caucccuugu gggugguuug gcugugcccu ggaguccuug aaaagggagu ccuucuuggg auccuggauc
1761 gccguaccga gauccgacac uggugcccu agcugaguag gcuaaggugu guuggguaa ccuggcuaag ggaaaucggc
     aaauuggccc cguaacuuug ggagaagggg ugccagccau gcggauggcu ggucgcagug acaggggggg cccgacuguu
1921 uaauaaaaac auagcuccua gcuagcccgu gagggugugu acuggggcg acaccugccc agugccggca cgugaagccc
     ugguucaacg gggugaagcg ccgguaaacg gcggggguaa cuauaacccu cuuaagguag cgaaaugccu ugccggauaa
2081 guaccggccu gcaugaaugg uugaacgagg ucccuacugu cccuagccag gaccuaguga agcugcuguu cuggugcaca
     agccagagac ucccaguggg aagcgaagac cccguagagc uuuacugcag ucugcuguug gggcuugguc augggguaugc
2241 aguguaggug ggaggcgucg augccauggu cgccaggcug ugguggaguc ggucaugaga caccaccuuc cugugacugu
     gucucuaacc ccauguuugu gggggacauc gguagauggg caguuuggcu ggggcggcac gcgcuugaaa ugguaucaag
2401 cgcgcccuaa ggucggcuca ggcgggacag agauccgcug uagaguguaa gggcauaagc cggcuugacu gugcuccuac
     uaguaggggg ugcaggugcg agagcagggc cuagcgaacc ccagaguccu cgucggugg ggccuggau gacagaaaag
2561 cuaccucggg gauaacuggg uggucgcagg caagagccca uaucgacccu gcggcuugcu acuucgaugu cgguucuuuc
     cauccugggu gugcagcagc acccaagggu gggguuguuc gcccauuaaa ggggaacgug agcugguuu agaccgucgu
2721 gagacagguu gguugcuauc uacugggagu gugugguugc cugaggggaa ggugguucca guacgagagg aacggaccgu
     cggcgccucu gguuuaccgg uuauccgagu ggguauugcc gggcggcuac gcgcuaugau uauaaaggcu gaaggcaucu
2881 aagccugagg uuuucccuga aaauaggugg cuuguggacu gcggguagaa gaccuguuug uugggcggg ggugugagcu
     ucgaggccug uuuugggccg aguuguuuag ccugccguuu ccaagguuuu uuguccccu
```

2. The expanded modes of the aligned sequences are

$$\begin{cases} H_{a'} = \{(121,20),(210,2),(318,20),(386,1),(567,4),(632,3),(696,3),\\ \qquad (766,9),(781,2),(1102,1),(1331,3),(1508,2),(1591,1),\\ \qquad (1676,1),(1733,3),(2288,2),(2337,5),(2541,1),(2878,1)\}\,,\\ H_{b'} = \{(0,2),(1192,7),(1421,1),(1817,1),(1887,12),\\ \qquad (2418,1),(2987,8),(3040,1),(3051,10)\}\,, \end{cases} \tag{4.40}$$

where (j_k, ℓ_k) are the position and length of insertion in sequence A' (or B'), respectively.

3. We find the shifting function of sequence A' (or B') as

$$L_{a'}(k)\,, \quad k = 1,2,\cdots,k_{a'}\,, \quad L_{b'}(k)\,, \quad k = 1,2,\cdots,k_{b'}$$

Table 4.6. Alignment output for Mc.vanniel and Mb.tautotr

```
   1 uaucuauuac ccuacccugg ggaauggcuu ggcuugaaac gccgaugaag gacgugguaa gcugcgauaa gccuaggcga
     --cuuuuuua ugccgucugg gggauggcuu ggcuugaguc gcugaugaag gccguggcaa gcugcgauaa gcccagggga
     ggcgcaacag ccuuugaacc uaggauuucc gaaugggacu u--------- ---------- -ccuacuuuu guaauccgua
     ggagcagcau ccuuggaucc ugggauugcc gaaugggacu ucccagccaa cccuucgggg uugugcuacu cccuguuaug
 161 aggauuggua acgcggggga uugaagcauc uuaguacccg caggaaaaga --aaucaacu gagauuccgu uaguagaggc
     gggaggggga acccgccgaa cugaaacauc uuaguaggcg gaggaagaga aagcaaauug cgacugccgu gaguaauggc
     gauugaacac ggaucagggc aaacugaauc ccuucgggga gauguggugu uauagggccu ucuuuucgcc uguugaga--
     gaaugaaagc ggugcaggac aaacugaacc ccuucgcagu gauguguugg gggauguggu guugucgauc ggugcguaug
 321 ---------- --------aa agcugaaguu gacuggaacg ucacacuaua gagggugaaa gucccg-uaa gcgcaaucga
     ggggugccgg gugugguggug uugaacuugg gcuggaaugc ccgggccgua gaggguuaaa gccccguaga ugcccaugcu
     uucagguuug aagugucccu gaguaccgug cguuggauau cgcgcgggaa uuugggaggc aucaacuucc aacucuaaau
     uggcucccug caccuuuccu gaguagcguc cauuggauau ugggcgugaa gcugggaggc aucgacuccu aauccuaaac
 481 acguuucaag accgauagcg uacuaguacc gcgagggaaa gcugaaaagc acccuuaaca gggugugaa aagagccuga
     acgucucaag uccgauagcg aacuaguacc gugagggaaa gcugaaaagu accccugaua gggguguga aagugccuga
     aacccag--- -guagguaug gaauggcgug gccccaaagg caacuguucu gaaggaaacc gucgcaaggc gg---cugua
     aaccaggcgg ugacagcccg gcacggcaug gaaggaaugu ggcugccccu guaagaaacc augguaacau gggaguaugu
 641 cgaagaacag agccaggguu gcguccuccg uuucgaaaaa cgggccgggg agugua---u uguuguggcg agcuuaagau
     gugggugguu gaacaguguc gugucguccg ucuugaaaca cgggccaggg aguuuagugg uuguggcgag gcuaagaagu
     cuucacgauc gaaggcguag ggaaaccaac aaguccgcag aaucuu---- -----uaggg a--cggggguc uuaagggccc
     gugucgcuuu guagucguag ggaaaccgac agguccgcag cagccuuugu gcugugaggg acgggggucuu aauagggccu
 801 ggagucacag caauacgacc cgaaaccggg cgaucuaggc cggggcaagg ugaagucccu caauugaggg auggaggccu
     ggagucacag cucuaaaacc cgaagccggu cgaucuagcc cuggguaggg ugaagucgcu cuuacgagug auggaggccc
     gcagaguugu ugccguucga agcacucuuc ugaccucggu cuagggguga aaggccaauc gagcccggag auagcugguu
     gcaggggugu ugucgugcga aacauuccuc uaaccugggg uuagugguga aaggccaauc aaggccggug acagcugguu
 961 ccccucgaag ugacucucag gucagccaga guucagguag ucggcagggu agagcacuga uaagaugguu aggggaagaa
     ccacccgaaa uggcucguag gccagccuga cuggagauag guggcggggu agagcacuua uuggguguuu agggggagag
     auuccucgcuguuuugucaa  acuccgaacc ugucgucgcc guaggcucug agugagggca ua-cggggua agcuguaugu
     aucccucggcauccuguaaa  acuccgaacu cgucaccguc guugaagguu ggagucaggg gcgcggggua agccuguguc
1121 ccgagacgggaauagccgag  acuuggguua aggccccuaa augccgauua agugugaaca cgaagggcgu ccuuggucua
     ccgagagaggaacaacucag  acuggggguua agguccuaa augccggcua agucuaaggg ggucuuuggc cc-------u
1281 agacagcagg gagguuggcu uagaagcagc cacccuuuaa agagugcgua acagcucacc ugucgagauc aagggccccg
     agacaauggg aaggugggcu uagaagcagc cauccuuuaa agaguucgua acagaucacc caucgagguc aaaggcaccg
     aaaauggacg gggcuaaauc ggcugccgag acccaaaggg caccgcaagg u---gauccc cguagggggg cguucugcga
     aaaauggagg ggaauuaagc cggcuaccga uaccucagag caccacuggu guggugggucu uguagggugg cguccgguug
1441 gggcagaagu ucggcuguga agucgagugg accucguaga aaugaagauc ccgguaguag uaacagcaua agugggguga
     ggguugaagu gggggcguga gcuccugugg acccggcugg aaugaggauc cugguaguag u-agcagcga agugaggugu
     gaaucccccac cgccgaaggg gcaaggguuc cacagcaaug uuugucagcu gugggguaagc cgguccua-- acucucgagg
     gaauccuuac cgccggaggg gcuaggguuc cuuggcaaug uucgucagcc aaggguuagu cgguccuaag gccgugggua
1601 uaacuccuuu gagaggaaag ggaaacaggu uaauauuccu gugccaucua gauacgcgug gcaacacaag g-uuaguuuc
     auguccauuu uggucgaaag gguaacgggu uaauauuccu guacggucca gguacuugcg gugacgcugg guugggcuuc
     caacgcuucu ggguaggcug aguguucuug ucuggacauu caagcuuaua
     ugacgcuuug ggguaggcug agcgggauuu ucguccuguu uaaggguuga
1761 aguccgggga gaguuguaau aacgag-aac cggaugaaag agugaugagc ucuccguuag gagaguucgg ccgaucucug
     agccugggga gagccguaau ggcgagaacc auggugaagg ccugaauagc caucccuugu ggguggguuug gcugugcccu
     gag---cccg ugaaaaggga acuagcaagg auucuagaug uccguaccca gaaccgacac uggugccccu aggugaguau
     ggaguccuug aaaagggagu ccuucuuggg auccuggauc gccguaccga gauccgacac uggugccccu agcugaguag
1921 ccuaaggcgu agcgggauga aucuagucga gggaagucgg caaauugguu ccguaacuuc gggagaagga gugccaguga
     gcuaagg-ug uguuggggua accuggcuaa gggaaaucgg caaauuggcc ccguaacuuu gggagaaggg gugccag---
     ucuuguuuaa auaugggauc gcuggucgca gugaccaggg agguccgacu guuuaauaca aacauagguc uuagcgagcc
     ---------c caugcggaug gcuggucgca gugacagggg gggcccgacu guuuaauaaa aacauagcuc cuagcuagcc
2081 ugaaaaggug uguacuaagg ccgacgccug cccagugcug guacgugaac cccgguucca accgggcgaa gcgccaguaa
     cgugagggug uguacugggg gcgacaccug cccagugccg gcacgugaag cccugguuca acggggugaa gcgccgguaa
     acggcggggg uaacuauaac ccucuuaagg uagcgaaauu ccuugucggg caaguuccga ccugcaugaa uggcguaacg
     acggcggggg uaacuauaac ccucuuaagg uagcgaaaug ccuugccgga uaaguaccgg ccugcaugaa ugguugaacg
2241 agaccuccac uguccccgac uagaauccgg ugaaccuacc auuccggcgc aaaggccgga gacuuccagu gggaagcgaa
     aggucccuac ugucccuagc caggaccuag ugaagcugcu guucuggugc acaagccaga gacucccagu gggaagcgaa
     gaccccgugg agcuuuacug cagccugucg uuggggcaug guugugagug uacaguguag gugggagcca ucgaaacc--
     gaccccguag agcuuuacug cagucugcug uuggggcuug gucauggua ugcaguguag gugggaggcg ucgaugccau
```

Table 4.6. (continued)

```
2401 uuuucgccag gaaaggugga ggcgauccug ggacaccacc cucucau--- --gaccaugu uccucacccu uuuaggggac
     ggucgccagg cuguggugga gucggucaug agacaccacc uuccugugac ugugucucua accccauguu uguggggac
     accgguaggu gggcaguuug gcuggggcgg uacccuccua aaaaugcauc aggagggccc caaagguugg cucaagcggg
     aucgguagau gggcaguuug gcuggggcgg cacgcgcuug aaauggua-u caagcgcgcc cuaaggucgg cucaggcggg
2561 ucaggacucc gcuguugagu guaagggcaa aagccagccu gacuuuguug ccaacaaaac gcaacgaaga ggcgaaagcc
     acagagaucc gcuguagagu guaagggcau aagccggcuu gacugugcuc cuacuaguag ggggugcagg ugcgagagca
     gggccuaacg a-accccugu gccucacuga ugggggccag ggaugacaaa aaagcuaccc cggggauaac agaguugucg
     gggccuagcg aaccccagag uccucgucgg ugggggccug ggaugacaga aaagcuaccu cggggauaac uggguggucg
2721 cgggcaagag cccauaucga ccccgcggcu ugcuaccucg augucgguuu uucccauccu gggucugcag caggacccaa
     caggcaagag cccauaucga cccugcggcu ugcuacuucg augucgguuc uuuccauccu gggugugcag cagcacccaa
     gggugggggcu guucgcccau uaaaggggau caugagcugg guuuagaccg ucgugagaca gguugguugc uaucugcugg
     gggugggguu guucgcccau uaaaggggaa cgugagcugg guuuagaccg ucgugagaca gguugguugc uaucuacugg
2881 auguguuagg cugucugagg gaaagguggc ucuaguacga gaggaacggg ccgucggcgc cucuagucga ucgguugucu
     gagugugugg uugccugagg ggaagguggu uccaguacga gaggaacgga ccgucggcgc cucugguuua ccgguuaucc
     gacaaggcac ugccgagcag ccacgcgc-c aagagauaag agcugaaagc aucuaagcuc gaaauucauc cugaaaauaa
     gagugggguau ugccgggcgg cuacgcgcua ugauuauaaa ggcugaaggc aucuaagccu gagguuuucc cugaaaauag
3041 gacagccguu uccuucggga acgagaacuc ccguagaaga cggguuugau aggcuagggg uguacgcauc aagguucuuc
     guggcuugug gacugcgggu agaagaccug uuuguugggg cgggggugug agcuucg--- -----aggcc uguuuugggc
     cgagauguuc agcccgcuag uacuaacagu cucgagagau aauuuaggca u
     cgaguuguuu agccugccgu uuccaagguu -uuuuguccc u--------- -
```

as follows:

k	1	2	3	4	5	6	7	8	9	10	11	12	13	14	15
ℓ	20	2	20	1	4	3	3	9	2	1	3	2	1	1	3
$L_{a'}(k)$	0	20	22	42	43	47	50	53	62	64	65	68	70	71	72
k	16	17	18	19	20	21	22	23	24	25	26	27	28	29	30
ℓ	2	5	1	1		2	7	1	12	1	8	1	1	10	
$L_{a'}(k)$	75	77	82	83	84	0	3	10	11	23	24	32	33	43	53

in which $L_{a'}(k_{a'}+1) = L_{a'}$ is the sum of the virtual symbols in sequence A', and $L_{b'}(k_{b'}+1) = L_{b'}$ is the sum of the virtual symbols in sequence B'. Then

$$n_{a'} = n_a + L_{a'} = n_{b'} = n_b + L_{b'} \ .$$

It follows that the expanded mode of the sequences (A, B) has the following form:

$$\begin{cases} K_a = \{(121,20),(190,2),(296,20),(344,1),(524,4),(585,3),(646,3), \\ \qquad (713,9),(719,2),(1038,1),(1266,3),(1440,2),(1521,1), \\ \qquad (1605,1),(1661,3),(2216,2),(2263,5),(2465,1),(2797,1)\}\,, \\ K_b = \{(0,2),(1189,7),(1411,1),(1806,1),(1864,12), \\ \qquad (2395,1),(2963,8),(3008,1),(3018,10)\}\,. \end{cases} \tag{4.41}$$

4. With the help of (4.41), we find that the mutation mode of sequences (A, B) is of the form

$$H_{(a,b)} = (H_a, H_b) =$$
$$\begin{cases} (0,-2),(121,20),(190,2),(296,20),(344,1),(524,4),(585,3), \\ (646,3),(713,9),(719,2),(1038,1),(1189,-7),(1266,3),(1411,-1), \\ (1440,2),(1521,1),(1605,1),(1661,3),(1806,-1),(1864,-12), \\ (2216,2),(2263,5),(2395,-1),(2465,1),(2797,1),(2963,-8), \\ (3008,-1),(3018,-10)\ , \end{cases} \tag{4.42}$$

in which type-III mutation occurs at position i_k in sequence A if ℓ_k is positive, and type-IV mutation occurs at position i_k in sequence A if ℓ_k is negative. Therefore, we obtain the mutation shifting function as in Table 4.7.

Here the mutation mode of A from B is

$$K_{(b,a)} =$$
$$\begin{cases} (0,2),(119,-20),(208,-2),(316,-20),(384,-1),(565,-4), \\ (630,-3),(694,-3),(764,-9),(779,-2),(1100,-1),(1252,7), \\ (1322,3),(1470,1),(1498,-2),(1581,-1),(1666,-1),(1723,-3), \\ (1871,1),(1928,12),(2268,-2),(2317,-5),(2454,1),(2523,-1), \\ (2856,-1),(3023,8),(3060,1),(3069,10)\ , \end{cases} \tag{4.43}$$

where type-III mutation occurs at position i_k in sequence B if ℓ_k is positive, and type-IV mutation occurs at position i_k in sequence B if ℓ_k is negative.

Table 4.7. The mutation shifting function

k	1	2	3	4	5	6	7	8	9	10
i_k	0	121	190	296	344	524	585	646	713	719
ℓ_k	−2	20	2	20	1	4	3	3	9	2
$L_{a,b}$	0	−2	18	20	40	41	45	48	51	60

k	11	12	13	14	15	16	17	18	19	20
i_k	1038	1189	1266	1411	1440	1521	1605	1661	1806	1864
ℓ_k	1	−7	3	−1	2	1	1	3	−1	−12
$L_{a,b}$	62	63	56	59	58	60	61	62	65	64

k	21	22	23	24	25	26	27	28	29	30
i_k	2216	2263	2395	2465	2797	2963	3008	3018		
ℓ_k	2	5	−1	1	1	−8	−1	−10		
$L_{a,b}$	52	54	59	58	59	60	52	51	41	

Local Modification Operations of Sequence Alignment

Sequence (A', B') in Table 4.6 is the output of the SPA. This result is not the minimum penalty alignment, but we can use local modification operations to reduce its penalty. The result after local modification operations is listed in Table 4.8.

Using Table 4.8, we find that the overall penalty is reduced by 48. If we denote the sequences after the local modification operation from (A', B') as (C', D'), then their similarity is

$$r(C', D') = 0.6690 + \frac{48}{3061} = 0.6690 + 0.0157 = 0.6847.$$

Several calculations indicate that the similarity can be increased by 1–3% after the local modification operation for the SPA. Therefore, we can implement optimal alignment. The virtual symbols become dispersed after the local modification operation, so here, $k_{a'}, k_{b'}$ are increased, while the absolute value of ℓ_k is decreased.

Estimation of the Parameters of Sequence Mutation

We make the following estimation of the parameters of the mutation structure of (A, B):

1. Based on the result (A', B'), the probability of type-III or type-IV mutations is:

$$\begin{cases} \epsilon_3 = \dfrac{k_{a'}}{n_a} = \dfrac{19}{2977} = 0.0064 \,, \\ \epsilon_4 = \dfrac{k_{b'}}{n_a} = \dfrac{9}{2977} = 0.0030 \,. \end{cases}$$

   Then the average length of type-III or type-IV mutation and its parameter of geometric distribution are:

$$\mu_3 = \frac{84}{19} = 5.421 \,, \quad p_3 = \frac{1}{\mu_3} = 0.2260 \,,$$
$$\mu_4 = \frac{53}{9} = 5.889 \,, \quad p_4 = \frac{1}{\mu_4} = 0.1698 \,,$$

   respectively.

   Based on the result (C', D'), we find that the probability of type-III or type-IV mutation increases slightly, but the average length of type-III or type-IV mutation is reduced slightly (the parameter of geometric distribution increases slightly). Therefore, the question of whether to base the estimation on results (A', B') or on (C', D') should be analyzed further.

Table 4.8. Local modification operations of the RNA sequences (A', B') shown in Table 4.6

Beginning position	Original result of alignment	Result of local modification operation	Reduced penalty
1	uauc —cu	uauc –cu–	1
211	—aa aagc	aa— aagc	2
314	ugaga—— gcguaugggg	u–g–a–g—— gcguaugggg	2
326	———————— ——aaagcugaag gccgggugugu gguguugaacuu	aaag–cug–aag– ———————— gccgggugugu gguguugaacuu	3
387	–ua uac	ua– uac	2
568	——gua cggugac	—gu–a– cggugac	3
633	—cu gagua	—cu– gagua	1
765	uu————u cuuugugcugug	—uu———u– cuuugugcugug	3
782	—cg cggg	cg— cggg	2
1199	uaa –ua	uaa ua–	2
1329	gu—g uguggu	–gu–g– uguggu	3
1509	—a agg	a— agg	1
1732	ag—cc gaguccd	–ag–cc– gaguccd	4
1899	aaaua –ccau	aaaua ccau–	2
2290	–u ug	u– ug	1
2338	———gac gacugugu	gac——— gacugugu	3
2419	ucaggaggg –ucaagcgc	ucaggaggg ucaagcgc–	4
2532	–accccugug accccagagu	accccugug– accccagagu	4
2994	acgcau —aggc	acgcau aggc—	3
3049	uaauuuaggcauc cu————	uaauuuaggcauc ————c–u–	2

2. The statistical results for type-II mutation occurring in (A', B') are given as follows:

(ℓ_1, ℓ_2)	1	2	3	4
1	23	11	4	4
2	9	3	5	
3	5	2	1	
4	1			

(4.44)

where the figures in the square matrix indicate the number of type-II mutations with mutation mode (ℓ_1, ℓ_2). The final is result that a type-II mutation occurs 68 times. Thus, the intensity of type-II mutation is $\epsilon_2 = 68/2977 = 0.023$.
Based on (4.44), we find the joint probability distribution of (ℓ_1^*, ℓ_2^*) is:

$p_{i,j}$	1	2	3	4
1	0.34328	0.16418	0.05970	0.05970
2	0.13433	0.04478	0.07463	
3	0.07463	0.02985	0.01493	
4	0.01493			

(4.45)

where $p_{i,j} = P_r\{(\ell_1^*, \ell_2^*) = (i,j)\}$. Therefore, the average length of (ℓ_1^*, ℓ_2^*) and their parameters of geometric distribution are:

$$\mu_1 = 1.5075 \,, \quad p_1 = \frac{1}{\mu_1} = 0.6634 \,,$$

$$\mu_2 = 1.6866 \,, \quad p_2 = \frac{1}{\mu_2} = 0.5929 \,.$$

Utilizing (4.45), we calculate their mutual information as

$$I(\ell_1^*, \ell_2^*) = \sum_{i=1}^{4} \sum_{j=1}^{4} p_{i,j} \log \left(\frac{p_{i,j}}{p_i q_j} \right) = 0.0772 \,, \tag{4.46}$$

where $p_i = \sum_{j=1}^4 p_{i,j}$, $q_j = \sum_{i=1}^4 p_{i,j}$. Since the value of $I(\ell_1^*, \ell_2^*)$ is sufficiently small in practice, we conclude that ℓ_1^*, ℓ_2^* are independent random variables.

3. Based on (A', B') and the statistical result of type-II mutation, we find that the total error occurring in the region of type-II mutation is 162. Therefore, the total error induced by type-I mutation is $1023 - 84 - 53 - 162 = 724$, where 84 is the error caused by the virtual symbols inserted in sequence C, and 53 is the error caused by the virtual symbols inserted in sequence D. Therefore, we get the intensity of type-I mutation: $\epsilon_1 = 724/2977 = 0.2432$.
In summary, we find estimations for the parameters of 4 types of mutations acting on sequences (A, B) as shown below:

ϵ_1	ϵ_2	ϵ_3	ϵ_4	p_1	p_2	p_3	p_4
0.24	0.02	0.006	0.003	0.663	0.593	0.226	0.170

(4.47)

Based on the above analysis, we find that the mutation structures of gene sequences Mc.vanniel and Mb.tautotr agree with the stochastic mutation modes presented in Chap. 2. However, the issue of the statistical independence of the variables must still be addressed. These results offer a reference for analyzing the mutation and alignment of multiple sequences.

4.5 Exercises

Exercise 16. Choosing the Hamming matrix as the penalty matrix, write down the computational program of the SPA, and do the following calculations:

1. Align the sequences given in Sect. 4.4.2.
2. For Sect. 4.4.2, compare the SPA with the dynamic programming algorithm (introduced in Chap. 1) for the vital indices of alignment, such as the similarity and CPU time.

Exercise 17. Use the SPA to align the sequences produced by the simulation method in Chap. 2, and consider the following problems:

1. Analyze the relationship between the length of the sequences and the CPU time using the SPA.
2. Compare the similarity between the aligned sequence obtained from the SPA and that from using the dynamic programming algorithm, and give the statistics of the difference of similarity, on average.

Exercise 18. Let (A, B) be the sequences given in Table 4.5, and let (A', B') be the aligned sequences given in Table 4.6. Analyze and calculate the following:

1. In (4.21), one kind of modulus structure of (A', B') is given. With a similar argument, give two other forms of the modulus structure of (A', B').
2. Modify the results given in Table 4.6 using the local modification operation given in Table 4.8, such that the final result is (A'', B''). Write down the representation of (A'', B'').
3. Compare the similarity between sequences (A'', B'') and the sequences obtained by a dynamic programming algorithm.

Exercise 19. State the differences between sequential decision and self-adaptive regression decision, and write a program for the improved SPA using self-adaptive regression decision.

Exercise 20. According to the various types of local modifications given in Table 4.8, give a computational program for local modification operation.

Hints

For the SPA for pairwise alignment, it is better to write a program that includes the steps presented in Sect. 4.2. If you have difficulties, you may use the algorithm provided on our Web site [99], and then try programming by yourself.

5

Multiple Sequence Alignment

5.1 Pairwise Alignment Among Multiple Sequences

In previous chapters, the structural features of pairwise sequence alignment and the features of mutations were discussed. Based on these features, dynamic programming-based algorithms and the statistical decision-based algorithm (SPA) were presented. These algorithms are restricted to performing alignments for a pair of sequences. However, to apply these alignment methods to bioinformatics, they must be able to process a family of sequences (many more than two sequences) simultaneously. The alignment algorithms that can accomplish this are called multiple sequence alignment, or simply MA. When studying these alignments, pairwise alignment are the best reference. Therefore, we begin this chapter by discussing the structure resulting from mutations, as well as the structure by alignment of multiple sequences. As we attempt to develop MA, we discuss how to use pairwise alignment to process multiple sequences, and we consider what types of problems this raises.

5.1.1 Using Pairwise Alignment to Process Multiple Sequences

Both dynamic programming-based algorithms and statistical decision-based algorithms for pairs of sequences are fast and programmable. Until true MA are developed, researchers must use pairwise alignment methods to process multiple sequences. Many current MA software packages, such as Clustal-W [102] etc., are in fact based on pairwise alignment. That is, the pairwise alignment is an important component of these software packages. The use of pairwise alignment methods to process multiple sequences is important for a rough analysis of the common structure of multiple sequences. For example, we can analyze affinity relationships and evolution between each pair of sequences, based on pairwise alignment methods. This demonstrates that pairwise alignment of multiple sequences is important enough to consider specifically.

Let m be the number of sequences in a multiple sequence set; the computational complexity then ranges from $O(m^2n)$ to $O(m^2n^2)$ if we use pairwise alignment methods to process the set of sequences. For example, if we use the SPA, then the complexity is $O(m^2n)$, while the computational complexity is $O(m^2n^2)$ for a dynamic programming-based algorithm such as the Smith–Waterman algorithm. Therefore, if the size of the benchmark set is not overly large, pairwise alignment is acceptable in the sense of complexity. Of course, the use of pairwise alignment methods to build the homologous family of a sequence is only a stopgap measure, not the final goal.

5.1.2 Topological Space Induced by Pairwise Alignment of Multiple Sequences

We maintain the notations in (1.1) and (1.2) such that

$$\mathcal{A} = \{A_s = (a_{s,1}, a_{s,2}, \cdots, a_{s,n_s}), \quad s = 1, 2, \cdots, m\} \tag{5.1}$$

is a multiple sequence in which A_s is the sth gene sequence whose length is n_s, and $a_{s,j} \in V_4 = \{0, 1, 2, 3\}$ is the state space of nucleotides in the gene sequence, and m is its multiplicity. In addition, we still must introduce the following notations and point out some problems.

The Matrices Induced by Pairwise Alignment of Multiple Sequences

If $\mathcal{A}$ is a multiple sequence defined in (5.1), then for any $s, t \in M = \{1, 2, \cdots, m\}$, we find the result of pairwise alignment of (A_s, A_t) as follows:

$$(C_{s,t}, C_{t,s}) = ((c_{s,t;1}, c_{s,t;2}, \cdots, c_{s,t;n_{s,t}}), (c_{t,s;1}, c_{t,s;2}, \cdots, c_{t,s;n_{t,s}})) . \tag{5.2}$$

Then, $C_{s,t}, C_{t,s}$ is the expansion of (A_s, A_t) in which, $c_{s,t;j}, c_{t,s;j} \in V_5$ and $n_{s,t} = n_{t,s}$ is the common length of the sequences ($C_{s,t}$ and $C_{t,s}$).

We then obtain a matrix induced by pairwise alignment of multiple sequences as:

$$\bar{\mathcal{C}} = (C_{s,t})_{s,t=1,2,\cdots,m} , \tag{5.3}$$

in which $C_{s,t}$ is defined by (5.2). For simplicity, we refer to $\bar{\mathcal{C}}$ as the alignment matrix induced by multiple sequences $\mathcal{A}$, or the simpler alignment matrix. It is easy to find that changing the order of the pairwise alignments for the multiple sequences will result in a different matrix.

Definition 25. *Let $\bar{\mathcal{C}}$ be the alignment matrix induced by the multiple sequences $\mathcal{A}$. Let $T_{s,t}$ be the shifting mutation mode from A_t to A_s, and let $W = \{w(a,b),\ a, b \in V_5\}$ be the penalty function on V_5. Then:*

1. *$\bar{\mathcal{C}}$ is the minimum penalty alignment matrix if the expansion $(C_{s,t}, C_{t,s})$ of (A_s, A_t) has the minimum penalty score.*

2. *$\bar{C}$ is the uniform alignment matrix if $(C_{s,t}, C_{t,s})$ is the uniform alignment of (A_s, A_t) based on the mode $T_{s,t}$ for every pair $s, t \in M = \{1, 2, \cdots, m\}$.*

The definitions of the minimum penalty alignment and uniform alignment, as well as the relationship between these two kinds of pairwise alignments, are outlined in Chaps. 1 and 4, respectively. In this chapter, we focus on the case of minimum penalty alignment. We will discuss the uniform alignment case in Chap. 7.

Penalty Matrix Induced by Pairwise Alignment of Multiple Sequences

Let $\bar{C}$ be the alignment matrix induced by the multiple sequence $\mathcal{A}$. If the penalty function $W = w(a, b)$ defined on V_5 is given, then for any $s, t \in M$, we have two expansions $C_{s,t}$ and $C_{t,s}$ based on the pair of sequences A_s and A_t. The penalty score for the pair $C_{s,t}$ and $C_{t,s}$ is defined by:

$$w_{s,t}(\bar{C}) = w(C_{s,t}, C_{t,s}) = \sum_{j=1}^{n_{s,t}} w(c_{s,t;j}, c_{t,s;j}) \ . \tag{5.4}$$

Definition 26. *Let $\bar{C}$ be the alignment matrix induced by multiple sequences $\mathcal{A}$, and let W be the penalty function defined on V_5. The matrix*

$$\bar{W}(\bar{C}) = [w_{s,t}(\bar{C})]_{s,t=1,2,\cdots,m} \tag{5.5}$$

is then the penalty matrix induced by pairwise alignment of multiple sequences $\mathcal{A}$, where $w_{s,t}(\bar{C})$ is defined by (5.4). It is acceptable to simply call this the penalty matrix.

A fixed penalty matrix $\bar{W}^o = (w^o_{s,t})_{s,t=1,2,\cdots,m}$, is the minimum penalty matrix if each $w^o_{s,t}$ is the score of the minimum penalty alignment of (A_s, A_t).

For the sake of simplicity, we use $\bar{W}(\bar{C})$ to replace

$$\bar{W} = (w_{s,t})_{s,t=1,2,\cdots,m} \ .$$

Theorem 22. *If the multiple sequence $\mathcal{A}$ and the penalty function W on V_5 are given, then the minimum penalty matrix $\bar{W}^o$ is uniquely determined, and denoted by $\bar{W}^o = \bar{W}^o(\mathcal{A})$.*

Proof. It is sufficient to prove that the score $w^o_{s,t}$ of the minimum penalty alignment is uniquely determined for any sequence pair (A_s, A_t). To do this, we check the definition

$$w^o_{s,t} = \min\{w_{s,t} = w(C_{s,t}, C_{t,s}) \colon (C_{s,t}, C_{t,s}) \text{ is the alignment of } (A_s, A_t)\} \tag{5.6}$$

in which the set at the right-hand side of expression (5.6) has a lower bound. It follows that the minimum value is unique. Hence, the minimum penalty matrix $\bar{W}^o$ is also unique.

We then have the relationship

$$\mathcal{M} = \{\mathcal{A}, \bar{W}^o\} .$$

This is called the minimum penalty representation of pairwise alignment of multiple sequences. We will prove later that $\mathcal{M}$ forms a metric space.

Metric Space Defined on a Finite Set

The metric space is a fundamental concept in mathematics. To show that $\mathcal{M} = \{\mathcal{A}, \bar{W}^o\}$ is a finite metric space, we present the general definition of a metric space defined on a finite set.

Let $M = \{1, 2, \cdots, m\}$ be a finite set, and let $w_{s,t}$ be a function defined on $M \times M = \{(s,t)\colon s, t \in M\}$.

Definition 27. *A function $w_{s,t}$ defined on $M \times M$ is a measure (or metric or distance), if the following conditions hold:*

1. ***Nonnegative property**: $w_{s,t} \geq 0$ holds all $s, t \in M$ and $w_{s,t} = 0$ if and only if $s = t$.*
2. ***Symmetry property**: $w_{s,t} = w_{t,s}$ holds for all $s, t \in M$.*
3. ***Triangle inequality**: $w_{s,r} \leq w_{s,t} + w_{t,r}$ holds for all $s, t, r \in M$.*

If the distance function $w_{s,t}$ defined on a finite set M is given, then the M endowed with this distance forms a metric space, and it is called a finite metric space, or finite distance space.

The Fundamental Theorem of Minimum Penalty Alignment

Let $\mathcal{M}$ be the minimum penalty representation of pairwise alignment of multiple sequences defined as above. It is a finite metric space under the natural distance induced by the minimum penalty matrix $\bar{W}^o$, although this is not obvious. In fact, we can not build the relationship among

$$w(C_{s,t}, C_{t,s}) , \quad w(C_{s,r}, C_{r,s}) , \quad w(C_{t,r}, C_{r,t})$$

directly because the expansions $C_{s,t}$ and $C_{s,r}$ based on A_s and A_t, A_r are not unique. The fundamental theorem of the minimum penalty alignment is that the minimum penalty representation $\mathcal{M}$ is a finite metric space if the penalty matrix $\bar{W}^o = [w^o_{s,t}]_{s,t \in M}$ satisfies the three conditions defined above.

Theorem 23. *(Fundamental theorem of minimum penalty alignment.)*
Let $\mathcal{A}$ be a multiple sequence and $\bar{W}^o = (w^o_{s,t})_{s,t \in M}$ be the minimum penalty matrix of the multiple sequences $\mathcal{A}$ defined by (5.5) under a given penalty function $W = w(a,b)$, $a, b \in V_5$. Then, the $\mathcal{M}$ endowed with the natural distance induced by $\bar{W}^o = (w^o_{s,t})_{s,t \in M}$ is a finite metric space.

Proof. For simplicity, let the penalty function $w(a,b)$ be the Hamming matrix on V_5. Then, let $\bar{W}^o = (w^o_{s,t})_{s,t \in M}$ be the minimum penalty matrix based on the multiple sequences $\mathcal{A}$ defined by (5.5). We consider the natural distance induced by this minimum penalty matrix as follows: $d(A_s, A_t) = w^o(s,t)$. Using the definitions of the Hamming matrix and $\bar{\mathcal{C}}$, we find that $d(\cdot)$ satisfies both the nonnegative and symmetry properties. Therefore, we need only prove that $d(\cdot)$ satisfies the triangle inequality. Alternatively, we prove that $w^o_{s,r} \leq w^o_{s,t} + w^o_{t,r}$ holds for all $s,t,r \in M$. For an arbitrary three $s,t,r \in \{1,2,\cdots,M\}$, we may assume that these three subscripts are different from each other. For simplicity, we also omit the subscripts of the three vectors A_s, A_t, A_r. Let

$$Z = (z_1, z_2, \cdots, z_{n_z})\,, \quad Z = A, B, C\,, \quad z = a, b, c \tag{5.7}$$

be the uniform representation of the three sequences, and let

$$\begin{pmatrix} A' \\ B' \end{pmatrix}, \quad \begin{pmatrix} A^* \\ C^* \end{pmatrix}, \quad \begin{pmatrix} B^o \\ C^o \end{pmatrix} \tag{5.8}$$

be the minimum penalty alignments of all possible combined pairs in A, B, C. Thus, we alternatively prove that

$$w(B^o, C^o) \leq w(A', B') + w(A^*, C^*) \tag{5.9}$$

holds, by using the steps outlined below:

1. Following from the definition in (5.7), we know that the sequences A', A^* are the expansions of A, B', B^o are the expansions of B, and C^*, C^o are the expansions of C, with the corresponding expanded modes given as

$$\begin{cases} H'_a = (\gamma'_{a,1}, \gamma'_{a,2}, \cdots, \gamma'_{a,n_a}) = \left\{ \left(i'_{a,k}, \ell'_{a,k} \right), \ k = 1,2,\cdots,k'_a \right\}, \\ H^*_a = (\gamma^*_{a,1}, \gamma^*_{a,2}, \cdots, \gamma^*_{a,n_a}) = \left\{ \left(i^*_{a,k}, \ell^*_{a,k} \right), \ k = 1,2,\cdots,k^*_a \right\}, \\ H'_b = \left(\gamma'_{b,1}, \gamma'_{b,2}, \cdots, \gamma'_{b,n_b}\right) = \left\{ \left(i'_{b,k}, \ell'_{b,k} \right), \ k = 1,2,\cdots,k'_b \right\}, \\ H^o_b = \left(\gamma^o_{b,1}, \gamma^o_{b,2}, \cdots, \gamma^o_{b,n_b}\right) = \left\{ \left(i^o_{b,k}, \ell^o_{b,k} \right), \ k = 1,2,\cdots,k^o_b \right\}, \\ H^*_c = (\gamma^*_{c,1}, \gamma^*_{c,2}, \cdots, \gamma^*_{c,n_c}) = \left\{ \left(i^*_{c,k}, \ell^*_{c,k} \right), \ k = 1,2,\cdots,k^*_c \right\}, \\ H^o_c = (\gamma^o_{c,1}, \gamma^o_{c,2}, \cdots, \gamma^o_{c,n_c}) = \left\{ \left(i^o_{c,k}, \ell^o_{c,k} \right), \ k = 1,2,\cdots,k^o_c \right\}, \end{cases} \tag{5.10}$$

   where n_a, n_b, n_c are the lengths of sequences A, B, C, respectively.
2. Let $n'_a, n^*_a, n'_b, n^o_b, n^*_c$ and n^o_c be the lengths of the sequences A', A^*, B', B^o, C^* and C^o respectively, where $n'_a = n'_b$, $n^*_a = n^*_c$, $n^o_b = n^o_c$. Then, following from $(H'_a, H^*_a, H'_b, H^o_b, H^*_c, H^o_c)$, the decompositions of N'_a, N^*_a,

N'_b, N^o_b, N^*_c and N^o_c are as follows:

$$\begin{cases} N'_a = \left\{\delta'_{a,1}, \delta'_{a,2}, \cdots, \delta'_{a,2k'_a+1}\right\} , \\ N^*_a = \left\{\delta^*_{a,1}, \delta^*_{a,2}, \cdots, \delta^*_{a,2k^*_a+1}\right\} , \\ N'_b = \left\{\delta'_{b,1}, \delta'_{b,2}, \cdots, \delta'_{b,2k'_b+1}\right\} , \\ N^o_b = \left\{\delta^o_{b,1}, \delta^o_{b,2}, \cdots, \delta^o_{b,2k^o_b+1}\right\} , \\ N^*_c = \left\{\delta^*_{c,1}, \delta^*_{c,2}, \cdots, \delta^*_{c,2k^*_c+1}\right\} , \\ N^o_c = \left\{\delta^o_{c,1}, \delta^o_{c,2}, \cdots, \delta^o_{c,2k^o_c+1}\right\} , \end{cases} \tag{5.11}$$

where the intervals δ's are connected in order. If we let

$$\begin{array}{ll} \Delta'_{a,1} = \left\{\delta'_{a,1}, \delta'_{a,3}, \cdots, \delta'_{a,2k'_a+1}\right\} & \Delta'_{a,2} = \left\{\delta'_{a,2}, \delta'_{a,4}, \cdots, \delta'_{a,2k'_a}\right\} \\ \Delta^*_{a,1} = \left\{\delta^*_{a,1}, \delta^*_{a,3}, \cdots, \delta^*_{a,2k^*_a+1}\right\} & \Delta^*_{a,2} = \left\{\delta^*_{a,2}, \delta^*_{a,4}, \cdots, \delta^*_{a,2k^*_a}\right\} \\ \Delta'_{b,1} = \left\{\delta'_{b,1}, \delta'_{b,3}, \cdots, \delta'_{b,2k'_b+1}\right\} & \Delta'_{b,2} = \left\{\delta'_{b,2}, \delta'_{b,4}, \cdots, \delta'_{b,2k'_b}\right\} \\ \Delta^o_{b,1} = \left\{\delta^o_{b,1}, \delta^o_{b,3}, \cdots, \delta^o_{b,2k^o_b+1}\right\} & \Delta^o_{b,2} = \left\{\delta^o_{b,2}, \delta^o_{b,4}, \cdots, \delta^o_{b,2k^o_b}\right\} \\ \Delta^*_{c,1} = \left\{\delta^*_{c,1}, \delta^*_{c,3}, \cdots, \delta^*_{c,2k^*_c+1}\right\} & \Delta^*_{c,2} = \left\{\delta^*_{c,2}, \delta^*_{c,4}, \cdots, \delta^*_{c,2k^*_c}\right\} \\ \Delta^o_{c,1} = \left\{\delta^o_{c,1}, \delta^o_{c,3}, \cdots, \delta^o_{c,2k^o_c+1}\right\} & \Delta^o_{c,2} = \left\{\delta^o_{c,2}, \delta^o_{c,4}, \cdots, \delta^o_{c,2k^o_c}\right\} \end{array} \tag{5.12}$$

then we have

$$a'_{\Delta'_{a,1}} = a^*_{\Delta^*_{a,1}} = A , \quad b'_{\Delta'_{b,1}} = b^o_{\Delta^o_{b,1}} = B , \quad c^*_{\Delta^*_{c,1}} = c^o_{\Delta^o_{c,1}} = C , \tag{5.13}$$

and all components in the following vectors

$$a'_{\Delta'_{a,2}} , \quad a^*_{\Delta^*_{a,2}} , \quad b'_{\Delta'_{b,2}} , \quad b^o_{\Delta^o_{b,2}} , \quad c^*_{\Delta^*_{c,2}} , \quad c^o_{\Delta^o_{c,2}}$$

are the inserted symbol "–".

3. Following from the union $H''_a = H'_a \vee H^*_a$ of the expanded modes H'_a, H^*_a, we may expand sequence A to A'' under the mode H''_a. Then, A'' is actually the virtual expansion of both A' and A^*, whose extra regions are $H''_a - H'_a$ and $H''_a - H^*_a$, respectively. Therefore, we get the expanded modes on A' and A^* as $(H''_a - H'_a)_{a'}$ and $(H''_a - H^*_a)_{a^*}$, respectively. $(H''_a - H'_a)_{a'}$, $(H''_a - H^*_a)_{a^*}$ are then two different quadratic expansions of sequence A, whose evolution process is given as:

$$A \xrightarrow{H'_a} A' \xrightarrow{(H''_a - H'_a)_{a'}} A'' , \quad A \xrightarrow{H^*_a} A^* \xrightarrow{(H''_a - H^*_a)_{a^*}} A'' . \tag{5.14}$$

4. Since the lengths of sequences B', C^* are equal to the lengths of sequences A', A^*, respectively, the expansions of B', C^* under the expanded mode $(H''_a - H'_a)_{a'}$, $(H''_a - H^*_a)_{a^*}$ are denoted by B'', C'', respectively. Then, $(H''_a - H'_a)_{a'}$, $(H''_a - H^*_a)_{a^*}$ are the common expanded regions of pair A'' and B'' and pair A'' and C'' respectively. Hence, we find that

$$w(A'', B'') = w(A', B') , \quad w(A'', C'') = w(A^*, C^*) . \tag{5.15}$$

On the other hand, based on the definition of $w(A'', B'') = \sum_{j=1}^{n''_a} w(a''_j, b''_j)$, we derive the following relationships:

$$\begin{aligned} w(A'', B'') + w(A'', C'') &= \sum_{j=1}^{n''_a} \left[w\left(a''_j, b''_j\right) + w\left(a''_j, c''_j\right) \right] \\ &\geq \sum_{j=1}^{n''_a} w\left(b''_j, c''_j\right) = w(B'', C'') , \end{aligned} \tag{5.16}$$

where the inequality holds due to the following expression:

$$w\left(a''_j, b''_j\right) + w\left(a''_j, c''_j\right) \geq w\left(b''_j, c''_j\right) \quad \forall j = 1, 2, \cdots, n''_a ,$$

where $w(a, b)$ is a measurement defined on V_5, and n''_a is the length of the sequence A''.

5. In view of (5.15) and (5.16), we have the following inequality:

$$\begin{aligned} w(B^o, C^o) \leq w(B'', C'') &\leq w(A'', B'') + w(A'', C'') \\ &= w(A', B') + w(A^*, C^*) . \end{aligned} \tag{5.17}$$

Similarly, we can prove that

$$\begin{aligned} w(A^*, C^*) &\leq w(A', B') + w(B^o, C^o) , \\ w(A', B') &\leq w(A^*, C^*) + w(B^o, C^o) . \end{aligned}$$

Hence, the required triangle inequality relationship of $\bar{W}^o = (w^o_{s,t})_{s,t \in M}$ holds. This is equivalent to saying that $\bar{W}$ is a distance function defined on $\mathcal{A}$. This ends the proof.

Next, we denote the finite metric space with a minimum penalty matrix by $\mathcal{M} = \{M, W\}$, and we may call it the **metric space of pairwise alignment** for short, where $M = \{1, 2, \cdots, m\}$ is the subscript of $\mathcal{A}$, and $W = (w_{s,t})_{s,t=1,2,\cdots,m}$ is the minimum penalty matrix induced by the pairwise alignment of the multiple sequences $\mathcal{A}$ under a given penalty function. This metric space $\mathcal{M}$ is useful in the clustering of multiple sequences, and in the analysis of the evolution of multiple sequences. Clustering analysis is useful in many aspects of sequence analysis. We will discuss this further in later sections.

5.2 Optimization Criteria of MA

5.2.1 The Definition of MA

Using pairwise alignment to process multiple sequences is not a true multiple alignment approach, as several problems cannot be solved through use of this strategy. For example:

1. **To search for common stable regions of a family of sequences**. In other words, in determining the common region of many biological sequences, the pairwise alignment methods do not work.
2. **For an overview of the characteristics and trends of multiple sequences**. The stable regions of multiple sequences do not perfectly coincide. Frequently, there are sequences in a multiple sequence set such that the stable regions are different. This difference often cannot be found through pairwise alignment, only by MA.
3. **Analyze these types of mutation comprehensively**. Structure of the sequence before mutation and prediction problems. In the mutating processes, many important mutation types will occur, for example, independent mutation and transitional mutation, etc. To analyze these types of mutation comprehensively, and to predict the trend of changes, we must involve MA.

In conclusion, MA are vitally important tools in the analysis of the common structure of a family of sequences, and their usefulness is not limited to the research on mutations in and evolution of biological sequences. It is used comprehensively to solve bioinformatics problems, for example, as a main tool for predicting the secondary structure of proteins.

To create MA, we begin by building the optimization criteria of MA methods, and then attempt the optimization of MA.

5.2.2 Uniform Alignment Criteria and SP-Optimization Criteria for Multiple Sequences

Definition of Uniform Alignment of Multiple Sequences

Uniform alignment of a pair of sequences was addressed when we discussed mutation and pairwise alignment. We now generalize this concept to fit multiple sequences.

Let $\mathcal{A} = \{A_1, A_2, \cdots, A_n\}$ be a multiple sequence and $\mathcal{C} = \{C_1, C_2, \cdots, C_m\}$ be the alignment of $\mathcal{A}$. Typically,

$$A_s = (a_{s,1}, a_{s,2}, \cdots, a_{s,n_s}) , \quad C_t = (c_{t,1}, c_{t,2}, \cdots, c_{t,n'_t}) , \quad s, t \in M ,$$

$a_{s,j}$, $c_{t,j}$ are elements of V_4, V_5, respectively, and each C_s is virtual expansion of A_s.

Within the multiple sequence $\mathcal{A}$, every pair A_s, A_t are mutated sequences acted on by shifting and nonshifting mutations. Let $T_{s,t}$ be the mutation mode for A_s, A_t mutating to C_s, C_t, respectively, and let (C'_s, C'_t) be the compressed sequences of (C_s, C_t). If $(c_{sj}, c_{tj}) = (4, 4)$, then delete these two components from C_s, C_t, respectively, so that the rest of (C'_s, C'_t) is still the expansion of (A_s, A_t).

Definition 28. *Let $\mathcal{C}$ be the multiple expansion of $\mathcal{A}$. Then $\mathcal{C}$ is the uniform alignment of $\mathcal{A}$, if for every $s \neq t \in M$, the following conditions are satisfied:*

1. *For every expansion C'_s of A_s, the added part just consists of the regions resulting from type-III mutation so that A_s to A_t.*
2. *For every expansion C'_t of A_t, the added part just consists of the regions resulting from type-III mutation so that A_t to A_s.*

Calculation of Uniform Alignment of Multiple Sequences

In Sects. 3.1 and 3.2, we mentioned the mutation mode of multiple sequences and their envelope. If a multiple sequence $\mathcal{A}$ has only shifting mutations, then the uniform alignment $\mathcal{C}$ of $\mathcal{A}$ can be computed by the following steps:

1. Calculate the minimum envelope C_0 of $\mathcal{A}$.
2. For each s, since C_0 is the expansion of A_s, we compare C_0 and A_s. For the extra coordinates of C_0 relative to A_s, we replace them with "−", and then renew the sequence denoted by C_s. The collection of all renewed sequences $\mathcal{C} = \{C_s,\ s \in M\}$ is the uniform alignment of the multiple sequence.
3. If $\mathcal{A}$ is a multiple sequence involving both shifting and nonshifting mutations, then the minimum envelope C_0 involves type-I and type-II mutations, and C_0 relative to A_s can be divided into two parts, namely, the expansion and nonexpansion parts as follows:
$$C_0 = \left(c_{\Delta'_{s,0}}, c_{\Delta'_{s,1}}\right) ,$$
where $c_{\Delta'_{s,0}}$ is the expansion part and $c_{\Delta'_{s,1}}$ is the nonexpansion part of A_s.
4. The uniform alignment of a multiple sequence $\mathcal{A}$ is the result processed the following way: replace the corresponding coordinates in the region of $c_{\Delta'_{s,0}}$ by the elements of A_s, and replace the coordinates in the region of $c_{\Delta'_{s,1}}$ by the virtual symbol "−". The renewed multiple sequence is then the uniform alignment of $\mathcal{A}$.

Example 19. Let $\mathcal{A}$ be a triple of sequences given by:

$$\begin{cases} A_1 = \text{aactg()ggga[tagat]gguuuaacgta\{aauau\}accgt}\,, \\ A_2 = \text{aactg(gta)ggga[]gguuuaacgta\{aauau\}accgt}\,, \\ A_3 = \text{aactg(gta)ggga[tagat]gguuuaacgta\{\}accgt}\,. \end{cases}$$

Comparing these three sequences, we find the following mutation relationships:

Based on A_1, we insert gta after position 5 and delete tagat after position 9, so that A_1 mutates to A_2.

Also based on A_1, we insert gta after spot 5 and delete aauau after position 25, so that A_1 mutates to A_3.

Based on A_2, we insert tagat after position 12 and delete aauau after position 23, so that A_2 mutates to A_3.

Obviously, the triple sequence $\mathcal{A}$ has shifting mutations only. Therefore, each sequence in $\mathcal{A}$ can be mutated from another sequence in $\mathcal{A}$ by type-III and type-IV mutations.

The minimum envelope and maximum core are given by:

$$\begin{cases} C_0 = \text{aactg(gta)ggga[tagat]gguuuaacgta\{aauau\}accgt}, \\ D_0 = \text{aactg()ggga[]gguuuaacgta\{\}accgt}, \end{cases}$$

in which the data in parentheses, brackets, or braces are the deleted segments in A_1, A_2, A_3, respectively. We set

$$\begin{cases} C_1 = \text{aactg(----)ggga[tagat]gguuuaacgta\{aauau\}accgt}, \\ C_2 = \text{aactg(gta)ggga[-------]gguuuaacgta\{aauau\}accgt}, \\ C_3 = \text{aactg(gta)ggga[tagat]gguuuaacgta\{-------\} accgt}. \end{cases}$$

This renewed triple sequence is the uniform alignment of the triple sequence $\mathcal{A}$.

If A_1, A_2, A_3 have nonshifting mutations, they have no unified envelope and core. We construct an envelope and core with type-I and type-II mutations, and then construct the corresponding uniform alignment. For example, if

$$\begin{cases} A_1' = \text{aaatg()ggga[tagat]gguuuaacgta\{aauau\}accgt}, \\ A_2' = \text{aactg(gta)ggga[]gguaauucgta\{aauau\}accgt}, \\ A_3' = \text{aactg(gta)ggga[tagat]gguuuaacgtaaccgg}. \end{cases}$$

then besides the shifting mutation like the one in A_1, A_2, A_3, there are also type-I and type-II mutations in A_1', A_2', A_3'.

At position 3 of A_1', c was mutated to a, relative to A_2. This is a type-I mutation. At positions 16–19 of A_2', aauu was mutated to uuaa, relative to A_1. This is a type-II mutation. At the last position of A_3', the t was mutated to t, relative to A_1 and A_2. This again is type-I mutation.

After this preprocessing, we denote the envelope and core of A_1', A_2', A_3' by C_0, D_0, and let the uniform alignment be:

$$\begin{cases} C_1' = \text{aaatg(----)ggga[tagat]gguuuaacgta\{aauau\}accgt}, \\ C_2' = \text{aactg(gta)ggga[-------]gguaauucgta\{aauau\}accgt}, \\ C_3' = \text{aactg(gta)ggga[tagat]gguuuaacgta\{-------\}accgg}. \end{cases}$$

Problems in Uniform Alignment of Multiple Sequences

Based on the definition of uniform alignment of multiple sequences, we know that all of the shifting mutations among the multiple sequences can be determined by uniform alignment of multiple sequences, after which all the mutations between every pair in the multiple sequence can be determined. Thus, uniform alignment of multiple sequences is the ultimate goal. On the other hand, there are several difficult problems involved in uniform alignment of multiple sequences which must be solved, such as:

1. Example 19 is a special case that may be solved. For general cases, the calculation of the uniform alignment is too complex; so we must still find a systematic algorithm in order to solve it.
2. It is difficult to judge whether a MA is a uniform alignment or not. We cannot establish a unified indexing system to judge uniform alignment.

This shows that the uniform alignment of multiple sequences is simply an ideal optimization criterion, which is in reality difficult to perform. So, we must still find other optimization criteria.

SP-Criterion of MA

The SP-penalty functions of MA presented in (1.9), are frequently involved in current literature. The involved notations are stated as follows:

1. Let $w(a, b)$, $a, b \in V_5$ be the metric function defined on V_5, which is also called the difference degree, or penalty matrix. The most popular penalty matrices for DNA (or RNA) are the Hamming matrix, the WT-matrix, etc. The definition of the WT-matrix is presented in (1.7). Generally, a metric function $w(a, b)$, $a, b \in V_5$ should satisfy the three axioms, namely, nonnegativity, symmetry and the triangle inequality.
2. The SP-function is the function most frequently used as the penalty function for multiple sequences. The definition of the SP-function is presented in (1.9).
3. A generalized form of the SP-function is the weighted WSP-function defined as follows:

$$w_{\rm WSP}(\mathcal{C}) = \sum_{j=1}^{n'} \sum_{t>s} \sum_{s=1}^{m-1} \theta_{s,t} w(c_{s,j}, c_{t,j}) = \sum_{j=1}^{n'} \sum_{s=1}^{m-1} \sum_{t>s} \theta_{s,t} w(c_{s,j}, c_{t,j}) \;, \tag{5.18}$$

where $\theta_{s,t}$ is a weighting function.

The functions $w_{\rm SP}(\mathcal{C})$ and $w_{\rm WSP}(\mathcal{C})$ defined by (1.9) and (5.18) respectively, are both called SP-penalty functions for multiple sequences. Thus, multiple sequence alignment (MSA) may be formed as follows. For a multiple sequences $\mathcal{A}$, search for its expansion $\mathcal{C}_0$ with the minimum penalty (or the maximum similarity) under the given penalty function. Alternatively, solve the expansion $\mathcal{C}_0$ of $\mathcal{A}$ such that

$$w_{\mathrm{SP}}(\mathcal{C}_0) = \min\{w_{\mathrm{SP}}(\mathcal{C}) \colon \mathcal{C} \text{ is the expansion of } \mathcal{A}\} . \tag{5.19}$$

4. In addition, the SP-scoring function is also commonly used in current literature. We then let $w(a,b)$, $a,b \in V_5$ be a scoring function defined on V_5. Similarly, we have the scoring matrices for DNA (or RNA) sequences based on the Hamming matrix, or the WT-matrix. The scoring matrix based on the Hamming matrix is defined below:

$$W = [w_{\mathrm{WT}}(a,b)]_{a,b\in V_5} = \begin{pmatrix} & - & a & c & g & t(u) \\ - & 0 & -2 & -2 & -2 & -2 \\ a & -2 & 1 & 0 & 0 & 0 \\ c & -2 & 0 & 1 & 0 & 0 \\ g & -2 & 0 & 0 & 1 & 0 \\ t(u) & -2 & 0 & 0 & 0 & 1 \end{pmatrix} , \tag{5.20}$$

where $-, a, c, g, t(u)$ are $4, 0, 1, 2, 3$ respectively. The definition of the SP-scoring function for multiple sequences is then the same as (1.9) or (5.18). The corresponding optimization criterion is to find the maximum value in (5.19).

5.2.3 Discussion of the Optimization Criterion of MA

Establishing the Optimization Criteria for Multiple Sequences

As presented above, the uniform alignment criterion is reasonable but difficult to judge, while the index of the SP-criterion is easily calculated based on the result $\mathcal{C}$ although its rationality is yet to be demonstrated. In fact, many other optimization criteria for MA have been proposed. Therefore, we first must understand how to find the fundamental rules for judging the quality of specific optimization criteria. For this, we propose the following requirements: rationality, decidability, comparability (with the optimization solution) and helpfulness in calculating the optimization solution. We detail them as follows:

Requirement 1: rationality. Rationality here means whether or not the proposed criterion is related to multiple alignments. We need to know how to judge rationality.

Requirement 2: decidability. Decidability here means it directly and quickly decides the quality of an optimization criterion based on the alignment output. Obviously, a good criterion should be decidable and easy to calculate.

Requirement 3: comparability (with optimization solution). Comparability here means that the alignment result determining the proposed optimization criterion should be comparable with the optimization solution, or should determine the difference between the alignment result and the optimization solution.

Requirement 4: usefulness in optimizing the solution.

Based on the definitions of a uniform alignment criterion and the SP-criterion, we know that a uniform alignment criterion satisfies requirement 1, but it does not satisfy requirements 2 and 3. The SP-criterion satisfies requirements 1 and 2, but does not satisfy requirement 3. Therefore, by using the SP-criterion, we can only judge whether the alignment result is good or bad compared with another result. We cannot calculate the difference between the alignment result and the optimal solution. Later, we may find that the SP-criterion is easily calculated, although it does not really satisfy requirement 1.

Rational Conditions of the Optimization Criteria of MA

As mentioned above, the optimization conditions of MA should relate to the goal of MA. That is, to search for stable regions within multiple sequences and to determine the trend of mutation. Therefore, we should use the "concentration" of alignment results as a basic index.

In mathematics, there are several methods for measuring the relationship between various data. For example, distance, surface area and volume are familiar measurements. As well, the uncertainty of a random variable is a basic element in informatics. The probability distribution is an important factor when determining the uncertainty.

Besides the expressions of metric relations between the data, we also should consider their specific characteristics. For example, in the case of distance, it includes not only the formulas in Euclidean space, but also the three characteristics: nonnegativity, symmetry and the triangle inequality. For measurement, its vital characteristic is its additivity. The uncertainty also has particular characteristics that will be discussed later.

In order to establish the optimization conditions for MA, the concentration is chosen as a candidate index. We add the following conditions on the penalty function of MA.

Condition 5.2.1 Nonnegative property. For any MA $\mathcal{C}$, we always have $w(\mathcal{C}) \geq 0$, and the equality holds if and only if

$$\mathcal{C} = \mathcal{A}, \quad \text{and} \quad A_1 = A_2 = \cdots = A_m \ . \tag{5.21}$$

Expression (5.21) means that there are no virtual symbols "−" in any sequence.

Condition 5.2.2 Symmetry property. This means that the overall penalty function is invariant if we permute the order of the sequences in $\mathcal{C}$. Generally, let σ_1, σ_2 be two permutations defined on sets

$$M = \{1, 2, \cdots, m\}, \quad N' = \{1, 2, \cdots, n'\} \ ,$$

respectively, and let

$$\begin{cases} \sigma_1(\mathcal{C}) = \{A_{\sigma_1(s)}, \quad s = 1, 2, \cdots, m\}, \\ \sigma_2(\mathcal{C}) = \{\sigma_2(A_s), \quad s = 1, 2, \cdots, m\}, \end{cases} \tag{5.22}$$

where

$$\sigma_2(A_s) = (\sigma_2(a_{s,1}), \sigma_2(a_{s,2}), \cdots, \sigma_2(a_{s,n'})) \tag{5.23}$$

and $\sigma_2(a_{s,j}) = a_{s,\sigma_2(j)}$. The symmetry property means that

$$w[\sigma_1(\mathcal{C})] = w(\mathcal{C}), \quad w[\sigma_2(\mathcal{C})] = w(\mathcal{C}). \tag{5.24}$$

Condition 5.2.3 Maximum–minimum condition. This condition is used to describe the column without the virtual symbol "–". It means that the penalty score is maximum if a, c, g, and u occur in this column obeying uniform distribution, and the penalty score is minimum if only one of a, c, g, or u occurs in this column. Condition 5.2.3 reflects the requirement for uniformity. We use this to find positions that are invariant for all sequences.

Another requirement for uniformity is that the penalty score of the mixed sequence produced by two multiple sequences should be greater than the sum of the penalty scores of two single multiple sequences. For example, if

$$\mathcal{A}_1 = \{A_{1,1}, A_{1,2}, \cdots, A_{1,m_1}\}, \quad \mathcal{A}_2 = \{A_{2,1}, A_{2,2}, \cdots, A_{2,m_2}\}. \tag{5.25}$$

These two multiple sequences have no common elements, where

$$A_{\tau;s} = (a_{\tau;s,1}, a_{\tau;s,2}, \cdots, a_{\tau;s,n_{\tau,s}}), \quad \tau = 1, 2, \quad s = 1, 2, \cdots, m_\tau. \tag{5.26}$$

The mixed multiple sequence is

$$\mathcal{A}_0 = \{\mathcal{A}_1, \mathcal{A}_2\} = \{A_{1,1}, A_{1,2}, \cdots, A_{1,m_1}, A_{2,1}, A_{2,2}, \cdots, A_{2,m_2}\}. \tag{5.27}$$

We denote $\mathcal{A}_0 = \mathcal{A}_1 \otimes \mathcal{A}_2$, and the operation $\otimes$ is called the row superposition of multiple sequences.

Let $\mathcal{C}_\tau$ be the alignments of the multiple sequences $\mathcal{A}_\tau$, $\tau = 1, 2$, and let $\mathcal{C}_0 = \mathcal{C}_1 \otimes \mathcal{C}_2$ be defined in the same way as (5.27). Then, $\mathcal{C}_0$ is the alignment of $\mathcal{A}_0$.

Condition 5.2.4 Convexity of row superposition. Let $\mathcal{C}_\tau$ be the alignment of multiple sequences $\mathcal{A}_\tau$, $\tau = 1, 2$, then $\mathcal{C}_0 = \mathcal{C}_1 \otimes \mathcal{C}_2$ satisfies the following inequality:

$$w(\mathcal{C}_0) \geq \frac{m_1}{m_1 + m_2} w(\mathcal{C}_1) + \frac{m_2}{m_1 + m_2} w(\mathcal{C}_2). \tag{5.28}$$

If $m_1, m_2 > 0$, then the equality in (5.28) holds if and only if there is no "–" in $\mathcal{C}_0$ and if the probabilities of finding a, c, g, t in each column of $\mathcal{C}_1$ and $\mathcal{C}_2$ are the same.

Let the jth column vector of $\mathcal{C}_0, \mathcal{C}_1$, and $\mathcal{C}_2$ be

$$c_{0;\cdot,j} = \begin{pmatrix} c_{1;\cdot,j} \\ c_{2;\cdot,j} \end{pmatrix} , \quad c_{\tau;\cdot,j} = \begin{pmatrix} c_{\tau;1,j} \\ c_{\tau;2,j} \\ \vdots \\ c_{\tau;m_\tau,j} \end{pmatrix} , \quad \tau = 1, 2 . \tag{5.29}$$

Then, the equality in Condition 5.2.4 holds if and only if there is no "−" occurring in both $c_{1;\cdot,j}$ and $c_{2;\cdot,j}$ and the probabilities of finding a, c, g, t are the same.

Conditions 5.2.1–5.2.4 are the basic requirements for uncertainty or concentration. We should pay attention when comparing them with other relations (i.e., distance, measurement, etc). These basic relations frequently form an axiomatic system in the mathematical sense. Different axiomatic systems lead to different branches of disciplines, and different branches which have different data structure relations. For example, the difference between Euclidean geometry and non-Euclidean geometry is the fifth postulate (the axiom of parallels).

Besides the uncertainty requirements, we still have several additional requirements. We add the following three conditions:

Condition 5.2.5 The invariance of the penalty function. This means the "penalty function" for MA should be a penalty function for pairwise alignment if it was restricted to a pair of sequences. The popular penalty functions for pairwise alignment include the generalized Hamming function, the WT-matrix, etc.

Condition 5.2.6 The number of virtual symbols "−" is a minimum. If there exists one row of $\mathcal{C}$ where all the elements are "−", i.e., if

$$c_{\cdot j} = \begin{pmatrix} c_{1,j} \\ c_{2,j} \\ \vdots \\ c_{m,j} \end{pmatrix} = \begin{pmatrix} - \\ - \\ \vdots \\ - \end{pmatrix} ,$$

then $\mathcal{C}$ is definitely not the minimum penalty expansion of $\mathcal{A}$.

The row superposition operation $\otimes$ for multiple sequences is defined in expression (5.27). Similarly, we define the column superposition operation for multiple sequences. Let

$$\mathcal{A}_1 = \{A_{1,1}, A_{1,2}, \cdots , A_{1,m}\} , \quad \mathcal{A}_2 = \{A_{2,1}, A_{2,2}, \cdots , A_{2,m}\}$$

be two multiple sequences with the same multiplicity, where

$$A_{\tau,s} = (a_{\tau;s,1}, a_{\tau;s,2}, \cdots , a_{\tau;s,n_{\tau,s}}) , \quad \tau = 1, 2 .$$

Then, the column superposition of two multiple sequences is defined as

$$\mathcal{A}_0 = \mathcal{A}_1 + \mathcal{A}_2 = (A_{0,1}, A_{0,2}, \cdots , A_{0,m})^T , \tag{5.30}$$

in which

$$A_{0,s} = (A_{1,s}, A_{2,s}) = (a_{1;s,1}, a_{1;s,2}, \cdots, a_{1;s,n_{1,s}} a_{2;s,1}, a_{2;s,2}, \cdots, a_{2;s,n_{2,s}}) . \quad (5.31)$$

For the alignment $\mathcal{C}_1, \mathcal{C}_2$ of $\mathcal{A}_1, \mathcal{A}_2$, we can also define the column superposition operation $\mathcal{C}_0 = \mathcal{C}_1 + \mathcal{C}_2$. If $\mathcal{C}_1, \mathcal{C}_2$ are the alignments of $\mathcal{A}_1, \mathcal{A}_2$, then $\mathcal{C}_0$ is the alignment of $\mathcal{A}_0$.

Condition 5.2.7 Convexity of column superposition. If $\mathcal{C}_\tau$ is the alignment of the multiple sequence $\mathcal{A}_\tau$, $\tau = 1, 2$, then $\mathcal{C}_0 = \mathcal{C}_1 + \mathcal{C}_2$ satisfies

$$w(\mathcal{C}_0) = w(\mathcal{C}_1) + w(\mathcal{C}_2) \quad (5.32)$$

These additional conditions are also the natural requirement for MA.

We can easily verify that the SP-penalty function defined by expression (1.9) does not satisfy the Conditions 5.2.4 and 5.2.7. Therefore, we may arrive at some unreasonable results such as; for example,

$$C_{\mathrm{SP}}\left(\left(\begin{matrix} - \\ - \\ \vdots \\ - \\ a \end{matrix}\right)\right) = C_{\mathrm{SP}}\left(\left(\begin{matrix} a \\ a \\ \vdots \\ a \\ - \end{matrix}\right)\right) = m - 1 .$$

In addition, if

$$\mathcal{C}_0 = \begin{pmatrix} a \\ a \\ c \\ c \end{pmatrix} , \quad \mathcal{C}_1 = \begin{pmatrix} a \\ c \end{pmatrix} , \quad \mathcal{C}_2 = \begin{pmatrix} a \\ c \end{pmatrix} , \quad (5.33)$$

then following from Condition 5.2.4, the penalty functions of $\mathcal{C}_0, \mathcal{C}_1, \mathcal{C}_2$ must be the same. Nevertheless, for the SP-function ($w_{\mathrm{SP}}(a, b)$ is assumed to be the Hamming matrix), we have

$$w_{\mathrm{SP}}(\mathcal{C}_0) = 4 , \quad \text{while} \quad w_{\mathrm{SP}}(\mathcal{C}_1) = w_{\mathrm{SP}}(\mathcal{C}_2) = 1 . \quad (5.34)$$

Obviously, the conclusion in (5.34) does not satisfy Condition 5.2.7. Generally, which conclusion, that of the conclusion of the SP-function or that of Condition 5.2.4, is more reasonable is a good question for discussion.

5.2.4 Optimization Problem Based on Shannon Entropy

The goal of alignment is to keep the corresponding components of multiple sequences as consistent as possible while minimizing the number of virtual symbols "−". Using Conditions 5.2.1–5.2.7, we may find that the penalty function of MA is actually a measure of the complexity or uncertainty of multiple sequences. Since Shannon entropy is a natural measurement to describe the uncertainty, it allows us to use the concept of information to describe the optimization criteria. We now introduce some pertinent notations and properties.

Notations for MA $\mathcal{C}$

Let $\mathcal{C}$ be the alignment of the multiple sequence $\mathcal{A}$, then we introduce the following notation to describe the structure of $\mathcal{C}$:

1. For simplicity, we may assume that the row lengths of $\mathcal{A}$ and $\mathcal{C}$ are the same. Then

$$n_1 = n_2 = \cdots = n_m = n\,, \quad n'_1 = n'_2 = \cdots = n'_m = n'\,.$$

   Let $a_{\cdot,j}, a_{i,\cdot}$ be the row vector and column vector of $\mathcal{A}$, respectively, and let $c_{\cdot,j}, c_{i,\cdot}$ be the row vector and column vector of $\mathcal{C}$, respectively.
2. Let

$$\chi_z(c_{s,j}) = \begin{cases} 1\,, & \text{if } c_{s,j} = z\,, \\ 0\,, & \text{otherwise} \end{cases}$$

   be the indicator function, here $c_{s,j}, z \in V_5$, and let

$$f_{j,z}(\mathcal{C}) = \sum_{s=1}^{m} \chi_z(c_{s,j})\,, \quad j = 1, 2, \cdots, n'\,, \quad z \in V_5 \tag{5.35}$$

   be the frequency distribution function of the value of each component in the column vector of the multiple sequence, then obviously we find that

$$f_{j,z} \geq 0\,, \quad \sum_{z=0}^{4} f_{j,z}(\mathcal{A}) = m$$

   holds for any $z \in V_5$, $j = 1, 2, \cdots, n'$.
3. Let $\theta_j(\mathcal{C}) = f_{j,4}(\mathcal{C})$, and let

$$p_{j,z}(\mathcal{C}) = \frac{f_{j,z}(\mathcal{C})}{m}\,, \quad j = 1, 2, \cdots, n'\,, \quad z = 0, 1, 2, 3, 4 \tag{5.36}$$

   be the frequency distribution function of the jth column of the multiple sequence, then $\sum_{z=0}^{4} p_{j,z}(\mathcal{C}) = 1$ holds; here $f_{j,z}(\mathcal{C}) p_{j,z}(\mathcal{C})$ are actually functions of $c_{j,z}$.
4. In the definitions of (5.35) and (5.36), we may omit the notation $\mathcal{C}$ sometimes, so

$$p_{j,\cdot}(\mathcal{C}) = p_{j,\cdot} = (p_{j,0}, p_{j,1}, p_{j,2}, p_{j,3}, p_{j,4})\,.$$

   We then define a function as follows:

$$w_{HG}(\mathcal{C}) = \sum_{j=1}^{n'} HG(c_{\cdot,j}) = \sum_{j=1}^{n'} [H(p_{j,\cdot}) + G(\theta_j)]\,, \tag{5.37}$$

   where

$$HG(c_{\cdot,j}) = H(p_{j,\cdot}) + G(\theta_j), \quad j = 1, 2, \cdots, n'$$

   Then $HG(c_{\cdot,j})$ is called the HG function of $\mathcal{C}$, and $G(\theta_j)$ is a strictly monotonically increasing function with $G(0) = 0$.

Definition 29. *In the HG function of $\mathcal{C}$, if*

$$H(p_{j,\cdot}) = -\sum_{z=0}^{4} p_{j,z} \log p_{j,z} \tag{5.38}$$

is a Shannon entropy, then the function $w_{HG}(\mathcal{C})$ is called the S-function, or information-based penalty function, and is denoted by $w_S(\mathcal{C})$.

As this is a very important penalty function, the reader should keep it in mind as we will use it in the text below.

The Selection of the Monotonically Increasing Function $G(\theta)$

There are several possible selections for the monotonically increasing function $G(\theta)$, some of which are listed as follows:

1. Linear function: $G_m(\theta) = \frac{\theta}{g(m)}$, where $g(m)$ is a nondecreasing function of m ($g(m) \geq g(m')$ if $m > m'$) which does not depend on θ. In particular, $g(m)$ may be chosen as a constant with respect to m, and $G_m(\theta)$ is a common linear function of θ for the multiple sequences.
2. Power law function: for example, $G(\theta) = \theta^2$.
3. Logarithmic function: i.e., $G(\theta) = \log(1 + \theta)$, etc.

If we choose $w_S(\mathcal{C})$ as the penalty function to process the optimization problems for multiple sequences, then these types of optimal problems are called information-based optimal problems. As well, the corresponding optimization criteria are called information-based criteria.

Information-Based Criteria of Multiple Sequences

Theorem 24. *If $w(\mathcal{C})$ is an information-based function of MA given by (5.37) and (5.38), then Conditions 5.2.1–5.2.7 hold.*

Since the proof of this theorem is long, we will outline for the reader the role played by this theorem, before providing the proof. It follows from this theorem that the information-based penalty function defined by (5.37) and (5.38) is a penalty function satisfying Conditions 5.2.1–5.2.7. That is, this new penalty function is the best one among all penalty functions.

Proof. The proof of this theorem is divided into nine steps as follows:

1. Verifying that Conditions 5.2.1 and 5.2.2 are satisfied. This is trivial because $H(p_{j\cdot})$, $G(\theta_j)$ are nonnegative functions, and $w_S(\mathcal{C}) = 0$ holds if and only if
$$H(p_{j\cdot}) = 0\,, \quad G(\theta_j) = 0\,, \quad j = 1, 2, \cdots n' \tag{5.39}$$
holds. Furthermore, (5.39) holds if and only if (5.21) holds. In addition, based on the definitions of $w_S(\mathcal{C})$ and $H(p_{j\cdot})$, θ_j, we know that they are symmetric functions with respect to the subscripts s, j. Hence, the symmetric Condition 5.2.2 is true.

2. Verifying Condition 5.2.3. Condition 5.2.3 can be directly proved by the properties of Shannon entropy. This is because $H(p_{j\cdot})$ is maximal if $p_{j\cdot}$ obeys uniform distribution, while it is minimum if $p_{j\cdot}$ obeys binary distribution (in other words,

$$p_{j,z} = \begin{cases} 1\,, & \text{if } z = z_0\,, \\ 0\,, & \text{otherwise}\,, \end{cases}$$

for a fixed $z_0 \in V_4$).
3. Verifying Condition 5.2.4. Let $\mathcal{C}_0 = \mathcal{C}_1 \otimes \mathcal{C}_2$, then both $\mathcal{C}_1$ and $\mathcal{C}_2$ are subsets of $\mathcal{C}_0$. Following from (5.26) and (5.27), we have

$$\mathcal{C}_\tau = \{C_{\tau,1}, C_{\tau,2}, \cdots, C_{\tau,m_\tau}\}\,, \quad \tau = 0,1,2\,, \tag{5.40}$$

where $m_0 = m_1 + m_2$, and

$$C_{\tau,s} = (c_{\tau;s,1}, c_{\tau;s,2}, \cdots, c_{\tau;s,n'})\,, \quad \tau = 0,1,2\,, \quad s = 1,2,\cdots,m_\tau\,. \tag{5.41}$$

We then define

$$\begin{cases} f_{\tau,j}(z) = \sum_{s=1}^{m_\tau} \chi_z(c_{\tau;s,j})\,, \\ \theta_{\tau,j} = f_{\tau,j}(4)\,, \\ p_{\tau,j}(z) = \dfrac{f_{\tau,j}(z)}{m_\tau}\,, \end{cases} \tag{5.42}$$

where $\tau = 0,1,2$, $z \in V_5$, $j = 1,2,\cdots,n'$. We then find that the equations

$$\begin{cases} \theta_{0,j} = \theta_{1,j} + \theta_{2,j}\,, \\ p_{0,j}(z) = \mu_1 p_{1,j}(z) + \mu_2 p_{2,j}(z) \end{cases} \tag{5.43}$$

hold for all $z = 0,1,2,3,4$, and $j = 1,2,\cdots,n$, where $\mu_1 = \frac{m_1}{m_0}$, $\mu_2 = \frac{m_2}{m_0}$.
4. Verifying the convexity of $G(\theta)$. Following from formula (5.37) for the S-penalty function for multiple sequences $\mathcal{C}_0$, and the property that $G(\theta)$ is a monotonically increasing function, we obtain the inequality:

$$G(\theta_{0,j}) = \mu_1 G(\theta_{0,j}) + \mu_2 G(\theta_{0,j}) \geq \mu_1 G(\theta_{1,j}) + \mu_2 G(\theta_{2,j})\,. \tag{5.44}$$

5. Verifying the inequality $\mu_1 w_S(\mathcal{C}_1) + \mu_2 w_S(\mathcal{C}_2) \leq w_S(\mathcal{C}_0)$ in Condition 5.2.4. Let $h(p) = -p\log p$ be the entropy density function, then $h(p) = -p\log p$ is a convex function of the variable $p \in (0,1)$. In fact, $\mu_1 + \mu_2 = 1$ and

$$p_{0,j}(z) = \mu_1 p_{1,j}(z) + \mu_2 p_{2,j}(z)$$

imply the inequality:

$$\begin{aligned} -p_{0,j}(z)\log p_{0,j}(z) &= -[\mu_1 p_{1,j}(z) + \mu_2 p_{2,j}(z)]\log[\mu_1 p_{1,j}(z) + \mu_2 p_{2,j}(z)] \\ &\geq -\mu_1 p_{1,j}(z)\log p_{1,j}(z) - \mu_2 p_{2,j}(z)\log p_{2,j}(z)\,. \end{aligned} \tag{5.45}$$

Taking the sum of z by the two sides of expression (5.45), we have

$$\begin{aligned} H(p_{0;j,\cdot}) &= -\sum_{z=0}^{4} p_{0,j}(z)\log p_{0,j}(z) \\ &= -\sum_{z=0}^{4}[\mu_1 p_{1,j}(z) + \mu_2 p_{2,j}(z)]\log[\mu_1 p_{1,j}(z) + \mu_2 p_{2,j}(z)] \\ &\geq -\sum_{z=0}^{4}[\mu_1 p_{1,j}(z)\log p_{1,j}(z) + \mu_2 p_{2,j}(z)\log p_{2,j}(z)] \\ &= \mu_1 H(p_{1;j,\cdot}) + \mu_2 H(p_{2;j,\cdot}) \ . \end{aligned} \tag{5.46}$$

Furthermore, using expressions (5.44) and (5.46), we find that the inequality

$$\mu_1[H(p_{1;j,\cdot})+G(\theta_{1,j})]+\mu_2[H(p_{2;j,\cdot})+G(\theta_{2,j})] \leq H(p_{0;j,\cdot})+G(\theta_{0,j}) \tag{5.47}$$

holds for all $j = 1, 2, \cdots, n'$. Again, taking the sum over j on both sides of expression (5.47), we have the inequality

$$\begin{aligned} &\mu_1 w_S(\mathcal{C}_1) + \mu_2 w_S(\mathcal{C}_2) \\ &= \sum_{j=1}^{n'} \{\mu_1[H(p_{1;j,\cdot}) + G(\theta_{1,j})] + \mu_2[H(p_{2;j,\cdot}) + G(\theta_{2,j})]\} \\ &\leq \sum_{j=1}^{n'}[H(p_{0;j,\cdot}) + G(\theta_{0,j})] = w_S(\mathcal{C}_0) \ . \end{aligned}$$

Hence, the inequality $\mu_1 w_S(\mathcal{C}_1) + \mu_2 w_S(\mathcal{C}_2) \leq w_S(\mathcal{C}_0)$ in Condition 5.2.4 holds.

6. Verifying the sufficient condition for the equation in Condition 5.2.4. If $\theta_{0,j} = 0$ and the relationship

$$p_{1,j}(z) = p_{2,j}(z) = p_{0,j}(z) , \quad \forall z \in V_5 \tag{5.48}$$

holds for all $j = 1, 2, \cdots, n'$, then $\theta_{1,j} = \theta_{2,j} = 0$ and then $G(\theta_{1,j}) = G(\theta_{2,j}) = 0$ holds. Furthermore, following from (5.48), we have $H(p_{1;j,\cdot}) = H(p_{2;j,\cdot}) = H(p_{0;j,\cdot})$. It implies that

$$H(p_{0;j,\cdot}) = \mu_1 H(p_{1;j,\cdot}) + \mu_2 H(p_{2;j,\cdot}) \ .$$

Thus,

$$H(p_{0;j,\cdot}) + G(\theta_{0,j}) = \mu_1[H(p_{1;j,\cdot}) + G(\theta_{1,j})] + \mu_2[H(p_{2;j,\cdot}) + G(\theta_{2,j})] \ . \tag{5.49}$$

If we sum over j on both sides of (5.49), then we have

$$w_S(\mathcal{C}_0) = \mu_1 w_S(\mathcal{C}_1) + \mu_2 w_s(\mathcal{C}_2) \ ,$$

showing that the equality in expression (5.28) holds.

7. Verifying the necessary condition for the equation in Condition 5.2.4. If the equal sign in expression (5.28) holds and $m_1, m_2 > 0$, then we have $\theta_{0,j} = \theta_{1,j} = \theta_{2,j} = 0$ and (5.48) hold. Since $G(\theta)$ is a strictly monotonically increasing function and $\theta_{0,j} = \theta_{1,j} + \theta_{2,j}$, it follows that
$$G(\theta_{0,j}) \geq \mu_1 G(\theta_{1,j}) + \mu_2 G(\theta_{2,j})$$
holds. Furthermore, the equality holds if and only if $\theta_{0,j} = 0$. On the other hand, following from the strictly convex property of function $H(p_{0;j,\cdot})$, we have
$$H(p_{0;j,\cdot}) \geq \mu_1 H(p_{1;j,\cdot}) + \mu_2 H(p_{2;j,\cdot}) \ .$$
The equality holds if and only if expression (5.48) is true. If the equal sign in expression (5.28) holds and $m_1, m_2 > 0$, then $\theta_{0,j} = 0$ and expression (5.48) holds. In conclusion, the function $w_S(\mathcal{C})$ satisfies Condition 5.2.4.
8. Verifying Conditions 5.2.5–5.2.7. Since Condition 5.2.7 can be directly verified using the definition of the penalty function, we only check that Conditions 5.2.5 and 5.2.6 hold. For Condition 5.2.6, we may assume that the jth row of $\mathcal{C}$ is such that all the elements are "$-$" in the form
$$\begin{pmatrix} c_{1,j} \\ c_{2,j} \\ \vdots \\ c_{m,j} \end{pmatrix} = \begin{pmatrix} - \\ - \\ \vdots \\ - \end{pmatrix} .$$
We obtain a new multiple expansion $\mathcal{C}'$ by deleting this purely "$-$" column from $\mathcal{C}$, and then we have $C_S(\mathcal{C}) > C_S(\mathcal{C}')$. Therefore, $\mathcal{C}$ is definitely not the minimum penalty alignment, and Condition 5.2.6 holds. Since verifying Condition 5.2.5 is a long process, we do this in the next step.
9. For verifying Condition 5.2.5, on the one hand, we begin by calculating $HG(c_{1,j}, c_{2,j})$ defined in (5.37) in the case $m = 2$. We then have the following subcases:
   (a) If $(c_{1,j}, c_{2,j}) = (-,-)$, then $\theta_j = 2$. Therefore,
   $$H(p_{j,\cdot}) = 0 \,, \quad G(\theta_j) = G(2) \,, \quad HG(c_{1,j}, c_{2,j}) = G(2) \ .$$
   (b) If $(c_{1,j}, c_{2,j}) = (-, c)\ \forall c \in \{0,1,2,3\}$, then $\theta_j = 1$. Therefore,
   $$H(p_{j,\cdot}) = 0 \,, \quad G(\theta_j) = G(1) \,, \quad HG(c_{1,j}, c_{2,j}) = G(1) \ .$$
   (c) If $(c_{1,j}, c_{2,j}) = (c, c') \quad \forall c = c' \in \{0,1,2,3\}$, then $\theta_j = 0$. Therefore,
   $$H(p_{j,\cdot}) = 0 \,, \quad G(\theta_j) = G(0) = 0 \,, \quad HG(c_{1,j}, c_{2,j}) = 0 \ .$$
   (d) If $(c_{1,j}, c_{2,j}) = (c, c') \quad \forall c \neq c' \in \{0,1,2,3\}$, then $\theta_j = 0, p_{j,0} = p_{j,1} = 1/2$. Therefore,
   $$H(p_{j,\cdot}) = 1 \,, \quad G(\theta_j) = G(0) = 0 \,, \quad HG(c_{1,j}, c_{2,j}) = 1 \ .$$

As a result, we find the penalty matrix of $HG(c,c'), \forall c,c' \in V_5$ as follows:

$$w(c,c') = \begin{pmatrix} & a & c & g & u & - \\ a & 0 & 1 & 1 & 1 & G(1) \\ c & 1 & 0 & 1 & 1 & G(1) \\ g & 1 & 1 & 0 & 1 & G(1) \\ u & 1 & 1 & 1 & 0 & G(1) \\ - & G(1) & G(1) & G(1) & G(1) & G(2) \end{pmatrix} . \tag{5.50}$$

On the other hand, to get the penalty matrix for multiple sequences, we choose a function $G(\theta)$ such that $G(2) \geq G(1) \geq 1$. The penalty matrix $w(c,c')$ coincides with the generalized Hamming penalty matrix that is commonly used for pairwise alignment. This ends the proof of this theorem.

Discussion of the Converse Theorem 24

In Theorem 24, we proved that the information-based criterion satisfies Conditions 5.2.1–5.2.7 of the penalty function. We now consider the inverse proposition: what kind of conditions will imply the information-based function defined in (5.38). To solve this problem, we use the definition and properties of Shannon entropy.

Condition 5.2.8 The penalty function $w(\mathcal{C})$ is formed by HG defined by (5.37), $H(p_0, \cdots, p_4)$ is a continuous function of $(p_1, \cdots, p_4)$, and $G(\theta_j)$ is a strictly monotonically increasing function with $G(0) = 0$.

Condition 5.2.9 If $\mathcal{C}_0 = \mathcal{C}_1 \otimes \mathcal{C}_2$ is defined as in Condition 5.2.4, and function $H(\cdot)$ satisfies:

$$H(p_{0;j,\cdot}) = H(\mu_1) + \mu_1 H(p_{1;j\cdot}) + \mu_2 H(p_{2;j,\cdot}) . \tag{5.51}$$

where $h(p) = -p \log p - (1-p)\log(1-p)$.

Theorem 25. *If the penalty function $w(\mathcal{C})$ satisfies Conditions 5.2.1–5.2.3, 5.2.8, and 5.2.9, then $w(\mathcal{C})$ is definitely the information-based penalty function defined by (5.37) and (5.38).*

The proof of this theorem is detailed in many informatics books, for example, [23,88], etc. Therefore, we omit it here and refer the reader to other literature sources.

5.2.5 The Similarity Rate and the Rate of Virtual Symbols

Problems of the SP-Penalty (or Scoring) Function and the Information-Based Penalty Function

In previous sections, we defined two important penalty functions: the SP-penalty function and the information-based penalty function, which are frequently used to study MA. We also discussed their roles in the optimal analysis. However, these discussions were not in-depth enough for further study. We must study these two functions with respect to the following:

1. The comparability of the minimum penalty solution must be solved. In other words, we are unable to show a difference between the optimal solution and the minimum penalty solution based on these two functions.
2. The rate of virtual symbols proportional to the length of a sequence. Based on the results of MA, the optimization index for MA often involves the rate of virtual symbols, which will be defined later. The value of the SP-penalty function or the information-based function increases as the rate of virtual symbols increases. Conversely, the value of the SP-penalty function or information-based function decreases as the rate of the virtual symbols decreases. Determining the exact relationship between the rate of virtual symbols and the value of the penalty function is the problem to be discussed.
3. These two functions are unable to construct an optimally fast alignment. Therefore, the optimization criteria of MA similar to the fast MA still need to be discussed further. In this subsection, we focus on finding more optimization indices of MA besides the SP-penalty (or scoring) function and the information-based penalty function.

Similarity Rate

Let $\mathcal{A}$ be a given multiple sequence, so that we may obtain the minimum penalty matrix $\mathcal{B} = (B_{s,t})$ based on $\mathcal{A}$, and the output $\mathcal{C} = \{C_1, C_2, \cdots, C_m\}$. Based on these three elements, we have the following results:

1. A scoring matrix $W = (w_{s,t})_{s,t=1,2,\cdots,m}$ is induced by the matrix $\mathcal{B} = (B_{s,t})$ in the natural way: $w_{s,t} = w(B_{s,t}, B_{t,s})$.
2. A scoring matrix of MA $W' = (w'_{s,t})_{s,t=1,2,\cdots,m}$ is induced by result $\mathcal{C}$ in the natural way: $w'_{s,t} = w(C_s, C_t)$.

We then define the similarity rate as follows:

$$R(\mathcal{C}) = \frac{1}{m(m-1)} \sum_{s=1}^{m} \sum_{t \neq s} \frac{w'_{s,t}}{w_{s,t}} . \tag{5.52}$$

Since $w_{s,t}$ is the score of the minimum penalty alignment based on A_s, A_t, we have that $w'_{s,t} \leq w_{s,t}$ always holds. Hence, $R(\mathcal{C}) \leq 1$ holds. We define $\mathcal{C}$ as the optimal (or suboptimal) alignment of $\mathcal{A}$ if $R(\mathcal{C}) = 1$ (or $R(\mathcal{C}) \sim 1$). Therefore, the similarity rate describes the closeness between the optimal alignment and the minimum penalty alignment.

Rate of Virtual Symbols

The so-called rate of virtual symbols is the proportion of all virtual symbols "$-$" (or 4) in $\mathcal{C}$, namely,

$$P(\mathcal{C}) = \frac{\text{the total of virtual symbols "}-\text{" in } \mathcal{C}}{m \times n}, \tag{5.53}$$

where m is the multiplicity of $\mathcal{C}$, and n is the length of each sequence in $\mathcal{C}$.

In conclusion, the challenge of the optimization problem of MA is how to make the value $w_{\text{SP}}(\mathcal{C})$ and the similarity ratio $R(\mathcal{C})$ as large as possible while making the rate of virtual symbols $P(\mathcal{C})$ as small as possible. Or, how to make the rate of virtual symbols as small as possible while making the value $w_{\text{SP}}(\mathcal{C})$ and the similarity rate $R(\mathcal{C})$ as large as possible.

5.3 Super Multiple Alignment

With the above principle in mind, we developed a fast algorithm for MA known as the super multiple alignment (SMA). The associated software package was also developed by the Nankai University group, and is freely available to the public on the website (see Table 5.1). Next, we introduce the relevant materials of SMA.

5.3.1 The Situation for MA

In Sect. 1.1, we introduced the general situation for the algorithms of MA, and we discuss this issue in more detail at this point.

Definition of the MA

In 1982, the pairwise alignment problem had been primarily solved as the Smith–Waterman algorithm was validated. Since then, interest has turned to the question of how to get MA and how to improve the existing pairwise alignment. Almost all bioinformatics literature such as [64] involve MA.

MA is widely used in various fields. For example, to study biological evolution, researchers analyze structural changes based on the MA of special DNA sequences or protein sequences (such as mitochondrial DNA, cytochrome, *C. intestinalis*, etc.). To study the virus genome, MA is also used to get the evolution processes of specific viruses (such as SARS, HIV-1, and various tumors) [101]. As a result, *Paguma larvata* is identified as the source of the SARS virus based on the MA of 63 SARS genome sequences. In contrast, the article [101] used pairwise alignment rather than MA, and as a result, too much information was lost.

Another feature of MA is that the sizes of both the multiplicity m and the lengths of sequences are growing rapidly as work on this problem progresses.

It is common for a MA to involve hundreds of sequences which are hundreds of million base pairs in length. For example, there are 706 HIV-1 sequences in the GenBank 2004 edition (release 43); hopefully, the total number of HIV-1 sequences in all databases combined will exceed 1000. Therefore, there is great demand for fast algorithms of MA for the analysis of these large-scale data.

Progress of MA

The earliest MA algorithm is the MA software package [56], which extended the dynamic programming-based algorithm for pairwise alignment to the multiple cases by changing the penalty matrix to the multiple penalty tensor. The computational complexity of this algorithm is $O(n^m)$, so it is hard to compute as m, n increase. As a result, the scale of this algorithm is only $(m, n) = (7, 300)$. Progress on the improvement of MA is very slow, so it does not keep pace with the exponential speed of the data growth.

After this phase, the study of MA has been developing along two directions. One is to discuss the computational complexity of the solution with minimum penalty, which many publications consider to be a very difficult problem. It was called the first open problem in biological computing in [46], while refs. [15,36,106] call it the NP-hard and Max-Snp hard problem. Hence, it is difficult to achieve MA with minimum penalty theoretically. The MA problems become problems of computational complexity, as described in these publications.

On the other hand, interest in this problem is ongoing because of the importance of MA. Many algorithms, software packages and alignment results appear in the literature one after another. For example, BLAST and FASTA are both able to perform MA. Several specialized software packages, such as CLUSTAL-W/X, BioEdit, MulAlin, GCG, Match-Box, BCM, and CINEMA, etc. are all specific algorithms for MA. The common feature of these algorithms is that they are not concerned with minimum penalty solutions, but result in an increased scale of alignment. These algorithms achieve the suboptimal solutions to some degree, and get a large return for increasing the alignment scale. The alignment scale and the performance indices are shown in Table 5.2.

With MA emerging, the question of how to judge the quality of an algorithm becomes increasingly important. The four indices given in Sect. 1.1.3, namely, the utility range, alignment size, computational speed, and optimization index, are useful when judging the quality of an algorithm. In addition, the SP-penalty function, information-based penalty function, similarity rate and the rate of virtual symbols defined in (1.9), (5.37), (5.38), (5.52), and (5.53), respectively, should also be comprehensively considered if we want to judge the quality of a MA.

Features of the SMA

The purpose of this section is to present a fast algorithm, the so-called super MA (SMA) to fit large-scale MA. Several specific features of the algorithm can be summarized here:

1. **Wide applicability**. This algorithm may still lead to good results if the homology (similarity) between the multiple sequences is only slightly larger than 50%. For instance, we may get good alignment of the DNA sequences of the mitochondria of Primates, although the sequence homology for these sequences ranges from 55 to 90%. In fact, the homology ratio approaches 1, which exceeds our expectations.
2. **Large-scale**. Generally, the computational scale of the SMA is without limitation if a super computer is used. Even running this algorithm on a PC, the size limit of $n \times m$ is beyond 20 Mbp. We may get better results if the size $m \times n$ is less than 20 Mbp and if the homology for these sequences is larger than 80%.
3. **Fast**. On a PC with a 2.8 GHz processor, the alignment of $118 \times 30{,}000$ SARS sequences, takes 21 min; while the alignment of 706×8000 bp HIV-1 sequences takes 34 min. This is much faster than other algorithms.
4. **Highly superior to other algorithms based on three indices**. We compare this algorithm with others based on the following three optimization indices: the SP-scoring function, similarity ratio and ratio of virtual symbols. This algorithm is superior to the other algorithms in all three cases.

The SMA has been published on the Nankai University website [99], and computational service is also offered there. In addition, the alignments for the SARS sequences and HIV-1 sequences are also included on the website. [1]

5.3.2 Algorithm of SMA

For a given multiple sequence $\mathcal{A}$, in order to get its MA, we must first construct an algorithm. To construct an algorithm, we begin by formulating the computational principles.

Principles of MA

Principles of MA include the following:

1. Pairwise alignment. The most popular pairwise alignment include dynamic programming-based algorithms (i.e., the Smith–Waterman algorithm) and the statistical decision-based algorithm (i.e., SPA) [69, 90, 95].

[1] http://mathbio.nankai.edu.cn/database/exe/sma/PerformanceofSMA/SarsPredictbySMA.txt;
http://mathbio.nankai.edu.cn/database/exe/sma/PerformanceofSMA/HivGeneMatchCompare/

These two kinds of algorithms are easy to compute. Using a dynamic programming-based algorithm, we get the minimum penalty alignment with computational complexity $O(n^2)$, while we may get the suboptimal alignment with the computational complexity $O(n)$ if we use statistical decision-based algorithms. Therefore, we may use the dynamic programming-based algorithms if the lengths of the sequences are less than 10 kbp.

2. Modulus structure. Let (C_s, C_t) be the alignment of (A_s, A_t); then we describe all the virtual symbols in the sequence (C_s, C_t) by a mathematical formula referred to as the modulus structure or alignment mode. The modulus structure is a set of transformations and operations detailed in [89].
3. Clustering analysis of multiple sequences $\mathcal{A}$. Using the characteristics of $\mathcal{A}$ such as length function $n_s = ||A_s||$, $s = 1, 2, \cdots, m$, the scoring matrix of pairwise alignment of $\mathcal{A}$, etc., we construct the phylogenetic tree or the minimum distance tree. Both the phylogenetic and minimum distance trees are typical clustering methods in statistics and combination graph theory [35].

Algorithm of MA

Using the principles of MA, we construct the MA as follows:

Step 5.3.1 Preprocess the relevant parameters and data:

1. Let $M' = \{\mathbf{A}_1, \mathbf{A}_2, \cdots, \mathbf{A}_{2m-1}\}$ be the set of nodes in the clustering tree, where each node $\mathbf{A}_s \in M'$ is a subset of $\mathcal{A} = \{A_1, A_2, \cdots, A_m\}$. Specifically, $\mathbf{A}_s$ is a single-point set, namely, $\mathbf{A}_s = \{A_s\}$ if $s = 1, 2, \cdots, m$, and $\mathbf{A}_s$ is a set with at least two sequences if $s > m$. In some cases, we may simply use the following form:
$$M = \{1, 2, \cdots, m\}\,, \quad M' = \{1, 2, \cdots, m'\}\,, \quad m' = 2m - 1\,.$$
2. Let $G' = \{M', V'\}$ denote the graph associated with the clustering tree, in which V' is the set of edges in the clustering tree, which will be defined later.
3. Let $w(s, t)$, $s, t \in M$ be the clustering function that may be chosen in many ways, as follows:
   (a) If C_s, C_t is the minimum penalty alignment of A_s, A_t, then choose $w(s, t) = w(C_s, C_t)$.
   (b) Let C_s, C_t be the minimum penalty alignment of A_s, A_t, and let $n(C_s, C_t)$ be the total number of the virtual symbols in C_s, C_t. We choose $w(s, t) = n(C_s, C_t)$.
   (c) If the sequences A_s, A_t are not the same length, we choose $w(s, t) = |n_a - n_t|$.

We now only show the algorithm based on the choice of Step 5.3.1, procedure 3a, leaving analysis of the remaining cases up to the reader.

Step 5.3.2 With the notations defined in Step 5.3.1, we plant the clustering tree based on the multiple sequence $\mathcal{A} = \{A_1, A_2, \cdots, A_m\}$ as follows:

1. Let $M^{(k)} = \{s_1, s_2, \cdots, s_{m-k+1}\} \subset M'$ be the set of states at the kth clustering. It then satisfies the following conditions:
   (a) Each node s_i in $M^{(k)}$ corresponds to a subset of M, denoted by $\mathbf{A}_{s_i}^{(k)}$, here $s_{m-k+1} = m + k$.
   (b) $M^{(1)} = M = \{1, 2, \cdots, m\}$ is the set of states at the initial clustering. Thus, each node s corresponds to a single-point set $\{A_s\}$ if $s \le m$; and it corresponds to a set $\mathbf{A}_s$ with at least two points if $s > m$.
   (c) All the points of $M^{(k)}$ comprise a division of M. In other words, these subsets are mutually disjoint, and the union of them is M.
2. If the $M^{(k)}$ is found, we calculate

$$w_{s,t}^{(k)} = \min\left\{w(s', t'),\ s' \in \mathbf{A}_s^{(k)},\ t' \in \mathbf{A}_t^{(k)}\right\}, \quad s \neq t \in M^{(k)} . \tag{5.54}$$

Let $s'_0 \in \mathbf{A}_s^{(k)}$, $t'_0 \in \mathbf{A}_t^{(k)}$ be the pair of points satisfying $w(s'_0, t'_0) = w_{s,t}^{(k)}$, and let the pair s'_0, t'_0 be the closest nodes within $\mathbf{A}_s^{(k)}$ and $\mathbf{A}_t^{(k)}$. If there is a pair $s_0, t_0 \in M^{(k)}$ such that

$$w_{s_0,t_0}^{(k)} = \min\left\{w^{(k)}(s, t),\ s, t \in M^{(k)}\right\} , \tag{5.55}$$

then the set $M^{(k+1)}$ at the $(k + 1)$th clustering is defined by: Let $\mathbf{A}_{m+k}$ denote the union of $\mathbf{A}_{s_0}^{(k)}$ and $\mathbf{A}_{t_0}^{(k)}$, and keep the rest of the nodes invariant. Then, $(s_0, m + k), (t_0, m + k)$ are two edges on the clustering tree G', and $m + k$ is the clustering point of s_0, t_0.
3. Continuing this procedure, we may get the structure for each point of M' defined in Step 5.3.1, and we may also get all the edges in graph G' defined by Step 5.3.2, procedure 2. Finally, we may find the graph of clustering tree G'.

Step 5.3.3 Based on the clustering tree $G' = \{M', V'\}$ obtained by Steps 5.3.1 and 5.3.2, we construct the MA of $\mathcal{A}$ as follows. If r is the clustering point of s, t, then s, t correspond to the union of sets

$$\mathbf{A}_s = \{A_{s,1}, A_{s,2}, \cdots, A_{s,p_s}\}, \quad \mathbf{A}_t = \{A_{t,1}, A_{t,2}, \cdots, A_{t,p_t}\} , \tag{5.56}$$

in which $\mathbf{A}_r = \mathbf{A}_s \cup \mathbf{A}_t$, $\mathbf{A}_s \cap \mathbf{A}_t = \emptyset$, and $\mathbf{A}_s, \mathbf{A}_t$ both are subsets of $\mathcal{A}$. If we found the MA for $\mathbf{A}_s$ and $\mathbf{A}_t$, respectively, then we construct the MA for $\mathbf{A}_r$ in the following way:

1. Let

$$C_s = \{C_{s,1}, C_{s,2}, \cdots, C_{s,p_s}\}, \quad C_t = \{C_{t,1}, C_{t,2}, \cdots, C_{t,p_t}\} \tag{5.57}$$

be the MA for A_s and A_t, respectively, and let

$$H_s = \{H_{s,1}, H_{s,2}, \cdots, H_{s,p_s}\}, \quad H_t = \{H_{t,1}, H_{t,2}, \cdots, H_{t,p_t}\} \tag{5.58}$$

be the expanded modes that A_s, A_t mutates to C_s, C_t, respectively.

2. To cluster, let s', t' be the closest nodes within sets $\mathbf{A}_s$ and $\mathbf{A}_t$, then $A_{s'}, A_{t'} \in \mathcal{A}$. Let $(C_{s'}, C_{t'})$ be the pairwise alignment of $(A_{s'}, A_{t'})$, and let $(H_{s'}, H_{t'})$ be the corresponding expanded mode such that $(A_{s'}, A_{t'})$ mutates to $(C_{s'}, C_{t'})$.
3. Constructing the union modes based on H_s, H_t defined in (5.58) and $(H_{s'}, H_{t'})$ defined in Step 3.5.3, procedure 2, we have two modes as follows:

$$\begin{cases} H_s \vee H_{s'} = \{H_{s,1} \vee H_{s'}, H_{s,2} \vee H_{s'}, \cdots, H_{s,p_s} \vee H_{s'}\}, \\ H_t \vee H_{t'} = \{H_{t,1} \vee H_{t'}, H_{t,2} \vee H_{t'}, \cdots, H_{t,p_t} \vee H_{t'}\}. \end{cases} \tag{5.59}$$

Furthermore, we construct the new mode

$$H_r = H_s \vee H_{s'} \cup H_t \vee H_{t'}. \tag{5.60}$$

This H_r is then the expanded mode by which multiple sequences $\mathbf{A}_r$ mutate to C_r.

Step 5.3.4 Repeating Step 5.3.3 for each clustering point on the tree G' defined by Steps 5.3.1 and 5.3.2, we calculate the MA of each $\mathbf{A}_r$, and finally find the alignment $\mathcal{C}$ of the multiple sequence $\mathcal{A}$.

Step 5.3.5 Generally, the MA $\mathcal{C}$ obtained by Steps 5.3.1–5.3.4 is a suboptimal solution. In order to improve the optimization index of MA, we continue to align $\mathcal{C}$ through the following steps:

1. For each given $s' \in \{1, 2, \cdots, m\}$, let

$$\mathcal{C}_{s'} = \{C_1, C_2, \cdots, C_{s'-1}, C_{s'+1}, \cdots, C_m\}. \tag{5.61}$$

This is a sequence with multiplicity $(m-1)$, where the general form of the component is represented as follows:

$$C_s = (c_{s,1}, c_{s,2}, \cdots, c_{s,n_c}), \tag{5.62}$$

where n_c is the common length for all components. Next, let $M_{s'} = \{1, 2, \cdots, s'-1, s'+1, \cdots, m\}$ denote the set of subscripts of $\mathcal{C}_{s'}$, so that it is a $(m-1)$-ary set.

2. For each column in $\mathcal{C}_{s'}$, calculate its frequency distribution: $\bar{f}_j = (f_{j,c},\ c \in V_{q+1})$, in which, $f_{j,c}$ is the number of the elements of $\bar{c}_{s',j}$ whose value is c. Then, the transpose of this column $\bar{c}_{s',j}$ is

$$\bar{c}^T_{s',j} = (c_{1,j}, c_{2,j}, \cdots, c_{s'-1,j}, c_{s'+1,j}, \cdots, c_{m,j}). \tag{5.63}$$

The SP-penalty function of $\mathcal{C}_{s'}$ is

$$w_{\mathrm{SP}}(\mathcal{C}_{s'}) = \sum_{s<t \in M_{s'}} \sum_{j=1}^{n_c} w(c_{s,j}, c_{t,j}) = \frac{1}{2} \sum_{j=1}^{n_c} \sum_{c \neq c' \in V_{q+1}} f_{j,c} f_{j,c'} w(c, c') \tag{5.64}$$

and the SP-penalty functions of $\mathcal{C}_{s'}$ and $\mathcal{C}$ satisfy the following relationship:

$$w_{\mathrm{SP}}(\mathcal{C}) = \sum_{s<t \in M} \sum_{j=1}^{n_c} w(c_{s,j}, c_{t,j}) = w_{\mathrm{SP}}(\mathcal{C}_{s'}) + \sum_{t=1}^{m} \sum_{j=1}^{n_c} w(c_{s',j}, c_{t,j}) . \tag{5.65}$$

Let $w_{\mathrm{SP}}(C_{s'}, \mathcal{C}_{s'}) = \sum_{t=1}^{m} \sum_{j=1}^{n_c} w(c_{s',j}, c_{t,j})$ and choose the $s_0 \in M$ such that

$$w_{\mathrm{SP}}(C_{s_0}, \mathcal{C}_{s_0}) = \max\{w_{\mathrm{SP}}(C_{s'}, \mathcal{C}_{s'}) , \ s' \in M\} . \tag{5.66}$$

3. Delete these columns of $\mathcal{C}_{s'}$ if they are purely "$-$" and let $\mathcal{C}'_{s'}$ denote the rest of the multiple sequence. If $C'_{s'} = (c'_{s',1}, c'_{s',2}, \cdots, c'_{s',n'})$ is the expansion of $A_{s'}$, we define the penalty function of $C_{s'}$ and $\mathcal{C}'_{s'}$ as follows:

$$w(C'_{s'}, \mathcal{C}'_{s'}) = \sum_{t=1}^{m} \sum_{j=1}^{n'_c} w\left(c'_{s',j}, c'_{t,j}\right) , \tag{5.67}$$

in which $n'_c = \max\{n', n_c\}$.
4. Compute the alignment of A_{s_0} and $\mathcal{C}'_{s_0}$ under the penalty function in (5.67) with the dynamic programming-based algorithm. Let $\mathcal{C}''$ be the output, then $\mathcal{C}''$ is united by C''_{s_0} and $\mathcal{C}''_{s_0}$, where C''_{s_0} is the expansion of A_{s_0}, and $\mathcal{C}''_{s_0}$ is the expansion of $\mathcal{C}'_{s_0}$ by inserting an $(m-1)$-dimensional vector consisting of "$-$". According to (5.67), we can get the corresponding penalty matrix:

$$w(c, \bar{c}) = \begin{cases} \sum_{c'=0}^{4} f_{c'} w(c, c') , & \text{if } \bar{c}' \text{ is a column vector in } \mathcal{C}'_{s_0} , \\ m-1 , & \text{if } \bar{c}' \text{ is an } (m-1)\text{-dimensional vector} \\ & \text{filled by virtual symbols, and } c' \neq 4 , \\ 0 , & \text{if } c' = 4, \text{ and } \bar{c}' \text{ is an } (m-1)\text{-dimen-} \\ & \text{sional vector filled by virtual symbols} . \end{cases} \tag{5.68}$$

Under this penalty matrix, we may prove that $\mathcal{C}''$ is the optimal alignment of sequence A_{s_0} and $\mathcal{C}'_{s_0}$, and

$$w_{\mathrm{SP}}(\mathcal{C}) \geq w_{\mathrm{SP}}(\mathcal{C}'') . \tag{5.69}$$

5. Repeating Step 5.3.5, we continue until the SP-penalty score can no longer be reduced.

Remark 3. The above steps form just the outline for the SMA. It still needs to be adjusted according to specific cases of multiple sequences if we are constructing a program.

Table 5.1. Comparison of the size of multiple alignment

Software package or name of algorithm	Multiplicity restriction	Length restriction	Web page
SMA	No	No	http://mathbio.nankai.edu.cn /eversion/align-query.php
HMMER	No	No	http://hmmer.janelia.org/
POA	No	< 1 kbp	http://www.bioinformatics.ucla.edu/poa
MLAGAN	< 31	Unrestricted	http://genome.lbl.gov/vista /lagan/submit.shtml
ClustalW 1.8	< 500	Unrestricted	http://www.ebi.ac.uk/clustalw/
MuAlin	< 80	< 20 kbp	http://bioinfo.genopole-toulouse.prd.fr /multalin/multalin.html
MSA	< 8	< 800 bp	http://searchlauncher.bcm.tmc.edu /multi-align/multi-align.html
Match-Box	< 50	< 2 kbp	http://searchlauncher.bcm.tmc.edu /multi-align/multi-align.html

Table 5.2. Comparison of the optimization indices

Name of sequence	Scale of alignment	Software package or algorithm	CPU time (min)	SP-score	Similarity rate (%)	Rate of virtual symbols (%)
SARS	118×30 kbp	ClustalW 1.8	4740	9.7×10^7	99.97	0.40
SARS	Same	HMMER 2.2	381	1.0×10^8	99.93	0.47
SARS	Same	SMA	21	1.0×10^8	99.99	0.53
HIV1	706×10 kbp	HMMER 2.2	256	1.65×10^9	98.03	49.13
HIV1	Same	SMA	34	1.68×10^9	98.58	31.23

5.3.3 Comparison Among Several Algorithms

To show how well the SMA performs, we compare it with some popular MA with respect to the indices listed in Tables 5.1 and 5.2.

Remark 4. CPU time is defined as the time required for a PC with a 2.8 GHz processor to compute. The results in Tables 5.1 and 5.2 were obtained in 2004.

Following from Tables 5.1 and 5.2, we draw the following conclusions:

1. For size, SMA is the same as the HMMER 2.2 algorithm, but is far superior to other algorithms.

2. For speed, SMA is 8–18 times faster than the HMMER 2.2 algorithm, as well as 230 times faster than the ClustalW 1.8 algorithm.
3. For the SP-score index and similarity ratio index, SMA is slightly better than the HMMER 2.2 and ClustalW 1.8 algorithms.
4. For the ratio of virtual symbols index, SMA is far superior to the HMMER 2.2 algorithm if we consider the case of HIV1 because its rate of virtual symbols is less 18%. SMA is slightly inferior to HMMER 2.2 algorithm and ClustalW 1.8 if we use SARS as the benchmark set.

As a result we conclude that SMA is generally superior to other MA in terms of size, CPU time, similarity rate and rate of virtual symbols since HMMER 2.2 and ClustalW 1.8 are both the best among existing MA.

5.4 Exercises, Analyses, and Computation

Exercise 21. The metric relations distance, measurement (or probability), and uncertainty are frequently used in mathematics. Compare them, focusing on the aspects of content, definition and difference. For example:

1. Write down the objects they act upon.
2. Construct the basic requirements (axiom system) for these metrics.
3. Write down the expressions of these metrics (i.e., the formula).
4. Write down the definitions of these metrics and indicate in which fields they tend to be applied.

Exercise 22. Check whether or not the SP-penalty functions satisfy Conditions 5.2.1–5.2.7.

Exercise 23. Check whether or not the criterion of similarity rate satisfies Conditions 5.2.1–5.2.7.

Exercise 24. Download the data sets of SARS and HIV-1 from the Web [99], obtain the pairwise alignment using a dynamic programming-based algorithm and the SPA algorithm, respectively, and then analyze the results based on CPU time for pairwise alignment. Compute the matrix consisting of similarity rates based on the Hamming matrix.

Exercise 25. Download the ClustalW algorithm for MA from the Web [22]. Input the SARS and HIV-1 sequences, and compute the alignment output.

Exercise 26. According to the steps in Sect. 5.3, develop a program to obtain the SMA algorithm, and align the SARS and HIV-1 sequences.

Exercise 27. Prove that the expansion C_r obtained by Step 5.3.3 is just the alignment of the multiple sequence A_r.

Exercise 28. Continue Exercises 22 and 23 to analyze MA outputs for SARS and HIV-1 according to the following indices:

1. CPU time and rate of virtual symbols.
2. The SP-penalty function, the information-based function.

Hints

For SMA, we suggest that the reader write a program satisfying the steps presented in Sect. 5.2. If this proves too difficult, the reader may use the algorithm given on the Nankai University website [99], and then try to develop a program independently.

6

Network Structures of Multiple Sequences Induced by Mutation

As fast multiple alignment (MA) algorithms become a reality, analysis and application of their results becomes the central problem of genome research. In this book, we discuss the network structure theory of the multi-sequences induced by mutations.

6.1 General Method of Constructing the Phylogenetic Tree

6.1.1 Summary

One of the main purposes of making multiple alignments is to construct the phylogenetic tree. Looking at the MA results, we find that it is a set of sequences of the same length. If the result is correct, then this output is a kind of family file of these multiple sequences, containing all the connections among this family and the phylogenetic information on this family. Based on this family file, we may determine the evolutionary state of each sequence in this family. Generally, we use a topological tree to describe the connection among the multiple sequences, which is called a phylogenetic tree.

Tree is a class of spacial point-line graphs. The point-line graph is given by $G = \{M, V\}$, where $M = \{1, 2, \cdots, m\}$ are the points of the graph, and V is the set of all pairs of points in M. Each pair in V is seen as an arc. A point-line graph $G = \{M, V\}$ is called an undirected graph if the pairs (s, t), $(t, s) \in M$ are the same. Otherwise, it is a directed graph. These two types of point-line graphs will frequently appear in the following text. The point-line graphs theory is considered in many books, and it is not discussed further in this book.

There are many methods for constructing the phylogenetic tree. We will introduce these methods in this section as follows:

1. Distance-based methods (e.g., neighbor-joining). Any alignment result may be used to compute a distance matrix between these sequences. Based

on this distance matrix, we may produce the corresponding phylogenetic tree. The most popular methods are called UPGMA and neighbor-joining.
2. Feature-based methods (e.g., maximum parsimony method). This kind of method uses the features (characteristics) of the alignment outputs to construct the phylogenetic tree.
3. Probability-based methods (e.g., maximum-likelihood method and Bayes method). Using these methods to construct the phylogenetic tree, we should begin by constructing a probability model for the sequence mutation, and then construct the phylogenetic tree based on both the output and the probability model.

6.1.2 Distance-Based Methods

There are many distance-based methods for constructing the phylogenetic tree, and we only introduce two of these in this subsection, namely, UPGMA and neighbor-joining.

Unweighted Pair Group Method with Arithmetic Mean

Unweighted pair group method with arithmetic mean (UPGMA) [63,96] is the simplest of all clustering methods used to construct a phylogenetic tree. This method requires that the substitution velocity of the nucleotides or amino acids be uniform and unchanging through the entire evolution process. In other words, the molecular clock hypothesis holds. At each parent node, the branch lengths from the parent node to the two child nodes are the same.

The most intuitive clustering method used to construct the phylogenetic tree is the system clustering method. This method assembles the two nearest classes to a new class, into a cluster each time, until all the classes are assembled into one class. The algorithm is trivially developed by following the steps listed below:

1. Given an n-multiple nucleotide sequence or amino acid sequence, choose a distance function (e.g., using the Hamming distance function) and compute the evolution distance for every pair of sequences based on their pairwise alignment result, producing a distance matrix.
2. Regard each sequence as a class, then use the n initial classes as the leaf-nodes of the phylogenetic tree.
3. Using the distance matrix, search the two classes X, Y that are nearest, and then assemble X, Y into a new class Z, which is then the parent node of X, Y. The distances from node Z to X and to Y (that is, the branch lengths from Z to X and to Y) are the same, and equal to $d(X, Y)/2$. The total number of classes is then $n-1$.
4. Compute the distances from the new node Z to other nodes. Let K be the query node for the distance to be computed from K to Z. Since $d(X, K)$

and $d(Y, K)$ are collected in the distance matrix, we compute the distance $d(Z, K)$ by one of the following ways:

$$\begin{cases} d(Z,K) = \min\{d(X,K), d(Y,K)\} , \\ d(Z,K) = \max\{d(X,K), d(Y,K)\} , \\ d(Z,K) = (d(X,K) + d(Y,K))/2 . \end{cases}$$

We then find a new distance matrix.

5. Repeat steps 3 and 4 until all the classes are assembled into one. This clustering method is easy to operate. In fact, this procedure is simply a MA process, and the result involves making MA using the pairwise alignment algorithm, based on the multiple sequences.

UPGMA is used to construct the phylogenetic tree in a way similar to the system clustering method, the main difference being the formula used to compute the distance of classes. Using step 4 above to compute the distance between two classes, if the numbers of the sequences in the two classes are different, we have to compute the distance from the new cluster to all other clusters as a weighted average of the distances from its components:

$$d(Z,K) = \frac{n_X}{n_X + n_Y} D(X,K) + \frac{n_Y}{n_X + n_Y} D(Y,K) ,$$

where n_X and n_Y are the number of sequences in X and Y, respectively.

The Neighbor-Joining Method

The neighbor-joining method [81] is a distance-based method used to construct a phylogenetic tree. This method does not depend on the molecular clock hypothesis, and it can process large-size sequences quickly. It has therefore been a popular method for constructing phylogenetic trees up to now.

Neighbor-joining is also a clustering method. We can prove that the summation of all the branch lengths in the phylogenetic tree generated by this method is the smallest. The phylogenetic tree with the smallest sum of branch lengths is not unique, but this method produces only one.

The neighbor-joining method starts from a starlike structure, and collects all "neighbors" together to form a tree without roots as the output. For a set of N sequences, the computing steps are given as follows:

1. Compute the distance matrix of the N sequences with respect to some chosen metric.
2. Regarding each sequence as a node, the initial topological structure is starlike, as in the schematic representation shown in Fig. 6.1a.
3. For an arbitrary pair of nodes, we compute the sum of all branch lengths if we combine this pair of nodes as a new node. Let D_{ij} be the distance between sequences i and j, and this distance can be obtained from step 1;

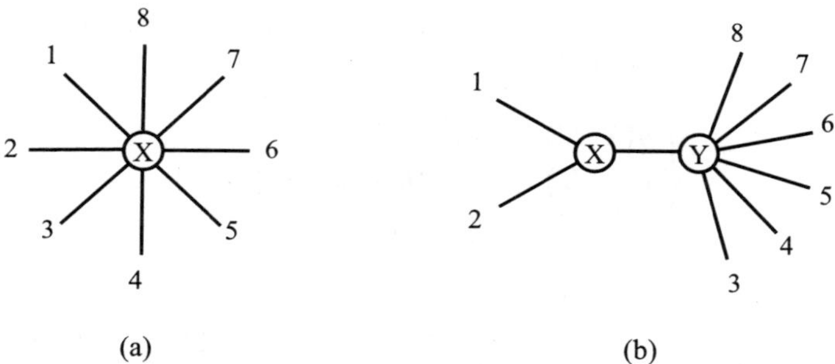

Fig. 6.1a,b. Neighbor-joining. **a** Initial starlike structure. **b** Treelike structure after nodes 1 and 2 have been joined. (From [81])

L_{ab} is the length between node a and node b, then the sum of the branch lengths of the starlike structure (Fig. 6.1a) is defined as follows:

$$S_0 = \sum_{i=1}^{N} L_{iX} = \frac{1}{N-1} \sum_{i<j} D_{ij} \,, \tag{6.1}$$

where X is the only inner node at the center of the starlike structure. The $\frac{1}{N-1}$ in formula (6.1) is due to the fact that each edge is counted $N-1$ times. We may assume that nodes 1 and 2 are joined. As in Fig. 6.1b, nodes 1 and 2 are seen as one class, and the other nodes as another class. The inner nodes are X and Y and the branch length L_{XY} between X and Y is defined by

$$L_{XY} = \frac{1}{2(N-2)} \left(\sum_{k=3}^{N} (D_{1k} + D_{2k}) - (N-2)(L_{1X} + L_{2X}) - 2 \sum_{k=3}^{N} L_{iY} \right) \,, \tag{6.2}$$

where the first term in parentheses is the sum of the lengths from the other nodes to nodes 1 and 2. The latter two terms are irrelevant to L_{XY} and should be subtracted because L_{XY} is counted $2(N-2)$ times in the first term in parentheses. Following Fig. 6.1b and the definition of the branch length, we have

$$L_{1X} + L_{2X} = D_{12} \,, \quad \sum_{k=3}^{N} L_{iY} = \frac{1}{N-3} \sum_{3 \le i < j} D_{ij} \tag{6.3}$$

and

$$S_{12} = L_{XY} + (L_{1X} + L_{2X}) + \sum_{k=3}^{N} L_{iY} \,. \tag{6.4}$$

Making use of (6.2) and (6.3), we have

$$S_{12} = \frac{1}{2(N-2)} \sum_{k=3}^{N} (D_{1k} + D_{2k}) + \frac{1}{2} D_{12} + \frac{1}{N-2} \sum_{3 \le i < j} D_{ij} \,, \tag{6.5}$$

in which D_{ij} are known. Therefore, following from (6.5), we may compute the sum of the branch lengths if nodes 1 and 2 are joined. Similarly, if an arbitrary pair of nodes are joined, we can compute the corresponding sum of the branch lengths.

4. Compare all sums of the branch lengths obtained in step 3, and choose this pair of nodes as the "neighbor" in case it minimizes the sum of branch lengths. We the find the topological structure shown in Fig. 6.1b if nodes 1 and 2 are joined. The branch lengths L_{1X} and L_{2X} are then computed as follows:

$$L_{1X} = (D_{12} + D_{1Z} - D_{2Z})/2\,, \quad L_{2X} = (D_{12} + D_{2Z} - D_{1Z})/2\,, \quad (6.6)$$

in which $D_{1Z} = \sum_{i=3}^{N} D_{1i}/(N-2)$ and $D_{2Z} = \sum_{i=3}^{N} D_{2i}/(N-2)$.

5. Compute the distance between the new node and other nodes. We may again assume that the new node is joined by nodes 1 and 2, and the distance between the new node and the jth old node is defined as

$$D_{(1-2)j} = (D_{1j} + D_{2j})/2\,, \quad j = 3, 4, \cdots, 8\,. \quad (6.7)$$

Therefore, the total number of outer nodes decreases from N to $N-1$, and inner nodes increase from 1 to 2.

6. Repeat steps 3–5 until the inner nodes become $N-3$. We then have a tree without a root, as required. To help the reader understand this method more easily, we give an example to illustrate how to use the neighbor-joining method to construct a phylogenetic tree.

Example 20. Let the distance matrix of the five species A, B, C, D, and E be

$$\begin{pmatrix} & A & B & C & D \\ B & 7 & & & \\ C & 8 & 5 & & \\ D & 11 & 8 & 5 & \\ E & 13 & 10 & 7 & 8 \end{pmatrix}.$$

We construct its phylogenetic tree using the neighbor-joining method.

Let us compute the sum of all branch lengths when two nodes are joined using formula (6.4). Then

$$\begin{pmatrix} (S) & A & B & C & D \\ B & 19.33 & & & \\ C & 20.67 & 20.67 & & \\ D & 21.00 & 21.00 & 20.33 & \\ E & 21.00 & 21.00 & 20.33 & 19.67 \end{pmatrix}.$$

From this matrix, we find that $S_{AB} = 19.33$ is a minimum. Thus, A and B are "neighbors" and we join A and B as a class, and then add an inner node X. The topological structure of the tree is shown in Fig. 6.2b. Using formulas (6.5) and (6.6), we find that L_{AX} and L_{BX} are 5 and 2, respectively. There are

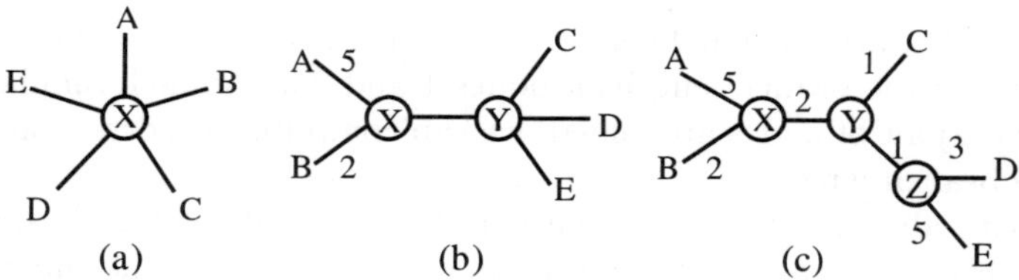

Fig. 6.2a–c. Constructing the phylogenetic tree using the neighbor-joining method. **a** Initial starlike structure. **b** Treelike structure after nodes A and B joined. **c** The complete tree without a root

then two inner nodes, so we continue the procedure. Following from formula (6.7) we find a new distance matrix as follows:

$$\begin{pmatrix} & A-B & C & D \\ C & 6.5 & & \\ D & 9.5 & 5 & \\ E & 11.5 & 7 & 8 \end{pmatrix} .$$

Repeating the above process, we obtain a new matrix of the sums of branch lengths:

$$\begin{pmatrix} (S) & A-B & C & D \\ C & 15.5 & & \\ D & 16 & 16 & \\ E & 16 & 16 & 15.5 \end{pmatrix} .$$

From the above matrix, we find that the sum of the branch lengths when $A-B$ and C are "neighbors" is the same as when D and E are "neighbors". If $A-B$ is seen as a node, then the topological structures of the trees for both cases are the same. Thus, let $A-B$ and C be "neighbors", and add a new inner node Z. The tree then has three inner nodes, and the minimum distance tree appears. Following from formulas (6.5) and (6.6), we get $L_{(A-B)Y} = 5.5$, $L_{CY} = 1$, $L_{DZ} = 3$ and $L_{EZ} = 5$. Furthermore, the lengths of the other branches are computed as:

$$L_{XY} = L_{(A-B)Y} - (L_{AX} + L_{BX})/2 = 5.5 - 3.5 = 2 ,$$
$$L_{YZ} = L_{CD} - L_{CY} - L_{DZ} = 1 .$$

This ends the procedure to construct the phylogenetic tree; the process is shown in Fig. 6.2.

6.1.3 Feature-Based (Maximum Parsimony) Methods

Feature-based methods often use the discrete features of data, for example, using alignment outputs for DNA or protein sequences to construct the phylogenetic tree. The most popular method is the maximum parsimony method, which uses features of DNA sequences to construct the phylogenetic tree.

These features of DNA sequences include the positions where the nucleotides differ. For positions where the nucleotides are the same for all sequences, the position does not join to construct the required phylogenetic tree if we use feature-based methods. However, they do join to construct the tree if we use distance-based methods. This is a major difference between feature-based and distance-based methods.

The Outline of the Maximum Parsimony Method

1. Perform the MA of the given multiple sequences, and obtain an output in which every sequence has the same length.
2. Based on the alignment output, we look for the informative positions. A position is defined as the informative position if at least two kinds of nucleotides occur with a high frequency in the column corresponding to this position. Otherwise, this position is a noninformative position. In the following example, the fifth, seventh, and ninth positions are informative positions marked with an asterisk, and the other positions are noninformative.

	1	2	3	4	5*	6	7*	8	9*
1	*A*	*A*	*G*	*A*	*G*	*T*	*G*	*C*	*A*
2	*A*	*G*	*C*	*C*	*G*	*T*	*G*	*C*	*G*
3	*A*	*G*	*A*	*T*	*A*	*T*	*C*	*C*	*A*
4	*A*	*G*	*A*	*G*	*A*	*T*	*C*	*C*	*G*

3. Construct the maximum parsimony phylogenetic tree based on the informative positions. We begin by giving all topological structures of possible phylogenetic trees for the sequences. For each of these trees, we let the informative positions be the leaf nodes, and we then predict their parent nodes based on the information of the leaf nodes, as well as giving the statistics of the differences between nucleotides within the neighbor nodes and computing the sum of the difference of nucleotides on the whole tree, which is called the length of the tree. We choose the tree with the minimum length as the estimation of the phylogenetic tree. For the above example, these four sequences may result in three possible trees without

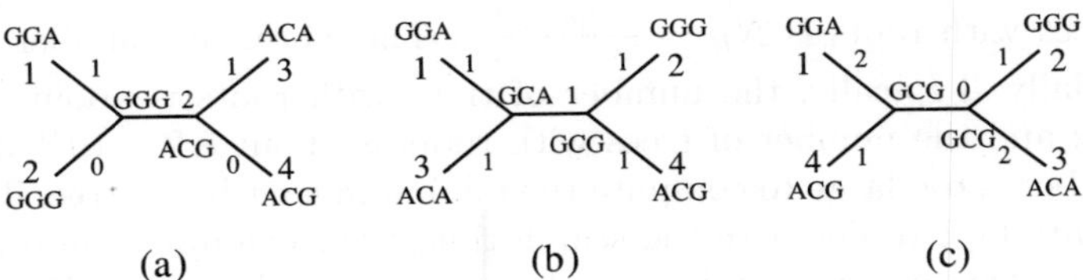

Fig. 6.3a–c. Using the maximal parsimony method to construct phylogenetic tree. **a** The topological structure of the first tree, whose length is 4. **b** The topological structure of the second tree, whose length is 5. **c** The topological structure of the third tree, whose length is 6

a root as shown as Fig. 6.3. For every possible tree, we compute the number of substitutions at the informative positions. We find the lengths of the three trees to be 4, 5, and 6, respectively. Therefore, we choose the tree in Fig. 6.3a as the estimation of the phylogenetic tree.

Calculation Using the Fitch Algorithm

In the above case, the parent nodes are easy to identify, as is the length of the tree. For the complex case where the tree has roots, then the length of the tree is calculated using the Fitch [30] algorithm as follows:

1. Give the range of each node. We define the range of the successor node as all the nucleotides occurring in the column corresponding to the successor node. For the inner nodes, the range is defined as the intersection of the ranges of the two successor nodes if it is not empty, or the union of the ranges of two successor nodes if their intersection is empty. Therefore, we may get the ranges of all the inner nodes and successor nodes.
2. Determine the value of each node. This process is opposite to the one above. We start from the value of the parent nodes to get the value of the successor nodes. For the root node, we choose an arbitrary value from its range as the value of the root node. For an inner node, if its range includes the value of its parent node, then this common value is defined as the value of this inner node. Otherwise, we select a value randomly from the range of this inner node as its value.
3. Determine the substitution times of the tree. The substitution times for the tree are defined as the total number of times the intersection set of the ranges of all the successor nodes generated in the first step is not empty.

Therefore, for a given tree with roots, we obtain the substitution times at each informative position according to the above three steps. The sum of the substitution times is the length of the tree.

We have outlined the process of constructing a phylogenetic tree using the maximum parsimony method. However, there remain some questions to be answered. First, if the number of species is too large, then the topological structures of the phylogenetic tree will generally be too high in number. For example, in trees with roots, when the number of species is $n \geq 2$, the number of trees with roots is $N_R = \frac{(2n-3)!}{2^{n-2}(n-2)!}$. Therefore, the number increases exponentially. Typically, the number of trees with roots is about 3.4×10^7 if $n = 10$; and the number of trees with roots is about 8.2×10^{21} if $n = 20$. This number is too large to compute the minimum length of a tree. Therefore, we must attempt to decrease the search times. For example, the branch and bound algorithm ensures a minimum length tree will be found. However, the time complexity of the algorithm is close to that of an exhaustive search algorithm in the worst case scenario. It is a time-consuming method. Heuristic search algorithms are another option. They highly reduce the search times but do not ensure the optimal solution will be found. Therefore, we consider using exhaustive search algorithms or the branch and bound algorithm only as long as the number of species is not excessive. It may be worth using heuristic search algorithms if the number of species is high. In addition, repeating this algorithm as the order of species changes will be helpful towards improving the quality of the result.

The second question is in regard to the probability of different nucleotide substitution in a true evolution process. For example, the number of transversion mutations is larger than the number of transition mutations in the real evolutionary process. This reminds us that the transition and transversion mutations are not equal. This question was not addressed in Fitch's algorithm, However, Sankoff's algorithm [82] offers a solution to this problem. The algorithm can deal with multiple features, and discusses the difference in probability corresponding to different features.

The maximum parsimony method uses the information on all the nucleotides at the informative positions to construct the phylogenetic tree. The advantages of this method are as follows: It uses the information on the alignment output completely. It obtains the information of ancestor sequences, and it does not show the difference between the nucleotides as is the case with the distance-based method. However, its disadvantages are also significant in that it does not use the information on the noninformative positions, its speed is much longer than that of distance-based methods, and the phylogenetic tree does not offer information about branch lengths. These weaknesses limit its applications.

6.1.4 Maximum-Likelihood Method and the Bayes Method

Among all the methods for constructing the phylogenetic tree, the maximum-likelihood method and the Bayes method are currently the most popular [2,28, 44,108,110,112]. These two methods are based on the use of probability theory to estimate the most probable topological structure of the phylogenetic tree. It allows different positions of sequences and different periods with evolution rate. It is the most credible method for constructing the phylogenetic tree. The well-known system analysis software programs PAML and MrBayes utilize the maximum-likelihood-based method and the Bayes inference-based method, respectively.

The Probability Models for Evolution

In Chap. 1, we introduced types-I, type-II, type-III, and type-IV mutations. For the conservation sequences, the probability that type-II, type-III, and type-IV happen in these sequences is small enough that we may ignore it. In other words, we will not consider this position if there is an insertion or deletion happening at this position. We consider type-II mutations to have the same effect as a type-I mutation occurring twice. For simplicity, we assume that only type-I mutations are occurring in these sequences.

We focus the discussion on DNA sequences and let the DNA sequence be of the following form:

$$A_t = (a_{t_1}, a_{t_2}, \cdots, a_{t_m}), \quad t \in R, a_{tj} \in \mathbf{Z}_4 , \tag{6.8}$$

where m is the length of the sequence. Let A_0 be the ancestor sequence and let A_t be the state that the ancestor sequence evolves to at the time t. The state at the jth position of A_t is considered a random variable ξ_{t_j}. For a given position j, the sequence $\{a_{t_j}, t \in [0, +\infty]\}$ is seen as a trail of the stochastic process $\{\xi_{t_j}, t \in [0, +\infty]\}$. We assume that the evolutions of the sequence are independent; in other words, that ξ_{t_j} is independent of j. That is to say that we only consider evolution at the jth position. For simplicity, we write $\{\xi_{t_j}, t \in [0, +\infty]\}$ as $\{\xi_t, t \in [0, +\infty]\}$.

We assume that the evolution process is a homogeneous Markov process, i.e., for any $t \geq 0$, the conditional probability

$$p_{YX}(t) = P\{\xi_{t+s} = Y \,|\, \xi_s = X\} \tag{6.9}$$

does not depend on $s (s \geq 0)$ where $p_{YX}(t)$ is the transition probability of ξ from state X to state Y after time t. Note that events at time t and at time s are independent, and following from the C-K equation, we obtain

$$p_{YX}(t+s) = \sum_{z \in Z_4} p_{YZ}(t) p_{ZX}(s) . \tag{6.10}$$

If we know the ancestor sequence of the Markov process, i.e., if ξ_0 is given, the process is unique if we get the transition probability matrix of the Markov process $P(t) = (p_{YX}(t))_{4\times4}$ where this transition probability matrix is the so-called substitution matrix. For example, to analyze the evolution of a protein, the PAM matrix and BLOSUM matrix are well-known, and these are the transition probability matrices we will discuss. The identifier numbers 0, 60, and 250 following the letters PAM in the matrices PAM0, PAM60, and PAM250 simply correspond to the t in the transition probability matrix $P(t)$, which is the evolution time.

To obtain $P(t)$, we assume that the following relationship holds:

$$\lim_{t \to 0^+} p_{YX}(t) = \delta(Y, X) = \begin{cases} 1, & Y = X , \\ 0, & Y \neq X . \end{cases} \tag{6.11}$$

This assumption indicates the probability that ξ was substituted in a very short time is 0, i.e., $P(0) = I$, where I is a 4×4 unit matrix. Let Q be the right derivative matrix of $P(t)$ at $t = 0$, then

$$Q = P'(0) = \lim_{t \to 0^+} \frac{P(t) - I}{t} , \tag{6.12}$$

namely,

$$P(\mathrm{d}t) = Q\mathrm{d}t + I . \tag{6.13}$$

From (6.13) we get

$$P(t+\mathrm{d}t) = P(t)P(\mathrm{d}t) \ . \tag{6.14}$$

From (6.14), we replace $P(dt)$ with $Qdt + I$ on the right side, to get

$$P(t+\mathrm{d}t) - P(t) = P(t)Q\mathrm{d}t \ ,$$

namely,

$$P'(t) = QP(t) \ . \tag{6.15}$$

Solving the differential equation, we find

$$P(t) = \mathrm{e}^{tQ} = I + \sum_{n=1}^{\infty} \frac{Q^n t^n}{n!} \ . \tag{6.16}$$

This is the transition probability matrix we require. Using this formula, we find that the transition probability matrix is uniquely determined by the right derivative matrix Q of $P(t)$ at $t = 0$ where Q is the so-called instantaneous transition probability matrix. If Q is symmetrical, then P is also symmetrical. This means the evolution process is reversible. If Q is an arbitrary matrix, then the formula (6.16) can be difficult to compute.

In practice, homogeneity, stationarity, and reversibility of Markov processes are all required. Homogeneity in the evolution process is equivalent to Q being independent over time. Stationarity in the evolution process means that the percentage of the nucleotides in the sequence is unchanged. Reversibility is obeyed when $\pi_X \mathit{\Pi}_{XY}(t) = \pi_Y \mathit{\Pi}_{YX}(t)$ holds, where π_X is the percentage of the nucleotide X in the sequence. This means that in theory we cannot distinguish a forward process from a reverse process. In a reversible process, we can diagonalize the matrix Q, i.e., it can be decomposed as $U \cdot \mathrm{diag}\{\lambda_1 t, \ldots, \lambda_4 t\} \cdot U^{-1}$, where $\{\lambda_1, \ldots, \lambda_4\}$ is the characteristic vector of Q. Thus, the formula (6.16) may be readily computed as follows:

$$P(t) = \mathrm{e}^{tQ} = I + \sum_{n=1}^{\infty} \frac{Q^n t^n}{n!} = U \cdot \mathrm{diag}\{\mathrm{e}^{\lambda_1 t}, \ldots, \mathrm{e}^{\lambda_4 t}\} \cdot U^{-1} \ . \tag{6.17}$$

The whole evolution process is determined with the computation of $P(t)$. This probabilistic model is supported by the three following suppositions:

1. This evolution process only involves type-I mutations.
2. The evolution processes at every pair of positions are independent.
3. The evolution process is an homogeneous, stationary, and reversible Markov process at each position.

In practice, the evolution process is not so ideal; insertions and deletions may happen although the sequences are conserved. These assumptions have little effect on the result.

The above evolution model is idealized, which tells us that the evolution process is determined by its initial state. In other words, the evolution process is determined by Q which is the right derivative of the transition probability matrix at time 0, or the instantaneous transition probability matrix. The matrix Q depends on the ancestor sequence. In practice, however, we know the present sequences, not the ancestor sequences. If we have the instantaneous transition probability matrix Q and the present sequence, we may predict the ancestor sequence and construct the entire phylogenetic tree.

Maximum-Likelihood Method for Constructing the Phylogenetic Tree

On one hand, the whole evolution process is determined by the instantaneous transition probability matrix Q according to the probabilistic evolution model. On the other hand, the probabilities of the phylogenetic tree may be computed if the topological structure of a phylogenetic tree is given. Therefore, for multiple sequences, we may use a maximum-likelihood method to get a maximum probability phylogenetic tree. This can be considered the maximum likelihood estimate of the true phylogenetic tree.

We assume that the probability of substitutions happening over an infinitesimal time interval Δt is $\lambda \Delta t$. Let the probability that the nucleotide mutates to X be p_X. Then, within Δt, the probability that X mutates to Y is

$$p_{XY}(\Delta t) = \begin{cases} 1 - \lambda \Delta t\,, & \text{if } X = Y\,, \\ \lambda \Delta t p_Y\,, & \text{otherwise}\,. \end{cases} \tag{6.18}$$

Following from the definition of $\delta(Y, X)$ given in the last section, we have

$$p_{XY}(\Delta t) = (1 - \lambda \Delta t)\delta(Y, X) + \lambda \Delta t p_Y\,. \tag{6.19}$$

Since the number of substitutions obeys the Poisson distribution, for a small t, $\mathrm{e}^{-\lambda t}$ is the probability that there is no substitution happening within $(0, t)$. Thus, the above formula can be corrected as follows:

$$p_{XY} = \mathrm{e}^{-\lambda t}\delta(Y, X) + (1 - \mathrm{e}^{-\lambda t})p_Y\,. \tag{6.20}$$

Generally, the distribution p is the stationary distribution of the Markov process if it is stationary. Based on the alignment output for multiple sequences, we may use the percentage of each nucleotide as the estimation of the stationary distribution. We may then evaluate the probability that nucleotide X mutates to Y within an interval $(0, t)$.

In conclusion, if multiple sequences are given, we can obtain the alignment output. At each position, we choose a proper parameter λ, and choose a topological structure of the tree and the sum of branch lengths, and then we may find the probability to generate the phylogenetic tree at this position. This routine is shown in Fig. 6.4.

There are four species on the phylogenetic tree without roots. The length of the branches is measured by the average numbers of nucleotides substituted at this position $\{v_i, i = 1, 2, 3, 4, 5\}$.

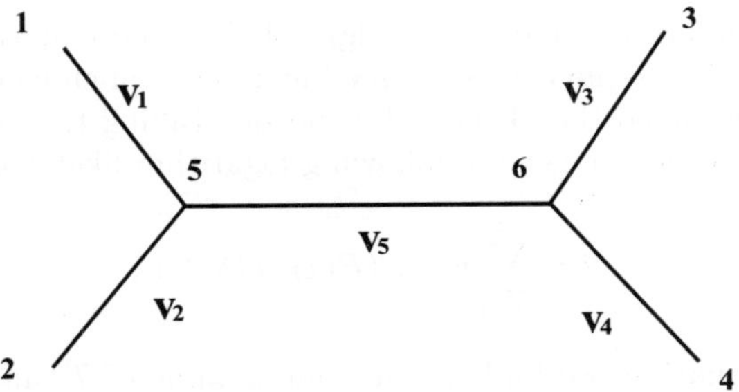

Fig. 6.4. The topological structure of a phylogenetic tree for four species. (From [108])

We may assume that the length of the alignment output for the four species is n where we ignore the insertion and deletion, i.e., neither type-III nor type-IV mutations happen. If there is an insertion or deletion at one position, the column corresponding to this position will be deleted. Let the nucleotides at the hth position of the MA output be $x_h = \{x_1, x_2, x_3, x_4\}^T$ and let $\{\pi_i, i = 1, 2, 3, 4\}$ be the stationary distribution of nucleotides, which can be approximated by the percentage of each nucleotide. Therefore, to generate the phylogenetic tree as in Fig. 6.4, the probability at position h is computed in the following way:

$$P(x_h, v) = \sum_{x_5=1}^{4} \sum_{x_6=1}^{4} \pi_{x_5} \left(P_{x_5x_1}(v_1)P_{x_5x_2}(v_2)P_{x_5x_6}(v_5) \times P_{x_6x_3}(v_3)P_{x_6x_4}(v_4)\right) . \tag{6.21}$$

If the molecular clock supposition holds, then the formula for the probability to construct the phylogenetic tree holds for any position. However, in most cases, the molecular clock supposition does not hold. The evolution speeds are different as the position is changed. That is, at different positions, the same branch length may not represent the same evolution time or the same substitution numbers. Therefore, λ is connected with the positions. As a result, Yang [108] proved that the distribution of λ is approximated by a Γ distribution. Let the value of λ at position h be λ_h so that the above formula can be written as

$$P(x_h, v|\lambda_h) = \sum_{x_5=1}^{4} \sum_{x_6=1}^{4} \left(\pi_{x_5}(P_{x_5x_1}(v_1\lambda_h)P_{x_5x_2}(v_2\lambda_h)P_{x_5x_6}(v_5\lambda_h) \right. \\ \left. \times P_{x_6x_3}(v_3\lambda_h)P_{x_6x_4}(v_4\lambda_h))\right) , \tag{6.22}$$

where P can be obtained from formula (6.22).

Furthermore, we assume that evolutions at different positions are independent. The probability that the whole sequence generates Fig. 6.4 is then computed by following formula:

$$P(X|T) = \Pi_{h=1}^{n} E(P(x_h, v|\lambda_h)) , \tag{6.23}$$

where the expectation value at the right side of the equation is under the condition λ_h, T is the phylogenetic tree including branch length information. This equation is called the likelihood equation. Taking the logarithm of both sides of the equation, we get the following logarithm likelihood equation:

$$l = \sum_{h=1}^{n} \log(E(P(x_h, v|\lambda_h))) . \tag{6.24}$$

In the above equation, we find the maximum value of T, and the maximum likelihood estimate of the phylogenetic tree.

Generally, nucleotide substitution involves not only stationary distribution, but also the percentages of transverse/transition mutations, and synonymous/nonsynonymous mutations. Currently, instantaneous transition probability matrices are commonly used. For example, the Jukes–Cantor model [49], F81 model, K2P model [52], HKY model, GTR model [100, 109], etc. all involve this matrix.

The maximum-likelihood method to construct a phylogenetic tree gives a probabilistic view of evolution. This model is superior to others. Especially in simulation research, this method is better than feature-based methods and distance-based methods. In different regions, we can choose different instantaneous transition probability matrixes. For example, in the region that code a protein, we may use the substitution model of a codon to construct the phylogenetic tree [34], while maximum likelihood methods would be time-consuming. For large size data, this method takes too long, or may not work at all.

The Bayes Method of Constructing the Phylogenetic Tree

The Bayes method of constructing the phylogenetic tree is based on the posterior probability distribution. We use the phylogenetic tree with the maximum posterior probability as the estimation of the true phylogenetic tree. Of course, we can use the Bayes formula to compute the $P(X|T)$ that is used in the maximum likelihood method as follows:

$$P(T_i|X) = \frac{P(X|T_i)P(T_i)}{P(X)} = \frac{P(X|T_i)P(T_i)}{\sum_{T_i} P(X|T_i)P(T_i)} , \tag{6.25}$$

where T_i is the topological structure and the branch lengths of some tree, X is a multiple sequence, and $P(X|T_i)$ is the conditional probability computed by formula (6.23).

Obviously, the posterior probability shown as (6.25) cannot be obtained through analytical approaches. The Monte Carlo method is a better tool to solve this problem. A popular method is the the Metropolis-Hastings method [37, 39, 62]; this is a Monte Carlo Markov chain (MCMC) method. It is outlined as follows:

1. Let T be the current state of the Markov chain. For the initial state, the selection of T is random.
2. Select a new state T' based on the transition probability matrix of the Markov chain. Generally, this state transition probability matrix is symmetrical. The probability from state T to T' is equal to that from state T' to T.
3. The probability that the new state is acceptable is computed as follows:
$$R = \min\left(1, \frac{P(X|T')}{P(X|T)} \times \frac{P(T')}{P(T)} \times \frac{q(T,T')}{q(T',T)}\right), \tag{6.26}$$
where q is the transition probability matrix of the Markov chain, and $\frac{q(T,T')}{q(T',T)} = 1$ if q is symmetric.
4. Generate a random number U in the open interval $(0,1)$. Then let $T = T'$ if $U \le R$, and keep the state T unchanged if $U > R$.
5. Repeat steps 2–4.

The distribution of T obtained from the above steps is the distribution of T in (6.25). We choose the maximum probability tree as the Bayes estimation of the real phylogenetic tree. Additionally, (6.26) is easy to compute because the large denominator is canceled. Therefore, in order to construct the phylogenetic tree, we choose this method when processing large-sized sequences.

6.2 Network Structure Generated by MA

The network structure generated by the MA outputs was proposed as a generalization of graphs and trees. We show that general theory of graphs and trees is perfectly suited to the analysis of the network structure generated by MA.

6.2.1 Graph and Tree Generated by MA

As above, let $\mathcal{A} = \{A_1, A_2, \cdots, A_m\}$ be a multiple sequence, and let $\mathcal{C} = \{C_1, C_2, \cdots, C_m\}$ be the alignment output. We then analyze the network structure generated by $\mathcal{C}$.

The Data Structure Generated by MA

The various data structures generated by the MA output $\mathcal{C}$ are defined as follows:

1. The distance matrix generated by MA is defined as:
Let $D = (d_{s,t})_{s,t=1,2,\cdots,m}$, where $d_{s,t} = d(C_s, C_t) = \sum_{j=1}^{n'} d(c_{s,j}, c_{t,j})$ and $d(c,c')$, $c, c' \in V_{q+1}$ be the distance function defined on V_{q+1}. Then $\hat{M} = \{M, D\}$ is a metric space.

Remark 5. The definition involves the alignment output $\mathcal{C}$, while it is not necessary for $\mathcal{C}$ to be the optimal alignment of $\mathcal{A}$.

2. Stable and unstable regions: A given j is the stable position if $c_{1,j} = c_{2,j} = \cdots = c_{m,j}$ holds. Otherwise, this position is an unstable position. A region is stable if all positions in this region are stable, and a region is unstable if all the positions in this region are not stable. Let Δ_0 and Δ_1 be the stable region and unstable region of $\mathcal{C}$, respectively.
The definition of a stable region and an unstable region can be generalized to the partial alignment case. Let M_0 be a subset of M, then $\mathcal{C}_0 = \{C_s, s \in M_0\}$ is the partial sequence of $\mathcal{C}$. With this new set, we may divide $N' = \{1, 2, \cdots, n'\}$, the set of positions of $\mathcal{C}$, into three parts as follows:
$$\begin{cases} \Delta_0(M_0) = \{j \in N,\ c_{s,j} = c_{s',j} \neq q,\ \forall s, s' \in M_0\}\,, \\ \Delta_1(M_0) = \{j \in N,\ \text{there is a pair } s \neq s' \in M_0,\ \text{such that } c_{s,j} \neq c_{s',j}\}\,. \\ \Delta_2(M_0) = \{j \in N,\ c_{s,j} = q,\ \forall s \in M_0\}\ , \end{cases} \tag{6.27}$$
then $\Delta_0(M_0)$, $\Delta_1(M_0)$ and $\Delta_2(M_0)$ are the stable region, unstable region and the insertion region of $\mathcal{C}_0$, respectively. Next, we let
$$g(M_0) = ||\Delta_0(M_0)||\,, \quad d(M_0) = ||\Delta_1(M_0)|| \tag{6.28}$$
be the lengths of the stable region and unstable region, respectively, for the partial alignment $\mathcal{C}_0$.
3. In the stable region $\Delta_0(M_0)$ and insertion region $\Delta_2(M_0)$,
$$\begin{cases} H_0(M_0) = \{(j, c_j),\ j \in \Delta_0(M_0)\}\,, & c_j \neq q\,, \\ H_2(M_0) = \{(j, c_j),\ j \in \Delta_2(M_0)\}\,, & c_j = q \end{cases} \tag{6.29}$$
are the modulus structures of the stable region and insertion region, respectively.
4. In the unstable region $\Delta_1(M_0)$,
$$H_1(M_0) = \{(j, c_{M_0,j}),\ j \in \Delta_1(M_0)\} \tag{6.30}$$
is the modulus structure of the unstable region, where $c_{M_0,j} = \{c_{s,j}, s \in M_0\}$.

These parameters reflect the data structure characteristics of mutation generated by multiple sequence alignments in different aspects. We can alternatively describe these structure characteristics using the network language. Let
$$\begin{cases} \tilde{\Delta} = \{(\Delta_0(M_0), \Delta_1(M_0), \Delta_2(M_0))\colon M_0 \subset M\}\,, \\ \tilde{\mathcal{H}} = \{(H_0(M_0), H_1(M_0), H_2(M_0))\colon M_0 \subset M\} \end{cases} \tag{6.31}$$
be the modulus structure of MA.

The Topological Tree Generated by MA

Above, we have shown that $\hat{M} = \{M, D\}$ generated by MA is a metric space. Following from the discussion of Sect. 6.1, we can generate different types of trees according to different data structures, as follows.

Minimum distance clustering tree, minimum distance tree, k-order tree, average minimum distance clustering tree, average minimum distance binary tree, average minimum distance binary colored arcs phylogenetic tree. The details of these trees can be found in [35].

The Phylogenetic Tree Generated by a Stable Region and an Unstable Region

In a phylogenetic tree $T' = \{M', V'\}$, let $e = 2m - 1$ be its root, let $T_t = \{M'_t, V'_t\}$ be the branch with root $t(m < t \leq 2m - 1)$, and let $w(e, t')$ be the sum of the lengths of all arcs from e to t. T' is then called the phylogenetic tree generated by a stable region and an unstable region of a multiple sequence if $w(e, t') = ||\Delta_0(M_t)|| + ||\Delta_2(M_t)||$, where M_t is the set of all leaves in T_t, and t' is the dual point of t.

For the phylogenetic trees generated by a stable region and an unstable region of multiple sequences, some properties can easily be found, namely:

1. For any $s \in M$, we always have that $w(e, s') = n$ holds, where n is the length of the MA output $\mathcal{C}$.
2. For any $t \in \{m+1, m+2, \cdots, 2m-1\}$ and $s \in M_t$, we always have that

$$w(e, t') = ||\Delta_0(M_t)|| + ||\Delta_2(M_t)|| \,, \quad w(t', s') = ||\Delta_1(M_t)||$$

   hold, where $\Delta_0(M_t)$ and $\Delta_1(M_t)$ are, respectively, the stable region and unstable region of multiple sequences M_t.
3. $w(e, e') = ||\Delta_0(M)||$ is the total length of the common stable region of the MA output. Let t_1, t_2 be the two successors of node e, the two branches generated by t_1, t_2 be T_{t_1}, T_{t_2}, and M_{t_1}, M_{t_2} be the sets of leaf nodes of T_{t_1}, T_{t_2}. The length of the arcs is then given as

$$\begin{cases} w(t_1, t'_1) = ||\Delta_0(M_{t_1})|| + ||\Delta_2(M_{t_1})|| - w(e, e') \\ \qquad\qquad = ||\Delta_0(M_{t_1})|| - ||\Delta_0(M)|| \,, \\ w(t_2, t'_2) = ||\Delta_0(M_{t_2})|| + ||\Delta_2(M_{t_2})|| - w(e, e') \\ \qquad\qquad = ||\Delta_0(M_{t_2})|| - ||\Delta_0(M)|| \,. \end{cases} \tag{6.32}$$

   Similarly, we get the lengths of arcs $w(t, t')$ of all $t \in \{m+1, m+2, \cdots, 2m-1\}$ in the phylogenetic tree T'.
4. If $s_1, s_2 \in M$ are the two leaf nodes on the phylogenetic tree T', and they have the same ancestor, then their arc lengths are the penalty function of the alignment sequences C_{s_1}, C_{s_2}. That is,

$$w(s_1, s'_1) = w(s_2, s'_2) = ||\Delta_1(s_1, s_2)|| = d(C_{s_1}, C_{s_2}) \,. \tag{6.33}$$

5. The triplet $T'(w) = \{M', T', w\}$ is called the colored arc graph of the phylogenetic tree T', where $w(t,t')$ or $w(s,s')$ are defined as in (6.33) or (6.32).
6. In the colored arc graph $T'(w)$ of the phylogenetic tree T', if we use the stable region and unstable region $\Delta_0(M_t)$, $\Delta_1(M_s)$ or the modulus structure of the stable region and unstable region $H_0(M_t)$, $H_1(M_s)$ to replace $w(t,t')$ and $w(s,s')$, this colored arc graph turns to the following two forms:
The colored arc graph of the stable region and unstable region is $T'(\Delta) = \{M', T', \Delta\}$ if we use $\Delta_0(t,t')$ or $\Delta_1(s,s')$ defined as in (6.31) or (6.30).
The colored arc graph of the stable region and unstable region is $T'(\mathcal{H}) = \{M', T', \mathcal{H}\}$ if we use the modulus structure $\mathcal{H}$ defined by (6.31), and $H_0(t,t')$ or $H_1(s,s')$ is defined by (6.29) or (6.28).

Minimum Unstable Region Phylogenetic Tree

In the above section, we have given the phylogenetic tree T' generated by the stable region and unstable region. It is simply called the phylogenetic tree T' of the stable region and unstable region. Let $w(T') = \sum_{s\in M} w(s,s')$; then it is the sum length of the unstable region of the phylogenetic tree T'.

Definition 30. *T_0' is called the minimum unstable region phylogenetic tree, if $w(T_0') \le w(T')$ holds for all other phylogenetic trees T'.*

The method of producing a minimum unstable region phylogenetic tree is similar to that for generating the minimum distance clustering tree. It can be clustered based on the length of the unstable region of the MA output. We will show this later with examples.

6.2.2 Network System Generated by Mutations of Multiple Sequences

Among the various topological trees generated from MA outputs, we use graphs and trees to express the connections between mutations and evolution. The modulus structure of the colored arc graph of the stable region and unstable region $T'(\mathcal{H}) = \{M', T', \mathcal{H}\}$ reflects the information of the MA output. However, some points are less clear for the description of these trees. For example, the combination relations of different sequences within the mutation region are still too complicated to be immediately understood. Therefore, we discuss them further. A network system generated by the mutations of multiple sequences is used to describe the mutation structure of the MA output through the colored arcs graph. To do this, we introduce the following notations.

Network System of Mutation

Let $M = \{1, 2, \cdots, m\}$ be the subscript set of a MA output $\mathcal{C}$, that is, each $i \in M$ corresponds to a sequence C_i. Then, graphs $G = \{M, V\}$ and $G' = \{M', V'\}$ are generated by MA output $\mathcal{C}$, in which, V is the arc set generated by the point pairs of M, and $\{M', V'\}$ is the extension of $\{M, V\}$ similar to that given by phylogenetic tree $T' = \{M', V'\}$. The network system generated by the MA output colors both points and arcs of the graph G or G'. Following from the metric relation w of MA output, two types of network structures may be generated respectively by the mutation region Δ and the modulus structure $\mathcal{H}$ as follows:

1. Topological network system generated by MA output: $G(W) = \{M, V, W\}$, in which w is the penalty function of the MA output defined by (6.33).
2. Mutation region network system: $G(\Delta) = \{M, V, \Delta\}$, in which Δ is the mutation region function of the multiple alignment output defined by (6.31).
3. Network system of mutation mode generated by multiple alignment output: $G(\mathcal{H}) = \{M, V, \mathcal{H}\}$, in which $\mathcal{H}$ is the modular function of the multi-sequence alignment given by (6.31).

These three network systems are called the network systems generated by the MA output, or simply the mutation networks. In the same way, we can define the graph G'. The purpose in researching the mutation network is to analyze the evolution relations of multiple sequences.

The Basic Mutation Types of Triple Sequences

Definition 31. *Let C_1, C_2, C_3 be a triple sequence in the MA output $\mathcal{C}$. Its basic types are stated as follows:*

1. *Orthogonal: Let δ_{12} and δ_{23} be the mutation regions induced in C_1, C_2, C_3. Then, H_{12} and H_{23} are orthogonal if $\delta_{12} \cap \delta_{23} = \emptyset$. We use the simpler form $H_{12} \perp H_{23}$ to represent the orthogonal relationship.*
2. *Overlapping: The triple sequences C_1, C_2, C_3 overlap if their mutations regions satisfy the following: $\delta_{12} = \delta_{13} = \delta_{23}$ and c_{1j}, c_{2j}, c_{3j} are different from each other for all $j \in \delta_{12}$.*

Theorem 26. *1. The orthogonal type is symmetric. Namely, if $H_{12} \perp H_{23}$ holds, then both $H_{23} \perp H_{12}$ and $H_{21} \perp H_{23}$ hold.*
2. $H_{12} \perp H_{23}$ holds if and only if $\delta_{12} \cap \delta_{23} = \emptyset$, in which δ_{ij} is the mutation region of H_{ij}.
3. If C_1, C_2, C_3 are overlapping, then $w_{12} = w_{13} = w_{23}$ holds.

It is easy to prove these three propositions, so we omit the proofs here.

The orthogonal type and overlapping type are the two extreme cases for mutations. In general, we frequently face mixed modes. Thus, we need the following decomposition theorem:

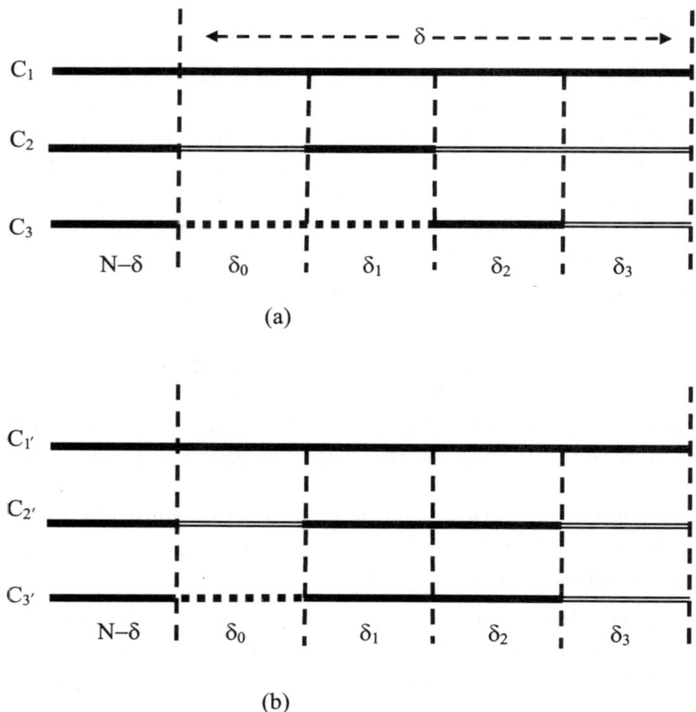

Fig. 6.5a,b. The decomposition of the mutation region of a triple alignment output

Theorem 27. *(The decomposition theorem of the triple alignment output.) Let C_1, C_2, C_3 be the alignment output of the triple sequence A_1, A_2, A_3. There is a new triple sequence $C_{1'}, C_{2'}, C_{3'}$ satisfying the following properties:*

1. *$C_{1'}, C_{2'}, C_{3'}$ are overlapping.*
2. *Mutation modes $H_{11'}, H_{22'}, H_{33'}$ are orthogonal to each other. As well, $H_{11'}$ and $H_{1'2}, H_{1'3}$, $H_{22'}$ and $H_{2'1}, H_{2'3}$, $H_{33'}$ and $H_{3'1}, H_{3'2}$ are all orthogonal.*

Remark 6. 1. In Fig. 6.5a, C_1, C_2, C_3 represent the alignment output of the triple sequence A_1, A_2, A_3, where $N - \delta$ is the stable region, in which the values of these three sequences are the same. δ is the unstable region, which can be decomposed to four subregions $\delta_1, \delta_2, \delta_3$, and δ_0 as shown in (6.34).
2. Figure 6.5b shows sequences $C_{1'}, C_{2'}, C_{3'}$ defined by (6.35).

Proof. Maintaining the notation given in Fig. 6.5, C_1, C_2, C_3 are the alignment output of triple sequences A_1, A_2, A_3 and the mutation region is δ. $N' - \delta$ is then the stable region. The unstable region can be decomposed into δ_{12},

δ_{23}, δ_{13}. These are the mutation regions of (C_1, C_2), (C_2, C_3), (C_1, C_3). Let

$$\begin{cases} \delta = \delta_{12} \cup \delta_{13} \cup \delta_{23}\,, \\ \delta_0 = \{j \in \delta \colon c_{1j}, c_{2j}, c_{3j} \text{ are not the same as each other}\}\,, \\ \delta_1 = \{j \in \delta \colon c_{1j} = c_{2j} \neq c_{3j}\}\,, \\ \delta_2 = \{j \in \delta \colon c_{1j} = c_{3j} \neq c_{2j}\}\,, \\ \delta_3 = \{j \in \delta \colon c_{2j} = c_{3j} \neq c_{1j}\}\,, \end{cases} \tag{6.34}$$

then $\delta_0, \cdots, \delta_3$ are four mutually disjoint regions. We denote the lengths of the four regions by $w_\tau = ||\delta_\tau||$, where $\tau = 0, 1, 2, 3$, respectively. Based on this decomposition, we construct new sequences $C_{1'}, C_{2'}, C_{3'}$ as follows. Let

$$c_{1'j} = \begin{cases} c_{2j}\,, & \text{if } j \in \delta_3\,, \\ c_{1j}\,, & \text{otherwise}\,, \end{cases} \qquad c_{2'j} = \begin{cases} c_{1j}\,, & \text{if } j \in \delta_2\,, \\ c_{2j}\,, & \text{otherwise}\,, \end{cases}$$

$$c_{3'j} = \begin{cases} c_{1j}\,, & \text{if } j \in \delta_1\,, \\ c_{3j}\,, & \text{otherwise}\,. \end{cases} \tag{6.35}$$

Then, the components of sequences $C_{1'}, C_{2'}, C_{3'}$ are different from each other in the region δ_0 but the same in the remaining regions. Therefore, it is the overlapping type. In addition, we analyze the mutation regions of sequences C_1, C_2, C_3 and $C_{1'}, C_{2'}, C_{3'}$ as follows. Since we then have

$$\delta_{11'} = \delta_3\,, \quad \delta_{2'2} = \delta_2\,; \quad \delta_{33'} = \delta_1\,,$$

and since regions δ_1, δ_2, δ_3 are mutually disjoint, it follows that $\{H_{11'}, H_{22'}, H_{33'}\}$ are orthogonal modulus structure. With the same reasoning, we may prove that the three groups of modes $H_{11'}$ and $H_{1'2}, H_{1'3}$; $H_{22'}$ and $H_{2'1}, H_{2'3}$; $H_{33'}$ and $H_{3'1}, H_{3'2}$ are orthogonal, respectively. Thus ends the proof.

Figure 6.6 shows the mutation relations between sequences C_1, C_2, C_3 and $C_{1'}, C_{2'}, C_{3'}$. The process by which C_1 mutates to C_2, C_3 can be decomposed, to where C_1 mutates to $C_{1'}$, and then $C_{1'}$ mutates to C_2, C_3. Therefore, Fig. 6.6 is called the network structure graph of the triple alignment output.

In Theorem 27, the triangle $\Delta(C_{1'}, C_{2'}, C_{3'})$ shrinks to a point if δ_0 is an empty set. If $C_{1'} = C_{2'} = C_{3'} = C_0$ are the same sequences, then

$$H_{10} \perp H_{20}\,, \quad H_{10} \perp H_{30}\,, \quad H_{20} \perp H_{30} \tag{6.36}$$

hold. The inverse proposition is also true, i.e., if there is a point C_0 making (6.36) true, then δ_0 must be an empty set.

Definition 32. *1. Under the conditions of Theorem 27, $C_{1'}, C_{2'}, C_{3'}$ are the orthogonal decomposition of the triple alignment output (C_1, C_2, C_3) if sequences $C_{1'}, C_{2'}, C_{3'}$ satisfy the theorem.*

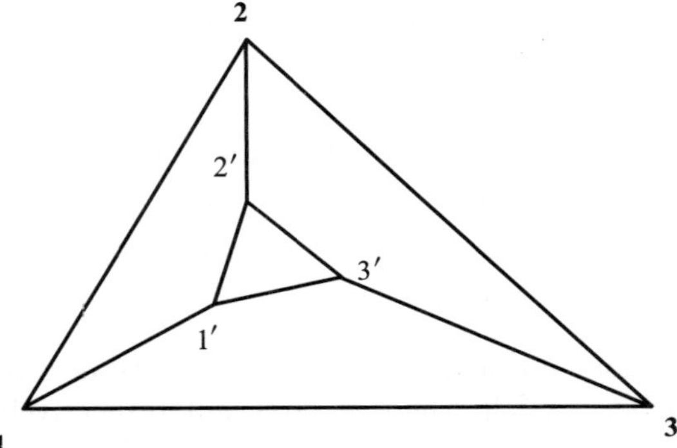

Fig. 6.6. The mutation network decomposition of a triple alignment output

2. *If there is a sequence* C_0 *such that (6.36) holds, then we say that* C_0 *makes the triangle* (C_1, C_2, C_3) *perfectly orthogonal. The triple sequences* (C_1, C_2, C_3) *can be made perfectly orthogonal if and only if* $C_{1'} = C_{2'} = C_{3'}$ *holds, where the mutation relationship of* C_1, C_2, C_3 *can be decomposed to the mutation relationship between* C_1, C_2, C_3 *and* $C_{1'}, C_{2'}, C_{3'}$.

The Mutation Network Tree Generated by a Binary Tree

In any book on graph theory, the reader can find the terms graph, tree, directed tree, node, arc, the extreme points of an arc, the starting point and end point of the directed arc, the root of a tree, and leaf all well-defined. Therefore, we do not repeat the definitions here. However, several new concepts are directly involved in the discussions presented in this book, which are defined as follows:

Definition 33. *1. For a mutation network* $\mathcal{E}$*, if each overlapping triangle is seen as a point, then the renewed mutation network* $\mathcal{E}'$ *is the reduction of* $\mathcal{E}$*.*

2. *A directed mutation network tree is a directed orthogonal mutation tree if any two arcs starting from any node are orthogonal.*
3. *An undirected mutation network tree is a perfectly orthogonal mutation tree if any two arcs with a common node are orthogonal.*

Theorem 28. *(The orthogonalization theorem of a mutation network tree.) For a given directed mutation network tree, there are some nodes such that the mutation network* $\mathcal{E}$*, which is generated by adding these nodes into the given tree, satisfies the following conditions:*

1. *If there are triangles in* $\mathcal{E}$*, they are overlapping triangles.*
2. *Let* $\mathcal{E}'$ *be the network induced by* $\mathcal{E}$ *in the case where each overlapping triangle is seen as a point, then* $\mathcal{E}'$ *is an orthogonal mutation tree.*

3. *Let $\Delta(a,b,c)$ be an overlapping triangle in the mutation network $\mathcal{E}$. Each arc with an extreme point a is then orthogonal to arcs ab, ac. Also, the same holds true for both b and c.*

Proof. For clarity, we follow Fig. 6.7 to give the proof as follows:

1. Figure 6.7a is the original undirected tree, where $G_1^{(0)} = \{M^{(0)}, V^{(0)}\}$ and $M^{(0)} = \{a,b,c,d,e\}$, $V^{(0)} = \{(a,b),(a,c),(b,d),(b,e)\}$. The virtual lines are 2-order arcs.
2. The orthogonalization starts from leaves a, c. Following from Theorem 27, there is an overlapping triangle $\Delta(a',b',c')$ which orthogonalizes (a,b,c). The modes $H_{aa'}, H_{bb'}, H_{cc'}$ are orthogonal to each other. If we reduce the network graph such that ab, ac, bc are seen as 2-order arcs, then we get Fig. 6.7b, and its mutation network tree is $G_1^{(1)} = \{M^{(1)}, V^{(1)}\}$, where

$$\begin{cases} M^{(1)} = \{a,b,c,d,e,a',b',c'\}, \\ V^{(1)} = \{(a,a'),(b,b'),(c,c'),(a',b'),(a',c'),(b',c'),(b,d),(b,e)\}. \end{cases}$$

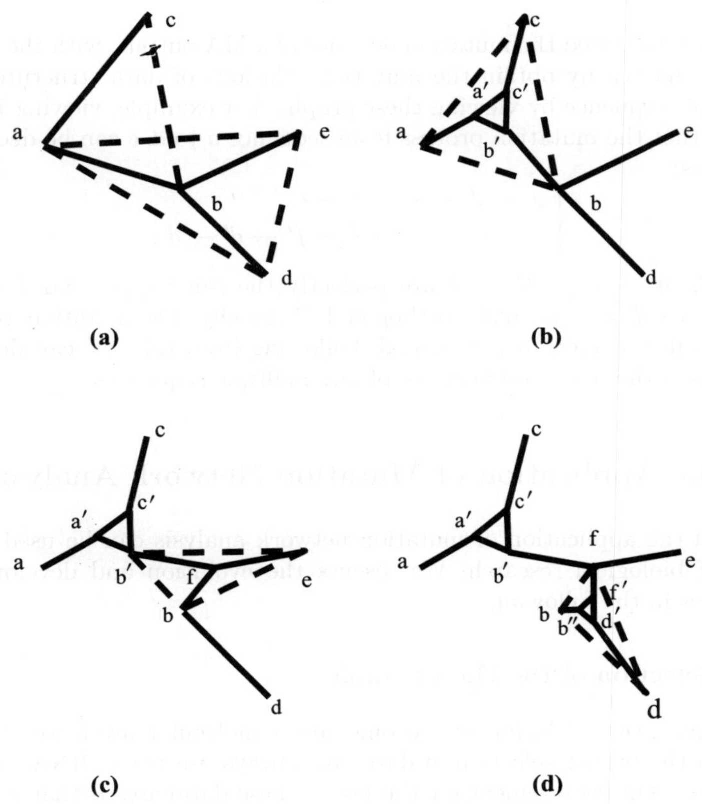

Fig. 6.7a–d. The orthogonalization procedure of a mutation network

3. Similarly to step 2, we orthogonalize triangle (b, b', e) in $G_1^{(1)}$. If this triangle is perfectly orthogonal, then Fig. 6.7c is obtained. Its mutation network tree is then $G_1^{(2)} = \{M^{(2)}, V^{(2)}\}$, where

$$\begin{cases} M^{(2)} = \{a, b, c, d, e, a', b', c', f\}, \\ V^{(2)} = \{(a, a'), (b, b'), (c, c'), (a', b'), (a', c'), (b', c'), (b', f), (b, f), \\ \qquad (f, e), (b, d)\} . \end{cases}$$

4. Continuing this procedure, we can do the orthogonalization procedure on $G_1^{(2)}$. Finally, we get $G_1^{(3)}$ as shown by Fig. 6.6d, where

$$\begin{cases} M^{(3)} = \{a, b, c, d, e, a', b', c', f, f', b'', d'\}, \\ V^{(3)} = \{(a, a'), (b, b'), (c, c'), (a', b'), (a', c'), (b', c'), (b', f), (f, f'), \\ \qquad (b'', f'), (f', d'), (b'', d'), (d', d), (b, b''), (f, e)\} . \end{cases}$$

The graph $G^{(3)}$ satisfies all the conditions in the theorem. Thus ends the proof.

We have introduced the mutation network of a MA output, with the intention that we may easily obtain the mutation relations of data structures among a multiple sequence by viewing these graphs. For example, viewing Fig. 6.7d, we find that the mutation process from sequence a to d, e can be decomposed as follows:

$$\begin{cases} a \to a' \to b' \to f \to e , \\ a \to a' \to b' \to f \to f' \to d' \to d , \end{cases}$$

in which, $a \to a' \to b' \to f$ are perfectly the same type, and $f \to e$ and $f \to f' \to d'$ are mutually orthogonal. Typically, the mutation process of each smaller segment is orthogonal. Following from this, we can deduce the relations of the mutation network of any multiple sequences.

6.3 The Application of Mutation Network Analysis

MA and the application of mutation network analysis can be used in many fields of biological research. We discuss the evolution and development of epidemics in the following.

6.3.1 Selection of the Data Sample

To examine the evolution of biosomes on a molecular level, we should begin with the proper selection of data. We always use DNA, RNA, or protein databases. The requirement for the use of these databases is that a sequence should have many homologous sequences in different biosomes. Research in biology indicates that many genes and proteins recur in many species. For example, chondriosome, cytochrome and cathepsin are found in many biosomes. In the process of selecting data samples, besides using existing data that may be obtained directly from databases such as GenBank, some special databases may also need to be tracked. Therefore, we need to design the data collection scheme before starting the sequencing. For example, to analyze the development of some epidemic or disease, we must design a good scheme for collecting the required data. Next, we choose chondriosome, SARS, and HIV-1, respectively, as examples to illustrate the procedure used to analyze the data. The explanation for the corresponding results is given below:

The Data Sample of Chondriosome

Biology research has revealed that chondriosome occurs in many biosomes. In GenBank, there are thousands of homologous sequences of chondriosome. To analyze the mutations, we select the ND1 gene coding region of 20 species of mammals as follows:

1. Bos taurus complete mitochondrial
2. Balaenoptera physalus mitochondrial, complete
3. Balaenoptera musculus mitochondrial DNA, complete
4. Phoca vitulina mitochondrial DNA, complete
5. Halichoerus grypus complete mitochondrial
6. Felis catus mitochondrion, complete
7. Equus caballus mitochondrial DNA, complete sequence
8. Rhinoceros unicornis complete mitochondrial
9. Rattus norvegicus mitochondrial
10. Homo sapiens mitochondrial DNA, complete sequence
11. Pan troglodytes mitochondrial DNA, complete sequence
12. Pan paniscus mitochondrial DNA, complete sequence
13. Gorilla gorilla mitochondrial DNA, comlete sequence
14. Pongo pygmaeus mitochondrial DNA, complete sequence
15. Pongo pygmaeus abelii mitochondrial
16. Hylobates lar complete mitochondrial DNA sequence
17. Didelphis virginiana complete mitochondrial
18. Macropus robustus complete mitochondrial
19. Ornithorhynchus anatinus mitochondrial DNA, complete
20. Mus musculus mitochondrial

SARS Sequences

In the spring of 2003, a SARS epidemic broke out in China. Research on the SARS virus has become an important problem in the fields of biology and medicine. In the GenBank database, new DNA sequences of SARS were continually announced. In September of 2003, an article published in *Science* [101] involved 63 DNA sequences of the SARS virus. As a result, this paper analyzed the evolution of the SARS epidemic from its onset to the metaphase and then to the mature phase. After September 2003, more new DNA sequences of the SARS virus were constantly being sequenced. As of September 2004, the total number of SARS virus sequences uploaded in the GenBank was 118. Their names and sources are shown in Table 6.1.

Remark 7. Under the "name" rubric, we only give the simpler name of the SARS coronavirus. For example, in number 4, we only use the name Sin850, while its full name is SARS coronavirus Sin850. Pagumalarvata is the Chinese southern Pagumalarvata. The CDC is CDC-200301157, Pagu. is Pagumalarvata, SH stands for Shanghai.

HIV-1 Virus Genome

The HIV-1 virus genome is the main type of AIDS virus. Besides HIV-1, there is HIV-2 along with other virus genomes of animals. Since HIV-2 appears in local districts, most studies of the AIDS virus genome focus on how to analyze the HIV-1 virus genome. In edition 2004/9 (release 43), the GenBank announced 706 sequences of HIV-1. The lengths of these sequences vary from 7,000 to 9,000 bp. Similarly to the SARS sequences, HIV-1 data contain both incomplete regions and nonsequenced regions. Therefore, we cannot adopt them mechanically. The nations and districts of origin for these 706 sequences of HIV-I are listed in Table 6.2.

6.3.2 The Basic Steps to Analyze the Sequences

The data samples we collected are a group of multiple sequences. Therefore, we process them by using various types of software packages to obtain a MA output. Let $\mathcal{A}$ be the multiple sequences consisting of the data samples, and let $\mathcal{A}'$ be its MA output.

The Procedure to Analyze the MA Output

1. Based on $\mathcal{A}'$, we compute the penalty (or scoring) matrix $W = (w_{s,t})$, modulus structure matrix $\mathcal{H} = (H_{s,t})$ and mutation region matrix Δ. Because the modulus structure matrix and the mutation region matrix $\mathcal{H}$, Δ are very complex, they may be considered to be parameters.
2. Based on the penalty matrix $W = (w_{s,t})$ to cluster the multiple sequences, we construct the minimum distance tree G_1, and then construct the k-order graph G_k and k-order mutation network $G_k(W)$.
3. Based on the minimum distance tree G_1 and mutation region matrix Δ, we orthogonalize the network, and give the corresponding graph for the orthogonal decomposition of the network.

Table 6.1. The names and numbered list of the 118 SARS sequences

No.	GenBank number	Name	Nation or district	No.	GenBank number	Name	Nation or district
1	NC-004718		Toronto	2	AY714217	CDC	USA
3	AY559097	Sin3408L	Singapore	4	AY559096	Sin850	Singapore
5	AY559095	Sin847	Singapore	6	AY559094	Sin846	Singapore
7	AY559093	Sin845	Singapore	8	AY559092	SinP5	Singapore
9	AY559091	SinP4	Singapore	10	AY559090	SinP3	Singapore
11	AY559089	SinP2	Singapore	12	AY559088	SinP1	Singapore
13	AY559087	Sin3725V	Singapore	14	AY559086	Sin849	Singapore
15	AY559085	Sin848	Singapore	16	AY559084	Sin3765V	Singapore
17	AY559082	Sin852	Singapore	18	AY559081	Sin842	Singapore
19	AY654624	TJF	Beijing	20	AY595412	LLJ-2004	Beijing
21	AY394850	WHU	Wuhan	22	AY274119	Tor2	Toronto
23	AY323977	HSR 1	Italy	24	AY291315	Frankfurt1	Germany
25	AY502932	TW9	Taiwan	26	AY502931	TW8	Taiwan
27	AY502930	TW7	Taiwan	28	AY502929	TW6	Taiwan
29	AY502928	TW5	Taiwan	30	AY502927	TW4	Taiwan
31	AY502926	TW3	Taiwan	32	AY502925	TW2	Taiwan
33	AY502924	TW11	Taiwan	34	AY502923	TW10	Taiwan
35	AY291451	TW1	Taiwan	36	AY390556	GZ02	Guangdong
37	AY395003	ZS-C	Guangdong	38	AY395002	LC5	Guangdong
39	AY395001	LC4	Guangdong	40	AY395000	LC3	Guangdong
41	AY394999	LC2	Guangdong	42	AY394998	LC1	Guangdong
43	AY394997	ZS-A	Guangdong	44	AY394996	ZS-B	Guangdong
45	AY394995	HSZ-Cc	Guangdong	46	AY394994	HSZ-Bc	Guangdong
47	AY394993	HGZ8L2	Guangdong	48	AY394992	HZS2-C	Guangdong
49	AY394991	HZS2-Fc	Guangdong	50	AY394990	HZS2-E	Guangdong
51	AY394989	HZS2-D	Guangdong	52	AY394987	HZS2-Fb	Guangdong
53	AY394986	HSZ-Cb	Guangdong	54	AY394985	HSZ-Bb	Guangdong
55	AY394983	HSZ2-A	Guangdong	56	AY394982	HGZ8L1-B	Guangdong
57	AY394981	HGZ8L1-A	Guangdong	58	AY394979	GZ-C	Guangdong
59	AY394978	GZ-B	Guangdong	60	AY508724	NS-1	Guangdong
61	AY463059	SH-QXC1	Guangdong	62	AY313906	GD69	Guangdong
63	AY310120	FRA	Italy	64	AY461660	SoD	Russia
65	AY485278	Sino3-11	Beijing	66	AY485277	Sino1-11	Beijing
67	AY345988	CUHK-AG03	Hong Kong	68	AY345987	CUHK-AG02	Hong Kong
69	AY345986	CUHK-AG01	Hong Kong	70	AY282752	CUHK-Su10	Hong Kong
71	AY357076	PUMC03	beijing	72	AY357075	PUMC02	Beijing
73	AY350750	PUMC01	Beijing	74	AY304495	GZ50	Hong Kong
75	AY304486	SZ3	Pagu.	76	AY427439	AS	Italy
77	AY283798	Sin2774	Singapore	78	AY278491	HKU-39849	Beijing
79	AY278489	GD01	Beijing	80	AY362699	TWC3	Taiwan
81	AY362698	TWC2	Taiwan	82	AY283797	Sin2748	Singapore
83	AY283796	Sin2679	Singapore	84	AY283795	Sin2677	Singapore
85	AY283794	Sin2500	Singapore	86	AY278741	Urbani	USA
87	AY351680	ZMY 1	Guangdong	88	AP006561	TWY	Taiwan
89	AP006560	TWS	Taiwan	90	AP006559	TWK	Taiwan
91	AP006558	TWJ	Taiwan	92	AP006557	TWH	Taiwan
93	AY278554	CUHK-W1	Hong Kong	94	AY348314	TaiwanTC3	Taiwan
95	AY338175	Taiwan TC2	Taiwan	96	AY338174	TaiwanTC1	Taiwan
97	AY321118	TWC	Taiwan	98	AY279354	BJ04	Beijing
99	AY278490	BJ03	Beijing	100	AY278487	BJ02	Beijing
101	AY297028	ZJ01	Beijing	102	AY278488	BJ01	Beijing
103	AY304488	SZ16	Pagu.	104	AY559083	Sin3408	Singapore
105	AY286320	ZJ01	Hangzhou	106	AY395004	HZS2-Bb	Guangdong
107	AY394988	JMD	Guangdong	108	AY394984	HSZ-A	Guangdong
109	AY394980	GZ-D	Guangdong	110	AY394977	GZ-A	Guangdong
111	AY463060	SH-QXC2	Shanghai	112	AY304494	HKU-66078	Hong Kong
113	AY304493	HKU-65806	Hong Kong	114	AY304492	HKU-36871	Hong Kong
115	AY304491	GZ60	Hong Kong	116	AY304490	GZ43	Hong Kong
117	AY304489	SZ1	Pagu.	118	AY304487	SZ13	Pagu.

CDC CDC-200301157, *Pagu.* Pagumalarvata, *SH* Shanghai

Table 6.2. The nations and districts for the 706 sequences of HIV-I

A	B	A	B	A	B	A	B	A	B	A	B
Botswana	72	Tanzania	41	Cameroon	82	South Africa	46	DR Congo	11	Senegal	7
Ethiopia	8	Nigeria	3	Zambia	2	Rwanda	1	Benin	1	Uganda	58
Kenya	45	Gabon	2	Central African	5	Chad	3	Niger	3	Mali	2
Finland	3	Belgium	11	France	23	Sweden	15	Greece	4	Belarus	2
Russia	3	Spain	14	Netherlands	14	Estonia	2	Britain	2	Germany	2
Ukraine	1	Norway	1	Taiwan	1	South Korea	2	China	17	Israel	1
India	15	Thailand	59	Ghana	3	Japan	4	Myanmar	9	Cyprus	2
Brazil	7	Uruguay	4	Argentina	26	Bolivia	2	Colombia	5	Australia	16
USA	43	Others	1								

A denotes the nation or district, *B* denotes the numbers of the sequenced genes

Analyzing the Biological Meaning of the Final Results

Based on the graph of the orthogonal decomposition of the network, we can construct a relationship of the mutations among sequences, and analyze the biological meaning. For the same biosome, there are many methods to collect the data sample, and for different data samples we may get different results. Therefore, we should analyze the biological meaning from several different angles.

Using the above general procedure, we next discuss several examples in biology and medicine. We will detail the content involved within the discussions.

6.3.3 Remarks on the Alignment and Output Analysis

The Mutation Analysis of Mammalian Mitochondrial Genome

1. The length of the mammalian mitochondrial genome is about 18 kbp. The length of the coding region ND1 is 900 bp. The length of its alignment output is 961 bp, as shown in [99].
2. The total length of the stable region of the multiple alignment output is $||N_0|| = 404$, the percentage is $\frac{404}{961} = 42.04\%$, proportional to the total length of the output. While we can readily produce the list of their modulus structure, we have omitted it for brevity.
3. Let $w(a, b)$ be the Hamming matrix, and let the penalty matrix be $w_{s,t} = \sum_{i=1}^{n} w(c_{s,i}, c_{t,i})$, where $s, t = 1, 2, \cdots, 20$ as shown in Table 6.3.
4. Based on the penalty matrix, we find the system clustering tree as shown in Fig. 6.8.

The Analysis of the SARS Virus Gene

1. The lengths of the 118 SARS sequences are about 18 kbp. We select 103 sequences which are well-sequenced. The length of the MA output is 29,908 bp. The result is shown in [99].

Table 6.3. The penalty matrix for the ND1 coding region

180																		
178	72																	
198	195	178																
192	192	175	29															
189	191	183	152	153														
184	192	186	174	173	180													
178	201	181	182	177	192	157												
250	243	238	255	255	246	246	258											
256	259	247	251	256	251	248	239	274										
256	262	245	270	267	265	246	243	269	94									
263	267	254	267	268	269	245	247	265	99	44								
257	264	251	259	256	258	251	237	279	103	111	116							
247	254	243	247	248	260	244	234	257	145	156	154	150						
248	259	244	250	250	267	249	232	253	146	157	156	156	60					
261	260	249	261	251	257	242	233	270	156	157	152	142	145	141				
264	257	244	270	267	253	261	254	286	300	292	299	298	308	304	325			
249	245	239	251	249	228	231	240	261	272	277	268	268	263	270	263	234		
259	285	265	259	261	265	263	260	307	306	312	304	302	309	304	316	273	246	
255	252	244	242	245	249	251	241	193	275	271	263	286	270	264	277	271	281	288

2. The SARS virus genome has high similarity because of the short time the disease has taken to develop and evolve. Except for a few sequences which may have sequencing errors, the sequence homology for most sequences is over 95%. In these 103 SARS sequences, we have determined their common stable region (at whose positions the nucleotides are invariant). The number of the positions in the common stable region is 26,924, which is 90.023% of the length of the sequence alignment output (29,908).
3. Analyze the unstable region of the MA output from different angles, including the head and tail of the SARS sequences. For the MA output, we can determine that the head comprised 20 positions and the tail comprised 43 positions. The percentages are 0.07% and 0.144%, respectively. In the head and tail part, the structure changes a great deal. The reason is that the start point and end point are both selected differently in sequencing. The distribution of the nucleotides in unstable positions can be denoted by $\bar{f}_i = (f_i(0), f_i(1), \cdots, f_i(4))$, where $f_i(z)$ is the number of nucleotides or inserting symbols z at position i. For example, $f_{19} = (1, 86, 0, 1, 15)$ means that that number of times that "a, c, g, t, and −" occur at position 19 of the 103 SARS sequences are 1, 86, 0, 1, and 15, respectively.
4. The penalty matrix $W = (w_{s,t})_{s,t=1,2,\cdots,103}$ follows from the multiple alignment output (shown in [99]), where $w_{s,t} = d_{\rm H}(A_s, A_t)$ is the Hamming distance between A_s and A_t.
5. Following from the penalty matrix W, we generate the phylogenetic tree, the minimum distance graph and the second-order structure graph. Construction of the network graph follows directly from these graphs.

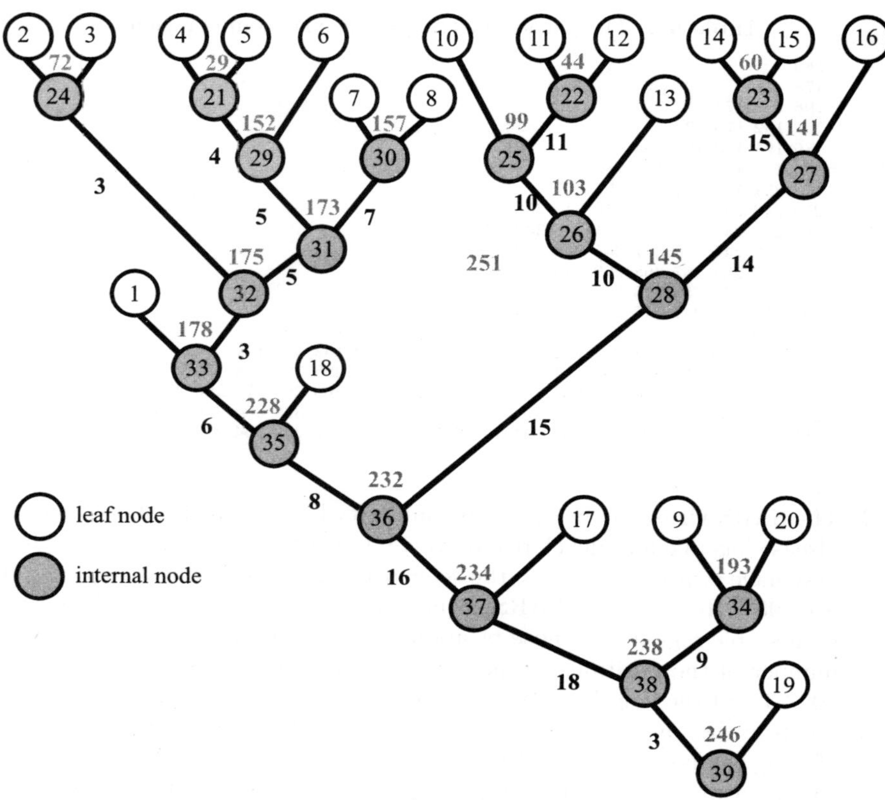

1. Bos taurus
2. Balaenoptera physalus
3. Balaenoptera musculus
4. Phoca vitulina
5. Halichoerus grypus
6. Felis catus
7. Equus caballus
8. Rhinoceros unicornis
9. Rattus norvegicus
10. Homo sapiens
11. Pan troglodytes
12. Pan paniscus
13. Gorilla gorilla
14. Pongo pygmaeus
15. Pongo pygmaeus abelii
16. Hylobates lar
17. Didelphis virginiana
18. Macropus robustus
19. Ornithorhynchus anatinus
20. Mus musculus

Fig. 6.8. The cluster tree generated by the multiple sequence alignment of the ND1 gene coding region of 20 sorts of mammals

The Network Graph Based on the SARS Sequences in Different Stages

In clinics, a disease is divided into many stages. SARS, as a particular disease, is also divided into an initial stage, a middle stage and a final stage. The SARS sequences change due to mutations as the stage or other conditions change.

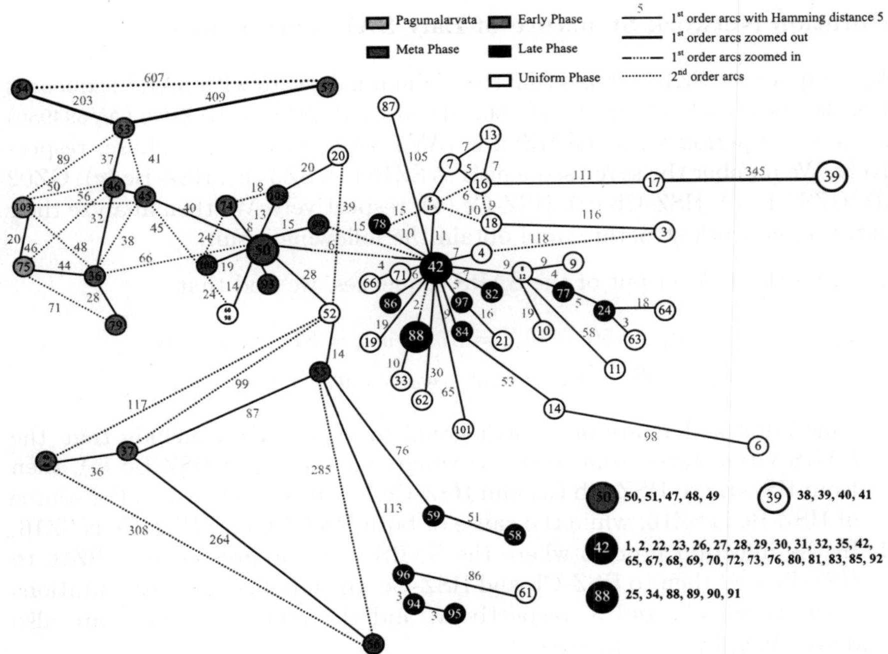

Fig. 6.9. The network graph based on the SARS sequences

To search the variance, we discuss the network graph based on the sequences collected at different stages. The discussion is detailed as follows:

1. For some sequences, for example, numbers 42, 50, and 51, the differences among them are very minor. They always come from the same district. It is useful to track their evolution processes (e.g., the time point for the onset, the development of the epidemic process, etc.).
2. Some sequences, e.g., numbers 5, 28, 76, and 93, form local clustering centers in the graph. These centers can be seen as sources of SARS in some districts.
3. Sequence 75 is the sequence of Pagumalarvata SZ3 (GenBank number: AY304486) (see [101]), the prevalent conclusion (including the conclusion in [101]) is that Pagumalarvata is the source of the SARS virus. However, based on the structure in Fig. 6.9, this conclusion can be challenged. If sequences 75, 36, and 47 were sequenced correctly, then 75 $\rightarrow$ 36 $\rightarrow$ 47, and double mutations happened at positions 48 and 68. If this conclusion is right, then the double mutations are the key causes of the SARS outbreak in 2003.

Mutation Network Structure of Ealy SARS Sequences

We compare the SZ16 (a) and SZ3 (b) of Pagumalarvata with SARS sequences HSZ-Bc (AY394994), GZ02 (AY390556), HSZ-Cc (AY394995), HSZ-Cb (AY394986) in the early period with the HZS2-E (AY394990) in the metaphase, respectively. We number the seven sequences as SZ16 (a), SZ3 (b), HSZ-Bc (c), GZ02 (d), HZS2-E (e), HSZ-Cb (u), HSZ-Cc (v), respectively. We then analyze their mutation network structure, and obtain the following result:

1. From the MA output of the SARS sequences, we find that

$$w_{ac} = 50\,, \quad w_{au} = 87\,, \quad w_{cu} = 37\,, \quad w_{cv} = 6\,,$$
$$w_{av} = 56\,, \quad w_{ac} + w_{cu} = w_{au}\,, \quad w_{ac} + w_{cv} = w_{av}\,.$$

   This implies that arc ac is orthogonal to cu, cv. We conclude that the SARS virus starts from SZ16 (a) (Pagumalarvata), to HSZ-Bc (c), then from HSZ-Bc to HSZ-Cb (u) and HSZ-Cc (v), respectively. i.e., the source of HSZ-Bc is SZ16, while the cause of both HSZ-Cb and HSZ-Cc is SZ16.
2. In the infection process where the SARS virus progresses from SZ16 to HSZ-Bc and then to HSZ-Cb and HSZ-Cc, the number of times mutations occur is 50, 37, and 3, respectively, and the mutation modes are also determined by the MA output.
3. The source of HSZ-Cb (u) and HSZ-Cc (v) is determined, we need only discuss the mutation structure of SZ16 (a), SZ3 (b), HSZ-Bc (c), GZ02 (d) and HZS2-E (e). This discussion is given below.

Remark 8. 1. Points a, b, c, d, e represent the five SARS sequences in the initial stage. Points f, g, h are the transitional sequences in orthogonal decomposition.
2. In the distance graph constructed by a, b, c, d, e, f, g, h nodes; thick lines are first-order arcs, and thin straight lines are second-order arcs. The numbers written on the sides of the lines represent the mutation errors.

The Analysis of the Network Structure Graph – Fig. 6.10

1. The triangles in the network structure, first-order arcs and second-order arcs are orthogonal. For example, in triangle $\Delta(a, b, h)$, the formula

$$|ab| = |ag| + |bg|\,, \quad |ah| = |ag| + |gh|\,, \quad |bh| = |bg| + |gh|$$

   holds.
2. For the 1st-order arcs in Fig. 6.10, the modulus structures are orthogonal to each other. For example, H_{ag}, H_{bg}, H_{hg} are mutually orthogonal.
3. In the SARS virus genome of Pagumalarvata, there are 20 mutation differences, and the mutation mode is H_{ab}. It may be decomposed orthogonally as $H_{ab} = H_{ag} + H_{bg}$.

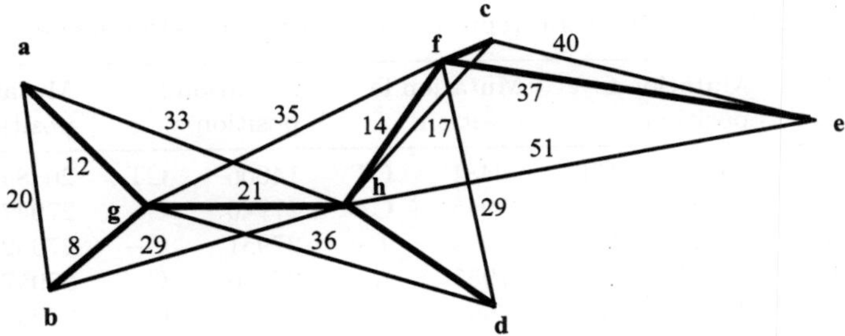

Fig. 6.10. The mutation network decomposition of SARS sequences in the initial stage

Table 6.4. The structural representation of the mutation mode H_{gh}

Mutation position	gh	Mutation position	gh	Mutation position	gh	Mutation position	gh	Mutation position	gh
1899	TG	3664	TC	6455	GA	69	CT	13882	TC
22216	AC	22317	AC	22615	CT	22974	AT	22997	GC
23356	CT	23531	CT	23641	TC	23768	GA	23802	TC
24221	GA	25340	AT	25562	AT	25598	TC	25682	GT
26464	AG								

4. When the SARS virus of Pagumalarvata infects human beings, the mutations of the genome consist of three parts: the first part is the mutation differences (i.e., H_{ag}, H_{bg}) of different Pagumalarvatas; the second part is the mutation differences (i.e., H_{hf}, H_{hc}) of different human beings; and the third part is the common mutation differences (i.e., H_{gh}) of human beings and Pagumalarvatas. We believe that the particular mutation H_{gh} is the key to how Pagumalarvata infects human beings. The mutation mode is shown in Table 6.4.

   Remark 9. The mutation position in the table is where the mutation happens. The capital letters are the nucleotides which mutate, e.g., 1899 TG means that the nucleotides in sequences g and h at the 1899th position of the alignment output are T and G, respectively.

5. After the SARS virus of Pagumalarvata infected human beings, many cases emerged. However, the SARS disease may break out only if the HZS2-E(e) virus occurs. Therefore, the mutation H_{fe} is the key to a SARS outbreak. The mutation modes are as shown in Table 6.5.

Remark 10. The data in Table 6.5 are defined the same way as those in Table 6.4.

Table 6.5. The structural representation of the mutation mode H_{fe}

Mutation position	fe	Mutation position	fe	Mutation position	fe	Mutation position	fe	Mutation position	fe
1196	CT	9406	CT	9481	CT	14606	CT	20884	AG
23873	GT	25028	GA	27945	CA	27946	C–	27947	T–
27948	A–	27949	C–	27950	T–	27951	G–	27952	G–
27953	T–	27954	T–	27955	A–	27956	C–	27957	C–
27958	A–	27959	A–	27960	C–	27961	C–	27962	T–
27963	G–	27964	A–	27965	A–	27966	T–	27967	G–
27968	G–	27969	A–	27970	A–	27971	T–	27972	A–
27973	T–	27974	A–						

Remark 11. The results listed in Tables 6.4 and 6.5 are mathematical results. They may be used as a reference for biology and medicine. Whether or not these results are correct must still be proved through observations and experiments.

The Alignment Output of the Sequences of HIV-1

Amongst the 706 HIV-1 sequences, we select 704 better sequences to be aligned. The lengths of the 704 HIV-1 sequences are within 7000–9000 bp. We produce the alignment output which is a 704 × 11,364 matrix. Because the 704 HIV-1 sequences refer to many nations or districts over a long time, we omit discussion of the alignment output.

6.4 Exercises, Analyses, and Computation

Exercise 29. Construct the phylogenetic tree and graph based on the penalty matrix in Sect. 6.3.3, according to the requirements listed below:

1. Minimum distance phylogenetic clustering tree, and the average minimum distance phylogenetic clustering tree
2. Directed and undirected minimum distance tree
3. Minimum distance two-order tree

Exercise 30. The ND1 gene coding region sequences of 20 species of mammals, and the MA outputs for 103 SARS sequences and 706 HIV-1 sequences are included on our Web site [99]. Construct the mutation network based on these datasets. Compute the stable and unstable regions for them, and represent them using modulus structure.

Exercise 31. Compute the similarity matrices of the MA outputs of the SARS sequences and HIV-1 sequences, and analyze the phylogenetic tree based on them. Also compute the following results:

1. Construct the phylogenetic clustering tree under the minimum distance.
2. Construct the first-order and second-order minimum distance undirected and directed topological distance trees, and represent the topological distance using colored arcs.
3. For the SARS sequences, construct phylogenetic trees using the maximum-likelihood method first, followed by the Bayes method.

Exercise 32. Perform MA based on the 8–12 earliest SARS sequences. Then, analyze the network structure based on the alignment output. Compute the following results:

1. Determine the stable and unstable regions, and express these using the modulus structure.
2. Construct the phylogenetic trees using minimum distance.
3. Construct first-order and second-order minimum distance undirected and directed topological distance trees, and express the topological distance using colored arcs.
4. Based on the first-order and second-order minimum distance undirected topological distance trees, perform orthogonal mutation network decomposition, and construct the graph of the orthogonal mutation network structure.
5. Based on the graph of the orthogonal mutation network structure, and using Pagumalarvata as the source of the disease gene, explain the gene mutation process and the path of the disease infection.

Exercise 33. Based on the MA outputs for the ND1 gene coding region sequences of 20 mammals, construct the phylogenetic tree according to the following typical requirements:

1. Using the characteristic value of the stable regions of MA outputs, construct the phylogenetic tree using the parsimony method.
2. Construct the phylogenetic tree using the maximum-likelihood method and the Bayes method.

Hint

Construct the phylogenetic tree for the maximum-likelihood method and the Bayes method, using the software packages Phylip [29], Paml [111], and MrBayes [44].

7

Alignment Space

In sequence alignment issues, the basic problem is the computation of the mutation error distance between two different sequences (minimum penalty or maximum score). In this chapter, we discuss the spaces created by alignment, called alignment space, in the sense of bioinformatics (although this view is applicable to many other fields, e.g., computer science, information and coding theory, cryptography, DNA computing, etc.). We do not discuss these other cases here, as they require expert knowledge in these fields.

7.1 Construction of Alignment Space and Its Basic Theorems

7.1.1 What Is Alignment Space?

Alignment space is a metric space of generalization errors. The generalization errors include the substitutions, insertions and deletions of symbols. These happen often in our daily life, for example, when writing, we may misspell a word, add an unnecessary word, or miss a word in a sentence. Therefore, generalization errors are frequently encountered in computer science, information theory and bioinformatics. The definitions and their consequences are different in different fields.

Alignment space is a very complicated nonlinear metric space. It differs from the common Euclidean space or Hamming space. The main characteristic of these two spaces is the measurement of the distance between two vectors of same length. They only measure the errors generated by substitutions. Therefore, these two spaces are linear and easy to process.

The earliest studies on generalization errors appeared in 1963, in which Levenshtein [55] defined several types of distances between two different sequences. One of them is the difference between the sum length of the two sequences and twice the length of the largest common subsequence.

We call this the L_2-distance. Another is defined by the minimal insertion/deletion/substitution operations that transform one sequence into another. In computer science, these data operations are called edit operations. We call this minimum number of operations the Levenshtein distance [68], the L_1-distance for short.

In 1974, Peter H. Sellers [85] employed the method of expansion sequences to define the evolutionary distance of two sequences, and to show that the space of sequences forms a metric space under evolutionary distance. Since the equivalence of evolutionary distance and Levenshtein distance [68] is not hard to prove, it may appear that the problem of the metric of generalization errors has been solved. However, one can find that in the proof [85] the triangle inequality of evolutionary distance is not strict. Since the structure of expansion sequences is complicated, it is difficult to describe clearly in several sentences. In this chapter, we introduce the modulus structure theory of augmented sequences and some different equivalent expressions. The modulus structure theory describes the relations and operations of different expansion sequences and shows that the operations form a Boolean algebra. Therefore, a stricter proof of Theorem 23 [85] is given. The data structure of sequences with generalization errors can then be characterized more clearly by the modulus structure theory.

Generalization errors are considered mutation errors in bioinformatics. This is one of the basic problems in bioinformatics, and we have described and discussed its function in the above text. We can say without exaggeration that the research on mutation errors is the central and essential problem in current molecular biology and bioinformatics.

In bioinformatics, the operation of seeking the mutation site is called the alignment operation on a sequence. The essential operation is to seek the minimum penalty alignment or maximum score alignment of different sequences. We call the penalty value of the minimum penalty alignment, the alignment distance of the two sequences. Alignment distance is equivalent to the evolutionary distance, so they are equivalent to Levenshtein distance. In this book, we explain the minimum penalty alignment and maximum score alignment of two nonequivalent sequences.

Because of the importance of mutations in biology, the alignment problem has been considered by many researchers from various angles; e.g., the Smith–Waterman algorithm is the dynamic programming-based method [95], the SPA algorithm is the statistic decision-based method [90], etc. The computational complexities are $O(n) - O(n^2)$. Following this research, we may quickly obtain the minimum penalty alignment of long sequences. The space consisting of all sequences with different lengths endowed with an alignment distance is called in this book an alignment space. We will then analyze its properties.

In this chapter, our intention is to discuss the data structure from the more popular and general points of view. For example, we discuss the properties and applications of the alignment space in the framework of general topological space.

7.1.2 The Alignment Space Under General Metric

Now we discuss how to measure the problem of generalization errors. There are many types of metric for measuring generalization errors; e.g., the Levenshtein distance, evolutionary distance, etc. The sequence alignment theory is very well-established, and its application in bioinformatics is very broad. Therefore, in this chapter we discuss in detail alignment distance and the corresponding alignment space. We also discuss the properties and applications of the alignment space under general conditions.

General Metric Space

Let V be a finite or infinite set. Let $V_+ = V \cup \{-\}$ be an expansion of V, which includes V and a virtual symbol "–".

Definition 34. *Let V_+ be the expansion of V and let $d_+(a,b)$ and $d(a,b)$ denote the distance functions defined on V_+ and V, respectively. If their distance functions are consistent, i.e., $d(a,b) = d_+(a,b)$ holds for all $a, b \in V$, then V is called the topological subspace of V_+, alternatively, V_+ is called the topological expansion of V.*

The main types of metric expansion space are as follows:

1. The finite set (discrete). In this case, $V = V_q = \{0, 1, \cdots, q-1\}$ is a finite set, $V_+ = V_{q+1} = \{0, 1, \cdots, q-1, q\}$. As q changes, the finite set has a different meaning. Typically, for $q = 4$, $V_q = \{a, c, g, t\}$ or $V_q = \{a, c, g, u\}$ form the familiar nucleotide table; and for $q = 20$, V_q is the familiar amino acid table.
   On a finite set V, the distance function $d(a,b)$ is represented by a distance matrix $D = (d(a,b))_{a,b \in V}$, e.g., the Hamming matrix of (1.6), or the WT-matrix of (1.7), etc. In this case, $\{V, D\}$ is a metric space.
2. The bounded infinite set. For example, $V = [0, q]$, $q > 0$ is a bounded interval, and a continuous set. There are many ways to express the distance between pairs of elements. For example, the mean square error (MSE) distance is: $d(a,b) = (b-a)^2$, absolute error distance is: $d(a,b) = |b-a|$, etc. Endowed with any one distance, $\{V, D\}$ is a metric space. Using any one of the above mentioned distance functions, the expansion distance function defined on the expansion space V_+ can be extended as follows:

$$d_+(a,b) = \begin{cases} d(a,b)\,, & \text{if both } a, b \neq \text{“–”}, \\ 0\,, & \text{if both } a, b = \text{“–”}, \\ q\,, & \text{if only one of } a, b \text{ is “–”, while } d(a,b) \text{ is the} \\ & \text{absolute error distance,} \\ q^2\,, & \text{if only one of } a, b \text{ is “–”, while } d(a,b) \text{ is the} \\ & \text{mean square error distance,} \end{cases} \tag{7.1}$$

then $\{V_+, D_+\}$ is a metric space.

3. The unbounded infinite set. For example, $V = (0, \infty)$ is an unbounded interval. Then the mean square error (MSE) distance, absolute error distance, etc., can be employed as the distance $d(a, b)$ on V. The extended distance on the expansion space V_+ can be defined as follows:

$$d_+(a,b) = \begin{cases} d(a,b)\,, & \text{if both } a, b \neq \text{“}-\text{”}\,, \\ 0\,, & \text{if both } a, b = \text{“}-\text{”}\,, \\ \delta + d(a,0)\,, & \text{if } a \in V \text{ and } b = \text{“}-\text{”}\,, \\ \delta + d(0,b)\,, & \text{if } b \in V \text{ and } a = \text{“}-\text{”}\,, \end{cases} \tag{7.2}$$

where $\delta > 0$ is a constant. Under this distance, we know that $\{V_+, D_+\}$ is a metric space.

The Problem of Sequence Alignment and the Definition of Alignment Space

The definition of pairwise alignment in this section is the same as that in Sect. 1.4.2. The slight difference is that here, V, V_+ are general metric spaces. For any two sequences $A = (a_1, a_2, \cdots, a_{n_a})$ and $B = (b_1, b_2, \cdots, b_{n_b})$ whose range of values is in V, the corresponding terms involved in pairwise alignment are stated as follows.

Definition 35. *1. If A and A' are two sequences in V and V_+, respectively, A' is the expansion of A if A remains after all the “$-$” in A' are deleted.*

2. If A, B are two sequences ranging into V, (A', B') is the alignment of (A, B) if A' and B' are the expansions of A and B, respectively, and their lengths are the same n'.
If (A', B') are the alignment of (A, B), and $d_+(a, b), a, b \in V_+$ is the distance function on V_+, then $d_+(A', B') = \sum_{i=1}^{n'} d_+(a_i, b_i)$ is defined as the distance between A' and B'.

3. Given a distance function d_+, (A', B') is the minimum penalty alignment of (A, B) if (A', B') are the alignment sequences of (A, B), and $d_+(A', B') \leq d_+(A'', B'')$ holds for any other alignment (A'', B''). $d_{\mathrm{A}}(A, B) = d_+(A', B')$ is called the alignment distance of (A, B). It is also referred to simply as the A-distance.

Based on Definition 35, we find the alignment distance uniquely for any pair A, B with values ranging on V, and this is a perfectly computational method for evaluating the distance in bioinformatics. For discrete or continuous bounded metric spaces, we may use the Smith–Waterman dynamic programming-based algorithm directly to compute the distance. For continuous unbounded metric space, the computation principle of the Smith–Waterman dynamic programming-based algorithm still may be used with a minor revision if we adapt the distance function defined by (7.2). Thus the corresponding software used in bioinformatics may still be used, albeit with revisions.

The elements in alignment space are the sequences with values ranging in V with different lengths. Let V^n be the set of all sequences with length n and whose values range in V, so that the alignment space V^* may be represented by $V^* = \bigcup_{n=1}^{\infty} V^n$. Typically, following from Definition 35, we know that for any $A, B \in V_q^*$, the A-distance is determined, and (V_q^*, d_{A}) is an alignment space or an A-space for short.

The Levenshtein Distance of Generalization Errors

We have mentioned the Levenshtein distance and evolutionary distance in the context of the alignment distance. In most cases, they are not equivalent. To understand how to process the generalization errors, we show the corresponding definitions and discuss their relationships. There are many ways to define the Levenshtein distance. Let A, B be two sequences with values ranging in V.

Definition 36. *1. (s, i, j) is an operation on A, which means that we change s many symbols in A, insert i many components, and delete j many components.*

2. B is an output of A under the operation (s, i, j) if A turns into B via the operation (s, i, j). Then (s, i, j) is called an operation from A to B.

3. (s_0, i_0, d_0) is a minimum operation from A to B if (s_0, i_0, j_0) is an operation from A to B, and $s_0 + i_0 + j_0 \leq s + i + j$ holds for any other operation (s, i, j) from A to B. Let $d_{\mathrm{L}_1}(A, B) = s_0 + i_0 + j_0$ be the L_1-distance of A, B, which is also called the evolutionary distance (or E-distance for short).

The L_1-distance can be also extended to the case where (s, i, j) has a different weight. If V is a general metric space with distance function $d(a, b)$, then the L_1-distance $d_{\mathrm{L}_1}(A, B)$ involves both the errors caused by s, i, j and the symbol of s, i, j.

The definition of L_2-distance for a sequence pair (A, B) is as follows:

Let $W = (w_1, w_2, \cdots, w_{n_w})$ be a sequence; if there is a subsequence $1 \leq i_1 < i_2 < \cdots < i_{n_w} \leq n_a$ such that $a_{i_j} = w_j$, $j = 1, 2, \cdots, n_w$ holds, then W is a subsequence of A. Then let

$$\rho(A, B) = \max\{|W| : W \text{ is the common subsequence of } A, B\} , \tag{7.3}$$

where $|W|$ is the length of sequence W. If W_0 is the common subsequence of A, B and $|W_0| = \rho(A, B)$, then W_0 is the largest common subsequence of A, B. The L_2-distance is defined as follows:

$$d_{\mathrm{L}_2}(A, B) = |A| + |B| - 2 \cdot \rho(A, B) . \tag{7.4}$$

The evolutionary distance is equivalent to the alignment distance, however, we discuss them in different ways. For the L-distance and the A-distance, we can describe their simple properties as follows:

If the penalty matrix d_+ is given, then for every pair of sequences A, B with values ranging on V, their L_1-distance, L_2-distance, and A-distance are uniquely determined. However, both the largest common subsequence W_0 of (A, B) and the minimum penalty alignment sequences (A', B') of (A, B) are not unique, and the minimum operation from A to B is not unique either.

The L_1-distance is equivalent to the A-distance. Nevertheless, the L_2-distance may not be equivalent to the A-distance. We illustrate this with the following example.

Example 21. If

$$\begin{cases} A = 00000111\,, \\ B = 11111000\,, \end{cases}$$

and if the penalty matrix is the Hamming matrix, we have

$$d_{\mathrm{A}}(A, B) = 8\,, \quad \rho(A, B) = 3\,, \quad d_{\mathrm{L}}(A, B) = 16 - 2 \times 3 = 10 \neq 8\,.$$

The Generalization of the Basic Theorem About the A-Distance

In Theorem 23, we proved that $\{V^*, D_A\}$ is a metric space if $\{V_+, D_+\}$ is a general metric space. The conclusion can be extended to the more general topological metric space.

Theorem 29. *If $\{V_+, D_+\}$ is a general metric space, then $\{V^*, D_A\}$ is a metric space, where the definitions of set V^* and the A-distance $d_{\mathrm{A}}(A, B)$ are the same as Definition 35. $d_{\mathrm{A}}(A, B)$ is the distance function defined on V^*, which means $d_{\mathrm{A}}(A, B)$ satisfies the three conditions of a distance function: nonnegativity, symmetry, and the triangle inequality.*

The proof is the same as that of Theorem 23. We can easily extend the proof of Theorem 23 to the general topological metric space $\{V_+, D_+\}$, so we do not need to repeat it here.

7.2 The Analysis of Data Structures in Alignment Spaces

For any pair of sequences $A, B \in V_q^*$, we may discuss structures beyond the A-distance. We will describe these other structures gradually.

7.2.1 Maximum Score Alignment and Minimum Penalty Alignment

The definitions of the maximum score alignment and the minimum penalty alignment have been mentioned before this section. We have explained that the maximum score alignment and the minimum penalty alignment are not the same in the general case. We now explain this in more detail.

The Uniqueness Problem of Optimal Alignment

In Example 2 we have shown that the minimum penalty alignment is not generally unique. The same example can also be used to explain the nonuniqueness of the maximum score alignment. Here, we explain why the minimum penalty alignment as well as the lengths of the alignment sequences may not be generally unique.

Example 22. If

$$\begin{cases} A = 00000111\,, \\ B = 11111000\,, \end{cases}$$

and if the penalty matrix is the Hamming matrix, then we have $d_{\text{A}}(A, B) = 8$. If we construct the sequences

$$\begin{cases} A_1 = 200000111\,, \\ B_1 = 111110002\,, \end{cases} \qquad \begin{cases} A_2 = 2200000111\,, \\ B_2 = 1111100022\,, \end{cases} \qquad \begin{cases} A_3 = 22200000111\,, \\ B_3 = 11111000222\,, \end{cases}$$

then

$$d_{\text{H}}(A, B) = d_{\text{H}}(A_1, B_1) = d_{\text{H}}(A_2, B_2) = d_{\text{H}}(A_3, B_3) = 8\,.$$

All of these are minimum penalty alignment sequences of (A, B), and their corresponding lengths are found as follows:

$$|A| = |B| = 8\,, \quad |A_1| = |B_1| = 9\,, \quad |A_2| = |B_2| = 10\,, \quad |A_3| = |B_2| = 11\,.$$

Thus, the lengths of minimum penalty alignment sequences may not be unique.

For a fixed pair (A, B), let $n_0(A, B)$ be the smallest length among the optimal alignment of (A, B), and we denote by $n_1(A, B)$ the largest length of the optimal alignment sequences of (A, B).

The Relationship Between the Maximum Score Alignment and the Minimum Penalty Alignment

Based on Definitions 3, we introduce the definition of the maximum score alignment sequence if $g_{\text{H}}(a, b) = 1 - d_{\text{H}}(a, b)$, $a, b \in V_5$. Following from Example 22, we find that the maximum score alignment and minimum penalty alignment are not the same. For example, here, (A_1, B_1) and (A_2, B_2) are the minimum penalty alignment sequences of (A, B), which do not represent its maximum score alignment sequences. (A_3, B_3) are both the minimum penalty and maximum score alignment sequences of (A, B).

Lemma 2. *The propositions below hold under the Hamming penalty and scoring matrix condition.*

1. *The minimum penalty alignment sequences* (A', B') *of sequences* (A, B) *are the maximum score alignment sequences of* (A, B) *if and only if the length* n' *of* (A', B') *is the largest length of the minimum alignment sequences of* (A, B).
2. *The maximum score alignment sequences* (A^*, B^*) *of sequences* (A, B) *are the minimum penalty alignment sequences of* (A, B) *if and only if the length* n^* *of* (A^*, B^*) *is the smallest length of maximum score alignment sequences of* (A, B).

Proof. Under the Hamming penalty and scoring matrix condition, for any alignment sequences (A', B') of (A, B) such that the equation $n' = d_{\rm H}(A', B') + g_{\rm H}(A', B')$ holds; if (A', B') are the minimum penalty alignment sequences, then only if n' is the maximal value do we have that (A', B') are the maximum score alignment sequences of (A, B). Proposition 1 of the lemma is correct.

Proposition 2 can be proved similarly.

Lemma 2 gives a necessary condition under which the minimum penalty alignment sequences are the maximum score alignment sequences. The following example expresses this: the maximum score alignment sequences are not always the minimum penalty alignment sequences.

Example 23. If

$$\begin{cases} A = 00001011\,, \\ B = 10111000\,, \end{cases}$$

then $A' = A$, $B' = B$ are the minimum penalty alignment sequences, $d_{\rm A}(A, B) = d_{\rm H}(A', B') = 5$, while the maximum score alignment sequences are

$$\begin{cases} A^* = 000010211222\,, \\ B^* = 222210111000\,, \end{cases}$$

maximum score $g(A^*, B^*) = 4(g(A', B') = 3)$. However, $d_{\rm H}(A^*, B^*) = 8$, so (A^*, B^*) are not the minimum penalty alignment sequences of (A, B).

7.2.2 The Structure Mode of the Envelope of Pairwise Sequences

The definitions of the envelope and core of multiple sequences, minimal (minimum) envelope and maximal (maximum) core, are given in Sect. 3.1. They are closely related to the data structure of alignment space. Let $A = (a_1, a_2, \cdots, a_{n_a})$, $B = (b_1, b_2, \cdots, b_{n_b})$ be two sequences with values ranging in V_q; let $C = (c_1, c_2, \cdots, c_{n_c})$, $D = (d_1, d_2, \cdots, d_{n_d})$ be the envelope and core of A, B. We now discuss the relationship between their structures.

The Structure Mode of the Envelope

If sequence C is the envelope of sequences A, B, then there must be two subsets Δ'_a and Δ'_b in N_c such that $c_{\Delta'_a} = A$ and $c_{\Delta'_b} = B$ hold. Δ'_a, Δ'_b are the sets of envelope sites. We discuss them as follows:

1. Let
$$\Delta'_0 = \Delta'_a \cap \Delta'_b\,, \quad \Delta'_1 = \Delta'_a - \Delta'_0,\; \Delta'_2 = \Delta'_b - \Delta'_0\,,$$
$$\Delta'_3 = N_c - (\Delta'_a \cup \Delta'_b) = N_c - (\Delta'_0 \cup \Delta'_1 \cup \Delta'_2)\,.$$
Then $\Delta'_0, \Delta'_1, \Delta'_2, \Delta'_3$ are four disjoint sets and their union is N_c.
2. If C is the minimal envelope of A and B, then Δ'_3 must be an empty set. We now discuss the structure under the minimal envelope condition.
3. For each $i \in N_c$, we define the structure functions $\tau_i = 0, 1, 2$, if $i \in \Delta'_0, \Delta'_1, \Delta'_2$ of sequence C. Obviously, the envelope C and the structure function $\bar{\tau} = (\tau_1, \tau_2, \cdots, \tau_{n_c})$ determine each other uniquely if sequences A and B are determined.

The structure function $\bar{\tau}$ is called the structure mode of the envelope C.

The Permutation of the Envelope

Let C be the envelope of sequences A and B. For any positions $i < j \in N_c$, permutated by c_i and c_j, sequence C becomes:

$$C' = (c_1, c_2, \cdots, c_{i-1}, c_j, c_{i+1}, c_{i+2}, \cdots, c_{j-1}, c_i, c_{j+1}, \cdots, c_{n_c})\,.$$

Let $C' = \sigma_{i,j}(C)$. If C' is also the envelope of A and B, then the permutation $\sigma_{i,j}$ is called an isotone permutation. Then C and C' are the equivalent envelopes of A and B.

Lemma 3. *In the structure mode $\bar{\tau}$ of the envelope C, if $\tau_i \neq \tau_{i+1} \in \{1, 2\}$, the permutation $\sigma_{i,i+1}$ must be the isotone permutation of C.*

The proof is obvious.

The Standard Structure Mode of the Envelope C

Definition 37. *If the structure mode $\bar{\tau}$ of the envelope C satisfies the following conditions, we may say that the envelope C has a standard structure mode:*

1. *Vector $\bar{\tau}$ can be decomposed into several alternating subvectors, let*
$$\bar{\tau} = ((\bar{e}_{0,1}, \bar{e}_{0,2}, \bar{e}_{0,0}), (\bar{e}_{1,1}, \bar{e}_{1,2}, \bar{e}_{1,0}), \cdots, (\bar{e}_{k_c,1}, \bar{e}_{k_c,2}, \bar{e}_{k_c,0}))\,, \tag{7.5}$$
where $\bar{e}_{k,\tau} = (\overbrace{\tau, \tau, \cdots, \tau}^{\ell_{k,\tau}})$ is a vector with a value $\tau \in \{0, 1, 2\}$ and length $\ell_{k,\tau}$.

2. In formula (7.5), the inequalities

$$\ell_{0,1} + \ell_{0,2} + \ell_{0,0} > 0 \quad \textit{and} \quad \ell_{k_c,1} + \ell_{k_c,2} + \ell_{k_c,0} > 0$$

hold. For any $k = 0, 1, \cdots, k_c$,

$$\ell_{k,3} > 0\,, \quad \ell_{0,1}, \ell_{k,2}, \ell_{k,0} \geq 0 \tag{7.6}$$

holds. If $\ell_{k,\tau} = 0$*, then vector* $\bar{e}_{k,\tau}$ *does not exist.*

Example 24. The following mode is a standard structure mode of an envelope.

$$\bar{\tau} = (11, 222, 0, 1111, 22, 00000, 2222, 0000000, 11111, 000, 111, 22, 0000, 111)\,,$$

where $k_c = 5$, and the value of $\ell_{k,\tau}$ is assigned as follows:

$$\begin{array}{llllll}
\ell_{0,1} = 2\,, & \ell_{0,2} = 3\,, & \ell_{0,0} = 1\,, & \ell_{1,1} = 4\,, & \ell_{1,2} = 2\,, & \ell_{1,0} = 5\,, \\
\ell_{2,1} = 0\,, & \ell_{2,2} = 4\,, & \ell_{2,0} = 7\,, & \ell_{3,1} = 5\,, & \ell_{3,2} = 0\,, & \ell_{3,0} = 3\,, \\
\ell_{4,1} = 3\,, & \ell_{4,2} = 2\,, & \ell_{4,0} = 4\,, & \ell_{5,1} = 3\,, & \ell_{5,2} = 0\,, & \ell_{5,0} = 0\,.
\end{array}$$

Using Lemma 3, we may conclude that all envelopes of sequences A and B must be equivalent to an envelope with a standard structure mode.

In the standard structure mode of the envelope, the sequence C may always be denoted by:

$$C = (c_{\delta_0}, c_{\delta_1}, \cdots, c_{\delta_{k_c}})\,, \tag{7.7}$$

where $\delta_k = (\delta_{k,1}, \delta_{k,2}, \delta_{k,3})$, and each $\delta_{k,\tau}$ is an integer vector of length $\ell_{k,\tau}$. These were arranged in order. Then, $c_{\delta_k} = (c_{\delta_{k,1}}, c_{\delta_{k,2}}, c_{\delta_{k,3}})$ is called a structure segment of envelope C.

In the structure segment c_{δ_k}, $c_{\delta_{k,3}}$ is called the last half-part of the structure segment, and $(c_{\delta_{k,1}}, c_{\delta_{k,2}})$ is called the first half-part of the structure segment.

The length of the last half-part of the structure segment c_{δ_k} in envelope C is $\ell_{k,1} + \ell_{k,2}$, where $\ell_{k,1}$ components are selected from A, and $\ell_{k,2}$ components are selected from B. Thus, the number of different selection methods is

$$\left(\frac{(\ell_{k,1} + \ell_{k,2})!}{\ell_{k,1}!\ell_{k,2}!} \right)\,.$$

The number of the envelopes which are equivalent to envelope C is

$$M(C) = \prod_{k=1}^{k_c} \left(\frac{(\ell_{k,1} + \ell_{k,2})!}{\ell_{k,1}!\ell_{k,2}!} \right)\,. \tag{7.8}$$

Properties of the Minimal (Minimum) Envelope

Let C be an envelope of multiple sequences $\mathcal{A}$ and let $\bar{\alpha} = (\alpha_1, \alpha_2, \cdots, \alpha_m)$ be the set of generation positions of envelope C, here $c_{\alpha_s} = A_s$, where $s = 1, 2, \cdots, m$ holds. If Δ is a subset of N_c, let $\alpha_s(\Delta') = \alpha_s \cap \Delta'$, then $c_{\alpha_s(\Delta')}$ must be a subsequence of A_s.

Lemma 4. *If C is a minimal (minimum) envelope of A and B, then $c_{\Delta'}$ must be a minimal (minimum) envelope of $c_{\alpha_s(\Delta')}$, $s = 1, 2, \cdots, m$.*

Proof. We prove this proposition by reduction to absurdity. If $c_{\Delta'}$ is not a minimal envelope of $c_{\alpha_s(\Delta')}$, where $s = 1, 2, \cdots, m$, then there exists a $j \in \Delta'$ such that $c_{\Delta''}$ is an envelope of $\bar{c}(\Delta')$, here $\Delta'' = \Delta' - \{j\}$. There now is a subset $\beta_s \subset \Delta''$ such that $c_{\beta_s} = c_{\alpha_s(\Delta')}$, where $s = 1, 2, \cdots, m$.

Since C is an envelope of (A, B), then $c_{(\Delta')^c}$ must be an envelope of

$$\bar{c}((\Delta')^c) = \left\{c_{\alpha_1((\Delta')^c)}, c_{\alpha_2((\Delta')^c)}, \cdots, c_{\alpha_m((\Delta')^c)}\right\} ,$$

where $\alpha_s((\Delta')^c) = \alpha_s \cap (\Delta')^c$, while $(\Delta')^c = N_c - \Delta'$. On the other hand, we obtain

$$A_s = c_{\alpha_s} = \left(c_{\alpha_s(\Delta')}, c_{\alpha_s((\Delta')^c)}\right) = \left(c_{\beta_s}, c_{\alpha_s((\Delta')^c)}\right) . \tag{7.9}$$

If we let $N'_c = \{1, 2, \cdots, j-1, j+1, \cdots, n_c\}$, then $\beta_s \cup (\Delta')^c \subset N'_c$. Following from (7.9), we find that $c_{N'_c} = (c_1, c_2, \cdots, c_{j-1}, c_{j+1}, \cdots, c_{n_c})$ is an envelope of (A, B). This contradicts the definition that C is a minimal envelope of (A, B). Therefore, the lemma is proved.

Theorem 30. *If C is a minimum envelope of A, B with a standard structure mode, and c_{δ_k} is a structure segment of C, then in the first half-part of the structure segment, any component in vector $c_{\delta_{k,1}}$ is different from that in $c_{\delta_{k,2}}$.*

Proof. We prove this proposition by reduction to absurdity. For a standard structure mode in (7.7), let

$$\begin{aligned} 1 = j_{0,1} \le j_{0,2} \le j_{0,0} \le j_{1,1} \le j_{1,2} \le j_{1,0} \le \cdots \\ \le j_{k_c,1} \le j_{k_c,2} \le j_{k_c,0} \le j_{k_c+1,1} = n_c , \end{aligned} \tag{7.10}$$

where $\ell_{k,\tau} = j_{k,\tau+1} - j_{k,\tau}$, $\tau = 1, 2, 3$. We denote $j_{k,4} = j_{k+1,1}$. Then $\ell_{k,\tau}$ satisfies formula (7.6).

If there is a structure segment c_{δ_k} such that the conclusion of the theorem is wrong, then there must be a $0 \le k \le k_c$ such that a component in $c_{\delta_{k,1}}$ is the same as that in $c_{\delta_{k,2}}$. Let j_1, j_2 satisfy $c_{j_1} = c_{j_2}$ and let $j_{k,1} < j_1 \le j_{k,2} < j_2 \le j_{k+1,0}$ hold. We construct a new sequence C' as follows:

$$C' = (c_1, c_2, \cdots, c_{j_1-1}, c_{j_{k,2}+1}, c_{j_{k,2}+2}, \cdots, c_{j_2-1}, c_j, \cdots, c_{n_c}) , \tag{7.11}$$

where $j = j_1 - 1 - j_{k,1} + j_2 - 1 - j_{k,2}$. C' is also the envelope of sequences A and B with a standard structure mode as follows:

$$\begin{cases} \delta'_{k',\tau} = \delta_{k,\tau} \,, \quad k' = 0, 1, \cdots, k-1 \,, \quad \tau = 1, 2, 3 \,, \\ \delta'_{k,1} = [j_{k,1} + 1, j_1 - 1] \,, \\ \delta'_{k,2} = [j_{k,2} + 1, j_2 - 1] \,, \\ \delta'_{k,0} = \{j\} = \{j_1 - 1 - j_{k,1} + j_2 - 1 - j_{k,2}\} \,, \\ \delta'_{k+1,1} = [j + 1, j + j_{k,2} - j_1] \,, \\ \delta'_{k+1,2} = [j + j_{k,2} - j_1 + j_{k,0} - j_2] \,, \\ \delta'_{k+1,0} = \delta'_{k,0} \,, \\ \delta'_{k+k',\tau} = \delta'_{k+k',\tau} \,, \quad k' = 2, 3, \cdots, k_c - k + 1 \,, \quad \tau = 1, 2, 3 \,. \end{cases} \tag{7.12}$$

However, the length of C' is $n_c - 1$. This contradicts the definition that C is the minimum envelope of A, B. The theorem is therefore proved.

7.2.3 Uniqueness of the Maximum Core and Minimum Envelope of Pairwise Sequences

The following example tells us that the maximum core and minimum envelope of pairwise sequences are not generally unique.

Example 25. 1. Let

$$\begin{cases} A = (11111011111) \\ B = (11110111111) \,, \end{cases}$$

then their maximum cores are as follows:

$$D = (1111011111) \,, \quad D' = (1111111111) \,.$$

Then $|D| = |D'| = 10$, and they are both the largest subsequence of A and B. In other words, they are both the maximum core of A and B. It follows that the maximum core of pairwise sequences is not always unique. In this example the minimum envelope $C = (111110111111)$ of A and B is unique.

2. Let

$$\begin{cases} A = (11111011110) \\ B = (111101111111) \,, \end{cases}$$

then the maximum core: $D = (1111011110)$ is unique, but the minimum envelopes are not unique, which is stated as follows:

$$C_1 = (11111011110111) \,, \quad C_2 = (11111011111011) \,,$$
$$C_3 = (11111011111101) \,, \quad C_4 = (11111011111110) \,.$$

These two examples show that maximum core and minimum envelope of pairwise sequences are not always unique. They are unable to determine each other.

7.2.4 The Envelope and Core of Pairwise Sequences

The Core Generated by the Envelope

If sequence C is the envelope of sequences A and B, then there are two subsets Δ'_a and Δ'_b in N_c such that $c_{\Delta'_a} = A$, and $c_{\Delta'_b} = B$. Moreover, the following properties hold:

1. Let
$$\Delta'_0 = \Delta'_a \cap \Delta'_b\,, \quad \Delta'_1 = \Delta'_a - \Delta'_0\,, \quad \Delta'_2 = \Delta'_b - \Delta'_0\,,$$
$$\Delta'_3 = N_c - (\Delta'_a \cup \Delta'_b) = N_c - (\Delta'_0 \cup \Delta'_1 \cup \Delta'_2)\ .$$
Then Δ'_0, Δ'_1, Δ'_2, Δ'_3 are four disjoint sets, and their union is N_c.
2. If sequence C is the minimal envelope of sequences A and B, then Δ'_3 must be an empty set.
In the following discussions, we always assume that C is the minimal envelope.
3. Let $D = c_{\Delta'_0}$ be the core of sequences A and B, it is referred to as the core of A and B generated by envelope C.
4. If sequence D is the core of A and B generated by the minimal envelope C, then $n_d = n_a + n_b - n_c$ holds.

The Modulus Structure of Core

If sequence D is the core of sequences A and B, then sequences A and B are both expansions of D. The modulus structures of expansions are denoted by

$$K_\tau = \{i_{\tau,0}, i_{\tau,1}, i_{\tau,2}, \cdots, i_{\tau,2k_\tau-1}, i_{\tau,2k_\tau}, i_{\tau,2k_\tau+1}\}\,, \quad \tau = a, b\ . \tag{7.13}$$

They satisfy the condition

$$0 = i_{\tau,0} \le i_{\tau,1} < i_{\tau,2} < \cdots < i_{\tau,2k_\tau-1} < i_{\tau,2k_\tau} \le i_{\tau,2k_\tau+1} = n_\tau\ . \tag{7.14}$$

This can generate small intervals

$$\delta_{\tau,k} = [i_{\tau,k} + 1, i_{\tau,k+1}]\,, \quad k = 0, 1, 2, \cdots, k_\tau\ .$$

If we let $\Delta_\tau = \bigcup_{k=1}^{k_\tau} \delta_{\tau,2k-1}$, then $a_{\Delta_a} = b_{\Delta_b} = D$ holds. Therefore,

$$||\Delta_a|| = \sum_{k=1}^{k_a} (i_{a,2k} - i_{a,2k-1}) = ||\Delta_b|| = \sum_{k=1}^{k_b} (i_{b,2k} - i_{b,2k-1})\ . \tag{7.15}$$

The Envelope Generated by the Core

Without loss of generality, we may construct the envelope C of A and B in the way shown in Fig. 7.1 based on the augmented mode H_a and H_b (from D to A and B, respectively) given in (7.15):

1. In Fig. 7.1, $k_a = k_b = 1$, hence
$$H_a = \{(i_{a,1}, \ell_{a,1})\}, \quad H_b = \{(i_{b,1}, \ell_{b,1})\} .$$
2. We now begin to compare the lengths of all small intervals. Let $\ell_{\tau,k} = i_{\tau,k+1} - i_{\tau,k}$, $\tau = a, b$, $k = 0, 1$, in which, $i_{\tau,0} = 1$, $i_{\tau,2} = n_\tau$, and $\tau = a, b$. The lengths of all small intervals in Fig. 7.1 satisfy the following relationship:
$$\ell_{a,0} < \ell_{b,0} < \ell_{a,0} + \ell_{a,1} < \ell_{b,0} + \ell_{b,1} < \ell_{b,0} + \ell_{b,1} + \ell_{b,2} = \ell_{a,0} + \ell_{a,1} + \ell_{a,2} .$$
3. In view of the above relationship, we cut the small intervals $\delta_{\tau,k}$ where $\tau = a, b$ and $k = 1, 3, 5$. Let
$$i_{d,0} = 0 < i_{d,1} = i_{a,1} < i_{d,2} = i_{b,1} < i_{d,3} = n_d .$$
Then we construct sequence C, letting
$$j_0 = 0, \quad j_1 = i_{a,1}, \quad j_2 = i_{a,1} + \ell_{a,1}, \quad j_3 = i_{b,1} + \ell_{a,1},$$
$$j_4 = i_{b,1} + \ell_{a,1} + \ell_{b,1}, \quad j_5 = n_c = n_a + \ell_{b,1} = n_b + \ell_{a,1} .$$
4. We then let
$$\begin{cases} \delta_{d,k} = [i_{d,k} + 1, i_{d,k+1}], & k = 0, 1, 2, \\ \delta_{c,k} = [j_k + 1, j_{k+1}], & k = 0, 1, 2, 3, 4, \end{cases}$$

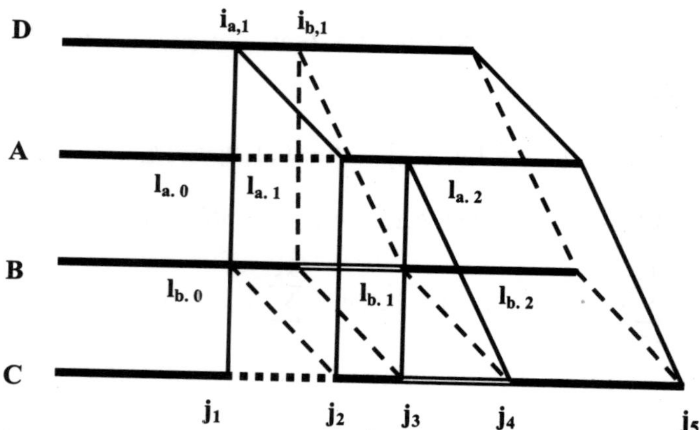

Fig. 7.1. Data relationship to find the envelope based on the core

and also

$$\begin{cases} c_{\delta_{c,2k}} = d_{\delta_{d,k}} , & k = 0,1,2 , \\ c_{\delta_{c,1}} = [a_{i_{a,1}+1}, a_{i_{a,1}+\ell_{a,1}}] , \\ c_{\delta_{c,3}} = [b_{i_{b,1}+1}, b_{i_{b,1}+\ell_{b,1}}] . \end{cases}$$

Then the sequence C is the envelope of sequences A and B generated by the core D.

The Theorem for the Relationship Between the Maximum Core and the Minimum Envelope

Based on the above discussion, we may prove a theorem that reflects the relationship between the maximum core and the minimum envelope.

Theorem 31. *1. If sequence D is the core of sequences A and B generated by minimum envelope C, then D must be the maximum core of sequences A and B.*

2. If sequence C is the envelope of sequences A and B generated by maximum core D, then C must be the minimum envelope of sequences A and B.

Proof. We prove proposition 1 by reduction to absurdity. If D is not the maximum core of sequences A and B then there must be a core D' of sequences A and B such that $n_{d'} > n_d$.

On the other hand, the envelope $C' = C(A, B; D')$ of sequences A and B can be generated by core D', so the lengths of sequences A, B, C', and D' satisfy the expression

$$n_{c'} = n_a + n_b - n_{d'} < n_a + n_b - n_d = n_c .$$

This contradicts the definition that C is the minimum envelope of sequences A and B. Consequently, $D = D(A, B, C)$ must be maximum core. Proposition 1 is proved.

Since proposition 2 can be proved similarly, we omit it here.

7.2.5 The Envelope of Pairwise Sequences and Its Alignment Sequences

The Alignment Sequences Generated by the Envelope of Pairwise Sequences

We discuss the sequences A, B, C, and D, where D is the core of A and B generated by envelope C of A and B, or C is the envelope of A, B generated by core D. The standard structure modes are given in (7.7), in which $\delta_k = (\delta_{k,1}, \delta_{k,2}, \delta_{k,3})$. Thus, the alignment sequences (A', B') may be obtained based on (A, B) in the following way:

$$a'_j = \begin{cases} c_j\,, & \text{when } j \in \Delta_a = \Delta_1 \cup \Delta_3\,, \\ q\,, & \text{when } j \in \Delta_2\,, \end{cases}$$
$$b'_j = \begin{cases} c_j\,, & \text{when } j \in \Delta_b = \Delta_2 \cup \Delta_3\,, \\ q\,, & \text{when } j \in \Delta_1\,, \end{cases} \tag{7.16}$$

in which $\Delta_\tau = \bigcup_{k=0}^{k_c} \delta_{k,\tau}$, where $\tau = 1, 2, 3$. Then, (A', B') are obviously the alignment sequences of (A, B), but not the optimal alignment sequences.

If sequence C is the minimum envelope of sequences A, B, then we construct the alignment sequences (A', B') of (A, B) as follows:

1. In the standard modulus structure (7.7) of the envelope of sequences A, B with C, we define

$$\begin{cases} \ell_{k,4} = \max\{\ell_{k,1}, \ell_{k,2}\}\,, & k = 0, 1, \cdots, k_c\,, \\ \ell_{k,5} = \min\{\ell_{k,1}, \ell_{k,2}\}\,, & k = 0, 1, \cdots, k_c\,. \end{cases} \tag{7.17}$$

   Furthermore, we find the interval sequence:

$$0 = j'_0 \le j'_1 < j'_2 < \cdots < j'_{2k_c} \le j'_{2k_c+1} = n_{c'}\,, \tag{7.18}$$

   in which $n_{c'} = \sum_{k=0}^{k_c} (\ell_{k,0} + \ell_{k,3})$ and

$$\begin{cases} j'_{2k+1} - j'_{2k} = \ell_{k,0}\,, & k = 0, 1, \cdots, k_c\,, \\ j'_{2k} - j'_{2k-1} = \ell_{k,3}\,, & k = 1, 2, \cdots, k_c\,. \end{cases}$$

   We then let $\delta'_k = [j'_k + 1, j'_{k+1}]$, where $k = 0, 1, \cdots, k_c$.
2. Based on $\delta'_k = [j'_k + 1, j'_{k+1}]$ and sequence C, we construct the segments $a'_{\delta'_{2k},\delta'_{2k+1}}$, $b'_{\delta'_{2k},\delta'_{2k+1}}$ of A' and B' for $k = 0, 1, \cdots, k_c - 1$ in turn. Let

$$a'_{\delta'_{2k+1}} = b'_{\delta'_{2k+1}} = c_{\delta_{k,3}}\,, \tag{7.19}$$

   and also

$$a'_{\delta'_{2k}} = \begin{cases} c_{\delta_{k,1}}\,, & \text{if } \ell_{k,0} = \ell_{k,1}\,, \\ \left(c_{\delta_{k,1}}, \overbrace{(4, 4, \cdots, 4)}^{\ell_{k,0}-\ell_{k,1}}\right), & \text{if } \ell_{k,0} > \ell_{k,1}\,, \end{cases}$$

   and

$$b'_{\delta'_{2k}} = \begin{cases} c_{\delta_{k,2}}\,, & \text{if } \ell_{k,0} = \ell_{k,2}\,, \\ \left(c_{\delta_{k,2}}, \overbrace{(4, 4, \cdots, 4)}^{\ell_{k,0}-\ell_{k,2}}\right), & \text{if } \ell_{k,0} > \ell_{k,2}\,. \end{cases} \tag{7.20}$$

Obviously, the sequences (A', B') induced by (7.19) and (7.20) are the alignment sequences of (A, B), and are called the alignment sequences generated by envelope C.

Theorem 32. *Under the Hamming score matrix, if sequence C is the minimum envelope of (A, B), then the alignment sequences (A', B') generated by envelope C must be the optimal alignment sequences of (A, B).*

Proof. We show this using reduction to absurdity. If the alignment sequences (A', B') generated by envelope C are not the optimal alignment sequences of (A, B), then there must be a pairwise sequences (A'', B'') such that (A'', B'') are the alignment sequences of (A, B) and $g_{\mathrm{H}}(A'', B'') > g_{\mathrm{H}}(A', B')$. We denote

$$\Delta_3'' = \{j \,|\, a_j'' = b_j'' \in V_4,\ j \in N''\}, \quad \Delta_3' = \{j \,|\, a_j' = b_j' \in V_4,\ j \in N'\} .$$

Then the sequence $D = a'_{\Delta_3'} = b'_{\Delta_3'}$ is the core of A and B generated by C, and

$$n_{d''} = ||\Delta_3''|| = g_{\mathrm{H}}(A'', B'') > g_{\mathrm{H}}(A', B') = ||\Delta_3'|| = n_d .$$

Now $D'' = a''_{\Delta_3''} = b''_{\Delta_3''}$ is the subsequence of A and B. If we denote D'' as the core, and construct the envelope C'' of A and B, then

$$n_{c''} = n_a + n_b - n_{d''} < n_a + n_b - n_d = n_c .$$

This contradicts the definition that C' is the minimum envelope of (A, B). Therefore, (A', B') must be the optimal alignment sequences of (A, B). The theorem is proved.

The Envelope Generated by Pairwise Alignment Sequences

If (A', B') are the alignment sequences of (A, B), then we define

$$\Delta_0' = \{i \in N' \,|\, a_i' = b_i' \in V_4\} , \tag{7.21}$$

and the region Δ_3' can be decomposed to small regions. We denote

$$0 = i_0 \leq i_1 < i_2 < i_3 < \cdots < i_{2k'} \leq i_{2k'+1} = n' , \tag{7.22}$$

and $\Delta_0' = (\delta_1', \delta_3', \cdots, \delta_{2k'-1}')$, where

$$\delta_k' = [i_k + 1, i_{k+1}] , \quad k = 0, 1, 2, \cdots, 2k' . \tag{7.23}$$

Therefore, we can use the alignment sequences (A', B') of (A, B) to construct the envelope C of (A, B). The computational steps are as follows:

1. We denote by ℓ_k, where $k = 0, 1, 2, \cdots, 2k'$, the length of the interval δ_k', and construct the sequence

$$C' = \left(a'_{\delta_0'}, b'_{\delta_0'}, a'_{\delta_1'}, a'_{\delta_2'}, b'_{\delta_2'}, a'_{\delta_3'}, \cdots, a'_{\delta'_{2k'-2}}, b'_{\delta'_{2k'-2}}, a'_{\delta'_{2k'-1}}, a'_{\delta'_{2k'}}, b'_{\delta'_{2k'}}\right) . \tag{7.24}$$

The length of sequence C' is then $n_{c'} = 2\sum_{k=0}^{k'} \ell'_{2k} + \sum_{k=1}^{k'} \ell'_{2k-1}$, in which $\ell'_k = ||\delta'_k|| = i_{k+1} - i_k$.
2. Deleting all of the components of C' whose value is 4, the rest of sequence C is the envelope of sequences (A, B), and C is generated by the alignment sequences (A', B'). In this case, the length of C is $n_c = n_{c'} - 2n' + n_a + n_b$.

Theorem 33. *Under the Hamming scoring matrix, if (A', B') are the optimal alignment sequences of (A, B), then the envelope of (A, B) generated by (A', B') must be the minimum envelope of (A, B).*

Proof. We use reduction to absurdity to prove this. If envelope C is not the minimum envelope of (A, B), then there must be an envelope C' of (A, B) such that $n_{c'} < n_c$ and C' has the standard structure mode as in (7.7). We denote this by

$$C' = \left(c'_{\delta'_0}, c'_{\delta'_1}, \cdots, c'_{\delta'_{k'_c}} \right) ,$$

in which, $\delta'_k = (\delta'_{k,1}, \delta'_{k,2}, \delta'_{k,3})$, and each $\delta'_{k,\tau}$ is a subscript set with the length $\ell'_{k,\tau}$. Following from (7.17) and (7.18), the core D' and alignment sequences (A'', B'') can be generated by envelope C'. The expression

$$n_{d'} = n_a + n_b - n_{c'} > n_a + n_b - n_c = n_d \tag{7.25}$$

holds, where n_d is the length of the core of A and B generated by envelope C. Since

$$g_{\mathrm{H}}(A'', B'') = ||D'|| = n_{d'} > n_d = ||D|| = g_{\mathrm{H}}(A', B') ,$$

this contradicts the supposition that (A', B') are the maximum score alignment sequences of (A, B). This contradiction shows that the theorem is true.

The Relationship Between the L₂-Distance and the A-Distance

Theorem 34. *Under the condition of Theorem 33, the inequality*

$$d_{\mathrm{L}_2}(A, B) \le |A| + |B| - 2n'(A, B) + 2d_{\mathrm{A}}(A, B)$$

holds for any $A, B \in V_2^*$. The necessary and sufficient condition for the equality sign is that the length of the maximum score alignment (A^*, B^*) of (A, B) be $n^* = n'(A, B)$, in which $n'(A, B)$ is the largest length of the minimum penalty alignment sequences of (A, B).

Proof. The proof can be extended by the conclusions of Theorem 29 and Lemma 2. Since

$$\begin{aligned} d_{\mathrm{L}_2}(A, B) &= |A| + |B| - 2\rho(A, B) = |A| + |B| - 2\rho^*(A, B) \\ &\le |A| + |B| - 2\rho'(A, B) = |A| + |B| - 2n' + 2d_{\mathrm{H}}(A', B') \\ &= |A| + |B| - 2n' + 2d_{\mathrm{A}}(A, B) , \end{aligned} \tag{7.26}$$

in which the equality sign is obtained from (7.12), the necessary and sufficient condition for the equality sign holding follows from Theorem 31. Thus the theorem is proved.

In Example 23 we had $d_{\mathrm{A}}(A,B)=5$, $n'(A,B)=|A|=|B|=8$, $\rho(A,B)=4$. Thus,

$$\begin{aligned} d_{\mathrm{L}_2}(A,B) &= |A|+|B|-2\rho(A,B) \\ &= 8 < |A|+|B|-2n'(A,B)+2d_{\mathrm{A}}(A,B) = 10 . \end{aligned}$$

The inequality in Theorem 32 is strictly true, thus, the sequences (A,B) do not satisfy the necessary and sufficient condition in Theorem 32.

7.3 The Counting Theorem of the Optimal Alignment and Alignment Spheroid

Above, we have mentioned that the optimal alignment sequences are not generally unique. The counting theorem of the optimal alignment is intended to determine the number of all the optimal alignments for a fixed sequence A and B. The alignment spheroid is the set of all the sequences whose A-distances arriving at a fixed sequence A are less than or equal to a constant.

7.3.1 The Counting Theorem of the Optimal Alignment

With the same notations as those used in the above section, for a fixed sequence A and B, let C be the minimum envelope, and let D be the maximum core. Then, C and D may be mutually determined. The structure mode of C is given in (7.7). Let

$$\tilde{\ell}=(\bar{\ell}_1,\bar{\ell}_2,\cdots,\bar{\ell}_{k_c}) \tag{7.27}$$

denote the lengths of every vector in (7.7), in which $\bar{\ell}_k=(\ell_{k,1},\ell_{k,2},\ell_{k,3})$, and $\ell_{k,\tau}$ is the length of vector $\delta_{k,\tau}$. We conclude the following:

1. The minimum envelope C may be generated if sequences A,B and their maximum core D are given. There are $M(C)$ envelopes equivalent to the minimum envelope C. The calculation of $M(C)$ is given by (7.8).
2. The minimum penalty alignment sequences (A',B') may be generated if sequences A,B and their maximum core D are given. Its standard mode is given by (7.18) and (7.19).
3. In the standard mode given by (7.18) and (7.19), each region $(a'_{\delta'_{2k}},b'_{\delta'_{2k}})$, undergoes the permutation of virtual symbols. The new sequences are also the maximum penalty alignment sequences of (A,B), and are called the equivalent sequences of (A',B'). The counting formula of the equivalent sequences of (A',B') is:

$$M(A,B;D)=\prod_{k=1}^{k_c}\left(\frac{\ell_{k,4}!}{\ell_{k,5}!(\ell_{k,4}-\ell_{k,5})!}\right) , \tag{7.28}$$

in which $\ell_{k,4}$, $\ell_{k,5}$ are defined by (7.17).

4. The maximum score alignment sequence of A and B (or the modulus structure of the maximum core) is not generally unique. If there are many maximum cores $D_1, D_2, \cdots, D_h$ in A and B, then the number of the maximum score alignment sequences for A and B is $M(A,B) = \sum_{h'=1}^{h} M(A,B;D_{h'})$.

Example 26. In Example 17, the minimum penalty alignment sequences of two RNA sequences are

$$\begin{cases} A': & \text{ugccuggcgg ccguagcgcg gugguccac cugaccccau gccgaacuca} \\ & \text{gaagugaaa ,} \\ B': & \text{—ccuaguga caauagcgga gaggaaacac ccguccc-au cccgaacacg} \\ & \text{gaaguuaag ,} \end{cases}$$

where the total penalty and the total score are 21 and 38, respectively. The maximum score alignment sequences are

$$\begin{cases} A'': & \text{ugccuggcgg ccguagcgcg -gugguccca ccugacccca ugccgaacuc} \\ & \text{agaagugaa a .} \\ B'': & \text{—ccuaguga caauagcg-g agaggaaaca cccguccc-a ucccgaacac} \\ & \text{ggaaguuaa g .} \end{cases}$$

In sequences (A'', B''), the maximum core is

$$D = \text{ccuag gcuag cgggg caccg cccau cccga accga aguaa ,}$$

and the total penalty and the total score are 22 and 39, respectively. Following from (7.28), it is easy to compute the number of maximum score alignment sequences as follows:

1. Based on (A'', B''), we find that the maximum core D is unique, but the modulus structure of maximum score alignment sequences is not unique.
2. We may obtain five modulus structures of maximum score alignment sequences of (A, B). Besides (A'', B''), we need only perform permutations of virtual symbols to get the others.
3. Following from (7.28), we uniquely obtain one equivalent pairwise sequence for each maximum score alignment sequence of (A, B). Thus, the total number of maximum score alignment sequences of (A, B) is 5.

7.3.2 Alignment Spheroid

To any sequence $A \in V_q^*$, we define the alignment spheroid with the center A and radius r as the following

$$O_{\mathrm{A}}(A, r) = \{B \in V_q^* : d_{\mathrm{A}}(A, B) \leq r\} \tag{7.29}$$

and let $S_A(A,r) = \{B \in V_q^* : d_A(A,B) = r\}$ denote the sphere of alignment spheroid $O_A(A,r)$. If the alignment spheroid is the Hamming matrix, then we call it the Hamming spheroid and denote it $O_H(A,r)$. Without loss of generality, we discuss the alignment spheroid under the condition $q = 2$.

As is well known in informatics, in a large n-dimensional vector space the properties of the Hamming spheroid are difficult to derive. An example, in space $V_2^{(n)}$, would be seeking the Hamming spheroid $O_H(A_i, k)$, $i = 1, 2, \cdots, m$ disjoint with maximum m and so on. Such problems still remain open in mathematics and informatics. Therefore, the properties of the alignment spheroid are difficult to analyze at present. In this book we only examine some simpler proteins.

The Run Decomposition of the Sequence

If A is a sequence on V_2^*, then A must be composed of vector segments with component values at either 0 or 1. These vector segments are called runs. A run is a 1-run if all components of the run are 1, and a 0-run if all components of the run are 0. The number of the runs is the run number; and the length of a run is the run length. A sequence expressed by runs is called the run decomposition of the sequence.

For example, the sequence $A = 0011101111100000000110$ may be decomposed as a 0-run and a 1-run, and the run number is 7. There are four 0-runs and three 1-runs amongst the seven runs and their corresponding run lengths are as follows:

$$\ell_1 = 2\,, \quad \ell_2 = 3\,, \quad \ell_3 = 1\,, \quad \ell_4 = 5\,, \quad \ell_5 = 7\,, \quad \ell_6 = 2\,, \quad \ell_7 = 1\,.$$

Then $n_a = \sum_{k=1}^{k_a} \ell_k$, in which, k_a is the run number and ℓ_k is the length of the kth run.

The Construction and Computation of an Alignment Spheroid with Radius $r = 1$

For a given sequence $A \in V_2^*$ and $r = 1$, the alignment spheroid is composed of those sequences satisfying

$$O_A(A,1) = \{A\} \cup S_H(A,1) \cup S_D(A,1) \cup S_I(A,1)\,, \tag{7.30}$$

in which $S_H(A,1), S_D(A,1), S_I(A,1)$ are the Hamming spheroid with radius 1, a deletion sphere, and an insertion sphere, respectively, and are defined as follows:

$$\begin{cases} S_D(A,1) = \{A_i = (a_1, a_2, \cdots, a_{i-1}, a_{i+1}, \cdots, a_{n_a}) \mid i = 1, 2, \cdots, n_a\}\,, \\ S_I(A,1) = \{A_i(\tau) = (a_1, a_2, \cdots, a_i, \tau, a_{i+1}, \cdots, a_{n_a}) \mid i = 0, 1, \cdots, n_a, \\ \qquad \tau = 0, 1\}\,, \end{cases} \tag{7.31}$$

in which $A = (a_1, a_2, \cdots, a_{n_a})$ is a fixed sequence. We explain the spheres $S_D(A,1), S_I(A,1)$ as follows:

1. The numbers of the sequences in $S_D(A,1)$ vary as sequence A changes. For example, the number $k_a = 2$ if $A = (11111000)$, especially $S_D(A,1) = \{(1111000), (1111100)\}$. Generally, $||S_D(A,1)|| = k_a$ holds.
2. The maximum of $||S_I(A,1)||$ is $2n_a$. Note that $||S_I(A,1)||$ varies as sequence A changes. For example, in the case $A = 11111000$, we have
$$S_I(A,1) = \{011111000, 101111000, 110111000, 111011000, 111101000, 111110000, 111111000, 111110100, 111110010, 111110001\}$$
3. Based on the definition of $S_D(A,1)$, we know that
$$||S_D(A,1)||$$
reaches the maximum n_a if the lengths of 0-runs and 1-runs are less than 2 in sequence A. For example, for the case $A = 010101$, we find that
$$S_D(A,1) = \{10101, 00101, 01101, 01001, 01011, 01010\} .$$
Therefore, $||S_D(A,1)||$ reaches $n_a = 6$.
4. The definition of $S_I(A,1)$ allows us to compute $||S_I(A,1)||$ as follows:
$$||S_I(A,1)|| = k_a + 2 + \sum_{k=1}^{k_a} (\ell_k - 1) = n_a + 2 . \quad (7.32)$$
Therefore, $||S_I(A,1)|| = 8 + 2 = 10$ if $A = 11111000$; and $||S_I(A,1)|| = 6 + 2 = 8$ if $A = 010101$. Then
$$S_D(A,1) = \{0010101, 0100101, 0101001, 10101010, 1010101, 0110101, 0101101, 0101011\} .$$
5. It is easy to prove that $S_{\mathrm{H}}(A,1) = n_a$. Thus, we have
$$||O_A(A,1)|| = 1 + ||S_{\mathrm{H}}(A,1)|| + ||S_D(A,1)|| + ||S_I(A,1)|| = 3 + 2n_a + k_a . \quad (7.33)$$
$S_{\mathrm{H}}(A,1)$, $S_D(A,1)$, $S_I(A,1)$ are three disjoint sets because the lengths of the sequences in sets $S_{\mathrm{H}}(A,1)$, $S_D(A,1)$, $S_I(A,1)$ are different.

The Recursion Construction and Computation of the Alignment Spheroid

If the alignment spheroid with radius $r = k$ is known, then we obtain the recursion construction and computation formula for the alignment spheroid with radius $r = k + 1$ as follows:

$$O_A(A, k+1) = \bigcup_{A' \in O_A(A,k)} O_A(A', 1) = O_A(A, k) \bigcup \left[\bigcup_{A' \in O_A(A,k)} S_A(A', 1) \right] . \tag{7.34}$$

Obviously, the sets $O_A(A', 1)$ for two different centers may overlap with each other as $A' \in O_A(A, k)$ ranges. The recursion is hard to compute, but may subsequently reduce the level of difficulty for the construction and computation of the alignment spheroid.

The Decomposition of the Alignment Spheroid According to the Lengths of Sequences

In the alignment sphere $S_A(A, k)$, the lengths of the sequences may range over $[n_a - k, n_a + k]$. Therefore, it might be decomposed as: $S_A(A, k) = \sum_{n=n_a-k}^{n_a+k} S(A, k; n)$, in which

$$S_A(A, k; n) = \{A' \in S(A, k) \,|\, |A'| = n\} ,$$
$$n = n_a - k, n_a - k + 1, \cdots, n_a, n_a + 1, \cdots, n_a + k , \tag{7.35}$$

where $S_A(A, k; n_a - k)$ is the set generated by deleting the k components of sequence A, $S_A(A, k; n_a + k)$ is the set generated by inserting k components of sequence A, and the other sets $S_A(A, k; n)$ are generated by substituting, deleting and inserting the components of sequence A.

7.4 The Virtual Symbol Operation in the Alignment Space

In the previous two sections, we referred to the permutation operation of the virtual symbol in alignment sequences. We now discuss it further. If sequences (A', B') are the alignments of (A, B), then A' and B' are defined on set V_+. If we perform the operation on the virtual symbols of A' and B', then we can get the virtual symbol operator. To understand alignment space in detail, we next analyze the virtual symbol operator.

7.4.1 The Definition of the Virtual Symbol Operator

The Virtual Symbol Operation on a Single Sequence

Let $A' = (a'_1, a'_2, \cdots, a'_{n'})$ be the expansion of sequence $A = (a_1, a_2, \cdots, a_{n'})$. Based on A', we may obtain A by deleting all the virtual symbols of A'. We obtain a new expansion of A if we continue to insert or delete some virtual symbols in A'. This operation is called the virtual symbol operation on A'.

If A'' is generated by sequence A' through the virtual symbol operation, we can also represent the relationship by modulus structure. Here, let

$$H^* = \{(j_k, \ell_k^*),\ k = 1, 2, \cdots, k_{a'}\} \tag{7.36}$$

be the modulus structure of the virtual symbol operation from A' to A'', in which $\ell_k^* \neq 0$, and (j_k, ℓ_k^*) means that we insert or delete ℓ_k^* components after the j_kth component of A', in which

$$0 = j_0 \leq j_1 < j_2 < j_3 < \cdots < j_{k-1} \leq j_k = n_{a'} . \tag{7.37}$$

The inserting or deleting operations depend on whether $\ell_k^* > 0$ or < 0. In addition, in the case $\ell_k^* < 0$, the $a'_{j_k+1}, a'_{j_k+2}, \cdots, a'_{j_k+\ell_k^*}$ are virtual symbols required to keep (7.36) well-defined. Let $A'' = \mathbf{H}^*(A')$, where $\mathbf{H}^*$ is the virtual symbol operator of A', with the following properties:

1. If A' is an expansion of A, and $\mathbf{H}^*$ is the virtual symbol operator of A', then $A'' = \mathbf{H}^*(A')$ is still an expansion of A.
2. If A' and A'' both are expansions of A, then there is a virtual symbol operator $\mathbf{H}^*$ of A' such that $A'' = \mathbf{H}^*(A')$.

The proofs of the two propositions are obvious, and hence we omit them here.

Example 27. In the set V_2, the components of value 2 are virtual symbols. The modulus structures of virtual symbol operation on the following sequences are shown as follows:

1. If
$$\begin{cases} A' = (11111\,22200\,00000) \\ A'' = (11111\,00000\,00) , \end{cases}$$
then $H^* = \{(5, -3)\}$.
2. If
$$\begin{cases} A' = (11111\,00000\,00) \\ A'' = (11111\,22200\,00000) , \end{cases}$$
then $H^* = \{(5, 3)\}$.
3. If
$$\begin{cases} A' = (11111\,21111\,11111) \\ A'' = (11111\,11112\,11111) , \end{cases}$$
then $H^* = \{(5, -1), (10, 1)\}$.

The Virtual Symbol Operation on Alignment Sequences

Let (A', B') be the alignments of (A, B), and $\mathbf{H}_a^*, \mathbf{H}_b^*$ be the virtual symbol operators of A' and B' to A and B, respectively. Let their modulus structures be

$$\begin{cases} H_a^* = \left\{ \left(j_{a,k}, \ell_{a,k}^* \right), \; k = 1, 2, \cdots, k_{a'} \right\} , \\ H_b^* = \left\{ \left(j_{b,h}, \ell_{b,h}^* \right), \; h = 1, 2, \cdots, k_{b'} \right\} . \end{cases} \tag{7.38}$$

It should be noted that the fact that $\mathbf{H}_a^*, \mathbf{H}_b^*$ are the modulus structures of the virtual symbol operators of A' and B', respectively, does not imply that $(\mathbf{H}_a^*, \mathbf{H}_b^*)$ are always the modulus structures of the virtual symbol operators of (A', B'). We add special conditions that are defined as follows.

Definition 38. $(\mathbf{H}_a^*, \mathbf{H}_b^*)$ *are the virtual symbol operators on the alignment sequences* (A', B') *of* (A, B) *if their modulus structures satisfy the following conditions:*

1. *H_a^*, H_b^* are the modulus structures of the virtual symbol operators of A' and B', and satisfy the expression $L_a^* = L_b^*$, where*

$$L_a^* = \sum_{k=1}^{k_{a'}} \ell_{a,k}^* , \quad L_b^* = \sum_{k=1}^{k_{b'}} \ell_{b,k}^* . \tag{7.39}$$

2. *In $(j_{a,k}, \ell_{a,k}^*)$, all $a'_{j_{a,k}+1}, a'_{j_{a,k}+2}, \cdots, a'_{j_{a,k}, \ell_{a,k}}$ are not virtual symbols if $\ell_{a,k}^* > 0$. Similarly, all $b'_{j_{b,h}+1}, b'_{j_{b,h}+2}, \cdots, b'_{j_{b,h}+\ell_{b,h}}$ are not virtual symbols if $\ell_{b,h}^* > 0$.*
3. *All $a'_{j_{a,k}+1}, a'_{j_{a,k}+2}, \cdots, a'_{j_{a,k}+\ell_{a,k}}$ are virtual symbols if $\ell_{a,k}^* < 0$. All $b'_{j_{b,h}+1}, b'_{j_{b,h}+2}, \cdots, b'_{j_{b,h}+\ell_{b,h}}$ are virtual symbols if $\ell_{b,h}^* < 0$.*

The Properties of Virtual Symbol Operation of Alignment Sequences

For the virtual symbol operation $(\mathbf{H}_a^*, \mathbf{H}_b^*)$ on the alignment sequences (A', B') of (A, B), the following properties also hold:

1. If (A', B') is the alignment of (A, B), and $(\mathbf{H}_a^*, \mathbf{H}_b^*)$ are the virtual symbol operators on (A', B'), then $(A'', B'') = (\mathbf{H}_a^*(A'), \mathbf{H}_b^*(B'))$ is also the alignment of (A, B).
2. If $(A', B'), (A'', B'')$ are all the alignment sequences of (A, B), then there must be a virtual symbol operator $(\mathbf{H}_a^*, \mathbf{H}_b^*)$ of (A', B') such that $(A'', B'') = (\mathbf{H}_a^*(A'), \mathbf{H}_b^*(B'))$.

The proofs of these two propositions are obvious and hence are omitted here.

7.4.2 The Modulus Structure of the Virtual Symbol Operator

In Chap. 3, we introduced the modulus structure theory for sequence mutations and alignments. Also, in the above section, we introduced the modulus structure for the virtual symbol operator. Now, we discuss the relationships between them.

The Properties of the Modulus Structure of the Virtual Symbol Operator for a Single Sequence

If A' is an expansion of A, then based on the discussion in Chap. 3, we infer that the transformation from A to A' is uniquely determined by modulus structure $H = \{(i_k, \ell_k)$, where $k = 1, 2, \cdots, k_a\}$. Furthermore, if A'' is generated by the virtual symbol operation on A', then with the same discussion as in the above section, we find that the transformation from A' to A'' is determined by the modulus structure $H^* = \{(j_k, \ell_k^*)$, where $k = 1, 2, \cdots, k_{a'}\}$. The operators $\mathbf{H}, \mathbf{H}'$ induced by the modulus structures H, H^* satisfy $A' = \mathbf{H}(A)$, and $A'' = \mathbf{H}'(A')$.

On the other hand, based on the proposition (2) of the virtual symbol operation for a single sequence, we conclude that A'' is the expansion of A. Therefore, there is a modulus structure H'' and its corresponding operator $\mathbf{H}''$ such that $A'' = \mathbf{H}''(A)$. The properties of the modulus structure of the virtual symbol operator for a single sequence are mainly the relations between H, H^* and H''. We discuss them below:

1. In the definitions of H, H^* and H'', we first note that H and H'' are the modulus structures of the sequence determined by A, while H^* is the modulus structure of the virtual symbol determined by A'. Here, position j_k in H^* has values in $\{1, 2, \cdots, n_{a'}\}$. Therefore, $a'_{j_k} \in V_+$ means a'_{j_k} can be valued in V or valued as "–".
2. Following from the relationship between A and A', we define

$$i'_k = j_k - L_{j_k} = j_k - \sum_{j=1}^{j_k - 1} \kappa_-(a'_j) \tag{7.40}$$

where

$$\kappa_-\left(a'_j\right) = \begin{cases} 1, & \text{if } a'_j \text{ is the virtual symbol "–"}, \\ 0, & \text{otherwise} . \end{cases}$$

Based on this, we define the modulus structure as follows:

$$H' = \{(i'_k, \ell'_k),\ k = 1, 2, \cdots, k_{a'}\} \tag{7.41}$$

in which i'_k is defined by (7.40), and $\ell'_k = \ell^*_k$.
3. Let $H'' = \{(i''_k, \ell''_k),\ k = 1, 2, \cdots, k''_a\}$ be the modulus structure from A to its expansion A'', and I, I', I'' be the position sets of H, H' and H'', then we define set $I'' = I \cup I' = \{i''_1, i''_2, \cdots, i''_{k''}\}$, where

$$\ell''_{k''} = \begin{cases} \ell_k , & \text{when } i''_{k''} = i_k \in I \text{ but } i''_{k''} \text{ is not in } I' , \\ \ell'_k , & \text{when } i_{k''} = i_k \in I \text{ but } i_{k''} \text{ is not in } I' , \\ \ell_k + \ell'_{k'} , & \text{when } i''_{k''} = i_k = i'_{k'} \in I \cap I' . \end{cases} \tag{7.42}$$

4. From the definitions of H' and H^* we see in (7.42) that $\ell'_k > 0$ must apply to the second case and $\ell_k + \ell'_{k'} > 0$ must apply to the third case; so $\ell''_k \geq 0$ is always true. Their relation is as follows:

$$A' = \mathbf{H}(A) , \quad A'' = \mathbf{H}''(A) = \mathbf{H}^*(A') . \tag{7.43}$$

The Properties of the Modulus Structure of the Virtual Symbol Operator for Alignment Sequences

The properties of the modulus structure of the virtual symbol operator for a single sequence can be easily extended to the case of alignment sequences. We elaborate on it below.

If sequences (A', B') are the alignment sequences of (A, B), the modulus structures of the virtual symbol operations and the operators of (A', B') are (H_a^*, H_b^*), $(\mathbf{H}_a^*, \mathbf{H}_b^*)$ respectively, then

$$(A'', B'') = (\mathbf{H}_a^*, \mathbf{H}_b^*)(A', B') = (\mathbf{H}_a^*(A'), \mathbf{H}_b^*(B')) . \tag{7.44}$$

We discuss the following properties:

1. The modulus structure of the alignment sequence (A', B') of (A, B) is: (H_a, H_b). The modulus structure of the virtual symbol operator for (A'', B'') with (A', B') is (H_a^*, H_b^*). Here, (A'', B'') given in (7.44) are also the alignment sequences of (A, B). Their modulus structure is denoted by (H''_a, H''_b).
2. Using the properties (7.41) of the modulus structure for the virtual symbol operator of a single sequence we can transform the modulus structure of (A', B') into the modulus structure (H'_a, H'_b) of (A, B), where

$$\begin{cases} H'_a = \left\{ \left(i'_{a,k}, \ell'_{a,k} \right), \; k = 1, 2, \cdots, k^*_{a'} \right\} , \\ H'_b = \left\{ \left(i'_{b,k}, \ell'_{b,k} \right), \; k = 1, 2, \cdots, k^*_{b'} \right\} , \end{cases} \tag{7.45}$$

and

$$\begin{cases} i'_{\tau,k} = j_{\tau,k} - L^*_{\tau,j_k} = j_{\tau,k} - \sum_{j=1}^{j_{\tau,k}} \kappa_-(\tau'_j) , & \tau = a, b , \\ \ell'_{\tau,k} = \ell^*_{\tau,k} , & \tau = a, b , \end{cases} \tag{7.46}$$

where

$$\kappa_-(\tau'_j) = \begin{cases} 1 , & \text{when } \tau'_j \text{ is valued "$-$"}, \\ 0 , & \text{otherwise} . \end{cases}$$

3. Similar to expression (7.42), we find the relation between (H_a, H_b), (H'_a, H'_b), and (H''_a, H''_b), so that the expression

$$\begin{cases} (A', B') = (\mathbf{H}_a, \mathbf{H}_b)(A, B) = (\mathbf{H}_a(A), \mathbf{H}_b(B)) , \\ (A'', B'') = (\mathbf{H}''_a(A), \mathbf{H}''_b(B)) = (\mathbf{H}_a^*(A'), \mathbf{H}_b^*(B)) \end{cases} \tag{7.47}$$

holds.

The Basic Types of Modulus Structure of the Virtual Symbol Operator for Alignment Sequences

Following from the definition of the modulus structure (H_a^*, H_b^*) of the virtual symbol operator for alignment sequences, we can determine its basic type as: $I_{j_a,j_b;\pm\ell}$, where ℓ is a nonnegative integer. This is the simpler form of $H_a^* = \{(j_a, \pm\ell)\}$, $H_b^* = \{(j_b, \pm\ell)\}$, where $j_a \neq j_b$. It means that after the sites j_a, j_b of sequences A', B', we insert or delete ℓ virtual symbols "−".

Obviously, any virtual symbol operator of alignment sequences can be decomposed to the product of several single symbol insertion and deletion types.

Example 28. Let

$$\begin{cases} A = (11111\,00000\,11111)\,, \\ B = (00111\,11000\,0011)\,, \end{cases}$$

and let

$$\begin{cases} A' = (11111\,22000\,00111\,11)\,, \\ B' = (00111\,11000\,00222\,11)\,, \end{cases} \qquad \begin{cases} A'' = (22111\,11000\,00111\,11)\,, \\ B'' = (00111\,11000\,0011222)\,, \end{cases}$$

then

$$\begin{cases} H_a = \{(5,2)\}\,, \\ H_b = \{(12,3)\}\,, \end{cases} \qquad \begin{cases} H_a^* = \{(0,2),(5,-2)\}\,, \\ H_b^* = \{(12,-3),(17,3)\}\,. \end{cases}$$

Furthermore,

$$\begin{cases} H_a' = \{(0,2),(5,-2)\}\,, \\ H_b' = \{(12,-3),(14,3)\}\,, \end{cases} \qquad \begin{cases} H_a'' = \{(0,2)\}\,, \\ H_b'' = \{(14,3)\}\,. \end{cases}$$

Then the decomposition of $(\mathbf{H}_a^*, \mathbf{H}_b^*)$ is shown as follows:

$$I_{0,17;2}\begin{pmatrix} A' \\ B' \end{pmatrix} = \begin{pmatrix} 221111\,12200\,00011\,111 \\ 00111\,11000\,00222\,1122 \end{pmatrix} = \begin{pmatrix} A_1' \\ B_1' \end{pmatrix}\,,$$

$$I_{7,12;-2}\begin{pmatrix} A_1' \\ B_1' \end{pmatrix} = \begin{pmatrix} 22111\,1100\,00011\,111 \\ 00111\,11000\,0021\,122 \end{pmatrix} = \begin{pmatrix} A_2' \\ B_2' \end{pmatrix}\,,$$

$$I_{14,15;1}\begin{pmatrix} A_2' \\ B_2' \end{pmatrix} = \begin{pmatrix} 22111\,11000\,00112\,111 \\ 00111\,11000\,00211\,222 \end{pmatrix} = \begin{pmatrix} A_3' \\ B_3' \end{pmatrix}\,,$$

$$I_{14,12;-1}\begin{pmatrix} A_3' \\ B_3' \end{pmatrix} = \begin{pmatrix} 22111\,11000\,00111\,11 \\ 00111\,11000\,00112\,22 \end{pmatrix} = \begin{pmatrix} A'' \\ B'' \end{pmatrix}\,.$$

Finally, we have

$$\begin{pmatrix} A'' \\ B'' \end{pmatrix} = \begin{pmatrix} \mathbf{H}_a^*(A) \\ \mathbf{H}_b^*(B) \end{pmatrix} = I_{14,12;-1} I_{14,15;1} I_{7,12} I_{7,12;-2} I_{0,17;2}\begin{pmatrix} A \\ B \end{pmatrix}\,.$$

Therefore, the decomposition

$$(\mathbf{H}_a^*, \mathbf{H}_b^*) = I_{14,12;-1} I_{14,15;1} I_{7,12;-2} I_{0,17;2}$$

holds.

7.4.3 The Isometric Operation and Micro-Adapted Operation of Virtual Symbols

For the sake of simplicity, we assume that for any case involving the matrix, we use the Hamming matrix.

The Definitions of the Isometric Operator and Micro-Adapted Operator for Virtual Symbols and Their Properties

Definition 39. *Let (A', B') be the alignment of (A, B), and (H_a^*, H_b^*) be the virtual symbol operators on (A', B') such that*

$$(A'', B'') = (\mathbf{H}_a^*(A'), \mathbf{H}_b^*(B')) .$$

Then, (H_a^*, H_b^*) is an isometric operator on (A', B') if $d_+(A'', B'') = d_+(A', B')$, and (H_a^*, H_b^*) is a micro-adapted operator on (A', B') if $d_+(A'', B'') < d_+(A', B')$.

Next, we discuss the properties of the isometric operator. Let (A', B') be the alignment, and the virtual symbol operator $(\mathbf{H}_a^*, \mathbf{H}_b^*)$. We introduce the following notations:

$$\begin{cases} j_m^* = \min\{j_{a,k}^*,\ k = 1, 2, \cdots, k_{a'},\ j_{b,k'}^*,\ k' = 1, 2, \cdots, k_{b'}\} , \\ j_M^* = \max\{j_{a,k}^* + \ell_{a,k}^*,\ k = 1, 2, \cdots, k_{a'},\ j_{b,k'}^* + \ell_{b,k'}^*,\ k' = 1, 2, \cdots, k_{b'}\} . \end{cases} \tag{7.48}$$

(j_m^*, j_M^*) is the value range of the virtual symbol operator. Based on (j_m^*, j_M^*), we can determine the subsequences of A' and B' as follows:

$$\begin{cases} A'_{mM} = \left(a'_{j_m^*+1}, a'_{j_m^*+2}, \cdots, a'_{j_M^*}\right) , \\ B'_{mM} = \left(b'_{j_m^*+1}, b'_{j_m^*+2}, \cdots, b'_{j_M^*}\right) . \end{cases} \tag{7.49}$$

Let $L_0 = \sum_{k=1}^{k_{a'}} \ell_{a,k} = \sum_{k=1}^{k_{b'}} \ell_{b,k}$ be the total number of insertions or deletions of the virtual symbol operator of the alignment sequences. Based on these notations, we can formulate the basic property theorem for the isometric operator as follows.

Theorem 35. *If $(A', B'), (A'', B'')$ are both the optimal alignments of (A, B), then there is an isometric operator $(\mathbf{H}_a^*, \mathbf{H}_b^*)$ such that $A'' = \mathbf{H}_a^*(A'), B'' = \mathbf{H}_b^*(B')$.*

On the other hand, if (A', B') are the optimal alignment sequences of (A, B), and $(\mathbf{H}_a^*, \mathbf{H}_b^*)$ is the isometric operator on (A', B'), then $(\mathbf{H}_a^*, \mathbf{H}_b^*)$ is also an optimal alignment, in which $A'' = \mathbf{H}_a^*(A')$ and $B'' = \mathbf{H}_b^*(B')$.

Proof. The proof of the theory is obvious. Because (A', B'), (A'', B'') are both the alignment sequences of (A, B), then making use of the properties of the virtual symbol operator of alignment sequences, we find the isometric operator $(\mathbf{H}_a^*, \mathbf{H}_b^*)$ such that $A'' = \mathbf{H}_a^*(A')$, $B'' = \mathbf{H}_b^*(B')$. Since (A', B'), (A'', B'') are both the alignments of (A, B), this implies that $d_+(A'', B'') = d_+(A', B')$ holds, meaning that (H_a^*, H_b^*) is an isometric operator on (A', B').

The converse proposition is obtained using the optimal alignment (A', B') and the definition of the isometric operator $(\mathbf{H}_a^*, \mathbf{H}_b^*)$.

The Basic Types of Isometric Operators

In Sects. 7.4.2 and 7.4.3, we have stated the basic types $I_{j_a,j_b;\ell}$ of isometric operators. We now discuss their properties.

Theorem 36. *In order for the basic type $I_{j_a,j_b;\ell}$ to be an isometric operator, the necessary and sufficient conditions in different cases are stated as follows:*

1. *In the case where $\ell > 0$ and $j_a < j_b - \ell$, then $I_{j_a,j_b;\ell}$ is an isometric operator if and only if*

$$d_{\mathrm{H}}\left(a'_{j_a,j_b}, b'_{j_a,j_b}\right) = d_{\mathrm{H}}\left(a'_{j_a,j_b-\ell}, b'_{j_a+\ell,j_b}\right) + 2\ell \,, \tag{7.50}$$

in which $c'_{i,j} = (c'_{i+1}, c'_{i+2}, \cdots, c'_j)$, and $c = a, b$.
By symmetry, in the case where $j_b < j_a - \ell$, $I_{j_a,j_b;\ell}$ is the isometric operator if and only if

$$d_{\mathrm{H}}\left(a'_{j_b,j_a}, b'_{j_b,j_a}\right) = d_{\mathrm{H}}\left(a'_{j_b+\ell,j_a}, b'_{j_b,j_a-\ell}\right) + 2\ell \,. \tag{7.51}$$

2. *In the case where $\ell < 0$ and $j_a < j_b + \ell$, $I_{j_a,j_b;\ell}$ is an isometric operator if and only if*

$$d_{\mathrm{H}}\left(a'_{j_a+\ell,j_b}, b'_{j_a+\ell,j_b}\right) = d_{\mathrm{H}}\left(a'_{j_a-\ell,j_b-\ell}, b'_{j_a,j_b}\right) - 2\ell \,. \tag{7.52}$$

By symmetry, in the case where $j_b < j_a+\ell$, $I_{j_a,j_b;\ell}$ is the isometric operator if and only if

$$d_{\mathrm{H}}\left(a'_{j_b-\ell,j_a}, b'_{j_b-\ell,j_a}\right) = d_{\mathrm{H}}\left(a'_{j_b,j_a}, b'_{j_b-\ell,j_a-\ell}\right) - 2\ell \,. \tag{7.53}$$

Proof. The proof of this theorem follows from Fig. 7.1. For the case $\ell > 0$ and $j_a < j_b - \ell$, the proof is demonstrated below.

In Fig. 7.2, (A', B') are the alignment sequences of (A, B). Operation $I_{j_a,j_b;\ell}$ inserts ℓ virtual symbols after position i_a of A' to obtain A'', and inserts ℓ virtual symbols after position j_b of B' to obtain B''. Then A'', B'' just act in the region $(j_a, j_b + 1)$ and do not involve the other regions. Therefore,

$$d(A'', B'') = d\left(a''_{0,j_a}, b''_{0,j_a}\right) + d\left(a''_{j_a,j_b+\ell}, b''_{j_a,j_b+\ell}\right) + d\left(a''_{j_b+\ell,n''}, b''_{j_b+\ell,n''}\right)$$

holds, in which case $n'' = n' + \ell$. We then have

$$d(A'', B'') = d\left(a'_{0,j_a}, b'_{0,j_a}\right) + d\left(a''_{j_a,j_b+\ell}, b''_{j_a,j_b+\ell}\right) + d\left(a'_{j_b,n'}, b'_{j_b,n'}\right) \tag{7.54}$$

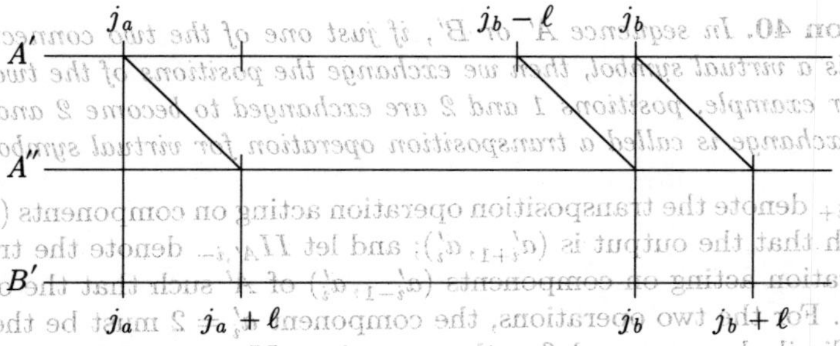

Fig. 7.2. $I_{j_a,j_b;\ell}$. The isometric operator

Thus,

$$d\left(a''_{j_a,j_b+\ell}, b''_{j_a,j_b+\ell}\right) = d\left(a''_{j_a,j_a+\ell}, b''_{j_a,j_a+\ell}\right) + d\left(a''_{j_a+\ell,j_b}, b''_{j_a+\ell,j_b}\right) + d\left(a''_{j_b,j_b+\ell}, b''_{j_b,j_b+\ell}\right) \ .$$

Since both $a''_{j_a,j_a+\ell}$ and $b''_{j_b,j_b+\ell}$ are virtual symbol vectors, following from Definition 38 we know that not every component in vectors $b''_{j_a,j_a+\ell}$ and $a''_{j_b,j_b+\ell}$ is a virtual symbol. Thus, we obtain the expression

$$d\left(a''_{j_a,j_a+\ell}, b''_{j_a,j_a+\ell}\right) = d\left(a''_{j_b,j_b+\ell}, b''_{j_b,j_b+\ell}\right) = \ell$$

and

$$d\left(a''_{j_a+\ell,j_b}, b''_{j_a+\ell,j_b}\right) = d\left(a'_{j_a,j_b-\ell}, b'_{j_a+\ell,j_b}\right)$$

Therefore, we have

$$d\left(a''_{j_a,j_b+\ell}, b''_{j_a,j_b+\ell}\right) = \ell_a + \ell_b + d\left(a'_{j_a,j_b-\ell}, b'_{j_a+\ell,j_b}\right) .$$

Substituting this result into (7.54), we find that (7.50) holds.

With the same arguments, we may prove the other cases, so that the theorem is proved in its entirety.

Remark 12. For the isometric operator, we should note that the product U_1U_2 may be an isometric operator, although both U_1 and U_2 are not isometric operators. For instance, in example 28, $I_{14,15;1}$, $I_{14,12;-1}$ are both not isometric operators, but their product is an isometric operator. In fact, $I_{14,12;-1}I_{14,15;1}$ is the same as the permutation below:

$$I_{7,12;-1}I_{14,15;1}\begin{pmatrix}A'_2\\B'_2\end{pmatrix} = I_{14,12;-1}\begin{pmatrix}22111\,11000\,00112\,111\\00111\,11000\,00211\,222\end{pmatrix}$$

$$= \begin{pmatrix}22111\,11000\,00111\,11\\00111\,11000\,00112\,22\end{pmatrix}$$

We define another type of virtual symbol operation as follows.

Definition 40. *In sequence A' or B', if just one of the two connected components is a virtual symbol, then we exchange the positions of the two components, for example, positions 1 and 2 are exchanged to become 2 and 1. This kind of exchange is called a transposition operation for virtual symbols.*

Let $II_{A',i+}$ denote the transposition operation acting on components (a'_i, a'_{i+1}) of A' such that the output is (a'_{i+1}, a'_i); and let $II_{A',i-}$ denote the transposition operation acting on components (a'_{i-1}, a'_i) of A' such that the output is (a'_i, a'_{i-1}). For the two operations, the component $a'_i = 2$ must be the virtual symbol. Similarly, we may define the operations $II_{B',j+}$ and $II_{B',j-}$.

Theorem 37. *The operation $II_{A',j\pm}$ is an isometric operator if and only if $b'_j = b'_{j\pm1}$. Similarly, the operation $II_{B',j\pm}$ is an isometric operator if and only if $a'_j = a'_{j\pm1}$.*

The proof of this theorem is obvious.

7.5 Exercises, Analyses, and Computation

Exercise 34. Prove the following propositions:

1. Give an example showing that Lemma 3 holds.
2. Prove Theorem 29 and proposition 2 of Theorem 31.
3. Prove formula (7.8).
4. Prove propositions 1–4 in Sect. 7.2.4.

Exercise 35. Based on Example 26, perform the following computations:

1. Compute the maximum core D of (A'', B'') and discuss its uniqueness.
2. Give the alignment modulus structure of (A'', B'') and the other four types of modulus structures.
3. Compute the five maximum scoring alignments of (A, B).

Exercise 36. For the pairwise alignment in Exercise 13 in Chap. 3:

1. Construct the maximum core and minimum envelope, and give the modulus structure.
2. Determine the standard mode of the minimum envelope generated by the modulus structure of the maximum core (may not be unique) and construct all the minimum envelopes from this.
3. Use the counting theorem of the minimum envelope to explain the envelopes generated by task 2 and all the minimum envelopes of these sequences.
4. Use the minimum envelope of the pairwise sequences to construct their minimum penalty alignment sequences.

Exercise 37. Based on the minimum penalty alignment sequences in the Exercise 3 in Chap. 1 with the Hamming matrix, construct the maximum core and minimum envelope of the pairwise sequences.

Exercise 38. Based on Exercise 35, discuss the counting problem of the optimal alignment sequences of pairwise sequences. That is to say, estimate the number of all the optimal alignment sequences of the pairwise sequences.

Part II

Protein Configuration Analysis

8

Background Information Concerning the Properties of Proteins

Proteins are among the most significant biological macromolecules which perform the key functions involved in life processes. The activity of a protein is not only related to its chemical components, but is also affected by its three-dimensional configuration. Our main focus in this chapter is the topic of protein configuration analysis. First, we introduce some background concerning the physical properties of proteins.

8.1 Amino Acids and Peptide Chains

Proteins are formed by different amino acids, of which 20 types are commonly found in nature. Different amino acids are arranged in a specific sequence to form the primary structure of proteins, and their three-dimensional configuration defines the three-dimensional structure of the protein. Parts of these segments are called chains. Some specific motifs formed by peptide chains (for example, α-helices, β-sheets) determine what is called the secondary structure of the proteins. We introduce first the chemical components and related notations of these 20 commonly occurring amino acids.

8.1.1 Amino Acids

Amino acids are the basic structural units of proteins. The chemical components of an amino acid refer to the types, amounts, and interactional structure of the atoms that it consists of. Therefore, different amino acids have distinct chemical components, three-dimensional configurations and physical chemistry characteristics. To properly analyze protein structures, we need some basic information on amino acids.

Fixed Parts and Movable Parts of Amino Acid Components

The chemical components of an amino acid comprise two parts: the "fixed part" and the "movable part." The "fixed part" is the common part of the 20

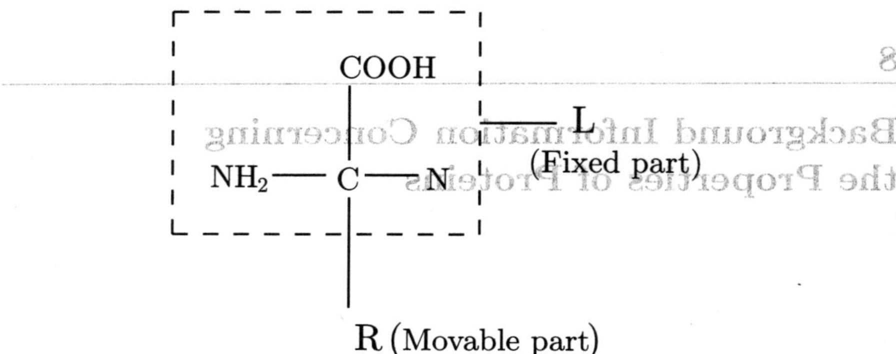

Fig. 8.1. The fixed part and the movable part of amino acid components

amino acids, consisting of an amine group (NH_2), a carboxyl group (COOH), and a carbon atom (C_A). When several carbon atoms are contained in an amino acid, where C_α is a specific one, it is called an α-carbon in molecular biology, and is denoted by A in this book.

The "movable part" of an amino acid refers to the various component elements of different amino acids. In molecular biology, the "fixed part" of an amino acid is denoted by L while the "movable part" (also called the "side chain" of the amino acid) is denoted by R. Here, L is formed by fixed elements, while the formation of R is not fixed, or distinctive for different amino acids (Fig. 8.1).

Names, Codes, and Chemical Components of Amino Acids

The names, codes, chemical structures, and physical chemistry characteristics of the 20 commonly occurring amino acids will be used frequently in this book. These are summarized in the following figures and tables.

The names, three-letter codes, and one-letter codes of the 20 commonly occurring amino acids are shown in Table 8.1.

In a database, we usually denote Gln and/or Glu by a three-letter code Glx, or a one-letter code Z; similarly for Asn and/or Asp we use Asx or a one-letter code B; and undetermined amino acids are given the symbol X.

Physical Chemistry Characteristic Indices of the Commonly Occurring 20 Amino Acids

The physical chemistry properties of an amino acid include: chemical type, molecular weight, volume, frequency, hydrophobicity, polarity, electron property, specific volume, dissociation degree, etc. Some key properties of the 20 commonly occurring amino acids are listed in Table 8.2.

Table 8.1. Names, three-letter codes, and one-letter codes of the 20 commonly occurring amino acids

No.	Name	Three-letter code	One-letter code	No.	Name	Three-letter code	One-letter code
1	Alanine	Ala	A	11	Leucine	Leu	L
2	Arginine	Arg	R	12	Lysine	Lys	K
3	Asparagine	Asn	N	13	Methionine	Met	M
4	Aspartic acid	Asp	D	14	Phenylalanine	Phe	F
5	Cysteine	Cys	C	15	Proline	Pro	P
6	Glutamine	Gln	Q	16	Serine	Ser	S
7	Glutamic acid	Glu	E	17	Threonine	Thr	T
8	Glycine	Gly	G	18	Tryptophane	Trp	W
9	Histidine	His	H	19	Tyrosine	Tyr	Y
10	Isoleucine	Ile	I	20	Valine	Val	V

8.1.2 Basic Structure of Peptide Chains

Figure 8.2 shows the common expression of the protein/peptide chain structure. The horizontal line is called the "backbone," while the sequence

$$\mathrm{R} = \{\mathrm{R}_1, \mathrm{R}_2, \cdots, \mathrm{R}_n\} , \tag{8.1}$$

is called the "side chain" of the peptide chain, and R_j is the movable part of the jth amino acid. If R_j is the name of an amino acid, then R in (8.1) is the primary structure sequence of the protein.

Furthermore, an oxygen atom is preserved in each amino acid residue, so an oxygen atom sequence exists within a peptide chain:

$$\mathrm{O} = \{\mathrm{O}_1, \mathrm{O}_2, \cdots, \mathrm{O}_n\} . \tag{8.2}$$

In this book, we call oxygen sequences **O** the oxygen chains of the peptide chains, where O_j is the oxygen atom of the jth amino acid. Apart from side

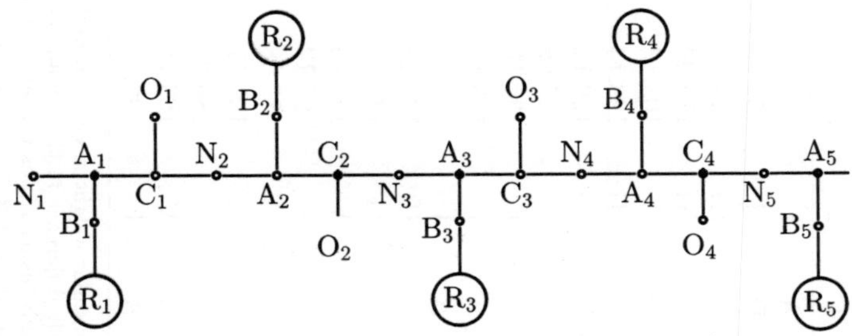

Fig. 8.2. Structural connection of backbone and side chains in common peptide chains

Table 8.2. Properties of the 20 commonly occurring amino acids

One-letter code	Category 1[a]	Category 2[b]	Molecular weight (D)	Volume ($Å^3$)	Surface area ($Å^2$)	Specific volume (mL/g)	pK_a	FR (%)	Hydrophobicity factors K-D	Eisenberg	Meek	F-P	Wolfenden
A	A	I	71.08	88.6	115	0.748		7.9	1.80(7)	0.25(7)	0.20(8)	0.31(10)	1.49(6)
R	D	II	156.20	173.4	225	0.666	12	5.4	−4.0(20)	−1.8(20)	−1.3(20)	−1.0(20)	−19.9(20)
N	B	III	114.11	117.7	160	0.619		4.1	−3.5(15)	−0.64(16)	−0.69(15)	−0.60(16)	−0.97(16)
D	C	III	115.09	111.1	150	0.579	4.5	5.4	−3.5(15)	−0.72(18)	−0.72(16)	−0.77(18)	−10.9(19)
C	B	IV	103.13	108.5	135	0.613	9.3	1.5	2.5(5)	0.04(9)	0.67(3)	1.54(5)	1.24(8)
Q	B	III	128.14	143.9	180	0.674		3.9	−3.5(13)	−0.69(17)	−0.74(17)	−0.22(15)	−9.4(14)
E	C	III	129.12	138.4	190	0.643		6.7	−3.5(15)	−0.62(15)	−1.1(18)	−0.64(17)	−10.2(17)
G	B	I	57.06	60.1	75	0.632		7.0	0.4(8)	0.16(8)	0.06(9)	0.0(13)	2.4(2)
H	D	II	137.15	153.2	195	0.670	6.2	2.3	−3.2(14)	−0.40(14)	−0.0(10)	−0.13(12)	−10.3(18)
I	A	I	113.17	166.7	175	0.884		5.9	4.5(1)	0.73(1)	0.74(1)	1.8(2)	2.15(4)
L	A	I	113.17	166.7	170	0.884		9.6	3.8(3)	0.53(4)	0.65(5)	0.17(4)	2.28(3)
K	D	II	128.18	168.7	200	0.789	10.4	5.9	−3.9(19)	−1.1(19)	−2.0(19)	−0.99(19)	−9.5(15)
M	A	IV	131.21	162.9	185	0.745		2.4	1.9(6)	0.26(6)	0.71(2)	1.23(6)	1.48(9)
F	A	V	147.18	189.9	210	0.774		4.0	2.8(4)	0.61(2)	0.67(3)	1.79(3)	−0.76(7)
P	A	V	97.12	122.7	145	0.758		4.8	−1.6(11)	−0.07(11)	−0.44(14)	−0.72(9)	5.40(1)
S	B	IV	87.08	89.0	115	0.613		6.8	−0.8(10)	−0.26(13)	−0.34(13)	−0.04(14)	−5.06(11)
T	B	IV	101.11	116.1	140	0.689		5.4	−0.7(9)	−0.18(12)	−0.26(12)	−0.34(11)	−4.88(10)
W	A	V	186.21	227.8	255	0.734		1.1	−0.9(11)	0.37(5)	0.45(7)	2.25(1)	−5.88(12)
Y	B	V	163.18	193.6	230	0.712	9.7	3.0	−1.3(12)	−0.02(10)	−0.22(11)	−0.96(8)	−6.11(13)
V	A	I	99.14	140.0	155	0.847		6.8	4.2(2)	0.54(3)	0.61(6)	1.22(7)	1.99(5)

The molecular weight is given in units of 1.07×10^{-24}g. The results may vary using different samples in amino acid frequency statistics, and there would be a little difference in the amounts obtained. The unit in the table is a percentage. pK_a denotes the dissociation constant of the side chains [65]. More specifically, using K_a as the equilibrium constant for the dissociation reaction of an acid, $pK_a = -\log_{10} K_a$. For example, the dissociation constant of the amide group are 6.8–7.9, while those of the carboxyl group are 3.5–5.1. The last five columns are the hydrophobicity factors of each amino acid under different statistical means and grades (in *parentheses*).

FR the frequency rate found using the Swiss-Prot'2006 Database.

[a] *A* nonpolar, *B* polar and uncharged, *C* electronegative, *D* electropositive

[b] *I* fatty hydroxyl, *II* alkaline group, *III* acid group, *IV* hydroxyl and sulfur, *V* aryl and loop

chain R_j and oxygen chain O_j, the remaining part

$$\mathbf{L} = \{(N_1, A_1, C_1), (N_2, A_2, C_2), \cdots, (N_n, A_n, C_n)\} , \tag{8.3}$$

is the backbone of the peptide chain, where (N_j, A_j, C_j) is an α-carbon and carbon atom in the fixed part of the jth amino acid, which appear alternately. In Fig. 8.2, B_i is the C_B atom of the ith amino acid.

We see from the above that proteins are made up of one or more peptide chain(s), while peptide chains are sequences composed of amino acids. This is the primary structure of the proteins. Within peptide chains, every amino acid and its atoms have a relevant three-dimensional configuration and form various geometrical structures. This is the tertiary structure or high level structure, called in general the three-dimensional structure of the proteins. The target of this book is to investigate the characteristics of the primary, secondary (motifs), and three-dimensional structures of proteins.

8.2 Brief Introduction of Protein Configuration Analysis

In this section, we introduce briefly the research status on protein configuration analysis, including protein structure databases, related issues in the study of three-dimensional structures of proteins and the bioinformatics issues emphasized in this book.

8.2.1 Protein Structure Database

There are many types of protein structure databases, which can be classified as follows.

Database of Protein Common Information

Protein common information databases involve information on primary structure sequence, and other information concerned with the commentary. The commentary deals with other aspects of the proteins, such as the origin, class, function, domain, key words and the connection to other databases, etc. The most popular ones are:

- PIR (Protein Information Resource) [9]: http://pir.georgetown.edu/
- Swiss-Prot [8]: http://expasy.org/sprot/

By December 2005, the number of proteins collected in PIR and Swiss-Prot exceeded 100,000 each.

Databases of Protein Structures

The main characteristic of protein structure databases is that they contain information regarding the three-dimensional structure of proteins. It can be as specific as providing the three-dimensional location of each atom. The most popular ones are:

- PDB (Protein Data Bank) [13]: http://www.rcsb.org/pdb/
- CSD (Cambridge Structural Database) [1]: http://www.ccdc.cam.ac.uk/prods/csd.html

Other Types of Databases

There are many other types of protein database. According to incomplete statistics, there are 120 well-known ones, of many types, including:

1. Databases concerning proteins together with peptides, such as TrEMBL [71] and GenPept database [11]. These are databases of proteins translated by nucleic acids.
2. Databases that derive from the original databases (such as PIR [9], Swiss-Prot [8], PDB [13]), such as a database of the similar proteins, for instance, PIR-ASDB is a database of similar proteins in PIR. PIR-ALN contains the sequence pairs whose similarity error is below 55% after sequence alignment [97].
3. Protein secondary structure databases and structure classification databases. For instance, the DSSP database [50] is a protein secondary structure database built upon PDB, while the SCOP (Structural Classification of Proteins) database [59] contains the protein domains, and classifies protein structure by seven hierarchies.
   The database used frequently in this book, the PDB-Select database [41], is obtained by deleting homologous proteins (or peptide chains) from the PDB database. Its characteristic is that each individual protein then has more independence.
4. Relatively specialized databases, such as a database of protein functions, classifications, enzymes, etc., which are not elaborated on here.

8.2.2 Brief Introduction to Protein Structure Analysis

For protein structure analysis, we mainly discuss the following topics in this book: sequence alignment of the protein primary structure and three-dimensional structure; prediction of protein secondary structure; in-depth analysis of protein three-dimensional particle systems; configuration analysis of the protein three-dimensional structure, and semantic analysis of the protein primary structure. We briefly introduce these topics here:

Sequence Alignment of the Protein Primary Structure and Three-Dimensional Structure

The sequence alignment of the protein primary structure and three-dimensional structure is an extension of Part I of this book. In the primary structure sequence alignment, the main difference between DNA sequences and protein sequences is that the data set is not V_4 but V_{20}, which represents the set of the 20 commonly occurring amino acids. As well, the penalty (or scoring) matrix between the 20 amino acids and the dummy notations is represented in terms of the PAM or BLOSUM matrix series.

In the protein three-dimensional structure sequence alignment, we mostly adopt in this book the sequence alignment represented by the protein backbone torsion angle. From these, we give a series of definitions and calculations concerned with protein three-dimensional structure similarity.

Prediction of Protein Secondary Structure

Many methods are used in the prediction of protein secondary structures, while in this book we mainly introduce only those concerning informatics and statistics. From this, we can see the difficulties which lie in the protein secondary structure commonly predicted using the PDB database. This restricts the accuracy of predicting the protein secondary structure, in view of informatics and statistics issues. Furthermore, in this chapter we discuss the relationship between the protein secondary structure and the torsion angles resulting from a particular sequence.

In-Depth Analysis of the Protein Three-Dimensional Particle System

If we take the atoms in a protein as a three-dimensional particle system, then a series of geometric calculations can be done on these particles, including an in-depth analysis of each particle in the system. Many approaches are applicable to the depth calculation of particles within proteins, such as hydrophobicity maps that are useful in biology, etc. In this book we use mathematical (originating from statistics) calculations. At the same time we discuss biological consequences of these calculations.

Calculations and Analysis of the Protein Three-Dimensional Configuration Structure

The discussion in this part forms the centrepiece of this book. Protein three-dimensional structure is tightly related to its functions. Therefore, the study of protein functions, such as virus analysis, rational drug design, etc. are all key problems investigated in bioinformatics.

Protein three-dimensional structure analysis covers many issues and structure types. For instance, in the SCOP database [5, 60, 66], the 7-hierarchy classification from the secondary structure jumping-off point i, is a discussion of protein structure and characteristics in view of the existing structure types. The emphasis in this book is on: protein secondary structure prediction, three-dimensional configuration alignment, analysis and calculations of the depth function and three-dimensional structure characteristics. These topics address the various properties of protein structures using different points of view, which will be helpful when further investigating structure-to-function relationships for proteins.

Characteristic Analysis of the Protein Three-Dimensional Structure Configuration

In recent years, much attention has been paid to characteristic analysis of the protein three-dimensional structure configuration [14, 42]. This problem mainly arose from the study of virus analysis and rational drug design. These two issues can be generalized to the ligand-receptor interaction, including the configuration characteristics and the causal requirements for interactions accompanied by configuration characteristics. Ligands can be represented by viruses, micromolecular and macromolecular drug entities that bind to receptors. Their interactions include ligand adsorption and penetration upon receptors, the analysis of which has to do with the interaction of molecules and atoms and the geometrical configuration. For instance, for a ligand to penetrate a receptor (such as those present in cell membranes), both the possibility and the driving force for binding are required. The possibility for binding refers to whether the ligand will be able to find its way to the receptor site, while the driving force refers to the free energy reduction upon binding of the ligand to the receptor.

Apart from three-dimensional structure prediction, many methods and approaches are conducted in protein three-dimensional structure investigations, which we generalize in this book to the following problems:

1. The comparison of the protein three-dimensional structural homology. This is to what extent the protein three-dimensional configurations could be called homologous; how to measure this similarity, and the definition of their comparison.
2. The in-depth analysis of the protein three-dimensional particle system. If we take the amino acids (or atoms of the amino acids) in a protein as spatially distributed particles, we can do depth calculations upon these particles. The definition of the depth varies and some of the problems there can be reduced to geometric calculations.
3. Characteristic analysis of protein three-dimensional configuration. This characteristic analysis has different aims, whether analyzing ligands or receptors. For ligands, we need only know the spatial configuration. The

characteristic structure needed for the analysis of receptors is relatively complicated. It requires the knowledge of voids and structural motifs inside the protein, types and sizes of cavities and grooves on the surface of the protein, etc. From these characteristic structures, the potential for a ligand-receptor interaction can be further evaluated.

These characteristic structural analyses involve a series of problems in geometric calculation, which will be discussed in detail in the following chapters.

8.3 Analysis and Exploration

Exercise 39. Describe the similarities and differences between the formation of bonds between different amino acids in *in vitro* experiments in the chemistry laboratory, and *in vivo* within living organisms.

Exercise 40. There is a relationship between crystallized proteins and proteins in living organisms (proteins in water or a buffer solution). The protein three-dimensional data in the PDB database refer to crystallized proteins. Does this have an impact on the analysis of protein function?

Exercise 41. The four-dimensional structure of a protein refers to its dynamic (or changing) three-dimensional structure when the protein is under different conditions. How can four-dimensional protein structural data be collected and studied?

Exercise 42. The genetic code table demonstrated how amino acids are encoded by triples of nucleotides. It shows a connection between two types of biological molecules. The biological process involved in producing amino acids requires a series of functions involving mRNA, tRNA, and rRNA, and as such is very complicated. Explain this process from the point of view of molecular movement.

Exercise 43. Protein three-dimensional structure analysis is the first step in investigating protein functions. The interactions between different proteins depend first on whether or not their configurations match, and also on whether different amino acids in the matched configurations can give rise to biochemical bonds and how strongly they may react. Discuss the effects that molecules of a drug and viruses may have on the function of proteins.

9

Informational and Statistical Iterative Analysis of Protein Secondary Structure Prediction

9.1 Protein Secondary Structure Prediction and Informational and Statistical Iterative Analysis

9.1.1 Protein Secondary Structure Prediction

The Anfinsen Principle of Protein 3D Structure Prediction

The Anfinsen (1972, 1973) principle [6] is the foundation of protein 3D structure prediction. It claims that all the information about a protein's 3D structure is contained in its primary structure (sequence). The main experimental basis for this is the fact that heating proteins in solution to deform their 3D structures causes these proteins to lose activity. However, when the temperature is lowered to its original value, the primary structures remain; and the 3D structure configuration resumes its original state. This experiment indicates that the primary structure of a protein determines its 3D structure.

In recent years, the Anfinsen principle for the protein 3D structure has been challenged [27, 84, 113]. Some experiments showed that the same primary structure may form diverse 3D structures under different conditions [27, 84, 113]. In other words, the same primary structure may lead to different 3D phase structures.

The Basic Problem of Protein Secondary Structure Prediction

As early as 1951, Pauling et al. proposed that protein partial segments can form special α-helix and β-sheet structures, which are examples of protein secondary structures. Thus the basic problem of protein secondary structure prediction (PSSP) is estimating from the primary structure which partial segments can form these special α-helix and β-sheet structures. Since then, it has been discovered that these secondary structures not only exist in proteins in great amounts, but are also closely related to the protein function. Thus, prediction of the secondary structure has been an important topic in protein structure investigation in the last two decades of the twentieth century. Since large amounts of data on protein 3D structure had been compiled, the study of this issue tends to diversification, and has led to a great deal of research (see the literature review in [79]). In this book we do not discuss these results again, but analyze the informational and statistical characteristics of this problem, to illustrate its current status.

General Status of the Protein 3D Structure Prediction

Protein 3D structure prediction is classified into several types. They are secondary structure prediction, secondary structure component prediction, and 3D structure prediction.

Secondary structure component prediction refers to the prediction of the α-helix and β-sheet proportions within a protein.

Protein 3D structure prediction aims to predict the 3D structure from the protein's primary structure. The key is the long-distance folding problem, and the commonly used methods are folding pattern classification, molecular dynamics calculations, partial peptide domain structure analysis, etc. Folding pattern classification aims to classify the types of protein 3D folding patterns, and make predictions based on these folding types. Molecular dynamics calculations determine the interactions between the atoms within proteins and build their potential of mean force; the stable 3D configuration is the state corresponding to the energy minimum of the potential field. The domain structure method builds corresponding databases and weight coefficients for longer (containing ten amino acids and above) peptide configurations, and then uses these coefficients to predict protein 3D configuration.

In recent years, although many papers on protein structure prediction have been published and much progress has been made, the overall effect is not ideal. Over the past 20 years, many methods of PSSP have been developed, but the accuracy rates have always been around 70% at best. The accuracy rate for 3D structure prediction has been even worse. For instance, in the most commonly used 3D structure classification, for which there is not a single preferred calculation method, the variance in 3D folding pattern estimation is huge. Because of the complexity of protein molecular structures, and since the atoms contained in a protein range in number from hundreds to tens of thousands, and the interaction force they produce is of many types (the van der Waals force, electrostatic, hydrogen bonds, etc.), the computational complexity of the interactions between atoms grows exponentially. Even if a massive supercomputer is used, it still cannot successfully deal with even the smallest peptide (such as one containing only 20 amino acids). Therefore, protein 3D structure prediction still has a long way to go before it can be considered reliable. In this chapter, we first analyze the statistical information characteristics of protein primary and secondary structure. The prediction algorithm is also involved.

Table 9.1. Primary and secondary structure sequences of protein 12E8H

```
EVQLQQSGAE VVRSGASVKL SCTASGFNIK DYYIHWVKQR PEKGLEWIGW IDPEIGDTEY VPKFQGKATM TADTSSNTAY
ooSSSSoooS SSSooooSSS SSSSSoooHH HSSSSSSSSo ooooSSSSSS SSoooooSSS oHHHoooSSS SSSooooSSS
4211110011 2113211111 1112331123 3111111110 3210111111 1133121111 1332333112 1113312111
LQLSSLTSED TAVYYCNAGH DYDRGRFPYW GQGTLVTVSA AKTTPPSVYP LAPGSAAQTN SMVTLGCLVK GYFPEPVTVT
SSSooooHHH oSSSSSSSSS oooooooooo oooSSSSSSo ooooooSSSS Soooooooooo ooSSSSSSSS SSooooooSSS
1113111332 2011121201 3210212321 1111111212 1011111111 2111212121 3112111123 2130321121
WNSGSLSSGV HTFPAVLQSD LYTLSSSVTV PSSTWPSETV TCNVAHPASS TKVDKKIVPR D
SHHHoooooS SSoooSSSoo SSSSSSSSSS oHHHoooooo SSSSSSHHHo SSSSSSoooo o
0002323321 1111111002 1121111120 1332333112 1111113312 1111112034 4
```

9.1.2 Data Source and Informational Statistical Model of PSSP

Data Source of Secondary Structure Prediction

The data source is the foundation of structure prediction. Protein structure databases are commonly used in secondary structure prediction, such as the PDB database, in which primary and secondary structure information for every classified protein is given, along with the 3D coordinates of each atom comprising the protein. Using this information, models and algorithms of PSSP can be established.

Data on more than 20,000 proteins and 70,000 peptide chains are contained in the PDB database version 2005. Because of the large numbers of homologous proteins in the PDB database, it is not suitable to use statistical analysis. Statistical analysis on the PDB database usually uses PDB-Select database, in which excess homologous sequences in the PDB database are deleted, and 3265 sequences are kept. Hence, it is a simplification of the PDB database.

The PDB database gives a clear indication of primary and secondary structure of every protein, which we express in an alternateive manner in dual-sequence form, as detailed below. For instance, for protein 12E8H (which has 221 amino acid residues), the primary and secondary structure sequences are shown in Table 9.1.

In the second line, the letters H, S, and o denote α-helix, β-sheet, and other structures, respectively. The third line expresses the torsion angle value of the protein backbone triangle, which will be further discussed later.

The Random Model of Secondary Structure Prediction

Secondary structure prediction is the prediction of the secondary structure status of each amino acid from the primary structure of the protein. In many protein databases, the relation between the primary and secondary structure is random, so we give their random relationship model as follows:

Let $\xi^{(n)} = (\xi_1, \xi_2, \cdots, \xi_n)$ be the primary structure sequence of a protein, where each ξ_τ, $\tau = 1, 2, \cdots, n$, represents the primary structure status viz. the name of the τth amino acid. Thus, the value of ξ_τ is in set $V_{20} = \{0, 1, \cdots, 19\}$, and n is the length of the protein.

Similarly, we set $\eta^{(n)} = (\eta_1, \eta_2, \cdots, \eta_n)$ to be the secondary structure of the protein, and the value of η_i is in set $\{1,2,3\} = \{\mathrm{H}, \mathrm{S}, \mathrm{o}\}$, then we call

$$\left(\xi^{(n)}, \eta^{(n)}\right) = ((\xi_1, \eta_1), (\xi_2, \eta_2) \cdots, (\xi_n, \eta_n)) \tag{9.1}$$

the protein primary–secondary joint structure random model (or joint structure random sequence), where $\tau = 1, 2, \cdots, n$ represents the order location of the sequence, ξ_τ, η_τ represent the primary and secondary structure status, respectively, at site τ of the protein. Hence the value of (ξ_τ, η_τ) is in $V_{20} \otimes V_3$, where $V_3 = \{0, 1, 2\} = \{\mathrm{H}, \mathrm{S}, \mathrm{o}\}$ is the set of protein secondary structure status. For ease of discussion, we introduce the following notations and terminologies:

1. In protein primary–secondary structure sequence $(\xi^{(n)}, \eta^{(n)})$, we denote as
$$\left(\xi_\tau^{(3)}, \eta_\tau^{(3)}\right) = ((\xi_\tau, \eta_\tau), (\xi_{\tau+1}, \eta_{\tau+1}), (\xi_{\tau+2}, \eta_{\tau+2})) \tag{9.2}$$
a tripeptide chain, which begins at site τ, and whose value is in the range $\{1, 2, \cdots, n-2\}$.
2. Set $a^{(3)} = (a_1, a_2, a_3)$, and $b^{(3)} = (b_1, b_2, b_3)$ to be the primary and secondary structure status vector of a protein of length 3, and $a_{\tau'} \in V_{20}$, $b_{\tau'} \in V_3$, $\tau' = 1, 2, 3$. Its corresponding dual-vector is denoted by
$$\begin{aligned}\left(a^{(3)}, b^{(3)}\right) &= ((a_1, a_2, a_3),\ (b_1, b_2, b_3)) \\ &= (s, t, r; i, j, k)\,, \quad s, t, r \in V_{20}\,, \quad i, j, k \in V_3\,. \end{aligned} \tag{9.3}$$
We call this the status vector of the tripeptide chain its primary–secondary structure.
3. The primary–secondary structure status vector of a tripeptide chain $(a^{(3)}, b^{(3)})$ can be considered to be a sample of a random vector $(\xi_\tau^{(3)}, \eta_\tau^{(3)})$; thus, we can define its probability distribution as follows:
$$\begin{aligned} p(s,t,r;i,j,k) = P_r\left\{\left(\xi_\tau^{(3)}, \eta_\tau^{(3)}\right) = (s,t,r;i,j,k)\right\}\,, \\ s,t,r \in V_{20}\,, \quad i,j,k \in V_3\,. \end{aligned} \tag{9.4}$$
These probability distributions can be obtained from the PDB or PDB-Select databases.
4. From the joint probability distribution in (9.4), we can obtain the conditional probability distribution, boundary distribution and conditional boundary distribution. For instance, the boundary distribution is
$$p(s,t,r) = \sum_{i,j,k=0}^{2} p(s,t,r;i,j,k)\,, \quad p(i,j,k) = \sum_{s,t,r=0}^{19} p(s,t,r;i,j,k)\,. \tag{9.5}$$

The corresponding conditional probability distributions are

$$p[(i,j,k)|(s,t,r)] = p(s,t,r;i,j,k)/p(s,t,r) \ , \tag{9.6}$$

etc.

5. From these probability distributions, all types of Shannon entropies and interaction information can be obtained, for instance, the joint Shannon entropy of $(\xi^{(3)}, \eta^{(3)})$ is

$$H\left(\xi^{(3)}, \eta^{(3)}\right) = -\sum_{i,j,k=0}^{2} \sum_{s,t,r=0}^{19} p(s,t,r;i,j,k) \log p(s,t,r;i,j,k) \ . \tag{9.7}$$

The conditional entropy of $\eta^{(3)}$ on $\xi^{(3)}$ is

$$H\left(\eta^{(3)}|\xi^{(3)}\right) = -\sum_{i,j,k=0}^{2} \sum_{s,t,r=0}^{19} p(s,t,r;i,j,k) \log p[(i,j,k)|(s,t,r)] \ . \tag{9.8}$$

The conditional mutual information of (η_1, η_3) on $(\xi^{(3)}, \eta_2)$ is

$$\begin{aligned} &I(\eta_1;\eta_3|\xi^{(3)},\eta_2) \\ &= \sum_{s,t,r=0}^{19} \sum_{i,j,k=1}^{3} p(s,t,r;i,j,k) \log \frac{p(i,k|s,t,r,j)}{p(i|s,t,r;j)p(k|s,t,r;j)} \ , \end{aligned} \tag{9.9}$$

where $p(i,k|s,t,r,j)$, $p(i|s,t,r;j)$, $p(k|s,t,r;j)$ are conditional probabilities derived from $p(s,t,r;i,j,k)$.

9.1.3 Informational and Statistical Characteristic Analysis on Protein Secondary Structure

Informational Characteristic Calculation on Protein Primary and Secondary Structure

We aim to predict a protein's secondary structure from its primary structure, thus we first analyze the conditional informational characteristics of the secondary structures on the primary structures. Our results are shown in Tables 9.2 and 9.3.

The data in Table 9.2 are results from conditional entropy, where data at the intersection of the first line and the first column represents $H(\eta_3|(\xi_1,\xi_2,\xi_3,\eta_1,\eta_2)) = 0.5798$, and data at the intersection of the second line and the first column represents $H(\eta_3|\xi_1,\xi_2,\eta_1,\eta_2) = 0.6807$.

The data in Table 9.3 are results from conditional mutual information, for instance, data in the first line, first column represents $I(\eta_1;\eta_2|(\xi_1,\xi_2,\xi_3,\eta_2)) = 0.31831$, while data in the second line, first column represents $I(\eta_1;\eta_2|(\xi_1,\xi_2,\eta_3)) = 0.31782$.

Table 9.2. Conditional entropy of protein primary and secondary structures

H	$k\|(i,j)$	$j\|(i,k)$	$i\|(j,k)$	$(j,k)\|i$	$(i,k)\|j$	$(i,j)\|k$
(s,t,r)	0.5798	0.2564	0.5451	1.1617	1.1712	1.1197
(s,t)	0.6807	0.3314	0.6293	1.2239	1.2559	1.2785
(s,r)	0.6636	0.3383	0.6363	1.2501	1.2965	1.3153
(t,r)	0.6564	0.3250	0.6670	1.2773	1.3042	1.3264
H	$j\|i$	$k\|i$	$i\|j$	$k\|j$	$i\|k$	$j\|k$
(s,t,r)	0.5819	0.9054	0.5914	0.6261	0.8634	0.5746
(s,t)	0.6845	1.0338	0.6659	0.7101	0.9877	0.6899
(s,r)	0.6692	0.9944	0.6444	0.6910	0.9311	0.6732
(t,r)	0.6482	0.9796	0.6214	0.6850	0.9318	0.6664
H	i	j	k	(i,j,k)		
(s,t,r)	1.2917	1.2822	1.3336	2.4534		
(s,t)	1.3667	1.3925	1.4534	2.7319		
(s,r)	1.3945	1.4000	1.4120	2.7273		
(t,r)	1.4146	1.3671	1.3927	2.7192		

Table 9.3. Conditional mutual information of protein primary and secondary structures

I	$(i;j)\|k$	$(i;k)\|j$	$(j;k)\|i$	$(i;j)$	$(i;k)$	$(j;k)$
(s,t,r)	0.31831	0.04626	0.32556	0.70031	0.42826	0.70757
(s,t)	0.31782	0.02939	0.35310	0.70805	0.41962	0.74333
(s,r)	0.34064	0.02736	0.33083	0.73081	0.41754	0.72101
(t,r)	0.33446	0.02866	0.32325	0.71890	0.41310	0.70769

Informational Characteristic Analysis on Protein Primary and Secondary Structure

Based on the above, we analyze the information transferring characteristics of each variable in protein primary and secondary structure as follows:

1. Hidden Markov property holds for the tripeptide chain sequences. We define tripeptide chain sequences as

$$\zeta_\tau = (\xi_\tau, \xi_{\tau+1}, \xi_{\tau+2}) , \quad \tau = 1, 2, \cdots, n-2 , \tag{9.10}$$

where $(\xi_1, \xi_2, \cdots, \xi_n)$ is the primary structure sequence of the protein. For its conditional mutual information,

$$I(\zeta_1; \zeta_3|\zeta_2) = I[\xi_1; \xi_4|(\xi_2, \xi_3)] = 0.0087 \approx 0 .$$

This indicates that when ζ_2 is fixed, ζ_1 and ζ_3 are nearly independent of each other. Thus the hidden Markov property holds for the tripeptide chain sequences.

2. Hidden Markov property holds for the tripeptide chain secondary structure. From the results in Table 9.2, we have

$$I[i,k|(s,t,r,j)] = I[\eta_1;\eta_3|(\xi_1,\xi_2,\xi_3,\eta_2)] = 0.04626 .$$

This indicates that when $(\xi_1,\xi_2,\xi_3,\eta_2)$ is fixed, η_1 and η_3 are nearly independent of each other. Thus, the hidden Markov property holds.
3. We see from the conditional entropy Table 9.2 that

$$H[\eta_1|(\xi_1,\xi_2,\xi_3)] = 1.2917 , \quad H[\eta_2|(\xi_1,\xi_2,\xi_3)] = 1.2822 ,$$
$$H[\eta_3|(\xi_1,\xi_2,\xi_3)] = 1.3336 .$$

Hence, out of the predictions of secondary structures η_1, η_2, η_3 separate from the primary structure (ξ_1,ξ_2,ξ_3) for the tripeptide chain, the best result is in η_2.
4. From

$$H[\eta_2|(\xi_2,\xi_3)] = 1.3671 , \quad H[\eta_2|(\xi_1,\xi_2,\xi_3)] = 1.2822 ,$$

we see that the result of the prediction can be improved by increasing the number of primary structures, but the effect will be minimal. From

$$H[\eta_2|(\xi_1,\xi_2,\xi_3)] = 1.2822 , \quad H[\eta_2|(\xi_1,\xi_2,\xi_3,\eta_3)] = 0.5746 ,$$
$$H[\eta_2|(\xi_1,\xi_2,\xi_3,\eta_1,\eta_3)] = 0.2564 ,$$

we see that the conditional entropy decreases sharply if we use more secondary structure information; specifically, the prediction of η_2 from $(\xi_1,\xi_2,\xi_3,\eta_1,\eta_3))$, is bound to be much better than using only η_2 for the prediction.

9.2 Informational and Statistical Calculation Algorithms for PSSP

9.2.1 Informational and Statistical Calculation f[illegible]PSSP

To establish informational and statistical calculation a[illegible]ms for PSSP, we must first classify the data of protein secondary struc[illegible] and build corresponding statistical calculation tables of prediction i[illegible]ation. Related discussions follow.

Data Classification

We base our discussion of the prediction problem on the [illegible]elect database. We denote its protein sequences as set Ω. This set can be divided in two subsets, Ω_1, Ω_2, called the training set and the validation set, respectively.

Their protein primary–secondary structures are denoted respectively by

$$\begin{cases} \Omega_1 = \{(A_1, B_1), (A_2, B_2), \cdots, (A_{m_1}, B_{m_1})\} , \\ \Omega_2 = \{(C_1, D_1), (C_2, D_2), \cdots, (C_{m_2}, D_{m_2})\} , \end{cases} \tag{9.11}$$

where A_s, C_t are the primary structure sequences of the two proteins, respectively, in databases Ω_1, and Ω_2, and B_s, D_t are the secondary structure sequences of the above two proteins s and t in databases Ω_1, and Ω_2. We then denote

$$Z_s = (z_{s,1}, z_{s,2}, \cdots, z_{s,n_s}) , \quad Z = A, B, C, D , \quad z = a, b, c, d ,$$

for their sequence expression. Using the PDB-Select database, we take $m_1 = 2765$, $m_2 = 500$.

Table of Conditional Probability Distribution

From the training set Ω_1, we calculate its conditional probability distribution table, the types of which are

$$\begin{cases} \text{Model I:} & p[i|(s,t,r)], p[j|(s,t,r)], p[k|(s,t,r)] , \\ \text{Model II:} & p[i|(s,t,r,j)], p[j|(s,t,r,i)], p[j|(s,t,r,k)], p[k|(s,t,r,j)] , \\ \text{Model III:} & p[i|(s,t,r,j,k)], p[j|(s,t,r,i,k)], p[k|(s,t,r,i,j)] , \end{cases} \tag{9.12}$$

where the tables of Model I are conditional probability distribution tables of primary structures on secondary structures, while the tables of Models II and III are conditional probability distribution tables of primary structures and some secondary structures on other secondary structures. The sizes of Models I, II, and III are 8000×3, $24{,}000 \times 4$, and $72{,}000 \times 3$ matrices, respectively. When Ω_1 is given, the joint probability distribution $p(s,t,r;i,j,k)$ is determined, and all these conditional probability distributions can be determined by the joint probability distribution $p(s,t,r;i,j,k)$.

Maximum Likelihood Estimate Prediction

1. Maximum likelihood estimate (MLE) prediction uses the tables of Model I, for instance, in $p[i|(s,t,r)]$, for every fixed $(s,t,r) \in V_{20}^{(3)}$, calculate the max $p[i|(s,t,r)]$ on $i = 1, 2, 3$, denoted by $i(s,t,r)$. Then

$$p[i(s,t,r)|(s,t,r)] = \max\{p[1|(s,t,r)], p[2|(s,t,r)], p[3|(s,t,r)]\} . \tag{9.13}$$

If the primary structure of the protein is $A = (e_1, e_2, \cdots, e_n)$, then its predicted secondary structure is

$$\begin{aligned} &\left(\hat{f}_2, \hat{f}_3, \cdots, \hat{f}_{n-2}\right) \\ &= (i(e_1,e_2,e_3), i(e_2,e_3,e_4), \cdots, i(e_{n-3},e_{n-2},e_{n-1}), i(e_{n-2},e_{n-1},e_n)) , \end{aligned} \tag{9.14}$$

where $(e_1, f_1), (e_2, f_2), \cdots, (e_n, f_n)$ are the protein primary–secondary structures, $\hat{f}_\tau$ is the prediction result of f_τ.

2. The most significant disadvantage of MLE prediction is that the joint information of primary and secondary structure in Table 9.2 is not used comprehensively. If each conditional distribution in (9.13) takes the values:

$$p[1|(s,t,r)] = 0.3\,, \quad p[2|(s,t,r)] = 0.3\,, \quad p[3|(s,t,r)] = 0.4\,,$$

then the result predicted from (9.13) is $i(s,t,r) = 3$. This type of prediction leads to large errors.

Using only the table of Model I in MLE prediction, the correct rate will not exceed 55%.

Threshold Series Prediction

In order to overcome the disadvantages of MLE prediction, we can adapt from statistics the threshold series prediction. Its essentials are listed below:

1. Choose the parameters θ_1, θ_2, θ_3 properly. The prediction can only be determined if the conditional probability distributions in Models I, II, and III are respectively greater than these parameters.
2. Using threshold series prediction, it is impossible to predict all the secondary structures at one time. Therefore, we need to use the threshold series prediction on conditional probability distributions in Models I, II, and III repeatedly, to reach the goal of predicting all the secondary structures. We present the algorithm in the next section.

9.2.2 Informational and Statistical Calculation Algorithm for PSSP

If $E = (e_1, e_2, \cdots, e_n)$ is the primary structure sequence of a protein, then we perform a recursive calculation using the table of conditional probability distributions (9.12) and threshold series prediction. We denote the secondary structure by $F = (f_1, f_2, \cdots, f_n)$, and the corresponding recursive algorithm as follows:

Step 9.2.1 Choose parameters $\theta_1, \theta_2, \theta_3 > 0.5$, and predict the secondary structure F from the primary structure E for the first time using the table of conditional probability distributions in (9.12). The main steps are:

1. For the fixed (e_p, e_{p+1}, e_{p+2}), calculate

$$p[f_\tau|(e_p, e_{p+1}, e_{p+2})]\,, \quad f_\tau = 0, 1, 2\,, \quad \tau = p, p+1, p+2\,.$$

2. If there exists $\tau \in \{p, p+1, p+2\}$, $f_\tau \in \{0, 1, 2\}$, such that: $p[f_\tau|(e_p, e_{p+1}, e_{p+2})] > \theta_1$, then f_τ is the secondary structure prediction result of (e_p, e_{p+1}, e_{p+2}) on the site τ.

3. We know from the characteristics of the conditional probability distribution that, for the same $\tau \in \{p, p+1, p+2\}$, there cannot be two $f_\tau \neq f'_\tau \in \{0,1,2\}$, such that $p[f_\tau|(e_p, e_{p+1}, e_{p+2})] > \theta_1$, and $p[f' + \tau|(e'_p, e_{p'+1}, e_{p'+2})] > \theta_1$ hold at the same time.
4. For the same $\tau \in \{p, p+1, p+2\}$, there may be two $p \neq p'$, such that

$$p[f_\tau|(e_p, e_{p+1}, e_{p+2})] > \theta_1 \ ,$$

and

$$p[f' + \tau|(e_{p'}, e_{p'+1}, e_{p'+2})] > \theta_1$$

hold for both, where $f_\tau, f'_\tau \in \{0,1,2\}$. If

$$p[f_\tau|(e_p, e_{p+1}, e_{p+2})] \geq p[f' + \tau|(e_{p'}, e_{p'+1}, e_{p'+2})] \ ,$$

then we have $\hat{f}_\tau = f_\tau$ as the prediction result for the first time.
5. Set N_1 to represent all the sites in the secondary structure prediction for the first time, that is

$$N_1 = \{\tau \in N \colon \text{Exists } p \in N, f_\tau \in \{0,1,2\}, \text{ such that } |p - \tau| \leq 2, \text{ and } p[f_\tau|(e_p, e_{p+1}, e_{p+2})] > \theta_1\} \ . \quad (9.15)$$

Then, for every site $p \in N_1$, there is a secondary structure prediction value f_p. We call the set

$$\mathbf{N}_1 = \{(p, f_p) \colon p \in N_1\} \quad (9.16)$$

the first-time prediction result of the protein secondary structure prediction.

Step 9.2.2 Based on the result obtained in Step 9.2.1, $\mathbf{N}_1$ is considered to be a known result, so we go on to the prediction for the sites in $N_1^c = N - N_1$. We denote one form of the conditional probability distributions of Models II and III as

$$\begin{cases} \text{Model II:} & p[f_\tau|(e_p, e_{p+1}, e_{p+2}, f_{\tau'})] \ , \quad \tau \neq \tau' \in \{p, p+1, p+2\} \ , \\ \text{Model III:} & p[f_\tau|(e_p, e_{p+1}, e_{p+2}, f_{\tau'}, f_{\tau''})] \ , \quad \tau, \tau', \tau'' \text{ different from} \\ & \text{each other, sphere } \tau, \tau', \tau'' \in \{p, p+1, p+2\} \ . \end{cases} \quad (9.17)$$

For site τ in set N_1^c, there exists $p \in N, \tau', \tau''$, such that:

1. There exists a conditional probability distribution for Model II, in (9.17) $p[f_\tau|(e_p, e_{p+1}, e_{p+2}, f_{\tau'})] > \theta_2$, or conditional probability distributions for Model III $p[f_\tau|(e_p, e_{p+1}, e_{p+2}, f_{\tau'}, f_{\tau''})] > \theta_3$.
2. In conditional probability distributions of Models II and III in Step 9.2.2, procedure 1, $\tau', \tau'' \in N_1$.
   Here, the combination that Step 9.2.2, procedures 1 and 2 both hold is denoted as (τ, f_τ), and we refer to it as the second-time prediction result. Following (9.16), we can get all the second-time prediction results

of the protein secondary structures similarly; $\mathbf{N_2} = \{(p, f_p), p \in N_2\}$, where N_2 is the collection of sites of the second-time prediction result of protein secondary structure.

Step 9.2.3 Based on the results of Steps 9.2.1 and 9.2.2, $\mathbf{N}_1\mathbf{N}_2$ are considered to be known data, so we go on to the prediction for the sites in $\mathbf{N}-\mathbf{N}_1-\mathbf{N}_2$. The corresponding steps are the same as those of Step 9.2.2, and the prediction result is $\mathbf{N}_3$.
Continuing like this, we arrive at a series of prediction results $\mathbf{N}_4, \mathbf{N}_5, \cdots$ etc. This operation continues until there is a $k > 0$, such that $\mathbf{N}_k$ is an empty set. If we denote $N_0 = \bigcup_{k'=1}^{k} N_{k'}$, then for every $p \in N_0$, there is a prediction result f_p.

Step 9.2.4 For the sites in $N_0^c = N - N_0$, we use the MLE prediction table in (9.13) to determine the prediction results for every $p \in N_0^c$. The secondary structure prediction result of all the sites in the protein is then obtained.

Step 9.2.5 Make predictions for all the proteins in the validation set Ω_2. We thus find the prediction result for every site $p \in \Omega_2$, denoted by f_p.
In the PDB-Select database, the secondary structure measurement result for all the amino acids in each protein is contained, denoted here by q_p. Prediction results such as the correct rate (or error rate) can then be compared. Obviously, the results are related to the parameters θ_1, θ_2, θ_3; so we denote its error rate by $e(\theta_1, \theta_2, \theta_3)$.

Step 9.2.6 Adjust the parameters θ_1, θ_2, θ_3 to minimize the error rate $e(\theta_1, \theta_2, \theta_3)$. The whole process of protein secondary structure prediction is then carried out. When the parameters θ_1, θ_2, θ_3 are fixed, the algorithm of protein secondary structure prediction (which is now fixed) is formed. We call this algorithm the informational and statistical threshold series prediction algorithm of protein secondary structure.

This algorithm is said to be ISIA.

9.2.3 Discussion of the Results

Prediction Results

For the $m = 3265$ proteins listed in the PDB database version 2005, there are 741,186 coterminous amino acids involved. We set the number of proteins in Ω_1, Ω_2 to be $m_1 = 2765$, $m_2 = 500$, containing 631,087 and 110,099 amino acids, respectively. We then consider Ω_1, Ω_2 to be two two-dimensional sequences of lengths 631,087 and 110,099 respectively, which is denoted by

$$\Omega_\tau = ((e_{\tau,1}, f_{\tau,1}), (e_{\tau,2}, f_{\tau,2}), \cdots, (e_{\tau,n_\tau}, f_{\tau,n_\tau})) , \quad \tau = 1, 2 , \tag{9.18}$$

where $n_1 = 631{,}087$ and $n_2 = 110{,}099$. For Ω_1, Ω_2 in (9.18), we distinguish the different proteins by list separators. Discussions on the calculations of these data follow:

1. From the training set Ω_1, we can find the joint frequency and joint frequency distribution table $p(s,t,r;i,j,k)$ of (9.4). The corresponding conditional probability distribution, the table of Models I, II, and III in (9.12) is then obtained.
2. If the above informational and statistical threshold series prediction is used, when $\theta_1 = \theta_2 = \theta_3 = 0.70$, the correct rate can be 4–5% higher than that obtained using MLE prediction. If the values of θ_1, θ_2, and θ_3 are adjusted constantly, the correct rate may be increased still further. However, the best prediction results have not yet been obtained. An overall introduction to the other algorithms in protein secondary structure software packages may be found in [79].
3. Secondary structure prediction is a complicated problem in the area of informational statistics. In the algorithms above, it is not only related to the choice of the parameters θ_1, θ_2, and θ_3, but also to the division Ω_1 and Ω_2 of the database Ω. Some sources in the literature set Ω_1 and Ω_2 to be the same as Ω, which will greatly increase the nominal prediction accuracy. However, in view of statistics, this is unreasonable, and therefore having it extended is meaningless.
4. Some of the secondary structure predictions add other protein information besides that contained in the PDB-Select database (such as information on the biological classification) in order to improve prediction accuracy. For example, the jackknife testing and multiple sequences alignment methods are used for this reason.

The Jackknife Test

The jackknife test uses a statistical testing method where:

1. $\Omega = \{1, 2, \cdots, m\}$ is the PDB-Select database, in which $i = A_i = (E_i, F_i)$, where

$$E_i = (e_{i,1}, e_{i,2}, \cdots, e_{i,n_i}) , \quad F_i = (f_{i,1}, f_{i,2}, \cdots, f_{i,n_i}) \tag{9.19}$$

   are the primary and secondary structure of protein i, respectively.
2. Ω_1 and Ω_2 are two sets of proteins, where $\Omega_2 = \{i\}$, and $\Omega_1 = \Omega - \Omega_2$. Ω_1 is the training set, and Ω_2 is the testing set.
3. We consider the set Ω_1, and give a two-dimensional sequence

$$\Omega_1 = ((e_{1,1}, f_{1,1}), (e_{1,2}, f_{1,2}), \cdots, (e_{1,n_0}, f_{1,n_0})) , \tag{9.20}$$

   where $n_0 = ||\Omega_1||$.
4. Using the calculations on Ω_1, the primary structure of protein Ω_2, and the predicted secondary structure of Ω_2, the prediction result of ISIA is

$$\hat{F}_i = \left(\hat{f}_{i,1}, \hat{f}_{i,2}, \cdots, \hat{f}_{i,n_i}\right) . \tag{9.21}$$

The error of the secondary structure prediction is

$$d\left(F_i, \hat{F}_i\right) = \sum_{i=1}^{n_i} \frac{1}{n_i} d_H\left(f_{i,j}, \hat{f}_{i,j}\right) , \tag{9.22}$$

where d_H is the Hamming distance.

5. Using the jackknife testing method for all $i \in \Omega$, one obtains prediction results for $\Omega_2 = \{i\}, i \in \Omega$. The error in the secondary structure prediction under the jackknife test is then

$$d_J(\Omega) = \sum_{i=1}^{m} \frac{n_i}{n_0} d\left(F_i, \hat{F}_i\right) . \tag{9.23}$$

Multiple Sequence Alignment

If we obtain a multiple sequence alignment (MSA) for $\hat{F}_i$ and

$$\Omega_{1,F} = \{F_1, F_2, \cdots, F_{i-1}, F_{i+1}, F_{i+2}, \cdots, F_m\} .$$

is the MSA result of $\hat{F}'_i$, we then obtain the error of the secondary structure prediction under jackknife testing and MSA is $d_{\mathrm{J,MSA}}(\Omega)$, given similarly by (9.22) and (9.23).

The error of the secondary structure prediction is $d_{\mathrm{J,MSA}}(\Omega) = 76.8\%$, when $\theta_1 = 0.70$, $\theta_2 = 0.85$, and $\theta_3 = 0.92$.

9.3 Exercises, Analyses, and Computation

Exercise 44. Obtain the protein secondary structure database Ω from PDB-Select at [99], and perform the following calculations:

1. Divide the database Ω into a training set Ω_1 and a validating set randomly, and set $m_1{:}m_2 = 5{:}1$.
2. On the training set Ω_1, calculate the statistical frequency and frequency distribution $n(s, t, r; i, j, k)$ and $p(s, t, r; i, j, k)$ of the tripeptide chain primary–secondary structure.
3. Calculate the conditional probability distribution of Models I, II, and III in (4.19) from the frequency distribution $p(s, t, r; i, j, k)$.
4. Calculate the MLE estimation table from the conditional probability distribution of Models I, II, and III.

Exercise 45. Based on Exercise 44, use the conditional probability distribution of Model I to do MLE on the protein sequences in Ω_2, then calculate the correctness rate.

Exercise 46. Based on Exercise 44, use the conditional probability distribution of Models I, II, and III and choose proper θ_1, θ_2, and θ_3 values to do threshold series estimation on the protein sequences in Ω_2, and then calculate the correctness rate.

Exercise 47. Changing the parameters θ_1, θ_2, and θ_3, compare the prediction results in Exercise 46, thereby determining the choosing of the best parameters and the correctness rate of the best prediction.

10

Three-Dimensional Structure Analysis of the Protein Backbone and Side Chains

It is known that the backbone of a protein consists of the atoms N, C_α, and C alternately, and any three neighboring atoms form a triangle. These coterminous triangles are called triangle splicing belts. We now discuss the structure and transformations of these triangles.

10.1 Space Conformation Theory of Four-Atom Points

The space conformation theory of four-atom points is the foundation of protein structure quantitative analysis. Atomic conformations of such clusters have been described in many ways in chemistry and biology. However, these descriptions have not yet been abstracted into mathematical language. In this chapter, we use geometry to abstract the theory into geometric relations of common space points, so that we may give the correlations and resulting formulas.

10.1.1 Conformation Parameter System of Four-Atom Space Points

The common conformation of four-atom space points refers to the structural relationship between the four discrete points a, b, c, and d in space. Their space locations are shown in Fig. 10.1. We now discuss their structural characteristics.

Basic Parameters of Four-Atom Points Conformation

For the four space points a, b, c, and d denote their coordinates in the Cartesian system of coordinates by

$$\boldsymbol{r}_\tau^* = \overrightarrow{oa}_\tau = (x_\tau^*, y_\tau^*, z_\tau^*) = x_\tau^*\boldsymbol{i} + y_\tau^*\boldsymbol{j} + z_\tau^*\boldsymbol{k}\,, \quad \tau = 1,2,3,4\,, \tag{10.1}$$

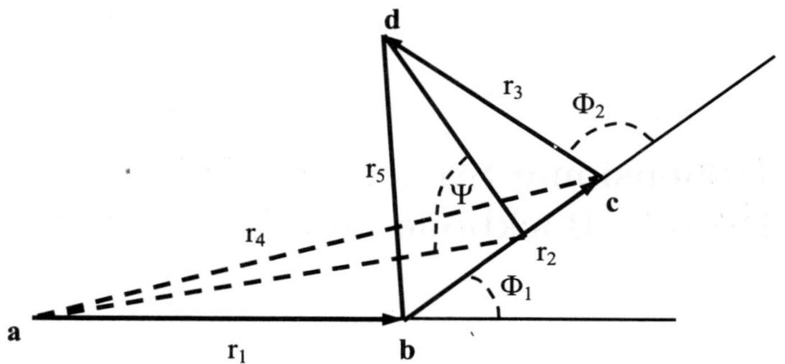

Fig. 10.1. Four-atom points conformation

where o is the origin of the coordinate system and $\boldsymbol{i}$, $\boldsymbol{j}$, $\boldsymbol{k}$ are the orthogonal basis vectors of the rectangular coordinate system. We introduce the following notations:

1. The vectors generated from the four space points a, b, c, and d are $\overrightarrow{ab}$, $\overrightarrow{bc}$, $\overrightarrow{cd}$, $\overrightarrow{ac}$, $\overrightarrow{bd}$, $\overrightarrow{ad}$, etc., denoted by $\boldsymbol{r}_1$, $\boldsymbol{r}_2$, $\cdots$, $\boldsymbol{r}_6$, respectively. Their coordinates as determined by (10.1) are

$$\begin{cases} \boldsymbol{r}_\tau = (x_\tau, y_\tau, z_\tau) = (x^*_{\tau+1} - x^*_\tau, y^*_{\tau+1} - y^*_\tau, z^*_{\tau+1} - z^*_\tau)\,, \quad \tau = 1,2,3\,, \\ \boldsymbol{r}_{\tau'} = (x_{\tau'}, y_{\tau'}, z_{\tau'}) = (x^*_{\tau'-1} - x^*_{\tau'-3}, y^*_{\tau'-1} - y^*_{\tau'-3}, z^*_{\tau'-1} - z^*_{\tau'-3})\,, \\ \qquad \tau' = 4, 5\,, \\ \boldsymbol{r}_6 = (x_6, y_6, z_6) = (x^*_4 - x^*_1, y^*_4 - y^*_1, z^*_4 - z^*_1)\,. \end{cases} \tag{10.2}$$

Their lengths are denoted by $r_1, r_2, \cdots, r_6$, where

$$r_\tau = |\boldsymbol{r}_\tau| = (x^2_\tau + y^2_\tau + z^2_\tau)^{1/2}\,, \quad \tau = 1,2,3,4,5,6\,. \tag{10.3}$$

2. We denote the angle between the vectors $\overrightarrow{ab}$ and $\overrightarrow{bc}$ by ϕ_1, and between the vectors $\overrightarrow{bc}$ and $\overrightarrow{cd}$ by ϕ_2. We call ϕ_1 and ϕ_2 the turn (bend) of the atomic points, and the formulas are obtained from the cosine theorem as

$$\phi_1 = \cos^{-1}\left(\frac{r_4^2 - r_1^2 - r_2^2}{2r_1r_2}\right)\,, \quad \phi_2 = \cos^{-1}\left(\frac{r_5^2 - r_2^2 - r_3^2}{2r_2r_3}\right)\,, \tag{10.4}$$

where cos is the cosine function, which has the domain $[0, \pi]$.

3. The triangles generated by the vectors $\overrightarrow{ab}$, $\overrightarrow{bc}$ and $\overrightarrow{bc}$, $\overrightarrow{cd}$ are denoted by $\delta(abc)$, $\delta(bcd)$, and the corresponding planes are denoted by $\pi(abc)$, $\pi(bcd)$, respectively. The normal vectors determined by planes $\pi(abc)$, $\pi(bcd)$ are denoted by

$$\boldsymbol{b}_1 = (x_7, y_7, z_7)\,, \quad \boldsymbol{b}_2 = (x_8, y_8, z_8)\,,$$

and their formulas are

$$\boldsymbol{b}_1 = \boldsymbol{r}_1 \times \boldsymbol{r}_2 = \begin{vmatrix} x_1 & y_1 & z_1 \\ x_2 & y_2 & z_2 \\ \boldsymbol{i} & \boldsymbol{j} & \boldsymbol{k} \end{vmatrix}, \quad \boldsymbol{b}_2 = \boldsymbol{r}_2 \times \boldsymbol{r}_3 = \begin{vmatrix} x_2 & y_2 & z_2 \\ x_3 & y_3 & z_3 \\ \boldsymbol{i} & \boldsymbol{j} & \boldsymbol{k} \end{vmatrix}, \tag{10.5}$$

where $\boldsymbol{r}_1 \times \boldsymbol{r}_2$ is the outer product of vectors $\boldsymbol{r}_1$, $\boldsymbol{r}_2$, while

$$\begin{vmatrix} x_1 & y_1 & z_1 \\ x_2 & y_2 & z_2 \\ \boldsymbol{i} & \boldsymbol{j} & \boldsymbol{k} \end{vmatrix}$$

is the third-order determinant.

4. The line of intersection of the planes $\pi(abc)$, $\pi(bcd)$ is bc, and the angle between them is denoted by ψ. We call ψ the torsion angle of the atom points. The formula describing it is readily found as

$$\psi = \cos^{-1}\left(\frac{\langle \boldsymbol{b}_1, \boldsymbol{b}_2 \rangle}{b_1 b_2}\right), \tag{10.6}$$

where b_1 and b_2 are the lengths of the normal vectors $\boldsymbol{b}_1$ and $\boldsymbol{b}_2$, respectively. The formula is the same as (10.3), while .

$$\langle \boldsymbol{b}_1, \boldsymbol{b}_2 \rangle = x_7 x_8 + y_7 y_8 + z_7 z_8 \tag{10.7}$$

is the inner product of vectors $\boldsymbol{b}_1, \boldsymbol{b}_2$. ψ is also defined on the domain $[0, \pi]$.

5. The mixed product of vectors $\boldsymbol{r}_1, \boldsymbol{r}_2, \boldsymbol{r}_3$ is defined as

$$[\boldsymbol{r}_1, \boldsymbol{r}_2, \boldsymbol{r}_3] = \langle \boldsymbol{r}_1 \times \boldsymbol{r}_2, \boldsymbol{r}_3 \rangle = \begin{vmatrix} x_1 & y_1 & z_1 \\ x_2 & y_2 & z_2 \\ x_3 & y_3 & z_3 \end{vmatrix}. \tag{10.8}$$

6. We denote

$$\vartheta = \vartheta(abcd) = \mathrm{sgn}\left([\boldsymbol{r}_1, \boldsymbol{r}_2, \boldsymbol{r}_3]\right) \tag{10.9}$$

as the mirror value (or chirality value) of $\boldsymbol{r}_1, \boldsymbol{r}_2, \boldsymbol{r}_3$, where

$$\mathrm{sgn}\,(u) = \begin{cases} +1, & \text{if } u \geq 0, \\ -1, & \text{otherwise} \end{cases}$$

is the sign function of u.

The mirror value (or chirality value) is a reflection of the chirality characteristics of vectors $\boldsymbol{r}_1$, $\boldsymbol{r}_2$, $\boldsymbol{r}_3$. That is, when $\vartheta > 0$, the three vectors $\boldsymbol{r}_1$, $\boldsymbol{r}_2$, $\boldsymbol{r}_3$ make a right-handed system, while if $\vartheta < 0$, $\boldsymbol{r}_1$, $\boldsymbol{r}_2$, $\boldsymbol{r}_3$ make a left-handed system.

Correlation of the Basic Parameters

From formulas (10.2)–(10.9), we obtain the parameter space for four-atom points:

$$\mathcal{E} = \{r_1, r_2, \cdots, r_6, \phi_1, \phi_2, \psi, \vartheta\} . \tag{10.10}$$

We denote

$$\mathcal{E}_1 = \{r_1, r_2, r_3, r_4, r_5, \psi, \vartheta\} , \quad \mathcal{E}_2 = \{r_1, r_2, r_3, \phi_1, \phi_2, \psi, \vartheta\} \tag{10.11}$$

to be the basic parameter space of the atom points, with the following properties:

1. Parameter systems $\mathcal{E}_1$ and $\mathcal{E}_2$ determine each other, since in the cosine theorem in (10.4), r_1, r_2, r_3, r_4, r_5 determines r_1, r_2, r_3, ϕ_1, ϕ_2, and vice versa.
2. Each parameter in parameter space $\mathcal{E}$ is invariant with respect to the coordinate system $\{o, \boldsymbol{i}, \boldsymbol{j}, \boldsymbol{k}\}$. That is, when the coordinate system undergoes a translation or rotational transformation, the value of each parameter in $\mathcal{E}$ remains the same. When the coordinate system undergoes a mirror reflection transformation, ϑ in $\mathcal{E}$ changes sign, while other parameters remain the same.
3. When the parameters in parameter space $\mathcal{E}_1$ or $\mathcal{E}_2$ are given, the configuration of the four-atom points is completely determined. That is, for two groups of four-atom points, if their parameters in parameter space $\mathcal{E}_1$ or $\mathcal{E}_2$ are the same, then after rigid transformations, the two groups of four-atom points are superposed.

Other Parameters in the Four-Atom Space of Point Configurations

We know from geometry that, in the four-atom space of point configurations, there are other parameters apart from the basic ones. For instance:

1. The area formula for the triangle determined by points a, b, c: $S = S(abc) = \frac{1}{2} r_1 r_2 \sin\phi_1$, or $S = [s(s - r_1)(s - r_2)(s - r_3)]^{1/2}$, where $s = \frac{1}{2}(r_1 + r_2 + r_3)$.
2. The volume formula of the tetrahedron determined by points a, b, c, d:
$$V = V(abcd) = \frac{1}{6} \left|[\boldsymbol{r}_1, \boldsymbol{r}_2, \boldsymbol{r}_3]\right| .$$
3. The formula for the relationship of the volume, surface area, and height of the tetrahedron determined by points a, b, c, d:
$$V(abcd) = \frac{1}{3} S(abc) h(abc) ,$$
where $h(abc)$ is the height from the bottom face $\delta(abc)$ to point d.

The formulas may vary under different conditions, which will not be described here.

10.1.2 Phase Analysis on Four-Atom Space Points

In the protein 3D structure parameter space $\mathcal{E}_1$, the values of r_1, r_2, r_3, r_4, and r_5 are relatively constant; thus, the main parameters affecting protein 3D structure configuration are ψ, ϑ. We focus our the analysis on these parameters.

Definition of the Phase of the Four-Atom Space Points

In parameter space $\mathcal{E}_1$, we have already given the definition of the mirror value ϑ. We call (ϑ, ϑ') the phase of a set of four-atom space points, where

$$\vartheta' = \begin{cases} +1 , & \text{if } 0 \leq \psi < \pi/2 , \\ -1 , & \text{if } \pi/2 < \psi \leq \pi . \end{cases}$$

The definition of phase is actually the value of the angle ψ in the four quadrants of the plane rectangular coordinate system. Here, when (ϑ, ϑ') takes the values $(-1, 1)$, $(-1, -1)$, $(1, -1)$, $(1, 1)$, the values of the angle ψ in the four quadrants of the plane rectangular coordinates system are 0, 1, 2, 3, respectively.

Definition of Types E and Z for Four-Atom Points

In the parameters of the four-atom points phase, we know that the mirror (or chirality) value is determined by parameter ϑ. We now discuss the definition of the parameter ϑ'. In biology and chemistry, the structural characteristics of four-atom points are usually distinguished by types E and Z, which are mathematically expressed as follows.

Let d' represent the projection of point d on plane $\pi(abc)$, then the four points a, b, c, and d' lie in the same plane. Let $\ell(bc)$ denote the line determined by points b and c.

Definition 41. *For the four space points a, b, c, d, if a and d' are on the same side of line $\ell(bc)$, we say that the four points a, b, c, d are of type E; while if a and d' lie on two different sides of the line $\ell(bc)$, then we say the four points a, b, c, d are of type Z.*

The type E and type Z structures of four-atom point configurations are shown in Fig. 10.2.

In Fig. 10.2, d' is the projection of d on plane ABC. In Fig. 10.2a, points a, d' are on the same side of line $\ell(bc)$; while in Fig. 10.2b, a, d' are on different sides of line $\ell(bc)$. They form type E and type Z, respectively.

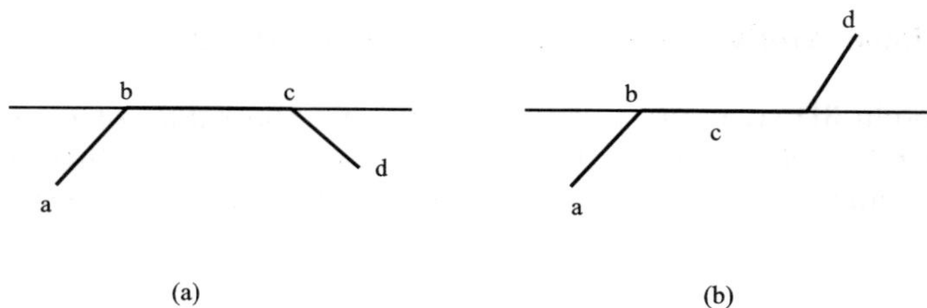

Fig. 10.2a,b. Types E and Z for four-atom point configurations

Determination of Type E and Type Z

We use geometry to determine the type E and type Z configuration of four-atom points. Let π_1 and π_2 represent planes determined by the three points a, b, c and b, c, d, respectively. Their normal vectors are $\boldsymbol{b}_1$ and $\boldsymbol{b}_2$, where the direction of the vectors is determined by the right-handed coordinate system, that is, their mixed product

$$[\boldsymbol{r}_1, \boldsymbol{r}_2, \boldsymbol{b}_1] , \quad [\boldsymbol{r}_2, \boldsymbol{r}_3, \boldsymbol{b}_2] > 0 .$$

Using the correlation of the normal vectors $\boldsymbol{b}_1$ and $\boldsymbol{b}_2$, we obtain the determining relationship for the structure type of the four-atom points conformation:

Theorem 38. *The determining relationship of the type E and Z structures of the four-atom points configuration is as follows: If $\langle \boldsymbol{b}_1, \boldsymbol{b}_2 \rangle < 0$, then the conformation of the four-atom points a, b, c, d is of type E, if $\langle \boldsymbol{b}_1, \boldsymbol{b}_2 \rangle > 0$, then the conformation of the four-atom points a, b, c, d is of type Z.*

Proof. We take the following steps to prove this proposition:

1. Because point d' is the projection of point d in plane Π_1, we set vector $\boldsymbol{r}_3' = \overrightarrow{cd'}$ to be the projection of $\boldsymbol{r}_3 = \overrightarrow{cd}$ in plane Π_1. If we denote the normal vector of plane c, d' d by $\boldsymbol{n} = \boldsymbol{b}_1 \times \boldsymbol{r}_2$, then $\boldsymbol{n}$ is in plane Π_1 and upright to $\boldsymbol{r}_2$, $\boldsymbol{b}_1$. At this time, $\boldsymbol{r}_1, \boldsymbol{r}_2, \boldsymbol{r}_3'$, and $\boldsymbol{n}$ are all in plane Π_1,

$$\begin{cases} \text{Iff } (\boldsymbol{r}_1 \cdot \boldsymbol{n})(\boldsymbol{r}_3' \cdot \boldsymbol{n}) > 0 , \text{ then } d, d' \text{ are on the same side of } \boldsymbol{r}_2 , \\ \text{Iff } (\boldsymbol{r}_1 \cdot \boldsymbol{n})(\boldsymbol{r}_3' \cdot \boldsymbol{n}) < 0 , \text{ then } d, d' \text{ are on different sides of } \boldsymbol{r}_2 , \end{cases} \tag{10.12}$$

   where $\boldsymbol{r}_1 \cdot \boldsymbol{n} = \langle \boldsymbol{r}_1, \boldsymbol{n} \rangle$ is the inner product of the two vectors.
2. In (10.12), $\boldsymbol{r}_3' \cdot \boldsymbol{n} = \boldsymbol{r}_3 \cdot \boldsymbol{n}$, and

$$(\boldsymbol{r}_1 \cdot \boldsymbol{n})(\boldsymbol{r}_3 \cdot \boldsymbol{n}) = \langle \boldsymbol{r}_1, \boldsymbol{b}_1 \times \boldsymbol{r}_2 \rangle \langle \boldsymbol{r}_3, \boldsymbol{b}_1 \times \boldsymbol{r}_2 \rangle . \tag{10.13}$$

   Then, from

$$\boldsymbol{b}_1 = \frac{\boldsymbol{r}_1 \times \boldsymbol{r}_2}{|\boldsymbol{r}_1 \times \boldsymbol{r}_2|}$$

and the outer product formula (10.5), we obtain

$$\boldsymbol{b}_1 \times \boldsymbol{r}_2 = \frac{1}{|\boldsymbol{r}_1 \times \boldsymbol{r}_2|}(\boldsymbol{r}_1 \times \boldsymbol{r}_2) \times \boldsymbol{r}_2 = \frac{1}{|\boldsymbol{r}_1 \times \boldsymbol{r}_2|}\left[(\boldsymbol{r}_1 \cdot \boldsymbol{r}_2)\boldsymbol{r}_2 - r_2^2 \boldsymbol{r}_1\right] .$$

Therefore,

$$(\boldsymbol{r}_1 \cdot \boldsymbol{n})(\boldsymbol{r}_3 \cdot \boldsymbol{n}) = \frac{1}{|\boldsymbol{r}_1 \times \boldsymbol{r}_2|^2}\left[(\boldsymbol{r}_1 \cdot \boldsymbol{r}_2)^2 - r_2^2 r_1^2\right] \cdot \left[(\boldsymbol{r}_1 \cdot \boldsymbol{r}_2)(\boldsymbol{r}_2 \cdot \boldsymbol{r}_3) - r_2^2(\boldsymbol{r}_1 \cdot \boldsymbol{r}_3)\right] .$$

Since for any two vectors $\boldsymbol{r}_1$ and $\boldsymbol{r}_2$, $(\boldsymbol{r}_1 \cdot \boldsymbol{r}_2)^2 \leq r_1^2 r_2^2$ always holds true,

$$\begin{aligned} \text{sgn}\left[(\boldsymbol{r}_1 \cdot \boldsymbol{n})(\boldsymbol{r}_3' \cdot \boldsymbol{n})\right] &= \text{sgn}\left[(\boldsymbol{r}_1 \cdot \boldsymbol{n})(\boldsymbol{r}_3 \cdot \boldsymbol{n})\right] \\ &= \text{sgn}\left[r_2^2(\boldsymbol{r}_1 \cdot \boldsymbol{r}_3) - (\boldsymbol{r}_1 \cdot \boldsymbol{r}_2)(\boldsymbol{r}_2 \cdot \boldsymbol{r}_3)\right] \end{aligned} \tag{10.14}$$

always holds.

3. On the other hand, we can also calculate $(\boldsymbol{b}_1, \boldsymbol{b}_2)$. The definition of $\boldsymbol{b}_\tau$ yields

$$(\boldsymbol{b}_1, \boldsymbol{b}_2) = \alpha(\boldsymbol{r}_1 \times \boldsymbol{r}_2, \boldsymbol{r}_2 \times \boldsymbol{r}_3) ,$$

where

$$\alpha = \frac{1}{|\boldsymbol{r}_1 \times \boldsymbol{r}_2| \cdot |\boldsymbol{r}_2 \times \boldsymbol{r}_3|} > 0 .$$

Then from the outer product formula, we obtain

$$(\boldsymbol{b}_1, \boldsymbol{b}_2) = \alpha\left[(\boldsymbol{r}_1, \boldsymbol{r}_2) \cdot (\boldsymbol{r}_2, \boldsymbol{r}_3) - r_2^2(\boldsymbol{r}_1, \boldsymbol{r}_3)\right] ;$$

therefore,

$$\text{sgn}\,(\boldsymbol{b}_1, \boldsymbol{b}_2) = -\text{sgn}\left[(\boldsymbol{r}_1 \cdot \boldsymbol{n})(\boldsymbol{r}_3' \cdot \boldsymbol{n})\right] . \tag{10.15}$$

From formula (10.15), we know that the conclusion of the theorem stands. Hence this theorem is proved.

Theorem 39. *The structure types E and Z of the four-atom points conformation can be determined by the value of the torsion angle ψ, by the relationship:*

$$\begin{cases} \textit{If } |\psi| \leq \dfrac{\pi}{2} & \textit{then the conformation of the four-atom points a,b,c,d} \\ & \textit{is of type } Z, \\ \textit{If } |\psi| > \dfrac{\pi}{2} & \textit{then the conformation of the four-atom points is of type } E , \end{cases} \tag{10.16}$$

where $0 \leq |\psi| \leq \pi$.

Proof. From the definition of the torsion angle (10.6), and comparing formula (10.16) with Theorem 39, we see that this theorem holds.

The Transformation Relationship for the Four-Atom Points Torsion Angle and Phase

In the definition of torsion angle (10.6), ψ is in the domain $[0, \pi]$. If it is combined with the mirror reflection value ϑ, ψ can be defined on $[0, 2\pi]$. We then take

$$\psi' = \begin{cases} \psi\,, & \text{if } \vartheta = -1\,, \\ 2\pi - \psi\,, & \text{if } \vartheta = 1\,. \end{cases} \tag{10.17}$$

At this time, ψ' is defined on $[0, 2\pi]$, and the value of ψ' in the four quadrants of the plane rectangular coordinate system coincides with the definition of (ϑ, ϑ').

10.1.3 Four-Atom Construction of Protein 3D Structure

In the construction of protein atoms, we have already introduced the structure of their backbones and side chains, from which various four-atom points of different types can be obtained. The important types that will be discussed are as follows.

The Types of Four-Atom Points Obtained from the Protein Backbones

1. Four-atom points obtained from the three atoms N, A, C in turn are of the three types: N, A, C, N′; A, C, N′, A′; C, N′, A′, C′, where N, A, C, N′, A′, C′ are the N, A, C atoms of two neighboring amino acids.
2. A series of four-atom points can be obtained by choosing an atom point in each amino acid, for example, A_t, A_{t+1}, A_{t+2}, A_{t+3} where $t = 1, 2, 3, \cdots$.

Four-atom points obtained from the side chains of the backbones are of the type N, A, C, O; N, A, C, B, etc.

Statistical Calculation for the Protein Backbone Four-Atom Points

There are 3190 proteins in the PDB-Select database (version 2005), on which we may perform statistical calculations for all the four-atom points of the linked amino acids. The results are as follows:

1. There are 739,030 linked amino acids in the 3190 proteins in the PDB-Select database (version 2005), and every amino acid consists of different atoms. The PDB-Select database gives the space coordinates of all the non-hydrogen atoms, and the total number of non-hydrogen atoms is about 5,876,820.

2. For the atoms N, A, C, N′, A′, C′, we denote r_1, r_2, r_3, r_4, r_5, r_6 to be the lengths of the vectors

$$\overrightarrow{NA}, \overrightarrow{AC}, \overrightarrow{CN'}, \overrightarrow{NC'}, \overrightarrow{AN'}, \overrightarrow{CA'} ,$$

respectively. We denote the torsion angle between plane $\pi(\text{NAC})$ and plane $\pi(\text{ACN}')$ by ψ_1, the torsion angle between plane $\pi(\text{ACN}')$ and plane $\pi(\text{CN}'\text{A}')$ by ψ_2, and the torsion angle between plane $\pi(\text{CN}'\text{C}')$ and plane $\pi(\text{N}'\text{A}'\text{C}')$ by ψ_3.
We calculate the mean, variance and standard deviation of each of the parameters $r_1, r_2, r_3, r_4, r_5, r_6, \psi_1, \psi_2, \psi_3$ for the 739,030 linked amino acids in the 3190 proteins in the PDB-Select database (version 2005). The results are given in Table 10.1.
We see from Table 10.1 that the values of the parameters r_1, r_2, r_3, r_4, r_5, r_6, and ψ_2 are quite constant, where the mean of ψ_2 is $\mu = 3.14716 = 180.320°$. This indicates that the four points A, C, N′, A′ are almost on the same plane.
3. The edge lengths of the four-atom points A_1, A_2, A_3, A_4 in the backbone are

$$r_1 = A_1A_2 , \quad r_2 = A_2A_3 , \quad r_3 = A_3A_4 , \quad r_4 = A_1A_3 , \quad r_5 = A_2A_4 .$$

The results of our statistical calculations for the torsion angle Ψ are shown in Table 10.2, where the torsion angle Ψ is the angle between the triangle $\delta(A_1A_2A_3)$ and $\delta(A_2A_3A_4)$, defined on $[0, 2\pi]$.
4. The phase distribution of ψ_1, ψ_2, ψ_3 and Ψ is shown in Table 10.3.

In the next section, we will analyze the four-atom structure of protein side chains. In the statistical analysis presented in this book, we include comprehensive statistics of all the amino acid parameters. If we analyzed each individual amino acid, the values of the related parameters would differ slightly, which we do not discuss here.

Table 10.1. Parameter characteristics of four-atom points N, A, C, N′, A′, C′ in different amino acids

	r_1	r_2	r_3	r_4	r_5	r_6
Mean (μ)	1.46215	1.52468	1.33509	2.45624	2.43543	2.43633
Variance (σ^2)	0.00018	0.00014	0.00021	0.00258	0.00070	0.00068
Standard deviation (σ)	0.01325	0.01198	0.01455	0.05080	0.02651	0.02608

	ψ_1	ψ_2	ψ_3
Mean (μ)	3.52116	3.14716	4.15748
Variance (σ^2)	3.77901	0.03136	1.84675
Standard deviation (σ)	1.94397	0.17709	1.35895

Table 10.2. Parameter characteristics of the four-atom points A_1,A_2,A_3,A_4 in different amino acids

	r'_1	r'_2	r'_3	r''_4	r''_5	ψ
Mean (μ)	3.81009	3.81007	3.80852	6.02735	6.02492	2.56770
Variance (σ^2)	0.00301	0.00308	0.00520	0.39028	0.39149	2.68689
Standard deviation (σ)	0.05490	0.05552	0.07210	0.62473	0.62569	1.63917

Table 10.3. Phase distribution table of ψ_1, ψ_2, ψ_3, and Ψ

Phase value	1	2	3	4
ψ_1	0.1625	0.3507	0.1926	0.3942
ψ_2	0.0014	0.5107	0.4863	0.0015
ψ_3	0.1155	0.0697	0.3156	0.4993
Ψ	0.3971	0.1937	0.2928	0.1164

10.2 Triangle Splicing Structure of Protein Backbones

There are two types of protein backbone structures. There are chains formed by three atoms N, A, C alternately, and chains made up of atoms A_1, A_2, $A_3, \cdots$. They are denoted respectively by

$$\begin{cases} \mathbf{L}_1 = \{(N_1, A_1, C_1), (N_2, A_2, C_2), \cdots, (N_n, A_n, C_n)\} , \\ \mathbf{L}_2 = \{A_1, A_2, A_3, \cdots, A_n\} , \end{cases} \tag{10.18}$$

where n is the length of the protein sequence, that is, the number of amino acids. We now discuss the structural properties of the protein backbone.

10.2.1 Triangle Splicing Belts of Protein Backbones

We can generally denote the sequences in (10.18) by

$$\mathbf{L} = \{Z_1, Z_2, Z_3, \cdots, Z_{n'-1}, Z_{n'}\} , \tag{10.19}$$

where n' is $3n$ or n, when $\mathbf{L}$ is $\mathbf{L}_1$ or $\mathbf{L}_2$, respectively. We denote the space coordinates of each point Z_j by (x_j^*, y_j^*, z_j^*), and $\mathbf{L}$ is the space structure sequence of protein backbones.

Parameter System for Protein Backbone 3D Structures

Similar to the structure parameter theory of four-atom points, we can construct the structure parameter for sequence $\mathbf{L}$. We introduce the relevant notations as follows:

1. If the three neighboring atoms in **L** are considered to form a triangle, then **L** would be formed from a series of conjoined triangles. These triangles are denoted by
$$\mathbf{L}_\delta = \{\delta_1, \delta_2, \delta_3, \cdots, \delta_{n'-2}\} , \tag{10.20}$$
where $\delta_i = \delta(Z_i Z_{i+1} Z_{i+2})$ is the triangle with vertices Z_i, Z_{i+1}, Z_{i+2}, and δ_i and δ_{i+1} are joined and have a common edge $Z_{i+1}Z_{i+2}$. We call $\mathbf{L}_\delta$ the triangle splicing belt of protein backbones.
2. For the triangle splicing belts of protein backbones, we introduce the parameter system as follows:
$$\begin{cases} r_i = |Z_i Z_{i+1}|, \ i = 1, 2, \cdots, n' - 1 , \\ r'_i = |Z_i Z_{i+2}|, \ i = 1, 2, \cdots, n' - 2 , \\ \phi_i \quad \text{angle between } \overrightarrow{Z_i Z_{i+1}} \text{ and } \overrightarrow{Z_{i+1} Z_{i+2}}, \ i = 1, 2, \cdots, n' - 2 , \\ \psi_i \quad \text{torsion angle of triangle } \delta_i \text{ and } \delta_{i+1}, \ i = 1, 2, \cdots, n' - 3 , \\ \vartheta_i \quad \text{mirror value of the four points } Z_i, Z_{i+1}, Z_{i+2}, Z_{i+3} . \end{cases} \tag{10.21}$$
This yields the parameter system of the triangle splicing belts of protein backbones:
$$\mathcal{E}_\delta = \{r_i, r'_{i'}, \phi_{i'}, \psi_{i''}, \vartheta_{i''}, \ i \in N'_1, \ i' \in N'_2, i'' \in N'_3\} , \tag{10.22}$$
where $N'_\tau = \{1, 2, \cdots, n' - \tau\}$, $\tau = 1, 2, 3$.
3. For the parameter system $\mathcal{E}_\delta$, similarly, we can divide it into two sets:
$$\begin{cases} \mathcal{E}_\delta^{(1)} = \{r_i, r'_{i'}, \psi_{i''}, \vartheta_{i''}, \ i \in N'_1, \ i' \in N'_2, \ i'' \in N'_3\} , \\ \mathcal{E}_\delta^{(2)} = \{r_i, \phi_{i'}, \psi_{i''}, \vartheta_{i''}, \ i \in N'_1, \ i' \in N'_2, \ i'' \in N'_3\} . \end{cases} \tag{10.23}$$
At this time, parameter systems $\mathcal{E}_\delta^{(1)}$ and $\mathcal{E}_\delta^{(2)}$ determine each other, and the 3D configuration of the protein backbone can be completely determined.

Phase Sequences of the Protein Backbone 3D Structure

Similar to the four-atom points, we can define the phase sequences of the protein backbone 3D structure. Here we take
$$\tilde{\vartheta} = \{(\vartheta_1, \vartheta'_1), (\vartheta_2, \vartheta'_2), \cdots, (\vartheta_{n'-2}, \vartheta'_{n'-2})\} , \tag{10.24}$$
where ϑ_i is defined in formula (10.22), while
$$\vartheta'_i = \begin{cases} +1 , & \text{if } |\psi_i| \le \pi/2 , \\ -1 , & \text{otherwise} , \quad i \in N'_3 . \end{cases} \tag{10.25}$$
Using $(\vartheta_i, \vartheta'_i)$ and formula (10.17), similarly, the value of ψ_i can be extended to $[0, 2\pi]$, and its values in the four quadrants of the plane rectangular coordinate

system can be determined. At this time, we take

$$\vartheta'' = \begin{cases} 0\,, & \text{if } \vartheta = -1,\ \vartheta' = 1\,, \\ 1\,, & \text{if } \vartheta = -1,\ \vartheta' = -1\,, \\ 2\,, & \text{if } \vartheta = 1,\ \vartheta' = -1\,, \\ 3\,, & \text{if } \vartheta = 1,\ \vartheta' = 1\,. \end{cases} \tag{10.26}$$

Table 9.1 gives the primary structure, secondary structure, and phase sequence of protein 12E8H, where the phase sequence of the protein is in the third line.

Plane Unwinding of the Triangle Splicing Belts of Protein Backbones

If the triangle splicing belt $\mathbf{L}_\delta$ of a protein backbone is given, its parameter system $\mathcal{E}_\delta^{(0)}$ will be determined by formula (10.23). We denote the parameter system of the plane unwinding formula of the protein backbone triangle splicing belt by

$$\mathcal{E}_\delta^{(0)} = \left\{r_i, r'_{i'}, \psi^0_{i''}, \vartheta_{i''},\ i \in N'_1, i' \in N'_2, i'' \in N'_3\right\}\ , \tag{10.27}$$

where $\psi^0_1 = \psi^0_2 = \cdots = \psi^0_{n'-2}$. The triangle splicing belt is then determined by $\mathcal{E}_\delta^{(0)}$

$$\mathbf{L}' = \{Z'_1, Z'_2, \cdots, Z'_{n'}\} \quad \text{or} \quad \mathbf{L}'_\delta = \left\{\delta'_1, \delta'_2, \cdots, \delta'_{n'-2}\right\}\ , \tag{10.28}$$

where $Z'_1, Z'_2, \cdots, Z'_{n'}$ are points in the fixed plane $\pi(Z_1Z_2Z_3)$,

$$\delta'_i = \delta\left(Z'_iZ'_{i+1}Z'_{i+2}\right)\ ,$$

and satisfy the conditions below:

1. Initial condition: $(Z'_1, Z'_2, Z'_3) = (Z_1, Z_2, Z_3)$.
2. Arc length condition: $|Z'_1Z'_2| = r_1$, $|Z'_2Z'_3| = r_2$, $|Z'_3Z'_4| = r_3$, $|Z'_1Z'_3| = r_4$, $|Z'_2Z'_4| = r_5$.

At this time, every protein backbone triangle splicing belt $\mathbf{L}_\delta$ can be unwound into a plane triangle belt $\mathbf{L}'_\delta$ by rigid rotations between the triangles δ_i and δ_{i+1}. Conversely, a plane triangle splicing belt $\mathbf{L}'_\delta$, using a rigid rotation between triangles δ'_i and δ'_{i+1} can become a protein backbone triangle belt $\mathbf{L}_\delta$. Rigid rotation occurs between δ_i and δ_{i+1} (or δ'_i and δ'_{i+1}), while the parameter structures of the other parts of the backbone triangle splicing belt $\mathbf{L}_\delta$ (or plane triangle splicing belt $\mathbf{L}'_\delta$) stay the same. The plane triangle splicing belt of the protein backbone is shown in Fig. 10.3.

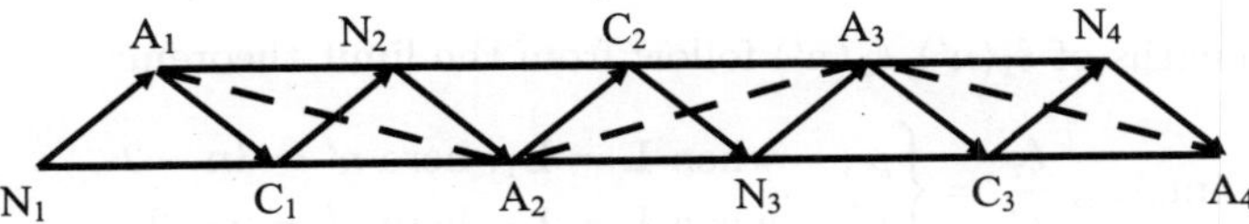

Fig. 10.3. Plane triangle splicing belt of the protein backbones

10.2.2 Characteristic Analysis of the Protein Backbone Triangle Splicing Belts

The Length and Width of the Protein Backbone Triangle Splicing Belts

If $\mathbf{L}_\delta$ is a fixed protein backbone triangle splicing belt, its parameter space will be $\mathcal{E}_\delta^{(1)}$. We can then define the edges and width of this triangle splicing belt as follows:

1. From vector groups

$$\begin{cases} L_1 = \{\boldsymbol{r}'_1, \boldsymbol{r}'_3, \cdots, \boldsymbol{r}'_{2n''-1}\} , \\ L_2 = \{\boldsymbol{r}'_2, \boldsymbol{r}'_4, \cdots, \boldsymbol{r}'_{2n''}\} , \end{cases} \tag{10.29}$$

where n'' is the integral value of $n'/2$. At this time, L_1, L_2 is formed by two groups of conjoined vectors, and they are parallel. They are named the upper edge and lower edge of the triangle splicing belts. Here we denote

$$\ell_1(n') = \sum_{i=1}^{n''} r'_{2i-1} , \quad \ell_2(n') = \sum_{i=1}^{n''} r'_{2i} \tag{10.30}$$

as the lengths of the upper and lower edge.

2. In triangle $\delta_i = \delta(Z_i Z_{i+1} Z_{i+2})$, we denote by h_i the height from the vertex Z_{i+1} to edge $Z_i Z_{i+2}$; its formula is:

$$h_i = 2S[\delta(Z_i Z_{i+1} Z_{i+2})]/|Z_i Z_{i+2}| = 2S_i/r'_i , \tag{10.31}$$

where S_i, $S[\delta(Z_i Z_{i+1} Z_{i+2})]$ are both the area of triangle $\delta_i = \delta(Z_i Z_{i+1} Z_{i+2})$, while $r'_i = |Z_i Z_{i+2}|$ is the length of the line $Z_i Z_{i+2}$.

3. We obtain from the calculation on the PDB-SEL database that the values of h_i are quite constant. In the calculation of triangle belts $\mathbf{L}_1, \mathbf{L}_2$, we obtain the means and variances of h_i as

	S_1	S_2	S_3	S'	h_1	h_2	h_3	h'
Mean (μ)	1.03572	0.90677	0.82643	6.64234	0.84362	0.74411	0.67951	2.24859
Variance(σ^2)	0.00074	0.00031	0.00048	0.72640	0.00134	0.00034	0.00047	0.22056
Standard deviation (σ)	0.02714	0.01768	0.02188	0.85229	0.03663	0.01835	0.02171	0.46964

where S_1, S_2, S_3, S' are the areas of triangles $\delta(\text{NAC})$, $\delta(\text{ACN}')$, $\delta(\text{CN}'\text{A}')$ and $\delta(\text{A}_1\text{A}_2\text{A}_3\text{A}_4)$, respectively, while h_1, h_2, h_3, and h' are relevant heights of these triangles.

4. The lengths of $\ell_1(n'), \ell_2(n')$ follow from the limit theorem:

$$\lim_{n' \to \infty} \frac{\ell_\tau}{n'} = \begin{cases} \mu\,, & \text{when } \mathbf{L} = \mathbf{L}_1, \text{ here } n' = 3n - 2\,, \\ \mu'\,, & \text{when } \mathbf{L} = \mathbf{L}_1, \text{ here } n' = n - 2\,, \end{cases} \tag{10.32}$$

where $\tau = 1, 2$, while

$$\begin{cases} \mu = \frac{1}{2}\mu(r_4 + r_5 + r_6) = \frac{1}{2}(2.45624 + 2.43543 + 2.43633) = 3.664\,, \\ \mu' = \frac{1}{2}\mu(r_4'') = \frac{6.02735}{2} = 3.0137\,. \end{cases}$$

Here, the data of $\mu(r_4), \mu(r_5), \mu(r_6), \mu(r_5'')$ are given by Tables 10.1 and 10.2.
Using the central limit theorem, the values of $\ell_1(n')$ and $\ell_2(n')$ can be estimated more precisely. We have validated these conclusions with a great amount of calculations on proteins.

The Relation Analysis of the Phase of the Protein Backbone Triangle Splicing Belt Between Secondary Structures

It has been mentioned above that the value of the protein backbone triangle splicing belt phase ϑ_i'' is closely correlated to the protein secondary structure. When $\vartheta_i'' = 0, 3$, four-atom points A_i, A_{i+1}, A_{i+2}, A_{i+3} form a Z type, while when $\vartheta_i'' = 1, 2$, four-atom points A_i, A_{i+1}, A_{i+2}, A_{i+3} form an E type. An appendix posted on the Web site [99] gives phase values and secondary structure sequences of all the proteins in the PDB-Select database. If we denote by M_1 the number of the amino acids in the database which take the secondary structure H or S, denote by M_2 the number of amino acids in the database which take the secondary structure H and phase value $\vartheta_i'' = 0, 3$, and denote by M_3 the number of amino acids in the database which take the secondary structure S and phase value $\vartheta_i'' = 1, 2$, then $\frac{M_2+M_3}{M_1} = 88.25\%$. From this, we conclude that the phase of the protein backbone triangle splicing belt is closely correlated to the secondary structures.

10.3 Structure Analysis of Protein Side Chains

We have stated in Chap. 7 that, for different amino acids, the component structure of the side chains may vary. We now discuss their configuration characteristics. We have pointed out before that the backbone of a protein consists of atoms N, C_α, C alternately, and they form a space triangle belt. Thus, the side chains of proteins can be considered to be components localized in this space triangle belt. We discuss first the configuration structure of an oxygen atom O and an atom C_B on the backbones, where the oxygen atom O is present in all amino acids, while atom C_B is present in all amino acids except the glycines.

10.3.1 The Setting of Oxygen Atom O and Atom C_B on the Backbones

Relevant Notation

1. We maintain the notation

$$\mathbf{L} = \{Z_1, Z_2, \cdots, Z_{3n}\} = \{N_1, A_1, C_1, N_2, A_2, C_2, \cdots, N_n, A_n, C_n\} \tag{10.33}$$

for the backbone of the protein, so the oxygen atom sequences and the side chain groups sequences are denoted by

$$\begin{cases} \bar{a} = \{O_1, O_2, \cdots, O_n\} , \\ \bar{a} = \{R_1, R_2, \cdots, R_n\} , \end{cases} \tag{10.34}$$

where O_i, R_i are the oxygen atom and the side chain group of the ith amino acid, respectively.

2. In the side chain group R, except for glycines, all the other amino acids contain C_B atoms, which are denoted by B atoms. Here, the tetrahedrons formed by the four-atom vertices N, A, C, O and N, A, C, B are V_O, V_B. Their shapes are shown in Fig. 10.4.
Figure 10.4 presents the structural relation of the atoms N, A, C, O, B, N′, where N′ is the nitrogen atom of the next amino acid. They form different tetrahedrons separately. For instance, $V_O = \{N, A, C, O\}$, $V_B = \{N, A, C, B\}$, where point B is usually on one side of the plane N A C, while point O can be on different sides of the plane N A C.

3. For the atom points N, A, C, O, B, N′, their coordinates are denoted separately by

$$\boldsymbol{r}_\tau^* = (x_\tau^*, y_\tau^*, z_\tau^*) , \quad \tau = 1, 2, 3, 4, 5, 6 . \tag{10.35}$$

We have already given the structural relations of the atom points N, A, C, N′ in the previous section, so we now discuss the relation between atoms O, B and atom points N, A, C, N′. We denote

$$\begin{cases} \boldsymbol{r}_\tau = \boldsymbol{r}_4^* - \boldsymbol{r}_\tau^* = (x_\tau, y_\tau, z_\tau) , \\ \boldsymbol{r}'_{\tau'} = \boldsymbol{r}_5^* - \boldsymbol{r}_\tau^* = (x'_{\tau'}, y'_{\tau'}, z'_{\tau'}) , \end{cases} \tag{10.36}$$

and their lengths are r_τ, r'_τ, $\tau = 1, 2, 3$.

4. In proteins, the tetrahedrons of different amino acids with regard to O, B atoms are denoted by $V_{i,O}$ and $V_{i,B}$. Similarly, we can define the relevant atom vectors $\boldsymbol{r}_{i,\tau}$, $\boldsymbol{r}'_{i,\tau}$, $\tau = 1, 2, 3$ to represent the vectors from atoms N_i, A_i, C_i to atom O_i and atom point B_i. Similarly to the definition in (10.36)

$$\begin{cases} \boldsymbol{r}_{i,\tau} = \boldsymbol{r}_{i,4} - \boldsymbol{r}_{i,\tau} = (x_{i,\tau}, y_{i,\tau}, z_{i,\tau}) , \\ \boldsymbol{r}'_{i,\tau} = \boldsymbol{r}_{i,5} - \boldsymbol{r}_{i,\tau} = \left(x'_{i,\tau'}, y'_{i,\tau'}, z'_{i,\tau'}\right) , \end{cases} \tag{10.37}$$

their lengths are $r_{i,\tau}$, and $r'_{i,\tau}$, $\tau = 1, 2, 3, i = 1, 2, \cdots, n$.

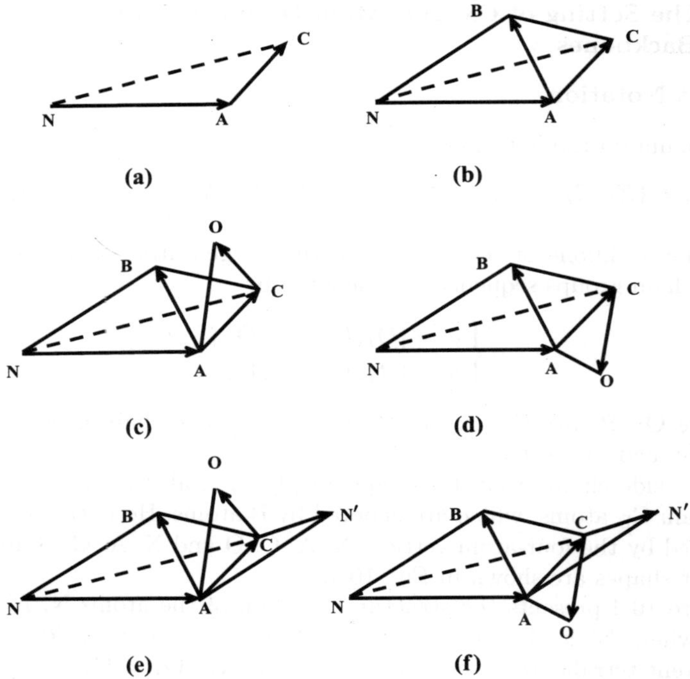

Fig. 10.4. Tetrahedrons $V_{\mathrm{O}}, V_{\mathrm{B}}$

Structural Analysis on Tetrahedrons $V_{\mathrm{O}}, V_{\mathrm{B}}$

When the protein backbone **L** is given, the locations of O_i are uniquely determined by the parameters $\{r_{2i}, r_{4i}, r_{6i}, \vartheta_i\}$, where ϑ_i is the mirror reflection value of the four-atom points A, C, N′, O, which has been defined in Chap. 8. Here we have the following formulas:

1. Formula for the volume of tetrahedron V_{O}:

$$V_{\mathrm{O}} = \frac{1}{6} \begin{vmatrix} x_1 & y_1 & z_1 \\ x_2 & y_2 & z_2 \\ x_3 & y_3 & z_3 \end{vmatrix}, \quad \text{its absolute value}. \tag{10.38}$$

2. Formula for the area of triangle $\delta(\mathrm{NAC})$ has been given in Sects. 10.1 and 10.2, which is denoted as $S = S(\mathrm{NAC})$. Here we introduce

$$h_{\mathrm{O}} = 3\frac{V_{\mathrm{O}}}{S}, \quad h_{\mathrm{B}} = 3\frac{V_{\mathrm{B}}}{S} \tag{10.39}$$

the height of tetrahedrons V_{O} and V_{B}, respectively, and V_{O} and V_{B} in (10.39) the volumes of the tetrahedrons. If we denote by $\vartheta_{\mathrm{O}}, \vartheta_{\mathrm{B}}$ the mirror reflection values of the tetrahedrons V_{O}, V_{B}, then we have

$$H_{\mathrm{O}} = \vartheta_{\mathrm{O}} h_{\mathrm{O}} , \quad H_{\mathrm{B}} = \vartheta_{\mathrm{B}} h_{\mathrm{B}} , \tag{10.40}$$

representing the directional volumes of tetrahedrons V_{O} and V_{B}, respectively. The formulas for the mirror reflection value ϑ_{O} of the four-atom points N, A, C, O are

$$\begin{cases} \vartheta_{\mathrm{O}} = \mathrm{sgn}\left[\left(\overrightarrow{\mathrm{NO}}, \overrightarrow{\mathrm{AO}}, \overrightarrow{\mathrm{CO}}\right)\right] = \mathrm{sgn}\left(\begin{vmatrix} x_1 & y_1 & z_1 \\ x_2 & y_2 & z_2 \\ x_3 & y_3 & z_3 \end{vmatrix}\right) , \\ \vartheta_{\mathrm{B}} = \mathrm{sgn}\left[\left(\overrightarrow{\mathrm{NB}}, \overrightarrow{\mathrm{AB}}, \overrightarrow{\mathrm{CB}}\right)\right] = \mathrm{sgn}\left(\begin{vmatrix} x_1' & y_1' & z_1' \\ x_2' & y_2' & z_2' \\ x_3' & y_3' & z_3' \end{vmatrix}\right) , \end{cases} \tag{10.41}$$

where $x_1, \cdots, z_3$ and $x_1', \cdots, z_3'$ are defined in (10.36), while

$$(\mathbf{r}_1, \boldsymbol{r}_2, \boldsymbol{r}_3) = \langle \boldsymbol{r}_1 \times \boldsymbol{r}_2, \boldsymbol{r}_3 \rangle = \begin{vmatrix} x_1 & y_1 & z_1 \\ x_2 & y_2 & z_2 \\ x_3 & y_3 & z_3 \end{vmatrix}$$

is the mixed product of the three vectors $\boldsymbol{r}_1$, $\boldsymbol{r}_2$, $\boldsymbol{r}_3$.

10.3.2 Statistical Analysis of the Structures of Tetrahedrons $V_{\mathrm{O}}, V_{\mathrm{B}}$

The statistical analysis of the structures of the tetrahedrons $V_{\mathrm{O}}, V_{\mathrm{B}}$ refers to the statistical analysis of the structural parameters of atoms O, B and N, A, C, N′. Using the PDB-Select database, we may perform statistical analysis on these structural parameters (Table 10.4).

Structure Analysis of the Type L and Type D Tetrahedron V_{B}

1. Among the 696,637 amino acids analyzed, the numbers of tetrahedrons of mirror value $+1$ or -1 are 37,682 and 658,955 respectively, thus the proportion of those with mirror reflection value $+1$ is 5.41%.
2. In the PDB-Select database, for tetrahedron V_{B}, the numbers of type L and type D tetrahedrons are 653,969 and 141, respectively, hence the proportion is about 10,000:2. Thus, type D tetrahedrons are rare.

Table 10.4. Statistics of the basic parameters of tetrahedrons V_O and V_B

	1	2	3	4	5	6	7	8	9	10
	R(NA)	R(AC)	R(CO)	R(ON′)	R(N′A′)	R(NC)	R(AO)	R(CN′)	R(OA′)	R(NAC)
Mean	1.46220	1.52459	1.22781	2.24328	1.46211	2.45620	2.40675	1.33654	2.35253	2.07139
Variance	0.00019	0.00016	0.00021	0.00147	0.00017	0.00260	0.00058	0.00237	0.16705	0.00296
Standard deviation	0.01371	0.01251	0.01454	0.03830	0.01300	0.05103	0.02400	0.04871	0.40872	0.05440
	11	12	13	14	15	16				
	R(ACO)	R(ACN′)	R(CN′A′)	ψ_1	ψ_2	ψ_3				
Mean	1.61603	1.81455	1.65359	0.38551	0.47297	3.15149				
Variance	0.00094	0.00294	0.00321	3.75848	2.73928	0.03761				
Standard deviation	0.03066	0.05421	0.05666	1.93868	1.65508	0.19392				
	1	2	3	4	5	6	7	8	9	10
	R(NB)	R(AB)	R(CB)	R(NB)	R(NAC)	R(ACB)		V(NACB)	S(NAC)	H(NACB)
Mean	2.44942	1.53138	1.43440	2.45479	2.07516	1.91471	−1.41089	−0.42143	1.03758	1.21819
Variance	0.00178	0.00060	0.19462	0.00257	0.00270	0.22472	0.88474	0.00059	0.00068	0.00240
Standard deviation	0.04217	0.02460	0.44115	0.05073	0.05201	0.47404	0.94061	0.02430	0.02600	0.04894
	11	12	13	14	15	16				
	R(ACO)	R(ACN′)	R(CN′A′)	ψ_1	ψ_2	ψ_3				
Mean	1.61603	1.81455	1.65359	0.38551	0.47297	3.15149				
Variance	0.00094	0.00294	0.00321	3.75848	2.73928	0.03761				
Standard deviation	0.03066	0.05421	0.05666	1.93868	1.65508	0.19392				

V_O is the volume of the tetrahedron $V(\text{NACO})$, $S(\text{NAC})$ is the area of the bottom triangle $\delta(\text{NAC})$, h_O is the height of the tetrahedron $V(\text{NACO})$ with regard to the bottom triangle $\delta(\text{NAC})$

10.4 Exercises, Analyses, and Computation

Exercise 48. Attempt the following calculation on the atoms of all the protein backbones in the PDB-Select database:

1. For every atomic sequence in the protein backbone $\mathbf{L}_1$ (see the definition in formula (10.18)), calculate the parameter sequences. The relevant parameters are defined in formula (10.21).
2. For every atomic sequence in the protein backbone $\mathbf{L}_2$ (see the definition in formula (10.18)), calculate the parameter sequences. The relevant parameters are defined in formula (10.21).
3. For every atomic sequence in the protein backbone $\mathbf{L}_2$ (see the definition in formula (10.18)), where A_i are replaced by N_i or C_i separately, calculate the parameter sequences. The relevant parameters are defined in formula (10.21).

Exercise 49. Based on the calculation in Exercise 48, provide the following discussion and analysis:

1. Analyze the stability of each parameter, and calculate its mean, variance and standard deviation.
2. In the protein backbone $\mathbf{L}_2$, when the atomic sequences are A_i, N_i, and C_i, respectively, compare the correlation of each parameter sequence.
3. It can been seen from the above analysis that the torsion angle sequences ψ_i and mirror sequences ϑ_i are unstable. Attempt a histogram analysis of the torsion angle sequences ψ_i and attempt to perform frequency analysis on the mirror sequences ϑ_i.

Exercise 50. From the calculation in Exercise 48, task 3, we obtain the torsion angle sequence ψ_i and mirror sequence ϑ_i for the atomic sequence A_i, and therefore obtain the phase sequence ϑ''_i (see the definition in formula (10.26)). Perform the following calculations:

1. Compare the statistical properties of the phase sequence ϑ''_i with the secondary structure sequence (see the definition and explanation in formula (9.1) and Table 9.1). For instance, the joint distribution of (ϑ''_i, η_i), properties such as the interprediction between ϑ''_i and η_i.
2. For the phase sequence ϑ''_i, there is also the problem of predicting ϑ''_i from the primary structure sequence ξ_i. Attempt to calculate the informational statistics table of the primary structure ξ_i and the phase sequences ϑ''_i using the method in Chap. 9, and predict the phase sequences from the primary structure sequences.

Exercise 51. Attempt the following calculations on the atoms of all the protein side chains in the PDB-Select database:

1. For each amino acid in the database, calculate each parameter for the tetrahedron structure $V_B = \{N, A, C, B\}$.

2. For the pairs of conjoined amino acids in the database, calculate each parameter for the five-atom structure $\{N, A, C, O, N'\}$, where N′ is the nitrogen atom of the next amino acid.
3. For the pairs of conjoined amino acids in the database, calculate each parameter for the six-atom structure $\{N, A, C, B, O, N'\}$, where N' is the nitrogen atom of the next amino acid.
4. For the pairs of conjoined amino acids in the database, calculate each parameter for the eight-atom structure $\{N, A, C, B, N', A', C', B'\}$, where N′, A′, C′, B′ are the corresponding atoms of the next amino acid.
5. For different amino acids and dipeptide chains, calculate the structural data for the atomic distances.

Exercise 52. Based on the calculation in Exercise 51, provide the following discussion and analysis:

1. Analyze the stability of each parameter, and calculate its mean, variance, and standard deviation. Also, discuss the similarities and differences between distinct amino acids.
2. From this, determine the structural characteristics of the related atoms, e.g., four points on the same plane, the unicity of the tetrahedron phase V_B.

Exercise 53. Based on the calculation in Exercises 51 and 52, provide the following discussion and analysis for different amino acids and dipeptide chains:

1. For different dipeptide chains, based on the calculation in Exercise 51, task 4, construct the second-rank topological structure map for eight-atom points $\{N, A, C, B, N', A', C', B'\}$.
2. For different amino acids and dipeptide chains, based on the calculation in Exercise 51, task 4, construct the second-rank topological structure map for all the atom points.
3. Together with the classification in Table 10.4, discuss the similarities and differences between the second-ranked topological structure map.

11

Alignment of Primary and Three-Dimensional Structures of Proteins

In the previous chapter, protein three-dimensional structures and a method for predicting protein secondary structures using informatics and statistics have been introduced. In this chapter, we focus on the issues of alignment algorithms for the three-dimensional structure of proteins.

11.1 Structure Analysis for Protein Sequences

In biology, sequence analysis refers to analyzing DNA (or RNA) sequences and protein sequences. In the above, we mainly focused our discussion on DNA (or RNA). Here we introduce structure analysis for protein sequences.

11.1.1 Alignment of Protein Sequences

Structure analysis for proteins is a huge area of active research. It can be divided into the following aspects.

Research on Protein Primary Structures

It is known that proteins are formed by binding different amino acid residues, and protein primary structures are the sequences consisting of different permutations of these amino acids. Thus, the study of protein primary structures concerns the sequence structures of these different amino acids' permutations. The issues are:

1. Structure analysis and alignment of protein sequences. Problems such as mutation and reorganization, etc. are relevant to our understanding of protein structures, and the evolution and classification of protein structures are also related. Thus, sequence alignment is still a basic problem.

2. Grammar and word-building analysis for protein sequence structures. If protein sequences are disassembled, many small protein sequence segments result. These segments are called characteristic chains in biology. If peptide chains are considered to be the basic building blocks for protein structure and function, they would correspond to words and phrases of a book. In order to get an overview of the structure and functions of proteins, one approach is to study the grammar and syntax. That is, to set up special databases (databases of grammar and syntax) of the peptide chains whose structure and function is known, and then search and compare the general structures of proteins in these databases, so that the general structures and functions of these proteins can be determined. This problem is still related to sequence alignment, where the segments are usually short in length and large in number.

Research on Protein Three-Dimensional Structures and Their Functions

Protein three-dimensional structure includes secondary structure and other higher level structures. From biological research, we know that the activity and function of a protein is related to its three-dimensional structure. If a protein is heated, it will be denatured and thus lose its activity. It is found biologically that, for the proteins whose primary structures are different, if they have the same three-dimensional structures, their functions may be similar. Therefore, the research on protein three-dimensional structures is closely related to that of their functions.

The topic of protein three-dimensional structures covers many aspects. They are:

1. Research on protein secondary structures. In biology, peptide chains with special structural characteristics are called secondary structures, such as α helices and β sheets. These structures are common units of protein general three-dimensional structures, and their combinations determine the functions of the proteins they make up. Therefore, the research on protein secondary structures is essential to that of the three-dimensional structures. The analysis of protein secondary structures focuses on their prediction, which is to confirm protein secondary structure characteristics according to their primary sequences.
   Many mathematical tools and methods have been used in the research on protein secondary structure prediction, while so far, the accuracy of the predictions is far from satisfactory. In the past ten years, the accuracy of the predictions of α helices and β sheets remained at around 70%. For some fixed proteins whose types are known, the accuracy of the prediction has improved. However, because the types are fixed, their definition of the prediction accuracy is not as good.
2. The main unresolved difficulty present in the analysis of protein three-dimensional structures is the long distance folding of the three-dimensional

structures. That is, under what condition will the protein three-dimensional structures begin to fold back, and to which part of the protein will they join. Although there is much research involved in studying this problem, no efficient method has yet been found.

3. Research on the relationship between protein three-dimensional structures and their functions. This problem has always been both of great interest and fraught with difficulties in bioinformatics. Since the formulation of the proposal of the proteome project, the research on the relationship between protein three-dimensional structures and their functions has focused on the interactions between different proteins and the relationship between these interactions and protein three-dimensional structures.

Generally speaking, the research on protein three-dimensional structures and the relationship between protein three-dimensional structures and their functions is still in an early stage of development. So far, large amounts of data have been accumulated and relevant information for medicine and biology is being drawn upon. However, there is still a long way to go and many concepts need to be developed until mature applications are carried out. These are the areas where mathematics will play an important role, although we will not discuss them in detail in this book.

11.1.2 Differences and Similarities Between the Alignment of Protein Sequences and of DNA Sequences

Above, we specified some of the problems in protein structure research. We know from existing methods that although some are just in the early stages, many are related to sequence alignment, and especially to multiple sequence alignment which plays an important role. However, some differences lie between the alignment of protein sequences and that of DNA sequences; thus the discussion in the above chapters does not directly translate to the computation of the alignment of protein sequences. The differences between the alignment of protein sequences and that of DNA sequences are described next.

Differences Between the Constitution of Protein Sequences and that of DNA Sequences

1. Proteins usually consist of the 20 commonly occurring amino acids, thus they are more complex than DNA which consists of only four kinds of nucleotides. In the alignment of DNA sequences, we assume the distributions of sequences follow a uniform distribution. This assumption is suitable for DNA sequences, but not for protein sequences. According to our statistical analysis and computation, even in the Swiss-Prot database, the statistics of single amino acids does not follow a uniform distribution. For the statistics on polypeptide chains, the differences between their distributions are even greater.

2. The length of a protein sequence is usually not large. For some long protein sequences, the lengths are in the thousands of residues. Thus in the pairwise alignment of sequences, an increase in the computational complexity will not be the main issue. Therefore, the advantages of using the SPA algorithm over the dynamic programming algorithm will not be immediately obvious. However, when it comes to multiple sequence alignment, the design of fast algorithms will still be a key issue.
3. The penalty functions for the alignment of DNA sequences are simpler, either in the Hamming matrix or the WT-matrix as they both involve the strong symmetry condition. The penalty functions used in the alignment of protein sequences are the PAM matrix series and BLOSUM matrix series, neither of which involves the strong symmetry condition, or even the symmetry condition. Therefore, it must be further examined whether the series of limit theorems given in this book will hold for the alignment of protein sequences.

Similarities Between the Alignment of Protein Sequences and that of DNA Sequences

The main similarity between the alignments of protein sequences and of DNA sequences is that the theory of their modulus structure analysis still applies. Therefore, the definition and algorithm of uniform alignment remains effective. These will be discussed in the following text.

11.1.3 The Penalty Functions for the Alignment of Protein Sequences

The penalty functions used in the alignment of protein sequences are the PAM and BLOSUM matrix series. We describe their origin and properties as follows.

Origin of the PAM and BLOSUM Matrix Series

In the alignment of DNA sequences, the Hamming matrix and the WT-matrix used as the penalty functions are relatively simple, and the settings are easy to understand. They both have the same characteristic of strong symmetry. In the Hamming matrix, the errors of the nucleotides occurring in the alignment are given the same penalty, while in the WT-matrix, a, g and c, u are differentiated by giving smaller penalties due to the errors. This is related to the double helix structure of DNA and the central dogma.

Because of the differences in protein sequences as compared to DNA sequences, the determination of their penalty function is more complex. In the PAM matrix series and the BLOSUM matrix series, it can usually be determined by statistical computation methods. Their values are the negative values of the penalty functions; the closer the amino acids are, the higher the score will be. Thus, we call them scoring matrices. The related steps of the computation are outlined below:

1. Choose the known homologous sequences to be statistical samples whose homologies have a specific biological definition. The PAM matrix series and BLOSUM matrix series are the results of the statistics applied to different types and scales.
2. Alignment with the selected homologous sequence samples and the result is constructed for a group of pairwise sequences.
3. Based on the statistics of the selected pairwise sequences, we compute their joint frequency distribution $p(a,b)$, $a,b \in V_q$, where q is equal to 20, 23 or else. The symbol $p(a,b)$ denotes the frequency of a,b occurring in the same position of the pairwise sequences after alignment.
4. According to the frequency distribution $p(a,b)$, $a,b \in V_q$, we calculate their mutual entropy density function:

$$k(a,b) = \log \frac{p(a,b)}{p(a)q(b)}\,, \quad a,b \in V_q\,, \tag{11.1}$$

where

$$p(a) = \sum_{b \in V_q} p(a,b)\,, \quad q(b) = \sum_{a \in V_q} p(a,b)\,.$$

5. Properly magnify the mutual entropy density function $k(a,b)$, and make it an integer, to obtain the scoring matrix.

Properties of the Relative Entropy Density Functions

In informatics, it is known that the average of the relative entropy density function is the Kullback–Leibler divergence.

$$D_0 = \sum_{a,b \in V_q} p(a,b)k(a,b) = \sum_{a,b \in V_q} p(a,b) \log \frac{p(a,b)}{p(a)q(b)}\,, \tag{11.2}$$

where $D_0 \geq 0$ definitely holds, and the equality holds if and only if

$$p(a,b) = p(a)q(b)$$

holds for any $a,b \in V_q$.

Besides the D_0 index, another function is frequently used in bioinformatics, namely,

$$D_1 = \sum_{a,b \in V_q} p(a)q(b)k(a,b) = \sum_{a,b \in V_q} p(a)q(b) \log \frac{p(a,b)}{p(a)q(b)}\,. \tag{11.3}$$

For D_1, the relationship

$$D_1 = - \sum_{a,b \in V_q} p(a)q(b) \log \frac{p(a)q(b)}{p(a,b)}$$

holds. Moreover, it is clear that $D_1 \leq 0$ holds.

This shows that D_0, and D_1 are two important indices that can be used to assess protein similarity. If we denote (A', B') as the alignment sequences for (A, B), then

$$(A', B') = ((c_1, d_1), (c_2, d_2), \cdots, (c_{n_c}, d_{n_c})) .$$

We denote as $p_{A,B}(c, d)$, where $c, d \in V_q$, the frequency of (c_j, d_j) taking the value (c, d), and then compute

$$\begin{cases} KL(C|D) = \sum_{c,d \in V_q} p_{C,D}(c, d) k(c, d) , \\ HL(C|D) = \sum_{c,d \in V_q} p_C(c) p_{b'}(d) k(c, d) , \end{cases} \quad (11.4)$$

where $k(c, d)$ is determined by the data of the PAM or BLOSUM scoring table, and

$$p_C(c) = \sum_{d \in V_q} p_{C,D}(c, d) , \quad p_{b'}(d) = \sum_{c \in V_q} p_{C,D}(c, d) .$$

If $KL(A|B)$ approaches or exceeds D_0, (A, B) can be considered similar. If $HL(A|B)$ is close to or less than D_1, (A, B) can be considered not similar.

However, in the sequence alignment, the scores for the segments of the aligned sequences should be raised in local areas. This will be further explained later in the book.

Several Typical Scoring Matrices in the Alignment of Protein Sequences

The following typical scoring matrices in the alignment of protein sequences are commonly used in various bioinformatics resources and websites. We enumerate them as follows:

1. PAM 70 matrix. The following is the PAM 70 matrix ["pam" version 1.0.6 (July 28, 1993)]. The main parameters are

   $$D_0 = 1.60 , \quad D_1 = -2.77 , \quad \text{Max} = 13 , \quad \text{Min} = -11 ,$$

   where Max and Min are, respectively, the maximum value and the minimum value of the matrix.
   The letters are the one-letter codes of the amino acids, "–" is a virtual symbol. B represents N or D; Z represents Q or E; X represents all kinds of amino acids.
2. BLOSUM 45 matrix. The data of the statistical samples is obtained from the following: blocks Database = /data/blocks-5.0/blocks.dat, $D_0 =$ 0.3795, $D_1 = -0.2789$.

	A	R	N	D	C	Q	E	G	H	I	L	K	M	F	P	S	T	W	Y	V	B	Z	X	-
A	5	-4	-2	-1	-4	-2	-1	0	-4	-2	-4	-4	-3	-6	0	1	1	-9	-5	-1	-1	-1	-2	-11
R	-4	8	-3	-6	-5	0	-5	-6	0	-3	-6	2	-2	-7	-2	-1	-4	0	-7	-5	-4	-2	-3	-11
N	-2	-3	6	3	-7	-1	0	-1	1	-3	-5	0	-5	-6	-3	1	0	-6	-3	-5	5	-1	-2	-11
D	-1	-6	3	6	-9	0	3	-1	-1	-5	-8	-2	-7	-10	-4	-1	-2	-10	-7	-5	5	2	-3	-11
C	-4	-5	-7	-9	9	-9	-9	-6	-5	-4	-10	-9	-9	-8	-5	-1	-5	-11	-2	-4	-8	-9	-6	-11
Q	-2	0	-1	0	-9	7	2	-4	2	-5	-3	-1	-2	-9	-1	-3	-3	-8	-8	-4	-1	5	-2	-11
E	-1	-5	0	3	-9	2	6	-2	-2	-4	-6	-2	-4	-9	-3	-2	-3	-11	-6	-4	2	5	-3	-11
G	0	-6	-1	-1	-6	-4	-2	6	-6	-6	-7	-5	-6	-7	-3	0	-3	-10	-9	-3	-1	-3	-3	-11
H	-4	0	1	-1	-5	2	-2	-6	8	-6	-4	-3	-6	-4	-2	-3	-4	-5	-1	-4	0	1	-3	-11
I	-2	-3	-3	-5	-4	-5	-4	-6	-6	7	1	-4	1	0	-5	-4	-1	-9	-4	3	-4	-4	-3	-11
L	-4	-6	-5	-8	-10	-3	-6	-7	-4	1	6	-5	2	-1	-5	-6	-4	-4	-4	0	-6	-4	-4	-11
K	-4	2	0	-2	-9	-1	-2	-5	-3	-4	-5	6	0	-9	-4	-2	-1	-7	-7	-6	-1	-2	-3	-11
M	-3	-2	-5	-7	-9	-2	-4	-6	-6	1	2	0	10	-2	-5	-3	-2	-8	-7	0	-6	-3	-3	-11
F	-6	-7	-6	-10	-8	-9	-9	-7	-4	0	-1	-9	-2	8	-7	-4	-6	-2	4	-5	-7	-9	-5	-11
P	0	-2	-3	-4	-5	-1	-3	-3	-2	-5	-5	-4	-5	-7	7	0	-2	-9	-9	-3	-4	-2	-3	-11
S	1	-1	1	-1	-1	-3	-2	0	-3	-4	-6	-2	-3	-4	0	5	2	-3	-5	-3	0	-2	-1	-11
T	1	-4	0	-2	-5	-3	-3	-3	-4	-1	-4	-1	-2	-6	-2	2	6	-8	-4	-1	-1	-3	-2	-11
W	-9	0	-6	-10	-11	-8	-11	-10	-5	-9	-4	-7	-8	-2	-9	-3	-8	13	-3	-10	-7	-10	-7	-11
Y	-5	-7	-3	-7	-2	-8	-6	-9	-1	-4	-4	-7	-7	4	-9	-5	-4	-3	9	-5	-4	-7	-5	-11
V	-1	-5	-5	-5	-4	-4	-4	-3	-4	3	0	-6	0	-5	-3	-3	-1	-10	-5	6	-5	-4	-2	-11
B	-1	-4	5	5	-8	-1	2	-1	0	-4	-6	-1	-6	-7	-4	0	-1	-7	-4	-5	5	1	-2	-11
Z	-1	-2	-1	2	-9	5	5	-3	1	-4	-4	-2	-3	-9	-2	-2	-3	-10	-7	-4	1	5	-3	-11
X	-2	-3	-2	-3	-6	-2	-3	-3	-3	-3	-4	-3	-3	-5	-3	-1	-2	-7	-5	-2	-2	-3	-3	-11
-	-11	-11	-11	-11	-11	-11	-11	-11	-11	-11	-11	-11	-11	-11	-11	-11	-11	-11	-11	-11	-11	-11	-11	1

	A	R	N	D	C	Q	E	G	H	I	L	K	M	F	P	S	T	W	Y	V	B	Z	X	-
A	5	-2	-1	-2	-1	-1	-1	0	-2	-1	-1	-1	-1	-2	-1	1	0	-2	-2	0	-1	-1	0	-5
R	-2	7	0	-1	-3	1	0	-2	0	-3	-2	3	-1	-2	-2	-1	-1	-2	-1	-2	-1	0	-1	-5
N	-1	0	6	2	-2	0	0	0	1	-2	-3	0	-2	-2	-2	1	0	-4	-2	-3	4	0	-1	-5
D	-2	-1	2	7	-3	0	2	-1	0	-4	-3	0	-3	-4	-1	0	-1	-4	-2	-3	5	1	-1	-5
C	-1	-3	-2	-3	12	-3	-3	-3	-3	-3	-2	-3	-2	-2	-4	-1	-1	-5	-3	-1	-2	-3	-2	-5
Q	-1	1	0	0	-3	6	2	-2	1	-2	-2	1	0	-4	-1	0	-1	-2	-1	-3	0	4	-1	-5
E	-1	0	0	2	-3	2	6	-2	0	-3	-2	1	-2	-3	0	0	-1	-3	-2	-3	1	4	-1	-5
G	0	-2	0	-1	-3	-2	-2	7	-2	-4	-3	-2	-2	-3	-2	0	-2	-2	-3	-3	-1	-2	-1	-5
H	-2	0	1	0	-3	1	0	-2	10	-3	-2	-1	0	-2	-2	-1	-2	-3	2	-3	0	0	-1	-5
I	-1	-3	-2	-4	-3	-2	-3	-4	-3	5	2	-3	2	0	-2	-2	-1	-2	0	3	-3	-3	-1	-5
L	-1	-2	-3	-3	-2	-2	-2	-3	-2	2	5	-3	2	1	-3	-3	-1	-2	0	1	-3	-2	-1	-5
K	-1	3	0	0	-3	1	1	-2	-1	-3	-3	5	-1	-3	-1	-1	-1	-2	-1	-2	0	1	-1	-5
M	-1	-1	-2	-3	-2	0	-2	-2	0	2	2	-1	6	0	-2	-2	-1	-2	0	1	-2	-1	-1	-5
F	-2	-2	-2	-4	-2	-4	-3	-3	-2	0	1	-3	0	8	-3	-2	-1	1	3	0	-3	-3	-1	-5
P	-1	-2	-2	-1	-4	-1	0	-2	-2	-2	-3	-1	-2	-3	9	-1	-1	-3	-3	-3	-2	-1	-1	-5
S	1	-1	1	0	-1	0	0	0	-1	-2	-3	-1	-2	-2	-1	4	2	-4	-2	-1	0	0	0	-5
T	0	-1	0	-1	-1	-1	-1	-2	-2	-1	-1	-1	-1	-1	-1	2	5	-3	-1	0	0	-1	0	-5
W	-2	-2	-4	-4	-5	-2	-3	-2	-3	-2	-2	-2	-2	1	-3	-4	-3	15	3	-3	-4	-2	-2	-5
Y	-2	-1	-2	-2	-3	-1	-2	-3	2	0	0	-1	0	3	-3	-2	-1	3	8	-1	-2	-2	-1	-5
V	0	-2	-3	-3	-1	-3	-3	-3	-3	3	1	-2	1	0	-3	-1	0	-3	-1	5	-3	-3	-1	-5
B	-1	-1	4	5	-2	0	1	-1	0	-3	-3	0	-2	-3	-2	0	0	-4	-2	-3	4	2	-1	-5
Z	-1	0	0	1	-3	4	4	-2	0	-3	-2	1	-1	-3	-1	0	-1	-2	-2	-3	2	4	-1	-5
X	0	-1	-1	-1	-2	-1	-1	-1	-1	-1	-1	-1	-1	-1	-1	0	0	-2	-1	-1	-1	-1	-1	-5
-	-5	-5	-5	-5	-5	-5	-5	-5	-5	-5	-5	-5	-5	-5	-5	-5	-5	-5	-5	-5	-5	-5	-5	1

3. BLOSUM 62 matrix. The data of the statistical samples are obtained from the following: blocks Database = /data/blocks-5.0/blocks.dat, $D_0 = 0.6979$, $D_1 = -0.5209$.
4. BLOSUM 80 matrix. The data of the statistical samples are obtained from the following: blocks Database = /data/blocks-5.0/blocks.dat, $D_0 = 0.9868$, $D_1 = -0.7442$.
5. PAM 30 matrix. The following is "pam" version 1.0.6 (28 July 1993). The main parameters are $D_0 = 2.57$, $D_1 = -5.06$, Max = 13, Min = −17.

11.1.4 Key Points of the Alignment Algorithms of Protein Sequences

Using the scoring matrix of the protein sequence alignment, the alignment of protein sequences can be carried out. The dynamic programming algorithm, SPA (which is based on statistical judgment), and the algorithms for multiple sequence alignment given above, are all based on the data for DNA (or RNA). Therefore, corresponding complementarity and modification should be made for protein sequences. The main points of modification of the related algorithms are illustrated as follows.

Dynamic Programming Algorithm for the Alignment of Protein Sequences

The basic steps of the dynamic programming algorithm remain the same. We compute with the PAM or BLOSUM matrix directly, following the discussion in Sect. 1.3.

Illustration of the SPA Algorithm

The key point of SPA is the computation of sliding window functions and the statistical estimations based on them. The sliding window functions are given by (4.10). They are

$$w(A,B;i,j,n) = w\left(a_{[i+1,i+n]}, b_{[j+1,j+n]}\right) = \frac{1}{n}\sum_{k=1}^{n} w(a_{i+k}, b_{j+k}) \,. \tag{11.5}$$

In the alignment of protein sequences, we only need to change the $w(a,b)$ function into the corresponding scoring function $v(a,b)$. In the statistical estimate formulas (4.14) and (4.16) used in SPA, the lack of bias and the consistency of these estimates can not be ensured, since complete randomness does not hold in the structure of protein sequences. This shows that SPA can still be used in the alignment of protein sequences, while more modifications and computations are needed to determine the position of the displacement mutation after the parameter estimation. Moreover, in SPA, further analysis is needed for the selection of parameters n, h, θ, θ', τ. The key problems in the alignment of protein sequences are: how to make use of the probability model and the limit properties of the stochastic sequence, and how to combine the origin of the PAM and BLOSUM scoring matrix series to achieve more systematic results.

Illustration of the Algorithm of the Alignment of Multiple Protein Sequences

The theory of sequence segments and module structures has nothing to do with the value of q in V_q, thus the definition of the uniform alignment and the algorithm for multiple sequence alignment can also be used in the alignment of multiple protein sequences.

Since the multiple sequence alignment given in this book is based on pairwise sequence alignment, and protein sequences are relatively short, the dynamic programming of their pairwise alignment is comparatively mature, and many websites and software packages can be used directly. Therefore, the alignment of multiple protein sequences can be implemented. First we get the pairwise sequence alignment by the dynamic programming algorithm, and then the sequence matrix and the module structure matrix of their pairwise sequence alignment are obtained. Finally, the alignment of multiple protein sequences can be implemented through the use of Steps 5.3.2 and 5.3.4 in the algorithm for multiple sequence alignment.

Further discussions are needed concerning whether the theory of the stochastic model of DNA sequences and performance analysis (such as the theory of complexity analysis and error analysis) can be used in the alignment and analysis of protein sequences. Some assumptions such as the complete randomness of the sequences, the limit properties of pairwise independent sequences, the characteristics of stochastic mutation flows, etc., need further demonstration and inspection. All of them require more discussion and study for the future development of the theory.

11.2 Alignment of Protein Three-Dimensional Structures

How to determine the similarity of the three-dimensional structures of different proteins, whether the method is reasonable and whether the expression of the similarity is valid are problems to be discussed and solved in the alignment of protein three-dimensional structures.

11.2.1 Basic Idea and Method of the Alignment of Three-Dimensional Structures

Alignment of Protein Three-Dimensional Structures

The traditional method of studying the homology of protein three-dimensional structures uses space movement and rotation [21, 51] to achieve the optimal fitting of every atom's point. This method is efficient for comparatively short local polypeptide chains (such as domain structures), but not so efficient for longer protein sequences. As movement and rotation are simple rigid transformations, they are not suitable for the complex comparison of protein three-dimensional structures. In this book, we try to use the sequence alignment algorithm to discuss the similarity of protein three-dimensional structures. Our main idea is outlined as follows:

1. The key to discussing protein three-dimensional structures is the three-dimensional structure of their backbones. Protein backbones can be considered to be spliced by a series of triangles, which are called the triangle splicing belts of protein backbones.
2. In the triangle splicing belt of a protein backbone, the length of each edge is relatively constant. The key to determining the configuration of a protein backbone is the dihedral angle between two neighboring triangles, and the value of the dihedral angle ranges within the interval $[-\pi, \pi)$ (or $[0, 2\pi)$). Thus the key to determining the configuration of a protein backbone is the sequence of the dihedral angles between two neighboring triangles. This sequence is called the dihedral angle sequence of a protein three-dimensional structure.
3. Homology alignment of protein three-dimensional structures can be carried out using the dihedral angle sequences of the protein three-dimensional structures, from which the determination of the homology of different protein three-dimensional structures can be obtained. Using these calculations, the homology of protein three-dimensional structures can be classified and searched.

We have illustrated above that the key to the determination of protein three-dimensional structures is the three-dimensional structure of their backbones, while the three-dimensional structure of the backbones can be spliced by a series of triangles. In the triangle splicing belt, the key to determining the configuration of the protein backbone is the dihedral angle sequence between two neighboring triangles, which are denoted as $\Psi = \{\psi_1, \psi_2, \cdots, \psi_{n-2}\}$, where ψ_i is the angle between triangles Δ_i and Δ_{i+1}, whose values range in the interval $V = [-\pi, \pi)$ (or $V = [0, 2\pi)$). We call Ψ the torsion angle sequence of this protein three-dimensional structure. In this chapter, the alignment of protein three-dimensional structures represented by their torsion angle sequences. We denote it as $V_+ = V \cup \{-\}$ and call it the expanding set of V.

Penalty Function of the Torsion Angle Sequences

Let $\Psi = (\psi_1, \psi_2, \cdots, \psi_{n_1})$, and $\Phi = (\phi_1, \phi_2, \cdots, \phi_{n_2})$ be the torsion angle sequences of two proteins, where ψ_i and ϕ_i are the radian angles whose values range in $(-\pi, \pi)$. We now give the definition of the penalty functions of the radian angle and the insertion symbol in $\psi', \phi' \in V_+$. We divide them into two types, discrete and consecutive. The definitions of their penalty functions are given as follows:

1. Penalty function of the discrete type. If the set V is divided into four parts $[0, \pi/2)$, $[\pi/2, \pi)$, $[-\pi, -\pi/2)$, $[-\pi/2, 0)$, these would be the four phases of Cartesian coordinates. If the value of angle ψ is determined by its phase, its range is given by $V_4 = \{0, 1, 2, 3\}$. The penalty function on

set $V_5 = \{0, 1, 2, 3, -\}$ will then be

$$d(\psi', \phi') = \begin{cases} |\psi' - \phi'| (\text{mod } 2)\,, & \text{if } \psi', \phi' \in V_4\,, \\ 2\,, & \text{if one of } \psi, \phi \text{ contains virtual symbol ``$-$''}, \\ & \text{but the other does not contain virtual} \\ & \text{symbol ``$-$''}, \\ 0\,, & \text{if both of the angles } \psi, \phi \text{ contain virtual} \\ & \text{symbol ``$-$''}. \end{cases} \tag{11.6}$$

2. Penalty function of the consecutive type. If the values of angles ψ', ϕ' range within V_+, the penalty function will be

$$d(\psi', \phi') = \begin{cases} |\psi' - \phi'|\,, & \text{if neither of the angles } \psi, \phi \text{ contains virtual} \\ & \text{symbol ``$-$''}, \\ \pi\,, & \text{if one of the angles } \psi, \phi \text{ contains virtual} \\ & \text{symbol ``$-$''}, \\ & \text{and the other does not contain virtual} \\ & \text{symbol ``$-$''}, \\ 0\,, & \text{if both of the angles } \psi, \phi \text{ contain virtual} \\ & \text{symbol ``$-$''}. \end{cases} \tag{11.7}$$

3. Denote by $\Psi'_s = \{\psi'_{s,1}, \psi'_{s,2}, \cdots, \psi'_{s,n'}\}$, $s = 1, 2$, two sequences whose lengths are the same and whose values range within either V_5 or V_+. The total value of the penalty function of Ψ'_1, Ψ'_2 is then defined as

$$d\left(\Psi'_1, \Psi'_2\right) = \sum_{i=1}^{n'} d\left(\psi'_{1,i}, \psi'_{2,i}\right)\,, \tag{11.8}$$

where $d(\psi'_{1,i}, \psi'_{2,i})$ is defined in (11.6) or (11.7) according to the discrete or consecutive type.
The definitions in (11.6) and (11.7) can be easily rewritten for scoring functions. For example,

$$w(\psi', \phi') = \begin{cases} 2 - d(\psi', \phi')\,, & \text{if } \psi', \phi' \in V_5\,, \\ \pi - d(\psi', \phi')\,, & \text{if } \psi', \phi' \in V_+\,. \end{cases} \tag{11.9}$$

Minimum Penalty Alignment of the Torsion Angle Sequences

We denote by $\Psi_s = \{\psi_{s,1}, \psi_{s,2}, \cdots, \psi_{s,n_s-2}\}$, $s = 1, 2$ the torsion angle sequences of two protein three-dimensional structures, where n_s, $s = 1, 2$ are the lengths of these two proteins. Here, Ψ'_1 and Ψ'_2 are two sequences whose values range within V_5 or V_+. We call (Ψ'_1, Ψ'_2) the alignment sequences of

(Ψ_1, Ψ_2) if sequences Ψ_1' and Ψ_2' are the same in length and they change into sequences Ψ_1, Ψ_2 after deleting symbols "–".

We call (Ψ_1^*, Ψ_2^*) the minimum penalty alignment sequences (or minimum penalty alignment) of (Ψ_1, Ψ_2), if (Ψ_1^*, Ψ_2^*) are alignment sequences of (Ψ_1, Ψ_2) and $d(\Psi_1^*, \Psi_2^*) \leq d(\Psi_1', \Psi_2')$ holds for any (Ψ_1', Ψ_2') which are the alignment sequences of (Ψ_1, Ψ_2).

For the penalty functions of fixed torsion angle sequences (Ψ_1, Ψ_2) and (11.8), their minimum penalty alignment sequences (Ψ_1^*, Ψ_2^*) can be obtained by the Smith–Waterman dynamic programming algorithm.

Measurement Index for the Homology of Protein Three-Dimensional Structures

From the minimum penalty alignment sequences (Ψ_1^*, Ψ_2^*) of torsion angle sequences (Ψ_1, Ψ_2), the measurement index for their similarity can be obtained. It is defined as follows:

1. Denote the total homology measurement index by the total value of the minimum penalty alignment $d(\Psi_1^*, \Psi_2^*)$ for the torsion angle sequences (Ψ_1, Ψ_2).
2. The total homology measurement index $d(\Psi_1^*, \Psi_2^*)$ can be decomposed as follows. Denote by k^* the total number of inserted symbols in sequences (Ψ_1^*, Ψ_2^*), and define

$$d_0(\Psi_1^*, \Psi_2^*) = d(\Psi_1^*, \Psi_2^*) - k^* \pi \tag{11.10}$$

   to be the measurement index of the local homology of the three-dimensional structures of proteins 1 and 2.
3. It is known from the alignment algorithm that the minimum penalty alignment (Ψ_1^*, Ψ_2^*) of torsion angle sequences (Ψ_1, Ψ_2) may be nonexclusive. We then take the maximum value k^* of the total number of inserted symbols in the minimum penalty alignment sequences, and $d_0(\Psi_1^*, \Psi_2^*)$ in formula (11.8) would be the maximum value of the local similarity measurement indices.

From this, we find that $[k^*, d_0(\Psi_1^*, \Psi_2^*)]$ are the two measurement indices for the homology of protein three-dimensional structures. They are the decomposition of the total homology measurement index $d(\Psi_1^*, \Psi_2^*)$.

11.2.2 Example of Computation in the Discrete Case

Analysis of Protein Three-Dimensional Structures with Multiple Models

In the PDB database, because of different measurements performed for its structure determination, a protein may have multiple models. For example,

Table 11.1. Spatial phase characteristic alignment for protein PDB1BF9

Model	Stable region of the torsion angles of amino acids	Phase	Atom site of origin and terminus of the stable region
1	222 020120111213333301133333300	11013223000	33–404
2	131 000320111233333201133220300	13013223020	607–979
3	122 020120111213333201133333300	11011203011	1181–1552
4	010 000120111213333201133333300	13011203011	1755–2126
5	010 020120111213333301133333300	11003203000	2329–2700
6	131 020120111213333201133333300	30101223012	2903–3274
7	113 020130111213333301133333300	31003223010	3477–3848
8	133 020130111213333201133333300	31011223013	4051–4422
9	001 020120111213333201133333300	13011223011	4625–4996
10	220 020120111233332301133332300	13013223020	5199–5570
11	111 000120111213333201133333300	13013223022	5773–6144
12	303 020120111213333301133333300	31011223030	6347–6718
13	303 020120111213333311123220300	13013223031	6921–7292
14	022 000120111213333301133333300	31011223032	7495–7866
15	201 020120111213333201133333300	11101223001	8069–8440
16	021 020130111213333201133333300	11011203021	8643–9014
17	002 020120111213333201133333300	30101223011	9217–9588
18	011 000130111233333201133333300	31011223001	9791–10162
19	000 020120111213333301133333300	11101203010	10365–10736
20	220 000120111213333201133333300	11011203021	10939–11310
21	102 020120111213333301133333300	31013223020	11513–11884
22	103 020130111233332300133333300	31001223002	12087–12458
23	202 020130 11121 333330113333300	30101223	12661–13032
	β-sheet (11121); Down α-helix (333330113333300); Bonding region		

protein PDB1BF9 has 23 models, and each model consists of 41 amino acids. All of their primary structures are the same, while their spatial coordinates are different (see file PDB1BF9, PDB database). For the backbones of these 23 models, we compute the value of their phase characteristics, shown in Table 11.1.

In Table 11.1, it is obvious that the phase characteristic sequences have bonding regions and β sheet regions. The bonding regions and α-helix regions of these 23 models can be visualized using Pymol, Rasmol, and other software on the data file PDB1BF9. We can see that the three-dimensional structures of their backbones are similar, especially in the stable regions.

In Table 11.2, the values of torsion angles range in the interval $V = [0, 2\pi)$. We can see that at sites 3–30, stable regions are formed.

Table 11.2. Torsion angle data for the 23 models of protein PDB1BF9

1	3.9	4.6	4.9	5.4	4.4	5.8	1.7	3.1	4.9	2.5	5.6	1.1	3.4	1.8	2.4	3.1	2.3	3.2	1.7	5.6
2	6.2	2.7	1.1	5.6	5.0	5.7	2.2	3.1	4.7	2.9	5.4	1.0	3.5	2.0	2.3	2.7	2.5	3.6	1.9	5.5
3	1.7	3.7	3.3	5.5	3.9	5.8	1.8	3.0	4.8	2.9	5.4	0.9	3.6	2.4	2.1	2.5	2.7	3.3	1.9	5.6
4	4.1	1.3	5.5	5.8	4.9	5.6	2.1	3.0	5.0	3.1	5.3	0.8	3.8	2.4	2.3	2.6	2.6	3.3	1.9	5.6
5	1.5	5.0	0.4	5.6	4.6	5.9	1.8	3.0	4.7	2.6	5.6	1.1	3.4	1.9	2.3	2.8	2.5	3.3	1.9	5.6
6	0.7	2.3	0.9	5.5	4.6	5.9	1.9	2.9	4.6	2.7	5.6	1.1	3.5	2.0	2.3	2.7	2.6	3.4	1.8	5.6
7	1.2	1.1	2.4	5.4	4.2	6.2	1.3	2.9	4.9	3.0	5.3	1.0	3.6	2.1	2.2	2.9	2.6	3.3	1.9	5.6
8	1.2	2.0	2.4	5.4	4.1	6.0	1.4	2.9	4.8	2.7	5.5	1.0	3.5	2.1	2.3	2.7	2.7	3.3	2.0	5.5
9	4.9	4.7	1.4	5.6	4.6	5.8	1.8	3.1	4.8	2.9	5.4	1.0	3.5	2.2	2.2	2.7	2.7	3.4	1.9	5.5
10	4.4	3.6	5.0	5.4	4.1	5.9	1.6	3.1	4.6	2.3	5.9	1.0	3.5	1.9	2.2	3.0	2.5	3.1	2.1	5.4
11	5.2	2.5	1.2	5.8	4.7	5.8	2.0	2.8	4.9	2.9	5.4	0.7	3.8	2.3	2.3	2.6	2.6	3.3	2.0	5.6
12	2.2	5.3	2.6	4.7	4.5	5.6	2.0	3.0	5.0	3.1	5.3	0.8	3.7	2.3	2.2	2.7	2.5	3.3	1.8	5.5
13	2.8	4.9	2.7	4.9	4.8	5.4	2.1	3.1	4.9	2.8	5.3	0.8	3.7	2.2	2.4	2.5	2.6	3.5	1.0	0.3
14	0.3	3.4	3.8	5.3	4.7	5.7	2.0	2.8	5.0	3.0	5.3	0.8	3.8	2.4	2.2	2.6	2.7	3.3	1.9	5.6
15	4.0	5.1	1.6	5.7	4.5	5.6	1.8	3.2	4.7	2.7	5.6	1.0	3.5	2.1	2.2	2.8	2.6	3.3	2.0	5.6
16	5.9	3.3	2.1	5.1	4.1	5.9	1.7	2.8	5.1	3.1	5.4	0.8	3.7	2.2	2.3	2.9	2.6	3.4	1.9	5.6
17	5.0	0.5	3.1	5.3	3.9	6.0	1.5	2.9	4.6	2.7	5.6	1.0	3.6	2.3	2.1	2.7	2.5	3.6	1.9	5.5
18	5.0	1.8	0.6	5.5	4.7	6.0	1.8	2.9	5.2	2.7	5.2	1.1	3.2	2.1	2.4	2.8	2.3	3.5	1.9	5.6
19	5.9	4.8	1.0	5.5	4.6	5.6	2.1	3.0	4.9	3.0	5.4	0.8	3.6	2.2	2.4	2.6	2.5	3.2	2.0	5.5
20	4.5	4.4	5.3	5.7	5.0	5.4	2.0	3.2	4.9	2.9	5.3	0.8	3.7	2.4	2.1	2.7	2.5	3.6	1.8	5.5
21	0.4	4.5	4.3	5.3	3.3	5.8	1.6	3.1	4.9	2.9	5.4	1.0	3.6	2.4	2.0	2.7	2.4	3.4	1.8	5.7
22	0.2	5.1	2.6	5.5	3.7	0.1	1.1	2.9	4.6	2.8	5.5	1.0	3.5	2.0	2.2	2.8	2.7	3.4	2.2	5.4
23	4.4	5.3	4.3	5.3	4.3	5.9	1.7	2.9	4.8	2.9	5.4	1.0	3.6	2.2	2.1	2.9	2.5	3.2	1.9	5.5
				|4							Stable region									|20

1	1.0	1.0	2.6	2.6	2.5	2.9	1.8	1.6	5.2	5.9	2.7	5.1	1.3	5.9	1.9	3.8	5.1	3.0
2	1.2	1.2	2.8	2.6	2.8	3.2	1.1	1.5	5.1	5.8	3.2	4.9	0.9	6.1	2.1	3.9	4.9	2.8
3	0.9	1.4	2.7	2.5	2.6	3.0	1.7	1.4	5.2	5.8	3.0	4.8	1.3	6.1	1.8	3.8	5.2	3.0
4	0.9	1.4	2.7	2.6	2.7	3.0	1.7	1.4	5.2	5.8	3.0	4.9	1.1	6.1	1.7	3.9	5.2	2.8
5	1.0	1.2	2.7	2.5	2.5	2.9	1.8	1.5	5.1	5.8	2.7	5.1	1.4	5.8	1.9	3.8	5.2	3.0
6	0.9	1.4	2.7	2.5	2.5	2.9	1.6	1.5	5.2	5.8	3.2	4.3	1.7	5.8	1.8	4.0	5.1	2.8
7	0.9	1.4	2.6	2.6	2.4	3.0	1.8	1.5	5.2	5.7	2.9	4.7	1.6	5.9	1.8	4.0	4.9	2.9
8	1.0	1.5	2.8	2.4	2.5	3.0	1.6	1.5	5.1	5.7	3.0	4.7	1.6	5.9	1.8	4.0	4.9	2.9
9	1.2	1.3	2.6	2.5	2.6	3.1	1.7	1.5	5.2	5.9	3.0	4.8	1.2	5.9	1.8	3.9	5.1	2.7
10	0.5	1.9	2.6	2.7	3.0	2.8	1.0	1.3	5.1	5.7	3.2	4.9	0.9	6.1	2.0	3.9	4.7	2.8
11	1.0	1.5	2.8	2.6	2.7	3.3	1.6	1.4	5.1	5.7	3.0	5.1	0.9	6.1	2.1	3.8	4.7	2.6
12	1.2	1.0	2.7	2.6	2.6	3.1	1.6	1.4	5.2	5.7	3.0	4.6	1.4	6.2	1.6	4.0	5.1	2.9
13	5.5	2.5	2.6	2.7	2.9	3.2	1.1	1.4	5.1	5.7	3.2	5.0	0.7	6.2	2.0	3.8	4.7	2.6
14	0.9	1.5	2.7	2.6	2.6	3.1	1.7	1.4	5.2	5.7	3.0	4.7	1.4	6.1	1.6	3.9	5.1	3.0
15	1.0	1.4	2.8	2.5	2.4	3.0	1.8	1.6	5.1	5.9	2.7	5.1	1.3	5.8	1.9	3.9	5.1	3.0
16	1.0	1.3	2.8	2.6	2.6	3.0	1.8	1.4	5.2	5.8	2.9	4.7	1.3	6.1	1.6	3.9	5.2	2.9
17	1.2	1.2	2.9	2.4	2.4	3.0	1.7	1.4	5.2	5.7	3.1	4.5	1.8	5.7	1.9	3.9	5.0	2.9
18	1.1	1.0	2.7	2.6	2.6	3.1	1.5	1.5	5.2	5.7	3.0	4.7	1.6	6.0	1.8	3.9	5.1	2.9
19	1.1	1.1	2.8	2.5	2.6	3.1	1.7	1.5	5.2	5.7	3.2	4.7	1.4	5.8	1.8	3.9	5.1	2.8
20	1.1	1.1	2.7	2.6	2.6	3.1	1.7	1.4	5.3	5.7	3.0	4.9	1.1	6.1	1.7	3.9	5.1	2.8
21	0.9	1.2	2.7	2.4	2.4	3.0	1.8	1.5	5.2	5.8	3.0	4.7	1.4	6.0	1.8	4.0	4.9	2.9
22	0.4	2.2	2.8	2.4	2.4	3.0	1.5	1.5	5.2	5.7	3.0	4.5	1.6	6.0	1.7	4.1	5.0	2.9
23	1.1	1.0	2.7	2.4	2.4	3.1	1.7	1.4	5.2	5.7	3.1	4.4	1.7	5.9	1.7	4.1	4.8	2.8
	|21				Stable region						|30							

Remark 13. Table 11.3 is a statistical table for the homology of the alignment of total spatial torsion angle phase sequences (all the torsion angles of each model) using the first 15 models of protein PDB1BF9. It was computed with the scoring function in formula (11.9).

Remark 14. Table 11.4 is the statistical table for the homology on the alignment of stable regions of the spatial torsion angle phases using the first 15 models of protein PDB1BF9. It was computed with the scoring function in formula (11.9).

Table 11.3. Statistical table for the homology of the alignment of total torsion angle phase sequences of protein PDB1BF9

Model	1	2	3	4	5	6	7	8	9	10	11	12	13	14
2	0.71													
3	0.85	0.68												
4	0.76	0.71	0.88											
5	0.88	0.66	0.80	0.83										
6	0.73	0.68	0.80	0.76	0.73									
7	0.83	0.66	0.78	0.73	0.85	0.80								
8	0.78	0.68	0.85	0.78	0.73	0.85	0.88							
9	0.80	0.73	0.88	0.90	0.78	0.83	0.76	0.83						
10	0.85	0.76	0.73	0.73	0.78	0.66	0.73	0.68	0.76					
11	0.80	0.83	0.80	0.85	0.78	0.80	0.78	0.78	0.85	0.78				
12	0.85	0.66	0.80	0.76	0.80	0.78	0.85	0.85	0.83	0.76	0.76			
13	0.73	0.73	0.68	0.68	0.68	0.61	0.68	0.66	0.76	0.71	0.71	0.78		
14	0.85	0.66	0.83	0.80	0.78	0.78	0.78	0.80	0.80	0.73	0.80	0.88	0.68	
15	0.83	0.66	0.83	0.78	0.80	0.85	0.76	0.78	0.88	0.71	0.78	0.80	0.68	0.76

Table 11.4. Statistical table for the homology of the alignment of stable regions of the torsion angle phases of protein PDB1BF9

Model	1	2	3	4	5	6	7	8	9	10	11	12	13	14
2	0.74													
3	0.96	0.78												
4	0.93	0.81	0.96											
5	1.00	0.74	0.96	0.93										
6	0.96	0.78	1.00	0.96	0.96									
7	0.96	0.70	0.93	0.89	0.96	0.93								
8	0.93	0.74	0.96	0.93	0.93	0.96	0.96							
9	0.96	0.78	1.00	0.96	0.96	1.00	0.93	0.96						
10	0.89	0.74	0.85	0.81	0.89	0.85	0.85	0.81	0.85					
11	0.93	0.81	0.96	1.00	0.93	0.96	0.89	0.93	0.96	0.81				
12	1.00	0.74	0.96	0.93	1.00	0.96	0.96	0.93	0.96	0.89	0.93			
13	0.81	0.78	0.78	0.74	0.81	0.78	0.78	0.74	0.78	0.74	0.74	0.81		
14	0.96	0.78	0.93	0.96	0.96	0.93	0.93	0.89	0.93	0.85	0.96	0.96	1.00	
15	0.96	0.78	1.00	0.96	0.96	1.00	0.93	0.96	1.00	0.85	0.96	0.96	0.78	0.93

We can see from Table 11.3 that the homologies between Model 1 and Model 2 or Model 5 are 0.71 and 0.88, respectively. It tells us that the homology between Model 1 and Model 5 is higher than that between Model 1 and Model 2.

Analysis of the Three-Dimensional Structures of Proteins Within the Same Family

The above three-dimensional structure analysis can be carried out within the same family. For example, in the analysis of the serpin ensemble, there is only one super-family in this fold, with only one family in this super-family, whose ID numbers in PDB are given as 7apiA, 8apiA, 1hleA, 1ovaA, 2achA, 9apiA, 1psi, 1atu, 1ktc, 1athA, 1antI, 2antI. These are from antitrypsin, elastase, inhibitor, ovalbumin, antichymotrypsin, and antitrypsin, antithrombin, respectively, in humans, horses, and cattle.

The result of their three-dimensional structure alignment is shown in Table 11.5. Except for protein pdb1ktc, the characteristic sequences of the three-dimensional structures of other protein backbones show comparatively high homologies.

Remark 15. Table 11.6 shows the multiple alignment result of the three-dimensional structure characteristic sequences of each serpin ensemble protein. Because of the complexity of the structure of protein PDB1KTCA, it is not listed in this table. Also, the value 4 stands for the virtual symbol "–".

Remark 16. In Table 11.7, A, B stand for the alignment result of the torsion angle phase sequences of the serpin ensemble protein PDB1ATHA and protein PDB8APIA, where 4 and – stand for the virtual symbol "–". The inserted symbols in the primary structures are determined by the corresponding positions of the insert symbols of the three-dimensional structure alignment.

Remark 17. We can see from Tables 11.7 and 11.8 that for the proteins PDB1ATHA and PDB8APIA, in the regions where the three-dimensional structures are homologous (such as sites 5–28, sites 32–67, etc.), the corresponding primary structures are quite different. It shows that, in proteins or peptide chains, primary structures which are quite different may generate similar three-dimensional structures.

11.2.3 Example of Computation in Consecutive Case

If the torsion angle ψ is considered a variable whose value ranges in the interval $(-\pi, \pi)$, Ψ_1 and Ψ_2 are two vectors who take values in $(-\pi, \pi)$. Their alignment can also be implemented by the dynamic programming algorithm. The statistical table of the average absolute deviation alignment of the torsion angle of the protein PDB1BF9 backbone is shown in Tables 11.9 and 11.10.

Remark 18. In Tables 11.9 and 11.10, each value is in radians. They are both statistical tables of the average absolute deviation alignment of the torsion angles for the first 15 models of protein PDB1BF9, where Table 11.9 shows the average absolute deviation of the torsion angles of all models, while Table 11.10 shows the average absolute deviation of the stable region of the torsion angle phase (from the fourth amino acid to the 31st amino acid) of each model.

Table 11.5. Multiple alignment result (homology) of the three-dimensional structure characteristics of the serpin ensemble protein family

	1ATHA	1PSI-	1ATU-	1ANTI	1ATTA	2ANTI	1OVAA	2ACHA	7APIA	8APIA	9APIA	1HLEA
1ATHA	1.00	0.64	0.55	0.61	0.59	0.58	0.59	0.58	0.60	0.60	0.62	0.58
1PSI-	0.64	1.00	0.69	0.57	0.57	0.58	0.61	0.66	0.67	0.66	0.67	0.59
1ATU-	0.55	0.69	1.00	0.54	0.48	0.51	0.56	0.59	0.60	0.58	0.62	0.56
1ANTI	0.61	0.57	0.54	1.00	0.62	0.65	0.56	0.57	0.55	0.56	0.59	0.58
1ATTA	0.59	0.57	0.48	0.62	1.00	0.63	0.56	0.54	0.53	0.55	0.54	0.56
2ANTI	0.58	0.58	0.51	0.65	0.63	1.00	0.55	0.57	0.55	0.57	0.56	0.58
1OVAA	0.59	0.61	0.56	0.56	0.56	0.55	1.00	0.59	0.60	0.60	0.62	0.65
2ACHA	0.58	0.66	0.59	0.57	0.54	0.57	0.59	1.00	0.84	0.82	0.84	0.72
7APIA	0.60	0.67	0.60	0.55	0.53	0.55	0.60	0.84	1.00	0.91	0.90	0.73
8APIA	0.60	0.66	0.58	0.56	0.55	0.57	0.60	0.82	0.91	1.00	0.89	0.74
9APIA	0.62	0.67	0.62	0.59	0.54	0.56	0.62	0.84	0.90	0.89	1.00	0.75
1HLEA	0.58	0.59	0.56	0.58	0.56	0.58	0.65	0.72	0.73	0.74	0.75	1.00

Table 11.6. Result of the alignment of the three-dimensional structure characteristics of the serpin ensemble protein family

1ATHA	4444444444	4444444444	4444444444	4001330300	0000000000	0000000000	0130213232	2000000000	0020213000	0000000444	4444444320	2211213000
1ANTI	4444444444	4444440213	2102133331	2122324400	0000000000	0000000000	0130213332	2000000000	0020013000	0000000444	4444412022	4112143000
1ATTA	3100211123	3203301331	1322121233	2301130200	0000000000	0000000002	4130013232	2000000000	0020013000	0000000444	4444444320	0210103000
2ANTI	4103003130	0302213330	2133330000	4443100200	0000000000	0000000004	4130213332	2000000000	0020213100	0000000444	4444444320	0213003020
1PSI-	4444444444	4444444444	4444444444	4444444400	0000000000	0000000004	4430013332	2000000000	0020013000	0000000444	4444412320	0014443000
1ATU-	4444444444	4444444444	4444444444	4444444400	0000000000	0000000000	0430233233	2000000000	0000011002	0000002444	4444410330	0014443000
7APIA	4444444444	4444444444	4444444444	4444444300	0000000000	0000000000	0430013232	2000000000	0020013000	0000000444	4444412330	0014443000
8APIA	4444444444	4444444444	4444444444	4444130200	0000000000	0000000004	4430013232	2000000000	0000213000	0000000444	4444412320	0014443000
9APIA	4444444444	4444444444	4444444444	4444444300	0000000000	0000000000	0430013232	2000000000	0020013000	0000000444	4444412320	0014443000
2ACHA	4444444444	4444444444	4444444444	4444444400	0000000000	0000000044	4430013210	2000000000	0020213000	0000000444	4444412320	0014443000
1HLEA	4444444444	4444444444	4444444444	4444444440	0000000000	0000000000	4431213232	2000000000	0000213000	0000000444	4444444320	0011443000
1OVAA	4444444444	4444444444	4444444444	4444444300	0000000000	0000000444	4431213332	2000000000	0000213000	0000000322	0211221000	0020003021
1ATHA	0000000000	0020034233	2334233204	4123334430	0000000003	2133333200	0200000000	0000000110	0222033000	0202313333	2233232330	3333002123
1ANTI	0000000000	0000120432	1233223233	3302333430	0000000003	0023333200	0200000000	0000000110	0202013000	1202313333	3233232330	3333002133
1ATTA	0000000000	0033334410	1123223323	3330233330	0000000003	2333333200	0200000000	0000000110	2312013020	0202313332	3233232330	1333002133
2ANTI	0000000000	0012324010	3333223233	3302133430	0000000003	2133333200	0200000000	0000000110	0303013120	1202303333	2233232330	1333002133
1PSI-	0000000000	0003314133	2333233333	3121334430	0000000003	2133233200	4300000000	0000000310	0313031133	0320123334	2234233330	1333002333
1ATU-	0000000000	0043310003	2332232233	3101324430	0000002003	3333221200	4300000000	0000000310	0203130133	0231143332	4232233320	3333123333
7APIA	0000000000	0021331014	2342323233	3302333430	0000000003	3333332202	4300000000	0000000110	0302030333	0324133323	4232232330	1333002023
8APIA	0000000000	0021331014	2342323233	2302133430	0000000003	3333332202	4300000000	0000000112	0202030333	0334133333	4233232330	1333002123
9APIA	0000000000	0021231014	2342323233	3302333430	0000000003	3333332202	4300000000	0000000310	0202030333	0324133333	4233232330	1333002123
2ACHA	0000000000	0201331014	2342323233	3302333430	0000000003	2333332200	4300000000	0000000110	0202030133	0234133323	4233232330	1333002121
1HLEA	0000000004	4421121114	2342323233	3302133430	0000000003	2333332200	0200000000	0000000110	0303013120	1302303332	3332232330	3333002133
1OVAA	2000000000	0211424103	2332334233	3302133430	0000000001	1203332200	0200000000	0000000110	0312013120	1202313332	3223232330	1333002123

Table 11.6. (continued)

```
1ATHA 3333344401 2332333333 2441232322 3430123332 3333110013 3333333001 4444443000 0002330000 0000042133 4232323423 3232333234 0000000124
1ANTI 3333310301 2333331010 3341223311 2333223330 3004414233 3334344411 2144443000 0002330000 0000024433 3212323233 2323232344 0000000102
1ATTA 3333304101 2333333323 2321233311 2333343331 1100133322 3330044414 4444443000 0002320222 0000220133 3232333123 3323224444 0000000122
2ANTI 2333304101 3333333111 0333223430 1421333233 3144430012 3332333044 2144443000 0002330000 0000024133 4232323443 3323233323 0000000122
1PSI- 2332014101 2333344332 2441233232 3420030133 3333310013 3332334411 2444442000 0002120000 0000230133 3232333441 0332232244 0000000122
1ATU- 3332110401 4333333303 4441033033 3420011033 2333310233 3332334411 2444443000 0002330000 0002012433 3410333331 0322232302 0000044102
7APIA 3332414141 2333323332 3441232322 3420031033 3333310112 3332334411 2444442000 0002120000 0000444423 2333323233 3123333323 0000000122
8APIA 3332414141 2333333332 3441232332 3420030133 2333310102 3332334411 2444442000 0002120000 0000444423 2133323233 3123233323 0000000122
9APIA 3332114101 4333333342 2441032332 3420030133 3333310013 3333334411 2444442000 0002320000 0000444423 2333323233 3323233323 0000000122
2ACHA 2332414141 2333323332 3441233233 2320030033 3233310012 3333334411 2444442000 0002320000 0000044423 3323213323 3323232323 0000000122
1HLEA 3332114101 2333333442 2441233122 3420030133 2333313201 3332333112 0110302000 0042330000 0002100044 1333323233 3123232322 0000000122
1OVAA 3332134001 2333333224 4441232323 3200311333 3333110013 3333330414 0244442000 0002120000 0002300244 1333323233 3323232333 0000000122

1ATHA 2000230023 2200133103 0332241332 3322233201 1104333233 2334123333 0244133222 0133332323 1032321130 32032
1ANTI 0002340023 2200321112 1212241333 3332233231 1233333123 3330333222 2003233232 3003241211 3243203403 44444
1ATTA 0002340023 2200030020 2233241332 3322231201 1333322223 3311001012 3224400333 3332300323 2113232030 04444
2ANTI 0002340023 2200321111 0312341333 2333233201 1121102332 3312333302 1322320133 3223230032 3213103203 03444
1PSI- 2000140423 2200031014 4332204332 2333331213 1312333333 3333103333 3220033323 3330032120 3103203024 44444
1ATU- 0000140012 2220443101 4333341333 3323331020 2310232303 0232301033 3320132323 3320032120 3103003024 44444
7APIA 2000140421 2200034440 0333320332 2333233201 1333222333 3334444444 4444444444 4444444444 4444444444 44444
8APIA 2000140423 2200034440 0333320332 2333233201 1332332323 3324444444 4444444444 4444444444 4444444444 44444
9APIA 0000140423 2200034440 0133321333 3333223201 1332233323 3334444444 4444444444 4444444444 4444444444 44444
2ACHA 2000140433 2200010410 0132441333 2323233201 1333223333 3334444444 4444444444 4444444444 4444444444 44444
1HLEA 0002340023 2200030410 0134421333 3322231201 1333222323 3334444444 4444444444 4444444444 4444444444 44444
1OVAA 0000140423 2200030340 2124421332 3322231201 0113330000 0000023012 3220032323 3330032320 3323204444 44444
```

Table 11.7. The alignment of spatial phase structures of protein 1ATHA and protein 8APIA

```
 |5         Similar region      28|      |32    Similar region        67|
A 0013303000 0000000000 0000000000 1302132322 0000000000 0202130000 0000004432 0221121300 0000000000 0002003423
B 4414302000 0000000000 0000000044 4300132322 0000000000 0002130000 0000001232 0001444300 0000000000 0002133101

A 3233423320 4412333430 0000000003 2133333200 0200000000 0000000110 0222033000 0202313333 2233232330 3333002123
B 4234232323 3230213330 0000000003 3333332202 4300000000 0000000112 0202030333 0334133333 4233232330 1333002123

A 3333301233 2333333212 3232233012 3332333311 0013333333 3001430000 0023300000 0000213324 3232323323 2343342300
B 3332111233 3333332312 3233232003 0133233331 0102333233 4411220000 0021200000 0004243421 3332323331 2323332300

A 0000012200 0230023220 0133103033 4221332433 2223320111 0433323323 3123333021 3322201333 3232310323 2113032032
B 0000012220 0010423220 0034404033 3320332233 3423320113 3233232343 3424444444 4444444444 4444444444 4444444444
```

Table 11.8. Primary structure alignment sequences for proteins 1ATHA and 8APIA

```
      |5                      28|   |32                                   67|
A ICTCIYRRRV WELSKANSRF ATTFYQHLAD SKNDNDNIFL SPLSISTAFA MTKLGACNDT LQQLME--VF KFDTISEKTS DQIHFFFAKL NCRLYRS-SK
B --D-HPTFNK ITPNLAEFAF SLYRQLAH-- -QSNSTNIFF SPVSIATAFA MLSLGTKADT HDEILEGLNF NLTE---IPE AQIHEGFQEL LRTLNQPDSQ

A LVSA-NRLFG --DKSLT-FN ETYQDISELV YGAKLQPLDF KENAEQSRAA INKWVSNKTE GRITDVIPSE AINELTVLVL VNTIYFKGLW KSKFSPENTR
B -LQ-LTTGNG LFLSEGLKLV DKFLEDVKKL YHSEAFTVNF -GDTEEAKKQ INDYVEKGTQ GKIVDLVKEL DRD-TVFALV -NYIFFKGKW ERPFEVKDTE

A KELFYKADGE SCSASMMYQE GKFRYRRVAE GTQVLELPFK GDDITMVLIL PKPE-KSLAK VEKELTPEVL QEWLDELEEM MLVVHMPRFR IE-DG-FSLK
B EEDFHVDQVT TVKVPMMKRL GMFNIQHCKK LSSWVLLMKY LGNATAIFFL --PDEGKLQH LENELTHDII TKF-L-E-NE DRRSASLHLP KLSITGTYDL

A EQLQDMGLVD LFSPEKSKLP GIVAEGRDDL -YVSDAF-HK AFLEVNEEGS E-AAASTAVV IAGRSLNLLP NRVTFKANRP FLVFIREVPL NTIIFMGRVA NPCV
B KSVLGQLGIT KVFS-NGADL SGV--T-EEA PLKLSKAVHK A-VLTIDEKG TEAAGAMF-L E-AIPM---- ---------- ---------- ---------- ----
```

Table 11.9. Statistical table of the average absolute deviation alignment of the torsion angles of the protein PDB1BF9 backbone

Model	1	2	3	4	5	6	7	8	9	10	11	12	13	14
2	0.365													
3	0.264	0.420												
4	0.328	0.386	0.293											
5	0.222	0.366	0.280	0.174										
6	0.293	0.390	0.284	0.320	0.270									
7	0.230	0.386	0.198	0.312	0.218	0.258								
8	0.269	0.432	0.210	0.375	0.298	0.214	0.179							
9	0.242	0.304	0.274	0.175	0.170	0.254	0.232	0.280						
10	0.298	0.410	0.337	0.511	0.432	0.412	0.366	0.364	0.411					
11	0.351	0.290	0.419	0.296	0.309	0.322	0.341	0.380	0.291	0.457				
12	0.245	0.405	0.313	0.332	0.321	0.366	0.264	0.323	0.376	0.403	0.385			
13	0.339	0.311	0.439	0.427	0.466	0.491	0.395	0.489	0.421	0.396	0.342	0.365		
14	0.262	0.395	0.255	0.317	0.329	0.282	0.244	0.227	0.352	0.400	0.276	0.236	0.372	
15	0.183	0.345	0.235	0.219	0.156	0.245	0.214	0.276	0.142	0.363	0.331	0.321	0.406	0.0344

Table 11.10. Statistical table of the average absolute deviation alignment of the torsion angles of the protein PDB1BF9 backbone

Model	1	2	3	4	5	6	7	8	9	10	11	12	13	14
2	0.316													
3	0.185	0.346												
4	0.199	0.276	0.181											
5	0.104	0.266	0.163	0.142										
6	0.140	0.244	0.156	0.132	0.069									
7	0.144	0.309	0.135	0.164	0.102	0.117								
8	0.142	0.300	0.139	0.200	0.125	0.117	0.101							
9	0.140	0.229	0.197	0.150	0.086	0.118	0.131	0.144						
10	0.292	0.382	0.215	0.382	0.319	0.312	0.286	0.235	0.321					
11	0.263	0.273	0.264	0.126	0.182	0.170	0.207	0.233	0.171	0.435				
12	0.141	0.312	0.145	0.113	0.160	0.155	0.166	0.174	0.172	0.320	0.218			
13	0.361	0.272	0.386	0.288	0.367	0.356	0.393	0.405	0.358	0.427	0.317	0.322		
14	0.172	0.320	0.192	0.086	0.178	0.163	0.164	0.189	0.188	0.336	0.153	0.116	0.294	
15	0.151	0.264	0.154	0.151	0.076	0.083	0.113	0.129	0.102	0.305	0.183	0.169	0.376	0.0197

Table 11.11. The error of pairwise alignment of the average absolute deviation of backbone torsion angles of different models of protein PDB1BF9

0.384																					
0.295	0.314																				
0.306	0.289	0.291																			
0.193	0.290	0.254	0.337																		
0.356	0.218	0.266	0.319	0.220																	
0.365	0.328	0.234	0.335	0.275	0.223																
0.356	0.277	0.187	0.360	0.267	0.151	0.103															
0.246	0.229	0.260	0.268	0.220	0.254	0.301	0.295														
0.256	0.332	0.339	0.340	0.371	0.392	0.437	0.395	0.346													
0.378	0.167	0.301	0.219	0.333	0.254	0.325	0.287	0.210	0.346												
0.316	0.357	0.215	0.308	0.274	0.324	0.252	0.282	0.245	0.447	0.365											
0.458	0.419	0.394	0.456	0.449	0.514	0.468	0.479	0.401	0.427	0.408	0.328										
0.322	0.291	0.157	0.247	0.317	0.265	0.254	0.243	0.258	0.348	0.249	0.222	0.425									
0.207	0.305	0.267	0.284	0.171	0.292	0.301	0.296	0.150	0.353	0.276	0.271	0.407	0.326								
0.326	0.249	0.194	0.289	0.287	0.252	0.231	0.212	0.203	0.355	0.223	0.247	0.467	0.165	0.267							
0.326	0.340	0.308	0.328	0.361	0.266	0.234	0.218	0.250	0.384	0.336	0.300	0.499	0.321	0.268	0.280						
0.302	0.238	0.357	0.242	0.302	0.207	0.270	0.251	0.221	0.380	0.232	0.341	0.541	0.309	0.299	0.242	0.261					
0.291	0.206	0.269	0.277	0.214	0.223	0.305	0.293	0.133	0.391	0.220	0.226	0.430	0.230	0.218	0.203	0.277	0.238				
0.227	0.265	0.271	0.165	0.298	0.350	0.428	0.412	0.180	0.301	0.258	0.260	0.401	0.225	0.253	0.280	0.324	0.269	0.206			
0.244	0.349	0.182	0.345	0.263	0.290	0.247	0.244	0.254	0.346	0.374	0.251	0.460	0.180	0.309	0.233	0.248	0.344	0.253	0.248		
0.386	0.381	0.282	0.468	0.315	0.298	0.229	0.236	0.291	0.375	0.411	0.304	0.453	0.303	0.295	0.290	0.248	0.377	0.300	0.414	0.242	
0.208	0.390	0.292	0.294	0.314	0.338	0.286	0.317	0.228	0.301	0.376	0.237	0.457	0.270	0.254	0.288	0.177	0.286	0.248	0.221	0.189	0.278

We see from the data in Tables 11.9 and 11.10 that the homology of the spatial phase characteristic sequences and the average absolute deviation of the torsion angles of the two protein backbones are generally consistent in the homology measurement index. That is, if the homology of spatial phase characteristic sequences is high, the average absolute deviation of the torsion angles would be low. For example, the homology of the characteristic sequences of the alignment between Model 1 and other models, and the average absolute deviation of their torsion angles are

Homology	0.71	0.85	0.76	0.88	0.73	0.83	0.78	0.80	0.85
Torsion angle deviation	0.365	0.264	0.328	0.222	0.293	0.230	0.269	0.242	0.298

Homology	0.80	0.85	0.73	0.85	0.83
Torsion angle deviation	0.351	0.245	0.339	0.262	0.183

.

The homology of the characteristic sequences of the alignment between Model 1 and other models in the stable region (from the fourth amino acid to the 31st amino acid) and the average absolute deviation of their torsion angles are given below

Homology	0.74	0.96	0.93	1.00	0.96	0.96	0.93	0.96	0.89
Torsion angle deviation	0.316	0.185	0.199	0.104	0.140	0.144	0.142	0.140	0.292

Homology	0.93	1.00	0.81	0.96	0.96
Torsion angle deviation	0.263	0.141	0.361	0.172	0.151

.

Based on these data, we may find that the homology and the average absolute deviation are inversely correlated. In most cases, the average absolute deviation would be low if the homology is high.

Table 11.11 contains the error of pairwise alignment of the average absolute deviation of the backbone torsion angles of the 23 models of protein PDB1BF9.

11.3 Exercises, Analyses, and Computation

Exercise 54. Find the data on the atomic spatial coordinates of the proteins whose ID numbers in the PDB database are 7apiA, 8apiA, 1hleA, 1ovaA, 2achA, 9apiA, 1psi, 1atu, 1ktc, 1athA, 1antI, 2antI. Compute the torsion angle sequences $\Psi_s = (\psi_{s,1}, \psi_{s,2}, \cdots, \psi_{s,n_b})$ for the atomic sequences in the backbone $\mathbf{L_1}$, where $s = 1, 2, \cdots, 12$ refer to the 12 proteins listed in this exercise.

Exercise 55. Decompose the torsion angle sequence Ψ_s into phase sequences $\bar{\vartheta}_s = (\vartheta_{s,1}, \vartheta_{s,2}, \cdots, \vartheta_{s,n_b})$. Implement the pairwise alignment by the dynamic programming algorithm, with the measure function given in formula (11.6), to obtain the pairwise minimum penalty alignment and the homology matrix of the pairwise alignment of these sequences.

Exercise 56. For the torsion angle sequences Ψ_s, implement the pairwise alignment by the dynamic programming algorithm with the measure function given in formula (11.7), to find the pairwise minimum penalty alignment and the homology matrix of the pairwise alignment of these sequences.

Exercise 57. Visualize the 3D structure of each protein in Exercise 54 using, for example, the Rasmol software package, and compare the relationship between the homologies and the configurations of different proteins. Visualize the 3D structures of the 15 models of protein PDB1BF9 in the same way, and compare the homology with Table 11.4.

12

Depth Analysis of Protein Spatial Structure

12.1 Depth Analysis of Amino Acids in Proteins

The depth of amino acids in a protein has been defined in many ways. For example, in biology, the amino acids, which come in contact with water molecules, form the surface of protein (zero depth) [54], so we may use this to determine the hydrophobic property of amino acids. Although these biological definitions have a clear physical sense, they lack a uniform computational methodology. For example, the concept of the hydrophobic factor of amino acids has been defined in many ways.

In mathematics, there are a series of definitions and calculational methods for the depth in a spatial particle system. How to apply these results and methods to the research of protein spatial structure will be discussed in this section.

12.1.1 Introduction to the Concept of Depth

We begin by introducing the basic mathematical concepts, definitions and generalization of depth in spatial particle systems.

Two Kinds of Definitions

There are two kinds of definitions of depth in mathematics, one is the generalization of the Tukey depth [103] in statistics, and the other is the accessible radius function theory in spatial particle systems:

1. The definition of Tukey depth comes from statistical theory. The median is a typical concept of "the deepest point" which is very important to statistics. The generalization of the median is the depth of one-dimensional data and high-dimensional data, and the definition of two-dimensional data is shown in the figure below. For one-dimensional data, they are well

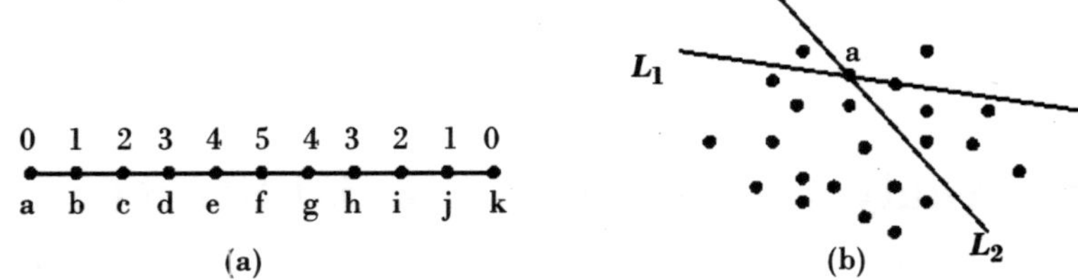

Fig. 12.1a,b. Depth of one-dimensional and two-dimensional data points

ordered, which implies that their depth can be quickly determined. However, for high-dimensional cases, the data are only semiordered. Therefore, we need methods to describe the depth. In modern statistical theory, there are a series of studies on both the definition and calculation of depth [58, 72, 80, 103, 114–116]. Since the concept of depth determines the relative position of the points in space, we can use it to analyze protein spatial structure.

Figure 12.1a shows the depth of one-dimensional data; where $a, b, \cdots, k$, are the points on a straight line, and the corresponding numbers are their depth. Figure 12.1b represents two-dimensional data, where we can see both interior points and exterior points, whose depth cannot be determined directly, but requires further statistical analysis.

In Fig. 12.1b, solid points represent all the points in a spatial particle system. L_1 and L_2 are straight lines passing through the point a, and these lines separate the set A into two parts. While rotating the line around point a, the corresponding fewest number of points that lie on one side of the line is the depth of point a in set A.

2. The accessible radius function theory in spatial particle systems is a biological concept of depth represented through the language of mathematics. It shows the largest possible radius within which each particle in A is in contact with the exterior. We will discuss this theory in detail below.

Protein Depth in Spatial Particle Systems

If we consider all proteins (or select parts of atoms) as a spatial particle system, then the spatial location of these atoms in the particle system can be used to analyze depth. In this book, we discuss the following issues:

1. On the basis of calculating depth, we can build a depth database of proteins. It means that we can calculate the depth of the amino acids (or C_α atoms) of all proteins (or peptides) in the PDB database, and build a depth database of proteins.
2. On the basis of the above depth database, we can do statistical depth analysis for different amino acids, such as depth tendency factor analysis, hierarchical analysis, and depth analysis of the peptide chain.

3. On the basis of depth analysis, we can do further analysis for special shapes of proteins, such as calculating the accessible radius function, calculating the interior structure of protein, and so on. We will discuss these issues in detail in the following text.

Mathematical Tools for Finding the Depth in Spatial Particle Systems

To analyze the depth of a spatial particle system, we use a variety of mathematical tools, such as the depth function algorithm, the geometric theory of convex polygons and hypergraph theory. We apply large-scale knowledge of analytic space geometry in the calculations.

12.1.2 Definition and Calculation of Depth

Before analyzing protein structure, we introduce the general theory of depth. In order to analyze protein structure, we give further discussion about both the calculation and properties of depth.

Definition of the Depth of a Discrete Set in 3D Space

Let

$$A = (a_1, a_2, \cdots, a_n) , \tag{12.1}$$

be a set of points in 3D space $\mathbf{R}^3$, and denote the three-dimensional coordinate of each point a_k by $\boldsymbol{r}_k = (x_k, y_k, z_k)$. Let π be a plane that contains the point a. Then, the plane cuts the set A into two parts, respectively denoted by $A_1(\pi, a)$ and $A_2(\pi, a)$, which are located on two different sides of π. Let $n_\tau(\pi, a)$ be the number of elements contained in the set $A_\tau(\pi, a)$, $\tau = 1, 2$. We always assume that $n_1(\pi, a) \le n_2(\pi, a)$.

Definition 42. *For a fixed point a and set A, we have the following definitions:*

1. *The depth of point a in set A is defined as*

$$s_A(a) = \min\{n_1(\pi, a) : \pi \in \tilde{\Pi}(a)\} , \tag{12.2}$$

in which $\tilde{\Pi}(a)$ is a plane that contains point a.

2. *Let π_0 be a plane that contains point a such that $n_1(\pi_0, a) = s_A(a)$, then we call π_0 a depth cut plane that contains point a.*
3. *If π_0 is a depth cut plane that contains point a, we call the side which contains $A_1(\pi_0, a)$ the outside of the depth cut plane. We call the other side the inside of the depth cut plane.*

Fast Calculation of Depth

In Definition 42, since $\tilde{\Pi}(a)$ is an infinite set, there is actually no way to calculate $s_A(a)$ in (12.2). Therefore, we first give some easily applied formulas for calculating depth.

Let $\pi(a;i,j)$ be a plane that contains three points (a, a_i, a_j), where $a_i, a_j \in A$. Then, the plane $\pi(a;i,j)$ cuts the set $A_\tau[\pi(a;i,j)]$, $\tau = 1,2$ into two parts, respectively denoted by A1 and A2, located on two different sides of plane $\pi(a;i,j)$. Let $n_\tau[\pi(a;i,j)]$, $\tau = 1,2$ be the number of elements contained in set $A_\tau[\pi(a;i,j)]$, $\tau = 1,2$, and always assume that $n_1[\pi(a;i,j)] \le n_2[\pi(a;i,j)]$.

For a given point a, and $a_i, a_j \in A$ that may change, let $s'_A(a)$ denote the minimum of $n_1[\pi(a;i,j)]$. Then the depth $s_A(a)$ of the point a in set A is just $s'_A(a)$.

Theorem 40. *The formula to compute the depth of point a in set A is given as follows:*

$$s_A(a) = \min\{n_1[\pi(a;i,j)] : a_i \ne a_j \in A, \text{ and } a, a_i, a_j \text{ are noncollinear}\} . \tag{12.3}$$

We will prove this theorem in Sect. 12.1.3. Thus, calculating the depth of point a can be simplified into calculating $n_1[\pi(a;i,j)]$, where $a_i, a_j \in A$. Therefore, the computational complexity is no more than $O(n^2)$, and we obtain the fast calculation of the depth of point a in set A.

The depth cut plane of point a can be restricted to a triangular area which is constructed by point a and two other points a_i, a_j in set A. Let $\pi(a, a_i, a_j)$ be a depth cut plane that contains point a. We then call triangle $\delta(a, a_i, a_j)$ the depth cut triangle of point a. Definitions of both the outside and inside of the depth cut triangle are determined by the corresponding outside and inside of the depth cut plane.

We can generalize the definition and fast calculation of depth to two-dimensional or higher dimensional point sets in the same way, so we will not repeat these here. Set A in Fig. 12.1 represents two-dimensional data. There might be many lines like L_1 and L_2 passing through point a, and they all cut the set A into two parts. The corresponding number of points are $n_1(L_1) = 2$, $n_2(L_1) = 19$, $n_1(L_2) = 7$, $n_2(L_2) = 15$, where L_1 is a cutting line. So the depth of a is $s_A(a) = 2$.

Remark 19. For the formula in Theorem 40, a may not be a point in set A. This means that for any point $a \in R^3$, formula (12.2) holds.

Several Properties of Depth

According to the definition of depth and the formula in Theorem 40, we immediately obtain the following properties:

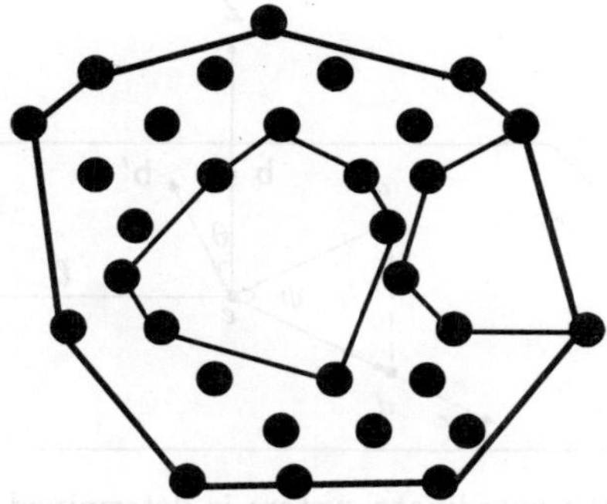

Fig. 12.2. The convex hull, groove, and cavity of a two-dimensional particle. The *thick line* is the convex closure $\Omega(A)$ of set A (their depths are zero), and the *thin lines* represent the interior groove and cavity of the convex closure $\Omega(A)$

1. If A is a fixed finite set and a is an arbitrary point in space R^3, then $s_A(a)$ is uniquely determined, and $0 \leq s_A(a) \leq \text{int}(n/2)$ is true, where $n = ||A||$, and $\text{int}(n/2)$ is the integral part of $n/2$.
2. If all points in A are coplanar, then the depth of each point in A is zero.
3. If Π_k is a depth cut plane of a_k, and $a \in A$ is in the plane Π_k, then $s_A(a) \leq s_A(a_k)$ holds. If $a \in A$ is in outside of Π_k (namely, it is in $n_1(\Pi_k)$), then $s_A(a) < s_A(a_k)$ holds.
4. Let $\Omega(A)$ be a convex closure of the set$A$; then $s_A(a) = 0$ if and only if a_k is on the boundary plane of $\Omega(A)$.

These properties can be easily deduced from Figs. 12.1 and 12.2.

Theorem 41. *Let s_A be the maximum of $s_A(a)$, $a \in A$. Then, s_A is called the deepest depth of set A. Denote by S_A the set of points in R^3 whose depth is s_A. Then, S_A is definitely a convex set, and this convex set does not contain any four points which are not coplanar in set A.*

We will prove this theorem in Sect. 12.1.4.

12.1.3 Proof of Theorem 40

Part One: Preliminary Nature

To prove Theorem 40, we must discuss the property of a plane turning around a straight line. We use the following symbols:

1. π is a plane in 3D space, whose normal vector is denoted by $\boldsymbol{b}$, with $\boldsymbol{b}$ and plane π being vertical. We denote by ℓ a directed line which lies in plane π; the direction of ℓ is arbitrarily selected, with its direction denoted by the vector $\boldsymbol{c}$.

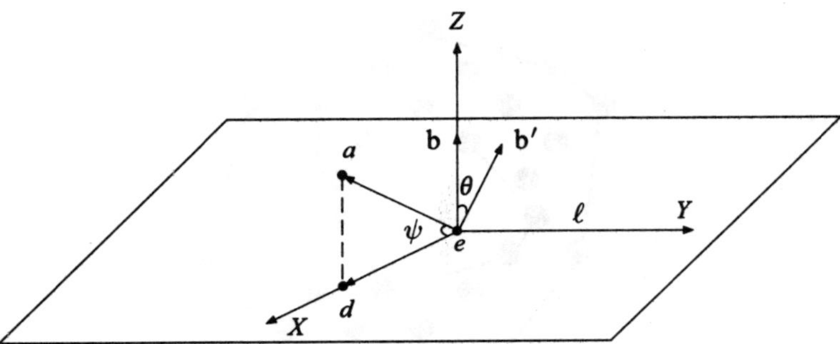

Fig. 12.3. Y-axis of the coordinate system is determined by the right-hand screw rule

2. Let a be a point which is not in plane π. We project a on to both plane π and line ℓ, and obtain two projective points denoted by d, e. Denote $\boldsymbol{b} = \overrightarrow{da}$, $\boldsymbol{d} = \overrightarrow{ed}$.
3. Let $\langle \boldsymbol{a}, \boldsymbol{b} \rangle$ be the inner product of $\boldsymbol{a}$ and $\boldsymbol{b}$; here the sign of $\langle \boldsymbol{a}, \boldsymbol{b} \rangle$ may be negative or positive. Define $\text{sgn}(\langle \boldsymbol{a}, \boldsymbol{b} \rangle)$, where
$$\text{sgn}\,(u) = \begin{cases} 1\,, & \text{if } u \geq 0\,, \\ -1\,, & \text{if } u < 0\,, \end{cases}$$
is the sign function of u.
4. Let $\psi = \angle aed$ be the angle between $\boldsymbol{a}$ and $\boldsymbol{d}$, then $-\pi/2 < \psi \leq \pi/2$. We provide the following rules: if $e = d$, which means that ae is perpendicular to plane π, then we let $\psi = \pm\pi/2$. if $\text{sgn}\,(\langle \boldsymbol{a}, \boldsymbol{b} \rangle) = +1$, $\psi > 0$; otherwise we let $\psi < 0$.
5. When plane Π is turning around line ℓ, we create a new plane Π', whose normal vector is denoted by $\boldsymbol{b}'$. The rotating dihedral angle, namely the angle θ between $\boldsymbol{b}'$ and $\boldsymbol{b}$, has range $|\theta| \leq \pi/2$. Its sign (positive or negative) is determined by the right-hand rule. We have the following properties.

Lemma 5. *If* $|\theta| < |\psi|$, *then*
$$\text{sgn}\,(\langle \boldsymbol{b}, \boldsymbol{a} \rangle) = \text{sgn}\,(\langle \boldsymbol{b}', \boldsymbol{a} \rangle)\ . \tag{12.4}$$

Proof. We begin to determine a rectangular coordinate system $\mathcal{E}$. Let e be the origin, and the X-axis and Z-axis are determined by vector $\overrightarrow{ed}$ and $\boldsymbol{b}$. The Y-axis is determined by the right-hand screw rule (Fig. 12.3).

Point a is in the plane XZ, $\psi = \angle aed$; without loss of generality, let $\boldsymbol{b} = (0, 0, 1)$, $|\overline{ea}| = 1$ then, $\boldsymbol{a} = (\cos\psi, 0, \sin\psi)$ and $\langle \boldsymbol{b}, \boldsymbol{a} \rangle = \sin\psi$.

When plane Π is turning around line ℓ with angle θ, then $\boldsymbol{b}' = (\sin\theta, 0, \cos\theta)$.
$$\langle \boldsymbol{b}', \boldsymbol{a} \rangle = \sin\theta\cos\psi + \sin\psi\cos\theta = \sin(\psi + \theta)$$

Since $|\theta| < |\psi|$ if $0 < \psi \le \frac{\pi}{2}$, $-\psi < \theta < \psi$ and $0 < \psi + \theta < 2\psi \le \pi$, we get $\mathrm{sgn}(\sin\psi) = \mathrm{sgn}[\sin(\psi+\theta)]$. If $-\frac{\pi}{2} \le \psi < 0$, $\psi < \theta < -\psi$ and $2\psi < \psi + \theta < 0$, we also find that $\mathrm{sgn}(\sin\psi) = \mathrm{sgn}[\sin(\psi+\theta)]$ holds. Thus, the theorem is proved.

Properties of a Plane Turning Around a Point

Lemma 5 suggests that the inner product changes if a plane is turning around a line. To discuss this in detail, we introduce the following notations.

Let $\boldsymbol{b}$, $\boldsymbol{b}'$, and $\boldsymbol{r}$ be three arbitrary vectors whose inner products are $\langle \boldsymbol{b}, \boldsymbol{r}\rangle$, $\langle \boldsymbol{b}', \boldsymbol{r}\rangle$. Let θ, ψ be the angle between $\boldsymbol{b}$ and $\boldsymbol{b}'$, $\boldsymbol{r}$, respectively, in which $0 \le \theta$, $\psi \le \pi$.

Lemma 6. *If θ and ψ satisfy either of the following conditions, then*

$$\mathrm{sgn}(\langle \boldsymbol{b}, \boldsymbol{r}\rangle) = \mathrm{sgn}(\langle \boldsymbol{b}', \boldsymbol{r}\rangle) . \tag{12.5}$$

Condition 1 *If $0 \le \psi \le \pi/2$, and $0 \le \theta \le \pi/2 - \psi \le \pi/2$.*
Condition 2 *If $\pi/2 \le \psi \le \pi$, and $0 \le \theta \le \psi - \pi/2 \le \pi/2$.*

Proof. For the sake of simplicity, in a rectangular coordinate system, we take

$$\boldsymbol{b} = (0,0,1)\,, \quad \boldsymbol{r} = (0, \sin\psi, \cos\psi)\,, \quad \boldsymbol{b}' = (\cos\phi\sin\theta, \sin\phi\sin\theta, \cos\theta)\,,$$

in which, ϕ is from $(0, 2\pi)$. Then

$$\mathrm{sgn}(\langle \boldsymbol{b}, \boldsymbol{r}\rangle) = \mathrm{sgn}[\cos\psi] = \begin{cases} +1, & \text{if } 0 \le \psi \le \pi/2\,, \\ -1, & \text{if } \pi/2 < \psi \le \pi \end{cases}$$

We calculate the value of $\mathrm{sgn}(\langle \boldsymbol{b}', \boldsymbol{r}\rangle)$. Here

$$\langle \boldsymbol{b}', \boldsymbol{r}\rangle = \sin\psi\sin\phi\sin\theta + \cos\psi\cos\theta\,. \tag{12.6}$$

To finish the proof of the lemma, we discuss the following two steps:

1. If $0 \le \psi \le \pi/2$, $0 \le \theta \le \pi/2 - \psi \le \pi/2$ (Fig. 12.4a), then $\cos\psi \ge 0$. Therefore, we need only prove $\langle \boldsymbol{b}', \boldsymbol{r}\rangle \ge 0$. Following from (12.6), we have

$$\langle \boldsymbol{b}', \boldsymbol{r}\rangle \ge \cos\psi\cos\theta - \sin\psi\sin\theta = \cos(\psi+\theta) \ge 0$$

   and since $\sin\psi, \sin\theta \ge 0, \sin\phi \ge -1, 0 \le \psi + \theta \le \pi/2$. This implies that $\langle \boldsymbol{b}', \boldsymbol{r}\rangle \ge 0$ holds. Thus the lemma is true under condition 1.
2. If $\pi/2 \le \psi \le \pi, 0 \le \theta \le \psi - \pi/2 \le \pi/2$ (Fig. 12.4b), then $\cos\psi \le 0$, so we need only prove $\langle \boldsymbol{b}', \boldsymbol{r}\rangle \le 0$. Following from (12.6), we have

$$\langle \boldsymbol{b}', \boldsymbol{r}\rangle \le \cos\psi\cos\theta + \sin\psi\sin\theta = \cos(\psi-\theta) \le 0\,.$$

   Since $\sin\psi, \sin\theta \ge 0, \sin\phi \le 1, \pi/2 \le \psi - \theta \le \pi$, we find that $\langle \boldsymbol{b}', \boldsymbol{r}\rangle \le 0$ holds. Additionally, $\cos\psi = 0$ implies that $\cos(\psi - \theta) = 0$. Thus the lemma is true under condition two, and the lemma is shown to be true in its entirety.

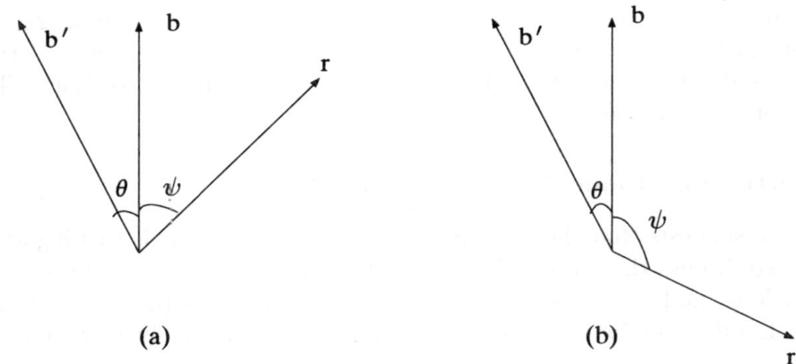

Fig. 12.4a,b. Proof of Lemma 6

Part Two: Proof of Theorem 40

Let $s'_A(a)$ be the depth defined by (12.3); obviously, we have that $s'_A(a) \geq s_A(a)$. Thus, we need only prove that $s'_A(a) \leq s_A(a)$ holds.

According to the definition of $s_A(a)$, we realize that, for an arbitrary point a, there must be a plane $\pi_0 \in \tilde{\Pi}(a)$, such that $n_1(\pi_0, a) = s_A(a)$. Three cases may occur on plane π_0:

1. *Case 1.* There are two points $a_i \neq a_j \in A$ such that a, a_i, a_j are not collinear.
   In this case, $\Pi_0 \in \tilde{\Pi}_0(a)$ implies $n_1(\pi_0, a) = s_A(a) \geq s'_A(a)$; thus we have that $s_A(a) = s'_A(a)$ holds.
2. *Case 2.* There is one point in plane π_0 such that $a_i \neq a \in \pi_0$, while the remaining points in A are neither in plane π_0, nor on line ℓ_{aa_i}, where ℓ_{aa_i} is a line determined by points a and a_i.
   Line ℓ_{aa_i} must be in plane Π_0, so we denote the set of points which are in A, while not in plane π_0, by A'.
   If $a_n \in A'$, let e_n and d_n denote the projection of a_n onto the line ℓ_{aa_i} and plane π_0, respectively, and let the sign of $\psi_n = \angle a_n e_n d_n$ be determined by the sign of $\langle \boldsymbol{a}_n, \boldsymbol{b} \rangle$. Moreover, let

$$\psi_0 = \min\{|\psi_n| \colon a_n \in A'\}$$

   and $a_{j_0} \in A'$ such that $|\psi_{j_0}| = \psi_0$. We then have the following conclusions:
   (a) $\psi_0 > 0$, otherwise $a_{j_0} \in \Pi_0$, which contradicts the definition of A'.
   (b) Point a_{j_0} must be in $A_{\Pi_0,2}$. Otherwise, we assume that point a_{j_0} is in $A_{\Pi_0,1}$, when Π_0 is rotating around line ℓ_{aa_i} with angle ψ_0, and we create a new plane Π' which is also through points a, a_i and point a_{j_0}. Several cases may arise according to the points in A':
      i. $\psi_n = \psi_0$, as shown in Fig. 12.5a, $\overline{e_n a_n}$ and $\overline{e_{j_0} a_{j_0}}$ have the same direction, so point a_n falls in plane Π'.

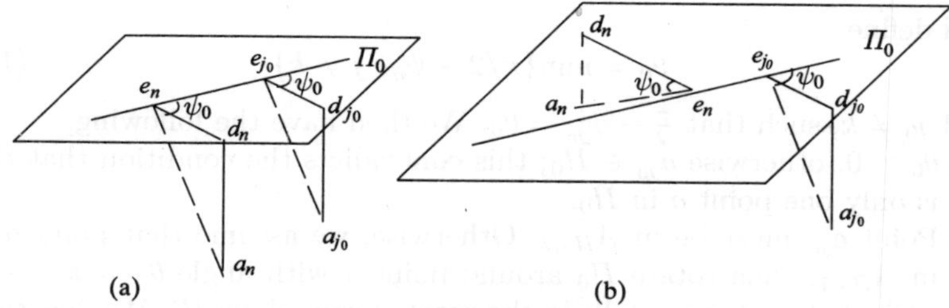

Fig. 12.5a,b. Proof of Theorem 40

ii. $\psi_n = \psi_0$, as shown in Fig. 12.5b, the angle formed by $\overline{e_n a_n}$ and $\overline{e_{j_0} a_{j_0}}$ is $\pi - 2\psi_0$.
Let the direction of $\overline{d_n a_n}$ and $\overline{e_n d_n}$ be the direction of the z-axis and x-axis, and let the origin be e_n. We then form a right cartesian coordinate system, and without loss of generality, we may set $\boldsymbol{b} = (0, 0, 1), |\overline{e_n a_n}| = 1$. This implies $\boldsymbol{a}_n = (\cos\psi_0, 0, \sin\psi_0)\, \boldsymbol{b}' = (\sin\psi_0, 0, \cos\psi_0)$, in which $0 < \psi_0 \leq \frac{\pi}{2}$. We then have

$$\langle \boldsymbol{b}\boldsymbol{a}_n \rangle = \sin\psi_0 \langle \boldsymbol{b}'\boldsymbol{a}_n \rangle = 2\sin\psi_0 \cos\psi_0$$

Thus, we have that $\operatorname{sgn}\langle \boldsymbol{b}\boldsymbol{a}_n \rangle = \operatorname{sgn}\langle \boldsymbol{b}'\boldsymbol{a}_n \rangle$ holds.

iii. In the case where $\psi_n = -\psi_0$, the situation is similar to that in case 2; two cases may occur and we have the same discussion.

iv. $|\psi_n| > \psi_0$. In this case, we may follow from Lemma 5 to get (12.4). Therefore, we have

$$n_1(\pi', a) < n_1(\pi_0, a) = s_A(a) \,.$$

This contradicts the definition of $s_A(a)$. Thus, point a_{j_0} must belong to $A_{\pi_0,2}$.

v. If a_{j_0} belongs to $A_{\Pi_0,2}$, then with the same arguments used in case 2, rotating Π_0 around line ℓ_{aa_i} with angle ψ_0, we get a new plane Π'. Here points a, a_i are still in plane Π', and the sphere is through point a_{j_0}. This is similar to case 2, the point a_n in A' may either fall in plane Π', or keep (12.5) stable, thus, we get

$$s'_A(a) \leq n_1(\Pi', a) \leq n_1(\Pi_0, a) = s_A(a) \,.$$

As a result, we find that $s'_A(a) = s_A(a)$ holds.

3. *Case 3.* The remaining points in A are not in Π_0. We denote $\boldsymbol{a}_n = \overline{aa_n}$, and denote the angle between $\boldsymbol{a}_n$ and $\boldsymbol{b}_0$ by ψ_n, where $0 \leq \psi_n \leq \pi$, and $\boldsymbol{b}_0$ is the normal vector of π_0. We now denote

$$\psi'_n = \begin{cases} \psi_n \,, & \text{if } 0 \leq \psi_n < \pi/2 \,, \\ \pi - \psi_n \,, & \text{if } \pi/2 \leq \psi_n \leq \pi \,, \end{cases} \tag{12.7}$$

and define

$$\theta_0 = \min\{\pi/2 - \psi'_n,\ j \neq k\} \tag{12.8}$$

and $j_0 \neq k$, such that $\frac{\pi}{2} - \psi'_{j_0} = \theta_0$. We then have the following:

(a) $\theta_0 > 0$, otherwise $a_{j_0} \in \Pi_0$; this contradicts the condition that there is only one point a in Π_0.

(b) Point a_{j_0} must be in $A_{\Pi_0,2}$. Otherwise, we assume that point a_{j_0} is in $A_{\Pi_0,1}$, then rotate Π_0 around point a with angle $\theta_0 = \pi/2 - \psi'_0$, such that vector $\boldsymbol{a}_{j_0}$ falls in the rotated new plane Π'. We denote the normal vector of plane Π' by $\boldsymbol{b}'$, and denote the angle between sphere $\boldsymbol{b}$ and sphere $\boldsymbol{b}'$ by θ_0. For any $j \neq k, j_0$, we have $\theta_0 \leq \pi/2 - \psi'_n$. Now, following from Lemma 6 and $a_{j_0} \in \Pi'$, we get $N_{\Pi',1} < N_{\Pi_0,1} = s_A(a)$, which contradicts the definition of $s_A(a)$. It follows that point a_{j_0} must belong to $A_{\Pi_0,2}$.

(c) If a_{j_0} is in $A_{\Pi_0,2}$, then similarly to the discussion above, we rotate Π_0 around point a with angle $\theta_0 = \pi/2 - \psi'_0$, such that vector $\boldsymbol{a}_{j_0}$ falls in the rotated new plane Π'.

By Lemma 6, we get

$$n_1(\pi', a) \leq n_1(\pi_0, a) = s_A(a) \ .$$

π is then also the depth cut plane of a, which comes back to case 2 and case 3. Therefore

$$s'_A(a) \leq n_1(\Pi', a) \leq n_1(\Pi_0, a) = s_A(a)$$

and $s'_A(a) = s_A(a)$ is true. The theorem follows.

12.1.4 Proof of Theorem 41

We prove this theorem in four steps:

1. **Step 12.1.1**. If $S_A = \{a\}$ is a set of a single point, then Theorem 41 is obviously true.
2. **Step 12.1.2**. If $S_A = \{a, b\}$, $a \neq b$ is a set of dual points, then for an arbitrary point c on the line segment ab, we have $s_A(c) = s_A$. We will prove this proposition by contradiction.
   Suppose $s_A(c) = s_A$ is not true; then by the definition of s_A, we get $a, b \neq c$, and $s_A(c) < s_A$. By the definition of depth, we realize that there must be a plane Π_c that is through point c, such that $n_1(\Pi_c) = s_A(c)$. Then two cases as follows may happen:
   (a) Line segment ab is in plane Π_c. By the definition of depth, we get $s_A(a), a_A(b) \leq n_1(\Pi_c) < s_a$, which contradicts the definition of points a, b, so this case will not happen.
   (b) Line segment ab is not in plane Π_c, then points a, b lie on both sides of plane Π_c. Without loss of generality, we let point a lie outside of plane Π_c. By the property of depth, we realize that $s_A(a) \leq n_1(\Pi_c) < s_a$ is

true, which contradicts the definition of point a, so this case will not happen either.

This shows that $s_A(c) < s_A$ is not true, so for any $c \in ab$, $s_A(c) = s_A$ is true.

3. **Step 12.1.3**. Similar to the proof in Step 12.1.2, we can prove that S_A must be a convex set. Thus, for any $a, b \in S_A$, and any point c on line segment ab, $s_A(c) = s_a$ must be true. So S_A must be a convex set.
4. **Step 12.1.4**. By the properties in Sect. 12.1.2, it is impossible to find any noncoplanar four points of set A in S_A. We do not give a detailed proof here. Thus, the theorem is proved.

12.2 Statistical Depth Analysis of Protein Spatial Particles

Statistical depth analysis of protein spatial particle requires the application of statistical analysis for depth indicators of spatial structure on all proteins in the PDB database. There are many types of depth indicators of protein spatial structure. In this book we choose several specific depth indicators for the statistical analysis.

12.2.1 Calculation for Depth Tendency Factor of Amino Acid

Depth Database of Protein

Using the method in Theorem 40 for the PDB database, we can build a depth database of proteins. Several important symbols are defined as follows:

Let

$$\mathcal{A} = \{A_1, A_2, \cdots, A_m\} \tag{12.9}$$

be all the protein sequences in the PDB database. For the sake of simplicity, we only discuss the proteins whose lengths (number of amino acids) are larger than 50. Here m is the number of proteins. A_i is the ith protein in the PDB database. We denote the primary structure sequence of protein A_i by

$$\boldsymbol{b}_i = (b_{i1}, b_{i2}, \cdots, b_{in_i}), \quad b_{ij} \in Z_{20},$$

in which $Z_{20} = \{1, 2, \cdots, 20\}$ represents the 20 common amino acids. We also denote the spatial structure sequence of protein A_i by

$$\boldsymbol{a}_i = (a_{i1}, a_{i2}, \cdots, a_{in_i}),$$

in which $a_{ij} = (x_{ij}, y_{ij}, z_{ij})$ represents the coordinate of C_α of the jth amino acid in protein A_i.

The data of sequence $\boldsymbol{b}_i$ and $\boldsymbol{a}_i$, $i = 1, 2, \cdots, m$, can both be obtained from the PDB database. Using these data, we can calculate the depth of each point in the sequence, and we denote $\boldsymbol{s}_i = (s_{i1}, s_{i2}, \cdots, s_{in_i})$, where $s_{ij} = s_{A_i}(a_{ij})$ is the depth of jth aminio acid (jth C_α atom) of the ith protein. We refer to

$$\mathcal{S} = \{(\boldsymbol{b}_i, \boldsymbol{a}_i, \boldsymbol{s}_i),\ i = 1, 2, \cdots, m\} \tag{12.10}$$

as the depth database of protein. This database is available on our Web site [99].

Definition of Depth Tendency Factor of Amino Acids

Through analysis of the depth database of protein, we realize that in protein spatial structures, distributions of the depths of different amino acids are obviously different. We call this phenomenon the **depth tendency factor of the amino acid**. There are various ways of defining the tendency factor. They will be discussed below:

1. **Zero depth tendency factor of amino acids**. The zero depth tendency factor of an amino acid is the proportion of each amino acid which appears on the surface of the protein (zero depth). We distinguish this proportion into two classes: absolute zero depth tendency factor and relative zero depth tendency factor:
   (a) Absolute zero depth tendency factor: This is the proportion of a certain amino acid in all the zero depth amino acids.
   (b) Relative zero depth tendency factor: For a certain amino acid, this is its proportion on the protein surface (zero depth).
2. **Deepest depth tendency factor of protein**. For each protein, there is one amino acid (sometimes several amino acids) whose depth is the largest. We call this (these) amino acid(s) the deepest point of the protein.
   The deepest depth tendency factor is defined as follows: the proportion of different amino acids in all of these amino acids that have the deepest depth.
3. **Average depth of amino acids**. This means the average depth of all the amino acids in the database. We refer to it as the average depth tendency factor of amino acid. We perform statistical calculations on protein chains (whose lengths are greater than or at least equal to 50) in the PDB database, and obtain their tendency factors as given in Table 12.1.

Remark 20. The symbols in Table 12.1 are explained as follows:

1. i is the sequence number of the amino acid, $a(i)$ is the total number of ith amino acids with zero depth, $n_0 = \sum_{i=1}^{20} a(i)$ is the total number of amino acids in zero depth, $n(i)$ is the total number of ith amino acids, $n = \sum_{i=1}^{20} n(i)$ is the total number of amino acids, $m = \sum_{i=1}^{20} c(i)$ is the total number of the deepest amino acids.
2. Thus, we have the following statistics:
   (a) $100n(i)/n$: percentage of frequency distribution of the amino acids

Table 12.1. Depth tendency factor of 20 common amino acids

One-letter code	0-depth F $a(i)$	Amino acid F $n(i)$	DPF $c(i)$	Amino acid F $100n(i)/n$	0-depth F $100a(i)/n_0$	0-depth RF $100a(i)/n(i)$	Grade of DP Rel. freq.	DPF $100c(i)/m$	Grade of DP
A	121175	718678	2881	8.202	7.477	16.861	i	8.127	i
R	79441	428179	1015	4.886	4.902	18.553	i	2.863	b
N	105583	390577	1426	4.457	6.515	27.033	h	4.023	b
D	157163	504074	1111	5.753	9.698	31.179	h	3.134	b
C	5766	128879	1108	1.471	0.356	4.474	b	3.126	b
Q	79403	327994	736	3.743	4.900	24.209	h	2.076	b
E	180862	553759	749	6.320	11.160	32.661	h	2.113	b
G	183580	675173	2693	7.705	11.328	27.190	h	7.597	i
H	30892	202744	971	2.314	1.906	15.237	i	2.739	b
I	25836	491050	4139	5.604	1.594	5.261	b	11.676	h
L	45885	763008	4298	8.708	2.831	6.014	b	12.125	h
K	151112	515808	644	5.887	9.325	29.296	h	1.817	b
M	17069	185310	697	2.115	1.053	9.211	b	1.966	b
F	18629	347024	2530	3.960	1.150	5.368	b	7.137	i
P	105018	405990	731	4.633	6.480	25.867	h	2.062	b
S	136614	527753	1511	6.023	8.430	25.886	h	4.263	b
T	99560	512565	2003	5.850	6.144	19.424	i	5.651	i
W	7397	130864	993	1.493	0.456	5.652	b	2.801	b
Y	21818	312638	1074	3.568	1.346	6.979	b	3.030	b
V	43717	619250	4047	7.067	2.698	7.060	b	11.417	h

0-depth F = zero depth frequency, *amino acid F* = amino acid frequency, *DPF* = deepest point frequency, *0-depth RF* = relative zero depth frequency, *grade of DP* = grade of the deepest point

(b) $100a(i)/n_0$: in all amino acids with zero depth, the percentage of ith amino acid (absolute zero depth tendency factor)
(c) $100a(i)/n(i)$: for the ith amino acid, percentage of zero depth (relative zero depth tendency factor)
(d) $100c(i)/m$: in all the deepest points of amino acids, the percentage of the ith amino acid (absolute deepest tendency factor)

Properties of Depth Tendency Factor

Based on Table 12.1, we can see that the depth tendency factor of an amino acid has the following characteristics:

1. In the view of the frequency and relative frequency of zero depth, the proportions of different amino acids on the surface of a protein are obviously different. The highest proportion is for the glutamic acid (E), whose proportion is as high as 32.66%, and the lowest proportion is for cysteine (C), whose proportion is only 4.47%.
2. Amino acids can also be classified by zero depth relative frequency. For example, the amino acid whose relative frequency is more than 20% (or 10–20%, or less than 10%) is said to be of the high (h) (or middle (i), or low (b)) surface tendency class, as shown in the table.
3. We can also use the depth frequency of the deepest point to classify amino acids. The amino acid whose depth frequency is above 10% (or 5–10%, or below 5%), is said to be of the high (h) (or middle (i), or low (b)) deepest point tendency class, as shown in the table.

12.2.2 Analysis of Depth Tendency Factor of Amino Acid

Relationship Between Depth Tendency Factors and Other Chemical and Physical Indices

We found that the depth tendency factors are comprehensively related to chemical and physical indices such as charge, polarity, chemical compound class, hydrophobicity, etc. (shown in Table 12.1). Correlative data are shown in Table 12.2.

The second row shown in Table 12.2 is the frequency of amino acids in the PDB database. The data of hydrophobicity comes from [26, 53, 61, 107]. The symbol +1 indicates electropositivity, −1 indicates electronegativity, and 0 indicates neutrality. In the column "Polarity" for dielectric polarity, +1 and −1 indicate polar and nonpolar properties, respectively.

In the column "Chemical compound class", $(-1,-1)$, $(1,-1)$, $(1,1)$ indicate the acyclic hydroxyl, alkaline, and acidic categories, respectively, $(0,0)$ indicates the hydroxy and sulfur-containing category, and $(-1,1)$ indicates the aromatic, and acclaimed category.

Using the data in Table 12.2, we can do data fitting on both the depth factor and the tendency factor. Let Y be the relative zero depth tendency

Table 12.2. Chemical and physical indices of amino acids and depth tendency factors

Code	Freq. (%) in PBD	Charge	Polarity	Chemical compound class	K-D	Eisenberg	Meek	F-P	Wolfenden	Absolute 0-depth TF	Relative 0-depth TF	Deepest depth TF
A	8.202	0	−1	(−1, −1)	1.80(7)	0.25(7)	0.20	0.31	1.49	7.477	16.861	8.127
R	4.886	1	−1	(1, −1)	−4.0(20)	−1.8(20)	−1.3	−1.0	−19.9	4.920	18.553	2.863
N	4.457	0	1	(1, 1)	−3.5(15)	−0.64(16)	−0.69	−0.60	−0.97	6.515	27.033	4.023
D	−5.753	1	−1	(1, 1)	−3.5(15)	−0.72(18)	−0.72	−0.77	−10.9	9.698	31.179	3.134
C	1.471	0	1	(0, 0)	2.5(5)	0.04(9)	0.67	0.54	1.24	0.356	4.474	3.126
Q	3.743	0	1	(1, 1)	−3.5(13)	−0.69(17)	−0.74	−0.22	−9.4	4.900	24.209	2.076
E	−6.320	1	−1	(1, 1)	−3.5(15)	−0.62(15)	−1.1	−0.64	−10.2	11.160	32.661	2.113
G	7.705	0	1	(−1, −1)	0.4(8)	0.16(8)	0.06	0.0	2.4	11.328	27.190	7.595
H	2.314	1	−1	(1, −1)	−3.2(14)	−0.40(14)	−0.0	−0.13	−10.3	1.906	15.237	2.739
I	5.604	0	−1	(−1, −1)	4.5(1)	0.73(1)	0.74	1.8	2.15	1.594	5.261	11.678
L	8.708	0	−1	(−1, −1)	3.8(3)	0.53(4)	0.65	0.17	2.28	2.831	6.014	12.125
K	5.887	1	−1	(1, −1)	−3.9(19)	−1.1(19)	−2.0	−0.99	−9.5	9.325	29.296	1.817
M	2.115	0	−1	(−1, 1)	1.9(6)	0.26(6)	0.71	1.23	1.48	1.053	9.211	1.966
F	3.960	0	−1	(0, 0)	2.8(4)	0.61(2)	0.67	1.79	−0.76	1.150	5.368	7.137
P	4.633	0	−1	(0, 0)	−1.6(11)	−0.07(11)	−0.44	−0.72	NM	6.480	25.867	2.062
S	6.023	0	1	(−1, 1)	−0.8(10)	−0.26(13)	−0.34	−0.04	−5.06	8.430	25.886	4.263
T	5.850	0	1	(−1, 1)	−0.7(9)	−0.18(12)	−0.26	−0.34	−4.88	6.144	19.424	5.651
W	1.493	0	−1	(0, 0)	−0.9(11)	0.37(5)	0.45	2.25	−5.88	0.456	5.652	2.801
Y	3.568	0	1	(0, 0)	−1.3(12)	−0.02(10)	−0.22	−0.96	−6.11	1.346	6.979	3.030
V	7.067	0	−1	(−1, −1)	4.2(2)	0.54(3)	0.61	1.22	1.99	2.698	7.060	11.417

TF = tendency factor, *NM* = no mention

Table 12.3. Table of data fitting of relative zero depth tendency factor of amino acid

Code	Act. val.	Reg. est.	Fit. err.
A	16.861	18.7888	−1.92777
R	18.553	19.0889	−0.53587
N	27.033	29.1475	−2.11446
D	31.179	30.6515	0.527526
C	4.474	6.92408	−2.45008
Q	24.209	22.7095	1.49953
E	32.661	32.2877	0.373308
G	27.190	25.1997	1.99029
H	15.237	10.1557	5.08133
I	5.261	5.41785	−0.156846
L	6.014	5.56178	0.452215
K	29.296	29.9323	−0.636279
M	9.211	12.3288	−3.1178
F	5.368	0.85603	4.51197
P	25.867	23.9452	1.92178
S	25.886	22.831	3.05504
T	19.424	20.0725	−0.648528
W	5.652	15.427	−9.775
Y	6.979	13.2924	−6.31337
V	7.060	7.57195	−0.511953

act. val. = actual value, *reg. est.* = regression estimate, *fit. err.* = fitting error

factor (in the 12th column), x_1, x_2 be the electric property and polarity (in the third and fourth columns), (x_3, x_4) be the data in the fifth column, and let $x_5, \cdots, x_9$ be hydrophobicity (in the sixth to tenth columns).

Thus, we can get their linear regression relationship through the formula:

$$\begin{aligned} \hat{Y} = 17.1434 &- 9.68877x_1 - 0.714291x_2 - 5.64749x_3 \\ &- 3.19165x_4 - 3.25787x_5 - 13.4022x_6 - 4.85877x_7 \\ &+ 1.92445x_8 + 1.05084x_9 \,. \end{aligned} \tag{12.11}$$

This fitted result is given in Table 12.3.

Furthermore, we obtain the mean square error, standard deviation, and mean error of fitting as follows:

$$\begin{cases} \text{mean square error } \sigma^2 = \sum_{i=1}^{20} p_i(y_i - \hat{y}_i)^2 = 6.29797 \,, \\ \text{standard deviation } \sigma = \sqrt{\sigma^2} = 2.50958 \,, \\ \text{mean error } e = \sum_{i=1}^{20} p_i|y_i - \hat{y}_i| = 1.74419 \,. \end{cases} \tag{12.12}$$

The errors in (12.12) are defined by percentage, where y_i is the actual zero depth tendency factor, $\hat{y}_i$ is the result which is fitted by chemical and physical indices of amino acid, and p_i is the frequency distribution of the amino acid in the PDB database.

Following from these data fitting results, we can see that the fitting error is not too large. Moreover, from fitting formula (12.12), we realize that the chemical and physical indices of amino acids have different effects on the relative zero depth tendency factor. For example, x_1, x_6 (charge of amino acids and the hydrophobic factor obtained from F-P and Eisenberg) have large effects, while x_2, x_8, x_9 (polarity of amino acids and the hydrophobic factor obtained from Wolfenden) have smaller effects. This differs from the conclusions in some publications which suggest that polarity is the major factor influencing the depth of amino acids in proteins.

Statistical Analysis of Both Length and Depth Indicators of Proteins

From the protein-depth database, we can perform statistical analysis of both length and depth indicators of proteins. The so-called depth indicators of proteins include:

1. The number of amino acids on the surface of a protein. For each protein, it is the total number of amino acids in zero depth.
2. The deepest depth of a protein, this means the depth of the deepest point in a protein.
3. Average depth of a protein, this means the average depth of all amino acids in a certain protein.

These three indicators relate to the length of the protein. We denote the length of the ith protein by $X(i) = n_i$, and denote three depth indicators of the ith protein by $Y_\tau(i)$, $\tau = 0, 1, 2$. In the PDB database, these indicators of proteins with different length are listed in [92]. Using the observation data $Y_\tau(i)$ and $X(i)$, we can perform regression analysis. Using the least-squares estimation, we get the linear regression equation of these data:

$$\hat{Y}_\tau(i) = \alpha_\tau X(i) + \beta_\tau\,, \quad \tau = 0, 1, 2\,, \tag{12.13}$$

in which the values of the regression coefficients α_τ and β_τ are defined by

τ	0	1	2
α_τ	0.05628	0.39897	0.06580
β_τ	31.80478	−11.82391	−2.30257

(12.14)

The regression errors of (12.14) are defined as follows:

τ	0	1	2
mean square error (σ_τ^2)	45.2454	97.0313	1.3734
standard error (σ_τ)	6.7265	9.8504	1.1719
absolute error (δ_τ)	5.1237	7.2439	0.8046

(12.15)

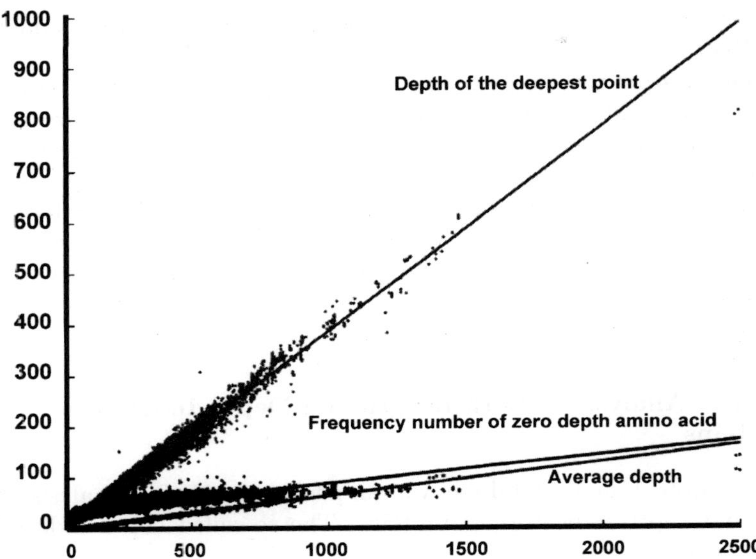

Fig. 12.6. The fitting relationship between protein length and depth map

where the definitions of mean square error, standard error, and absolute error are

$$\begin{cases} \sigma_\tau^2 = \dfrac{1}{m} \displaystyle\sum_{i=1}^{m} \left(\hat{Y}_\tau(i) - Y_\tau(i) \right)^2 , \\ \sigma_\tau = \sqrt{\sigma_\tau^2} , \\ \delta_\tau = \dfrac{1}{m} \displaystyle\sum_{i=1}^{m} \left| \hat{Y}_\tau(i) - Y_\tau(i) \right| . \end{cases}$$

Following from formulas (12.13)–(12.15), we realize that the fitting error of the average depth is very small. Therefore, we propose the following formula: the sum of the depths of all the amino acids in a protein, -2.30257, is in proportion to the square of the length of the protein. Also,

$$\sum_{j=1}^{n} s_j \sim 0.06580 \times n^2 + 2.30257 \times n \ , \tag{12.16}$$

in which n is the length of protein, and s_j is the depth of jth amino acid. The fitting relationship of formula (12.11) is shown in Fig. 12.6.

12.2.3 Prediction for Depth of Multiple Peptide Chains in Protein Spatial Structure

Using the protein depth database, we can predict the depth of each amino acid in a protein. In Table 12.1, the relative frequency of the zero depth is the probability of finding the amino acid on the surface of the protein. From Table 12.1, we can see that the probability that glutamic acid (E) is on the surface of a protein is 32.66%. To improve the accuracy of our forecasts, we can use multiple polypeptide chains on the surface of a protein.

Prediction of Tripeptide Chain

Let ABC be three linked amino acids. If one of them is on the surface of a protein, then we say that tripeptide chain ABC is on the surface of the protein. Let $n(\mathrm{A,B,C})$ be the number of tripeptides ABC on the surface of proteins in the PDB database, let $n_0(\mathrm{A,B,C})$ be the number of all tripeptides on the surface of proteins in the PDB database. Then $f_0(\mathrm{A,B,C}) = n_0(\mathrm{A,B,C})/n(\mathrm{A,B,C})$ is the zero depth tendency factor of the tri-peptide ABC on the surface of the protein. Table 12.4 shows all the tripeptide chains whose zero depth tendency factor satisfies $f_0(\mathrm{A,B,C}) > 75\%$.

Each tripeptide chain (for example, RED) consists of three amino acids. Frequency is the number of of tripeptides in the entire amino acid sequence whose length is greater than or at least equal to 50 in the PDB database, and the zero depth tendency factor has the value of $f_0(A,B,C)$.

Prediction of Tetrapeptide

The definition of the zero depth tendency factor of tetrapeptide ABCD on the surface of a protein is similar to the definition for a tripeptide. We list in Table 12.5 the tetrapeptides whose zero depth tendency factors are greater than 95%.

Following from Tables 12.4 and 12.5, we can get the probability of both tripeptides and tetrapeptides appearing on the surface of the protein. For example, from Table 12.5, we realize that tetrapeptide VYQR appears on the surface of the protein with probability 97.142%. Therefore, we can determine that the tetrapeptide VYQR must appear on the surface of the protein.

Remark 21. In Table 12.5, the zero depth tendency factor of a tetrapeptide has been given under two conditions, namely, the zero depth tendency factor of tetra-peptide must be more than 95%, and their frequency in the PDB database must be more than 200. We list those tetrapeptides whose zero depth tendency factors is more than 95%, with no restriction on frequency in [99].

We can also calculate the zero depth tendency factor of polypeptide chains in the PDB-Select database, for this may exclude the influence of homologous sequences. Since the result is similar, we do not list these.

Table 12.4. Tripeptide of zero depth tendency factor larger than 75%

Tripeptide	Freq.	0-depth TF	Tripeptide	Freq.	0-depth TF	Tripeptide	Freq.	0-depth TF	Tripeptide	Freq.	0-depth
RED	1459	77.039	NQS	894	75.727	NED	1589	75.267	NEE	1471	80.081
NEK	1367	78.346	NEP	1582	75.284	NHK	471	87.261	NKD	1400	83.500
NKH	316	76.582	NKP	828	75.362	DDP	1361	77.222	DQP	829	83.232
DEP	2148	75.418	DGK	1983	76.550	DKQ	941	77.258	DKH	690	75.942
DWP	322	83.229	CRN	573	79.755	CDP	298	78.859	CQE	341	82.404
CKM	229	92.576	QHQ	341	75.073	QWE	414	76.570	ENN	768	78.385
ENK	1013	78.479	EDQ	678	79.203	EDE	1838	77.257	EDK	2119	77.772
EDP	1729	78.716	EDS	1645	75.987	EED	1959	77.335	EEP	2856	76.890
EGD	2257	78.112	EKN	1130	75.398	EKD	1632	78.063	EKE	2504	75.039
GNK	2435	80.492	GEK	2645	77.769	GKE	2891	80.214	HEE	732	81.010
HEK	545	75.596	HKE	705	79.858	KNP	715	75.944	KDD	1017	83.087
KDQ	952	78.676	KQD	604	75.662	KED	1445	78.615	KGD	4048	75.148
KHK	824	83.859	KKD	1978	75.126	KPQ	625	76.160	MKP	622	77.491
PCM	131	76.335	PSK	2007	75.884	TQN	882	76.984	WPP	263	77.566

freq. = frequency, *TF* = tendency factor

Table 12.5. Tetrapeptides whose zero depth tendency factor is greater than 95%

Tetra-peptide	Freq.	0-depth TF	Tetra-peptide	Freq.	0-depth TF	Tetra-peptide	Freq.	0-depth TF	Tetra-peptide	Freq.	0-depth TF	Tetra-peptide	Freq.	0-depth TF
ARNP	381	97.375	AEDK	446	97.533	AKDQ	218	96.330	AKHG	312	95.192	APKD	205	97.560
ASCP	249	99.598	RNRW	418	100.000	RNKL	214	97.663	RNPT	391	96.163	RDPV	248	95.967
RQTW	226	100.000	RESG	275	96.000	RGDN	243	100.000	RKQQ	420	99.285	RPAD	234	96.581
RSGP	304	96.381	RYND	231	99.134	NAKP	248	97.580	NRNT	232	98.706	NRLI	423	96.453
NRSG	306	97.385	NRYN	213	98.591	NNKW	217	96.313	NEPD	288	97.916	NGPH	201	97.512
NHRE	264	96.969	NLSP	411	99.026	NPTQ	432	97.222	NSEF	229	97.379	NYWR	388	97.938
DRNQ	211	99.526	DDVH	201	95.024	DCHL	230	96.956	DEPT	231	95.670	DESA	263	95.437
DHGV	207	97.584	DLGH	205	95.121	DFHP	261	99.233	DFKE	243	96.707	DTNR	266	95.112
DWRK	395	99.240	DVNV	220	97.727	CRNR	437	98.398	CGQV	218	97.247	CPIN	214	99.532
QASG	285	96.140	QRES	318	95.597	QRVD	221	95.475	QNYW	585	98.803	QDYQ	304	100.000
QCRN	225	99.111	QESS	448	98.437	QGDL	201	95.024	QGDP	211	98.104	QPDR	227	97.356
QPSK	226	97.345	QSDI	208	95.192	QSGP	208	95.192	QTGA	235	95.319	QTWP	228	99.122
QWEG	267	97.378	QVDT	226	97.345	QVPP	203	98.029	EAEH	236	98.728	EDKT	678	99.852
EDWR	408	99.019	ECRE	211	97.630	EQKT	246	99.593	EQPS	229	96.943	EEPE	221	97.285
EESN	277	96.750	EGCV	205	98.536	EGSE	216	95.370	EGSK	206	98.058	EGTE	256	97.656
EKED	221	95.022	EPHS	224	95.535	ETYS	218	96.330	GRIW	215	98.139	GRKG	227	96.035
GRYN	234	97.863	GNED	457	96.498	GDNC	431	97.679	GDIK	332	97.590	GDPQ	203	97.044
GCGQ	225	98.222	GCIE	256	98.437	GQDA	201	97.512	GEKD	248	95.564	GETD	461	95.444
GHAK	349	96.848	GHRK	213	97.652	GIMK	304	97.039	GKCR	215	99.534	GKKL	311	96.141
GPTK	224	100.000	GTER	222	95.045	GTED	294	98.979	HREY	251	99.601	HCAN	274	99.270
HGCI	245	95.510	HLKD	302	96.357	HPFY	215	100.000	HVAP	230	95.217	INCL	209	100.000
ITDA	212	95.754	LQES	341	96.187	LKDC	255	98.431	LKHA	219	97.260	LSPS	432	97.222
KANR	409	97.310	KDCH	224	98.214	KCRN	210	99.523	KQAC	217	98.617	KQSD	213	99.061

Table 12.5. (continued)

Tetra-peptide	Freq.	0-depth TF	Tetra-peptide	Freq.	0-depth TF	Tetra-peptide	Freq.	0-depth TF	Tetra-peptide	Freq.	0-depth TF	Tetra-peptide	Freq.	0-depth TF
KGNE	456	98.903	KGER	237	95.358	KGHA	297	99.663	KGSK	322	95.031	KGWL	221	96.380
KHAH	203	99.014	KHGC	234	100.000	KIDK	208	96.153	KIFG	350	96.571	KKDG	209	96.650
KKEF	227	96.916	KKKD	227	98.678	KPNG	252	98.015	KPKW	305	100.000	KTAH	204	97.549
KTTP	281	98.576	KWRG	249	97.590	MNDT	244	98.770	MGNG	211	97.630	MKPK	315	95.873
FRDP	253	98.418	FNEP	262	97.709	FHPF	219	98.173	FTSL	220	95.909	PRQW	207	99.516
PNGE	227	96.035	PEGE	225	97.333	PHEE	214	100.000	PINC	242	98.347	PLSM	215	98.604
PKNK	236	95.338	PKWR	247	99.190	PSKR	208	95.673	PSKT	398	97.738	PTRG	305	98.032
PTQN	382	97.643	PTPG	324	99.074	PWTS	205	95.609	PYVV	242	97.520	SNHR	250	98.000
SCPI	218	99.082	SLDS	240	98.333	SLPK	232	95.258	SKEE	212	95.283	SKSA	233	95.708
SKWA	204	97.058	SKYC	218	97.706	SPSK	412	96.844	SPWT	207	99.033	SSKT	226	95.132
TRGD	328	98.780	TNRN	223	99.103	TNFN	224	95.535	TNSN	213	95.305	TDDA	487	96.509
TCGI	222	98.198	TQNY	388	96.391	TEPH	244	97.540	TGKC	213	100.000	TKGQ	248	95.161
TKIS	253	95.652	TPHE	271	98.892	TPFS	256	95.312	TWPY	262	100.000	WRKQ	382	100.000
WRSK	397	98.488	WLTI	257	97.276	WPYV	232	100.000	WTSS	217	96.774	WVGD	213	95.774
YADW	393	97.709	YEDK	365	98.904	YGLE	215	95.348	YWRS	410	96.341	VGCG	284	95.774
VPRE	233	95.708	VYQR	210	97.142									

TF = tendency factor

12.2.4 The Level Function in Spatial Particle System

Based on the definition and calculation of depth in the last section, we present the definition and calculation of the level function in a spatially distributed particle system. We denote the particle system by A, and denote all the zero-depth points in A by A_0. Let $\Omega(A) = \Omega(A_0)$ be the convex closure of the space particle system, and let its whole boundary plane be $\omega(A_0)$. Then, $\omega(A_0)$ can eventually be decomposed into a number of interconnected triangles. We call $\omega(A_0)$ the zero hierarchical plane of A, and A_0 is the point set on the zero hierarchical plane of A.

Let $A_1 = A - A_0$, and denote by A_{10} all zero depth points of the particle system A_1, where its corresponding convex closure is denoted by $\Omega(A_1) = \Omega(A_{10})$. We also denote its whole boundary plane by $\omega(A_1) = \omega(A_{10})$, which is called a first level plane of A, and A_{10} is the point set on a one level plane of A.

Similarly, we define $A_k = A_{k-1} - A_{k-1,0}$, and denote all zero depth points in the particle system A_k by $A_{k,0}$. Its convex closure is denoted by $\Omega(A_k) = \Omega(A_{k,0})$, and its whole boundary plane is denoted by $\omega(A_k) = \omega(A_{k,0})$, which we call the k-level plane of A. $A_{k,0}$ is a point on the k-level plane of A. Finally, we can get a k_0, such that $A_{k_0} = A_{k_0,0}$, and we call k_0 the largest level in the particle system A.

Using the notations above, if $a \in A_{k,0}$, then the level function at point a is k. Obviously, the definitions of level function and depth function are different. Level function has the following properties:

1. The group of set $A_{k,0}$, $k = 0, 1, \cdots, k_0$ is a subdivision of set A, they are disjoint, and their union is A, where $A_{0,0} = A_0$.
2. For any $k = 0, 1, \cdots, k_0 - 1$, the relationship $\Omega(A_{k-1,0}) \subset \Omega(A_{k,0})$ always holds.

Following the level function in a particle system, we may determine the structural characteristics of the particle system.

12.2.5 An Example

We use E. coli OmpF porin as an example to show the calculation process.

Depth and Level Function

This protein consists of 340 amino acids. Its primary structure, the coordinates of the C_α atom of amino acids, as well as their depth and level function, are shown in Table 12.6.

Remark 22. Coordinate (x, y, z) in Table 12.6 is the coordinate of the C_α atom of each amino acid.

Table 12.6. E. coli OmpF porin protein primary structure and coordinates of C_α atom

No.	Code	$C(x,y,z)$	D	L	No.	Code	$C(x,y,z)$	D	L	No.	Code	$C(x,y,z)$	D	L
1	A	100.516, 40.633, 38.037	6	3	2	E	102.617, 43.532, 36.772	5	2	3	I	105.790, 41.755, 35.780	0	0
4	Y	107.217, 44.778, 34.002	1	1	5	N	107.120, 48.575, 33.979	1	1	6	K	110.059, 50.246, 32.284	0	0
7	D	110.318, 53.037, 29.738	0	0	8	G	106.579, 53.601, 29.941	2	1	9	N	105.369, 50.182, 28.870	2	1
10	K	103.850, 48.105, 31.635	7	2	11	V	103.049, 44.467, 31.050	7	2	12	D	100.505, 42.693, 33.205	11	3
13	L	100.870, 38.907, 32.878	9	2	14	Y	98.075, 37.225, 34.829	15	4	15	G	96.507, 33.824, 35.283	18	4
16	K	94.238, 31.504, 37.256	24	4	17	A	94.210, 27.922, 38.414	6	2	18	V	90.663, 26.991, 39.232	16	4
19	G	90.117, 23.680, 40.996	2	1	20	L	86.490, 23.073, 40.118	12	3	21	H	83.821, 20.395, 40.368	5	2
22	Y	80.136, 20.149, 39.502	9	3	23	F	77.519, 18.166, 41.271	2	1	24	S	74.277, 17.120, 39.580	2	1
25	K	71.998, 14.075, 39.921	0	0	26	G	72.960, 11.001, 37.927	0	0	27	N	75.366, 11.695, 35.071	1	1
28	G	73.685, 15.106, 34.979	3	2	29	E	72.978, 15.269, 31.251	4	2	30	N	69.623, 16.609, 32.377	5	2
31	S	71.142, 19.123, 34.811	13	3	32	Y	71.683, 22.865, 34.374	32	4	33	G	75.454, 22.808, 34.361	40	5
34	G	76.455, 19.173, 34.201	20	3	35	N	78.335, 16.926, 36.575	7	2	36	G	82.024, 16.146, 37.069	3	1
37	D	85.500, 17.722, 37.284	3	1	38	M	85.569, 21.184, 35.739	17	4	39	T	89.088, 22.337, 36.588	11	3
40	Y	90.758, 24.873, 34.301	16	4	41	A	93.330, 27.625, 34.154	15	3	42	R	93.911, 30.973, 32.407	25	5
43	L	96.901, 32.907, 31.124	15	3	44	G	97.037, 36.294, 29.519	23	4	45	F	98.595, 39.703, 29.343	19	3
46	K	97.499, 43.279, 29.042	25	5	47	G	100.105, 45.761, 27.960	13	3	48	E	100.022, 49.537, 27.968	15	3
49	T	102.631, 51.846, 26.481	5	2	50	Q	102.440, 55.664, 26.574	3	1	51	I	103.089, 56.987, 23.111	0	0
52	N	103.026, 60.444, 24.743	0	0	53	S	100.781, 62.344, 27.154	0	0	54	D	97.433, 61.714, 25.446	2	1
55	L	98.094, 58.693, 23.318	2	1	56	T	98.335, 55.219, 24.745	8	2	57	G	98.739, 51.947, 22.923	3	1
58	Y	97.466, 48.786, 24.593	16	4	59	G	97.034, 45.128, 23.766	13	3	60	Q	95.502, 42.061, 25.395	30	5
61	W	95.411, 38.300, 24.950	21	4	62	E	93.631, 35.815, 27.211	36	6	63	Y	93.586, 32.036, 26.813	18	5
64	N	91.576, 29.301, 28.518	21	5	65	F	93.396, 26.073, 29.268	10	3	66	Q	91.162, 23.133, 30.243	11	3
67	G	92.340, 20.942, 33.068	6	2	68	N	89.638, 18.319, 33.081	6	2	69	N	91.085, 16.542, 30.049	4	1
70	S	93.370, 13.540, 29.974	0	0	71	E	96.772, 13.761, 28.428	0	0	72	G	95.439, 11.624, 25.615	0	0

Table 12.6. (continued)

No.	Code	$C(x,y,z)$	D	L	No.	Code	$C(x,y,z)$	D	L	No.	Code	$C(x,y,z)$	D	L
73	A	93.665, 12.692, 22.413	0	0	74	D	91.146, 14.860, 24.229	2	1	75	A	93.982, 16.977, 25.630	3	1
76	Q	92.940, 20.124, 23.696	4	2	77	T	89.189, 19.773, 24.201	7	3	78	G	88.135, 23.159, 25.446	12	4
79	N	91.320, 25.149, 25.015	9	4	80	K	90.473, 28.410, 23.267	12	5	81	T	91.617, 31.990, 22.889	15	4
82	R	89.183, 34.246, 24.755	45	6	83	L	90.417, 37.707, 23.827	54	6	84	A	92.973, 39.209, 21.509	19	4
85	F	93.028, 42.816, 20.411	20	4	86	A	94.921, 46.059, 20.189	7	2	87	G	94.171, 49.750, 20.231	8	3
88	L	95.050, 53.320, 21.037	3	2	89	K	93.281, 55.392, 23.682	14	3	90	Y	93.614, 59.131, 23.210	2	1
91	A	93.103, 62.118, 25.505	3	2	92	D	89.346, 62.617, 25.664	2	2	93	V	88.471, 61.583, 22.128	0	0
94	G	88.173, 57.993, 23.282	8	3	95	S	89.619, 54.569, 22.587	18	4	96	F	89.688, 52.362, 19.538	9	3
97	D	90.751, 48.752, 19.434	19	4	98	Y	90.261, 45.765, 17.179	10	3	99	G	90.387, 41.993, 17.407	11	3
100	R	88.616, 39.340, 19.423	37	5	101	N	86.728, 41.309, 22.055	67	6	102	Y	83.347, 41.596, 23.859	89	7
103	G	80.317, 42.873, 21.968	87	7	104	V	78.697, 46.125, 23.044	66	6	105	V	75.431, 44.574, 24.176	72	5
106	Y	77.621, 42.892, 26.756	94	7	107	D	78.093, 46.307, 28.350	66	5	108	A	74.505, 45.967, 29.556	53	5
109	L	74.578, 42.250, 30.350	76	6	110	G	77.737, 42.910, 32.305	78	6	111	Y	75.726, 44.192, 35.236	53	4
112	T	74.322, 40.757, 36.022	61	5	113	D	77.353, 38.646, 35.142	85	6	114	M	78.569, 38.541, 38.743	59	4
115	L	77.605, 35.067, 39.994	56	4	116	P	80.297, 32.859, 41.649	46	4	117	E	80.256, 30.366, 38.766	66	5
118	F	77.067, 30.270, 36.686	81	6	119	G	74.949, 33.180, 35.470	89	6	120	G	75.609, 35.344, 32.426	114	7
121	D	73.360, 32.932, 30.570	95	6	122	T	72.295, 35.265, 27.775	96	6	123	A	75.775, 36.376, 26.822	119	7
124	Y	76.673, 33.549, 24.425	98	7	125	S	80.166, 33.602, 22.921	86	6	126	D	80.410, 33.972, 19.141	54	5
127	D	76.819, 35.210, 19.053	60	6	128	F	76.815, 38.707, 17.511	48	5	129	F	76.849, 41.505, 20.139	71	6
130	V	76.190, 39.823, 23.516	98	7	131	G	79.481, 38.066, 24.207	111	7	132	R	83.033, 37.669, 22.980	96	7
133	V	83.324, 37.988, 19.200	52	6	134	G	85.841, 37.942, 16.410	20	4	135	G	86.789, 40.966, 14.340	8	2
136	V	84.858, 43.703, 16.115	28	4	137	A	86.112, 47.244, 15.800	11	3	138	T	85.180, 49.142, 18.905	32	5
139	Y	85.110, 52.825, 19.698	14	3	140	R	84.391, 53.966, 23.213	25	4	141	N	84.215, 57.503, 24.519	9	3
142	S	84.493, 58.569, 28.122	12	3	143	N	82.564, 61.144, 30.078	6	2	144	F	81.214, 62.427, 26.806	1	1
145	F	84.591, 63.793, 25.714	1	1	146	G	85.240, 64.688, 29.339	1	1	147	L	82.237, 67.038, 29.430	0	0

Table 12.6. (continued)

No.	Code	$C(x,y,z)$	D	L	No.	Code	$C(x,y,z)$	D	L	No.	Code	$C(x,y,z)$	D	L
148	V	79.942, 64.826, 31.501	0	0	149	D	81.243, 62.615, 34.314	2	1	150	G	79.750, 59.144, 34.149	6	2
151	L	78.331, 59.560, 30.660	3	1	152	N	79.878, 56.896, 28.448	13	3	153	F	79.031, 55.602, 25.001	10	3
154	A	80.545, 53.247, 22.433	19	4	155	V	79.842, 52.223, 18.874	8	2	156	Q	80.912, 48.835, 17.627	22	4
157	Y	81.224, 47.531, 14.138	3	1	158	L	80.923, 43.772, 13.791	15	3	159	G	82.680, 42.279, 10.821	1	1
160	K	81.207, 39.360, 8.892	0	0	161	N	82.487, 35.959, 10.035	3	1	162	E	81.495, 33.041, 7.797	0	0
163	R	82.590, 29.992, 9.748	1	1	164	D	81.984, 26.257, 9.393	0	0	165	T	79.047, 26.600, 11.754	6	2
166	A	75.826, 28.516, 11.665	12	3	167	R	76.184, 29.016, 15.426	28	4	168	R	79.627, 30.575, 15.282	24	4
169	S	78.908, 32.540, 12.089	13	3	170	N	77.708, 36.140, 11.799	17	3	171	G	77.141, 38.723, 9.063	4	2
172	D	78.288, 42.346, 9.431	0	0	173	G	76.546, 44.704, 11.773	2	1	174	V	76.678, 47.681, 14.105	2	1
175	G	75.911, 48.232, 17.748	14	3	176	G	76.358, 50.811, 20.460	12	3	177	S	75.879, 51.453, 24.144	20	4
178	I	75.225, 54.367, 26.475	5	2	179	S	75.858, 54.234, 30.196	13	3	180	Y	75.638, 56.733, 32.962	3	1
181	E	77.286, 55.987, 36.287	9	2	182	Y	76.517, 58.260, 39.263	0	0	183	E	76.900, 57.972, 43.059	0	0
184	G	76.686, 54.181, 42.932	2	1	185	F	74.009, 53.865, 40.286	1	1	186	G	74.395, 52.823, 36.719	8	2
187	I	71.939, 52.814, 33.923	0	0	188	V	72.702, 51.179, 30.536	11	3	189	G	71.098, 50.662, 27.207	2	1
190	A	72.806, 48.906, 24.337	19	4	191	Y	71.730, 47.577, 20.974	8	3	192	G	73.059, 46.010, 17.808	13	4
193	A	71.897, 44.329, 14.642	4	2	194	A	73.647, 42.445, 11.870	5	2	195	D	72.924, 40.318, 8.829	1	1
196	R	72.345, 36.615, 9.384	7	2	197	T	74.403, 34.595, 6.891	1	1	198	N	73.123, 32.031, 4.376	0	0
199	L	74.073, 29.116, 6.611	1	1	200	Q	72.240, 30.724, 9.495	8	2	201	E	69.169, 31.458, 7.411	2	1
202	A	69.473, 27.865, 6.250	0	0	203	Q	68.683, 26.494, 9.710	3	2	204	P	65.211, 25.004, 10.523	1	1
205	L	64.622, 27.028, 13.676	9	2	206	G	64.522, 30.778, 13.306	10	2	207	N	63.471, 32.869, 10.341	0	0
208	G	64.828, 36.239, 9.266	0	0	209	K	67.733, 37.978, 7.530	0	0	210	K	68.520, 40.391, 10.331	1	1
211	A	69.414, 39.430, 13.878	8	3	212	E	69.279, 42.177, 16.486	7	3	213	Q	69.897, 42.166, 20.214	30	4
214	W	69.359, 44.764, 22.875	19	4	215	A	69.285, 45.004, 26.632	22	4	216	T	69.331, 47.477, 29.459	7	2
217	G	70.841, 47.472, 32.876	16	3	218	L	70.455, 49.149, 36.207	0	0	219	K	72.667, 48.704, 39.213	7	2
220	Y	73.374, 49.984, 42.663	0	0	221	D	76.986, 49.221, 43.402	10	2	222	A	78.278, 50.929, 46.546	2	1
223	N	78.741, 50.247, 50.248	0	0	224	N	79.226, 46.466, 50.091	3	1	225	I	75.938, 46.290, 48.198	0	0

Table 12.6. (continued)

No.	Code	$C(x,y,z)$	D	L	No.	Code	$C(x,y,z)$	D	L	No.	Code	$C(x,y,z)$	D	L
226	Y	75.562, 45.068, 44.656	13	2	227	L	72.131, 45.035, 43.127	1	1	228	A	71.666, 44.775, 39.396	14	3
229	A	68.959, 43.983, 36.867	3	1	230	N	68.988, 43.339, 33.141	20	3	231	Y	66.255, 43.005, 30.584	5	2
232	G	66.613, 42.511, 26.867	11	3	233	E	65.142, 40.983, 23.730	6	2	234	T	66.439, 39.513, 20.481	18	3
235	R	65.340, 38.777, 16.936	5	2	236	N	66.671, 35.589, 15.357	16	3	237	A	69.830, 35.567, 17.500	41	5
238	T	69.766, 33.319, 20.561	59	6	239	P	70.999, 29.878, 19.454	51	5	240	I	68.593, 27.038, 20.294	38	5
241	T	68.154, 23.295, 19.770	19	4	242	N	65.284, 20.903, 19.834	6	3	243	K	66.731, 17.761, 21.452	3	2
244	F	63.842, 15.700, 20.178	0	0	245	T	63.597, 16.706, 16.523	0	0	246	N	67.393, 17.044, 16.655	1	1
247	T	66.947, 20.412, 14.982	3	2	248	S	68.876, 23.649, 15.576	10	3	249	G	68.739, 27.307, 14.702	19	3
250	F	67.875, 30.472, 16.551	32	4	251	A	64.957, 31.649, 18.650	23	3	252	N	62.831, 34.069, 16.671	8	2
253	K	62.280, 36.087, 19.847	3	2	254	T	63.345, 35.988, 23.521	11	3	255	Q	62.468, 37.964, 26.660	2	1
256	D	65.446, 37.988, 29.018	18	4	257	V	65.453, 38.946, 32.702	3	1	258	L	68.460, 38.802, 34.996	34	4
259	L	68.531, 39.871, 38.647	6	2	260	V	71.369, 39.934, 41.148	18	3	261	A	71.623, 40.705, 44.828	0	0
262	Q	74.956, 40.401, 46.617	9	2	263	Y	76.452, 41.778, 49.832	0	0	264	Q	80.162, 42.009, 50.685	5	2
265	F	80.947, 41.259, 54.311	0	0	266	D	84.150, 42.678, 55.693	0	0	267	F	85.685, 39.332, 56.668	0	0
268	G	85.527, 38.003, 53.103	2	1	269	L	82.180, 36.274, 52.790	1	1	270	R	80.186, 37.470, 49.792	7	2
271	P	76.828, 35.684, 49.493	3	1	272	S	74.879, 35.827, 46.236	12	2	273	I	71.246, 35.649, 45.244	0	0
274	A	69.958, 35.759, 41.674	11	2	275	Y	67.164, 35.047, 39.213	3	1	276	T	67.565, 34.221, 35.534	30	4
277	K	64.836, 33.767, 32.987	14	3	278	S	65.187, 33.533, 29.240	27	4	279	K	62.133, 32.676, 27.199	8	3
280	A	62.031, 32.049, 23.475	11	3	281	K	58.935, 33.099, 21.618	2	1	282	D	57.690, 32.128, 18.169	0	0
283	V	59.786, 28.956, 18.174	5	2	284	E	58.463, 27.066, 15.159	1	1	285	G	55.901, 24.460, 16.054	0	0
286	I	56.320, 25.244, 19.734	1	1	287	G	55.687, 28.900, 20.406	0	0	288	D	56.999, 29.918, 23.808	0	0
289	V	59.765, 27.956, 25.472	6	2	290	D	61.899, 28.667, 28.455	12	3	291	L	65.522, 28.374, 27.517	33	4
292	V	66.716, 29.218, 31.007	38	5	293	N	64.958, 29.612, 34.344	13	3	294	Y	66.200, 29.229, 37.909	7	2
295	F	67.156, 30.784, 41.201	1	1	296	E	70.673, 30.930, 42.493	10	2	297	V	72.116, 30.911, 45.950	0	0
298	G	75.854, 31.230, 46.197	10	2	299	A	78.746, 32.500, 48.267	7	2	300	T	82.353, 33.295, 47.520	15	3
301	Y	84.658, 33.608, 50.474	2	1	302	Y	87.718, 35.602, 49.488	6	2	303	F	90.758, 34.921, 51.630	0	0

Table 12.6. (continued)

No.	Code	$C(x,y,z)$	D	L	No.	Code	$C(x,y,z)$	D	L	No.	Code	$C(x,y,z)$	D	L
304	N	93.037, 37.607, 50.283	0	0	305	K	93.147, 39.072, 46.750	4	1	306	N	94.751, 35.917, 45.292	3	1
307	M	92.572, 33.263, 46.853	2	1	308	S	88.953, 32.339, 47.242	6	2	309	T	86.502, 29.510, 47.159	3	2
310	Y	82.858, 29.475, 46.285	13	3	311	V	79.799, 27.360, 46.296	5	2	312	D	77.203, 27.966, 43.668	23	3
313	Y	73.777, 26.426, 43.705	6	2	314	I	71.299, 26.463, 40.870	16	3	315	I	67.682, 25.815, 41.743	0	0
316	N	66.685, 24.804, 38.228	6	1	317	Q	63.162, 25.944, 37.477	0	0	318	I	63.012, 24.266, 34.080	4	1
319	D	60.458, 21.436, 34.073	0	0	320	S	60.724, 17.991, 32.525	0	0	321	D	58.035, 18.694, 29.983	0	0
322	N	60.354, 21.220, 28.263	3	1	323	K	59.616, 21.258, 24.531	3	1	324	L	63.137, 21.971, 23.341	11	2
325	G	64.311, 19.309, 25.747	8	2	326	V	66.215, 21.613, 28.055	23	4	327	G	67.675, 20.056, 31.212	17	3
328	S	65.639, 20.262, 34.407	6	2	329	D	67.990, 19.097, 37.194	3	1	330	D	69.974, 21.275, 39.558	3	1
331	T	73.746, 21.745, 39.630	10	3	332	V	76.235, 22.728, 42.311	5	2	333	A	79.674, 24.073, 41.611	17	4
334	V	82.402, 24.112, 44.226	3	2	335	G	85.576, 25.916, 43.305	13	3	336	I	88.826, 26.926, 44.913	1	1
337	V	90.689, 29.594, 42.968	11	3	338	Y	94.287, 30.765, 43.066	2	1	339	Q	95.302, 33.702, 40.876	9	3
340	F	97.768, 36.451, 40.152	7	2										

C = coordinate, D = depth, L = level

Table 12.7. Distribution of depth of each amino acid in protein E. coli OmpF porin

D	0	1	2	3	4	5	6	7	8	9	10	11	12	13	14	15
F	51	21	23	28	7	12	16	13	12	9	9	12	8	11	5	7
D	16	17	18	19	20	21	22	23	24	25	26	27	28	29	30	31
F	6	4	6	8	5	2	2	4	2	3	0	1	2	0	3	0
D	32	33	34	35	36	37	38	39	40	41	42	43	44	45	46	47
F	3	1	1	0	1	1	2	0	1	1	0	0	0	1	1	0
D	48	49	50	51	52	53	54	55	56	57	58	59	60	61	62	63
F	1	0	0	1	1	2	2	0	1	0	0	2	1	1	0	0
D	64	65	66	67	68	69	70	71	72	73	74	75	76	77	78	79
F	0	0	3	1	0	0	0	1	1	0	0	0	1	0	1	0
D	80	81	82	83	84	85	86	87	88	89	90	91	92	93	94	95
F	0	1	0	0	0	1	1	1	0	2	0	0	0	0	1	1
D	96	97	98	99	100	101	102	103	104	105	106	107	108	109	110	111
F	2	0	2	0	0	0	0	0	0	0	0	0	0	0	0	0
D	112	113	114	115	116	117	118	119								
F	0	0	1	0	0	0	0	1								

D depth, *F* frequency

Statistical Distribution of Depth

According to the discussion in Chap. 11, we can calculate the depth of all particles in the space particle system. Its distribution function describes the number of points with different depth, for example, for protein E. coli OmpF porin, its statistical distribution function is shown in Table 12.7.

In Table 12.7, the value of the depth ranges from 0 to 119. The number of points with different depths are listed in the table; for example, there are 51 zero-depth points, 21 one-depth points, and so on.

12.3 Exercises

Exercise 58. In the case of a plane, generalize the essential theorems and formulas presented in this chapter. For example:

1. For the plane particle system A, write out the definition of depth, then describe Theorem 40 and prove it.
2. Write a program for calculating the depth and level function of a point in plane particle system A.
3. Discuss the essential properties of the depth of the plane particle system A.

Exercise 59. Write a program for calculating the depth of three atoms N, C_α, C on the main chain of proteins in the PDB-Select database, and perform the following calculations:

1. Build three depth databases of the above three atoms: $\Omega_{\rm N}$, $\Omega_{\rm A}$, $\Omega_{\rm C}$. You may use the template of Table 12.6 and append depth indices for each amino acid.
2. Use these three depth databases of the above three atoms to calculate the depth tendency factor of different amino acids, such as the average depth tendency factor, the absolute zero depth tendency factor, the relative depth tendency factor and the deepest tendency factor, and so on.
3. Write out the forecast result for both tripeptides and tetrapeptides on the surface of a protein: write out the tripeptides on the surface of the protein with a probability of more than 75%, and the tetrapeptides on the surface of the protein with a probability of over 95%.

Exercise 60. Using the depth program in the plane particle system (Exercise 58), for the atoms N, C_α, C in the main chain of proteins in the PDB-Select database, write out the depth and level function of its projection on plane XY, XZ, YZ, and do the following statistical analysis:

1. Build a plane depth database of different amino acids and calculate their depth tendency factors, such as average depth tendency factor, absolute zero depth tendency factor, relative zero depth tendency factor and the depth tendency factor of the deepest point, and so on.
2. For both a tripeptide and a tetrapeptide, write out the forecast result of its projection on plane XY, XZ, YZ, write out the tripeptide whose projection on plane XY, XZ, YZ is on the surface of a protein with a probability of more than 75%, and the tetrapeptide whose projection on plane XY, XZ, YZ is on the surface of protein with a probability of more than 95%.

Exercise 61. On the basis of the calculation in task 1 of Exercise 58, compare the results for three atoms N, C_α, C, and discuss their relationships:

1. Compare the relationship of the depth tendency factors among different amino acids, such as the relationship among the depth tendency factors of atoms N, C_α, C.
2. Compare the relationship for predictions of polypeptide chains on the surface of a protein, for example, for atoms N, C_α, C, the data structure of the tetrapeptide which is on the surface of a protein with a probability of more than 95%.

Exercise 62. On the basis of task 1 in Exercise 58, calculate the depth and level database $\Omega'_{\rm N}$, $\Omega'_{\rm A}$, $\Omega'_{\rm C}$, where $\Omega'_{\rm N}$, $\Omega'_{\rm A}$, $\Omega'_{\rm C}$ is the database obtained by appending the level indices of each amino acid to the original database $\Omega_{\rm N}$, $\Omega_{\rm A}$, $\Omega_{\rm C}$.

Exercise 63. Prove properties 1–4 in Sects. 12.1.2 and 12.1.3.

13

Analysis of the Morphological Features of Protein Spatial Structure

13.1 Introduction

13.1.1 Morphological Features of Protein Spatial Structure

As mentioned in the previous chapter, there are a number of methods for studying protein spatial structure. In this chapter, we discuss the analysis of the morphological features of proteins. The morphological features of proteins include basic morphological features, such as sphericity, cylindricity, coffin shapes, umbrella shapes, etc., and interior features of proteins, such as interior cavities, grooves, channels, etc. In the area of drug design and virus analysis, these features have been redefined. We introduce the relevant issues as follows.

Interaction Between Ligands and Receptors

In the fields of drug design and virus analysis, a key issue is how to determine the interaction between a receptor and a ligand. This issue can be divided into two subissues, namely:

1. Morphological matches between ligands and receptors, such that the ligand can be bound, enter or transit through the receptor, and where these functions take place.
2. Whether they react if the ligand and receptor are matched; that is, whether they can be chemically bonded, and whether the biochemical functions of the protein are affected after they bond.

For the above two issues, the first one is a geometrical issue, while the second is a biological and chemical issue. For the first issue, there is an abundance of data for spatial locations of the atoms comprising the protein. This provides us with the basis for geometrical analysis. In this book we discuss only the first issue.

Geometrical Calculations of Ligands and Receptors

Geometrical calculations are routinely involved in many types of research, especially in the combinatiorial design of the shapes of objects manufactured in industrial production, which has led to the development of an integrated theory and computational algorithm [93, 98]. Many functions of biological macromolecules (such as proteins) relate to their morphology, which leads to research of the morphology of biological macromolecules becoming more and more important.

As we have mentioned above, proteins consist of large numbers of amino acids, and each amino acid consists of many different atoms, so that we may consider a protein as a spatially distributed particle system. This particle system can be described and researched from several angles, such as the secondary structure and spatial structure mentioned in the last three chapters of this book, as well as the morphological features which will be discussed below.

The essential morphological feature of a protein is that the particles which comprise the protein have a large scale and irregular distributions, and the known results are often too imprecise to be readily useful in drug design and virus analysis. Traditional geometric computational methods (such as the spline method, and particle distribution method) are not always effective. Therefore, we must explore new methods. The main goal of this chapter is to study the morphological features of protein spatial structures and to explore new mathematical theories, tools and methods. The involved issues include mathematical and geometric calculation:

1. The spatial polyhedron is a familiar geometrical object. If a polyhedron consists of large numbers of boundary planes, we need to use certain mathematical formulas and computations to describe and construct these polyhedrons. The concept of the hypergraph is the generalization of a graph with points and lines. It uses a family of sets to replace the arcs defined in graph theory. Using the hypergraph, we can describe the spatial polyhedron exactly.
2. In the previous chapter, we have discussed in detail the concept of depth, which is based on geometry. It combines the three concepts of the spatial polyhedron, the hypergraph and depth, which can create an essential tool for the study of a spatial particle system.
3. In the study of protein structures, in order to analyze their morphological features, the method of the small rolling sphere has already been used. We develop this method to a universal geometric calculation method and present several new concepts, such as the γ-accessible radius in a spatial particle system, which provides the biological concepts such as hydrophobicity with more precise mathematical definitions.
4. The question of how to effectively calculate and search the structural features of proteins which have different sizes and morphologies (Figs. 13.1, 13.2) is a new issue in geometrical calculations.

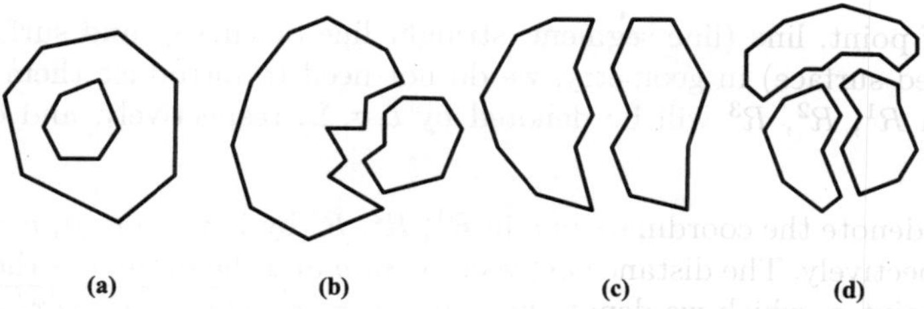

Fig. 13.1a–d. Spatial structure of the protein space

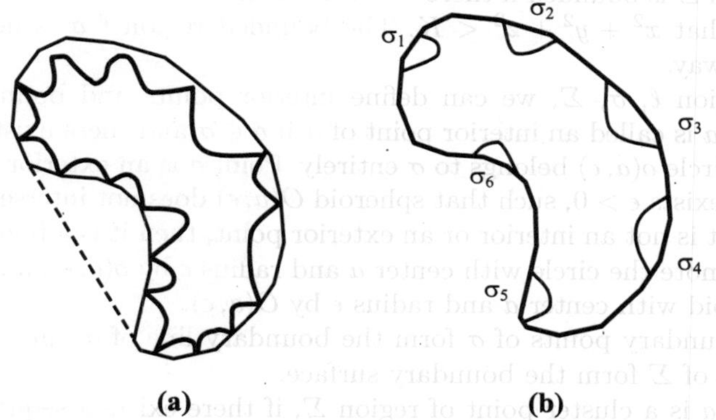

Fig. 13.2a,b. The rugged surface of a protein

Figure 13.1a represents a "cavity" in a protein, Fig. 13.1b represents the relationship between a ligand and a receptor, where the receptor has the structure of a "pocket" into which the ligand can enter, and Fig. 13.1c represents a "channel" in a protein, a tube-shaped structure in space. Figure 13.1d represents a protein consisting of a number of polymers forming autonomous domains.

In Fig. 13.2a, the bold lines represent the actual morphology of the surface of a protein; thin lines and broken lines represent the convex hull of the protein. Figure 13.2b shows grooves on the surface of the protein.

13.1.2 Several Basic Definitions and Symbols

In Chap. 12, we presented the definition and calculation of depth in a particle system, as well as the computation of both the convex hull and level. In this chapter, we add several new definitions and symbols.

Topological Structure of Spatial Set

We denote the one-dimensional, two-dimensional, and three-dimensional Euclidean spaces by R^1, R^2, R^3, respectively. Being very familiar with the concepts of point, line (line segment, straight line or curve), and surface (plane or curved surface) in geometry, we do not need to introduce them here. Regions in R^1, R^2, R^3 will be denoted by ℓ, σ, Σ, respectively, and defined as follows:

1. We denote the coordinate of a in R^1, R^2, R^3 by x, $\boldsymbol{r} = (x, y)$, $\boldsymbol{r} = (x, y, z)$, respectively. The distance between point a and the origin e is the modulus of point a, which we denote by $|ea| = r = |x|$, $\sqrt{x^2 + y^2}$, $\sqrt{x^2 + y^2 + z^2}$, where a is a point in R^1, R^2, R^3, respectively.
   Region Σ is bounded if there exists a constant $K > 0$, for any $(x, y, z) \in \Sigma$ such that $x^2 + y^2 + z^2 < K$. The bounded region ℓ, σ is defined in the same way.
2. In region ℓ, σ, Σ, we can define interior points and boundary points. Point a is called an interior point of σ if $a \in \sigma$ and there exists $\epsilon > 0$ such that circle $o(a, \epsilon)$ belongs to σ entirely. Point a is an exterior point of Σ if there exists $\epsilon > 0$, such that spheroid $O(a, \epsilon)$ does not intersect with Σ. If a point is not an interior or an exterior point, then it is a boundary point. We denote the circle with center a and radius ϵ by $o(a, \epsilon)$, and denote the spheroid with center a and radius ϵ by $O(a, \epsilon)$.
   All boundary points of σ form the boundary line of σ, and all boundary points of Σ form the boundary surface.
3. Point a is a cluster point of region Σ, if there exists a sequence $a_n \in \Sigma$, $n = 1, 2, \cdots$, such that $a_n \to a$ as $n \to \infty$, $a_n \to a$ means $|\boldsymbol{r}_n - \boldsymbol{r}| \to 0$, where $\boldsymbol{r}_n, \boldsymbol{r}$ is the corresponding coordinate vector of points a_n, a.
   Region Σ is closed if all its cluster points are in Σ.
4. Region Σ is connected, if for any two points $a, b \in \Sigma$, they can be connected with a curve C, $C \subset \Sigma$.

Region Σ is nondegenerate, if for any point a in Σ and any $\epsilon > 0$, there always exists a point $b \in O(a, \epsilon)$, such that b is an interior point of Σ.

A bounded, nondegenerate, closed connected region is called a natural region. For the case of a one-dimensional and two-dimensional region ℓ, σ, we can define its bounded, nondegenerate, closed connected and natural regions in a similar way.

Polygons and Polyhedrons

Obviously, the one-dimensional natural region is a line segment. We will now discuss the cases where σ, Σ are two-dimensional and three-dimensional natural regions, respectively:

1. In the two-dimensional natural region σ, if its boundary line consists of several line segments, we call this region a planar polygon.
2. In the three-dimensional natural region Σ, if its boundary surface consists of several planar polygons, we call this region a spatial polyhedron.

Based on Euclidean geometry, we can identify the following properties:

1. For any planar polygon σ, we can decompose it into several triangles $\delta_1, \delta_2, \cdots, \delta_k$. These triangles have no common interior points, and their union is σ. We can easily explain that this decomposition is not unique.
2. For any polyhedron Σ, its boundary surface consists of several polygons, so this boundary surface can be decomposed into several triangles $\delta_1, \delta_2, \cdots, \delta_k$. These triangles have no common interior points, and their union is the corresponding boundary surface. We call these triangles the boundary triangles of the polyhedron. Intersection lines of a noncoplanar boundary triangle are the edges of the polyhedron, and the endpoints of the edges are the vertices of the polyhedron.

Convex Polyhedron and Its Convex Hull

We continue discussing the natural region Σ (σ can be discussed in a similar way):

1. The spatial region Σ is a convex region, if for any $a, b \in \Sigma$, such that $\overline{ab} \subset \Sigma$ holds, where $\overline{ab}$ is a line segment with endpoints a, b.
2. The intersection of any convex regions is still a convex region. The intersection of all convex regions that contain Σ is the convex hull of Σ, denoted by $\Omega(\Sigma)$.
   If Σ is closed, then its convex hull $\Omega(\Sigma)$ must be closed, and we call $\Omega(\Sigma)$ the convex closure of Σ.
   If $\Omega(\Sigma)$ is a polyhedron, then $\Omega(\Sigma)$ is a convex polyhedron.
3. If $A = \{a_1, a_2, \cdots, a_n\}$ is a spatial particle system, the convex hull which is formed by set A is denoted by $\Omega(A)$. Obviously, $\Omega(A)$ is a spatial convex polyhedron.
4. If Ω is a spatial convex polyhedron, then its boundary surface is the convex hull of this polyhedron. This convex hull can be decomposed into several triangles, and these boundary surfaces can generate the edges and vertices of the polyhedron.

Hypergraph Determined by Polyhedron

The concept of the hypergraph is the generalization of a graph with points and lines [31]. If A is a point set, $V = \{V_1, V_2, \cdots, V_k\}$ is a group of subsets, and we call $G = \{A, V\}$ a hypergraph, the elements of A are the vertices of hypergraph G, and the elements of V are the arcs of hypergraph G, or simply the hyperarcs. Obviously, when V_i is a dual set, the hypergraph has become a common graph with points and lines. If we denote $||V_i|| = r$, then we call V_i the r-order arc. We denote all r-order arcs of G by $V^{(r)}$. If $\subset_{r=1}^{p} V^{(r)} = A$, then we call G a p-order hypergraph. Here we denote it by

$$G = \{A, V\} = \left\{A, V^{(1)}, V^{(2)}, \cdots, V^{(p)}\right\} , \tag{13.1}$$

which is called the decomposition of hypergraph G. A spatial polyhedron Ω can be represented by a hypergraph. In this case,

$$G = \left\{A_0, V^{(2)}, V^{(3)}\right\}$$

satisfies:

1. A_0 represents all the vertices of the spatial polyhedron, while $V^{(2)}$, which comprises the whole edges, is set of pair points.
2. $V^{(3)}$ are all the boundary triangles of the spatial polyhedron, where each $V_i = \{a_i, b_i, c_i\}$, $a_i, b_i, c_i \in A_0$ is a third-order subset of A_0.

Definition 43. *If hypergraph $G = \{A, V^{(2)}, V^{(3)}\}$ satisfies the above conditions 1 and 2, then the hypergraph G is determined by the polyhedron Ω. Thus G is a third-order hypergraph.*

An important polyhedron is the convex polyhedron. If $V_i \in V^{(3)}$, and the spatial plane determined by V_i is denoted by π_i, then Ω is a convex polyhedron if and only if for each $V_i \in V^{(3)}$ all points in A_0 are on the same side of π_i (or on π_i).

Symbols Used in this Part

1. δ is a spatial triangle, and $\boldsymbol{o}(\delta)$, $o(\delta), r(\delta)$ represent the circumcircle of triangle δ, circumcenter, and circumradius, respectively.
2. If a, b, c are three points in space, and $\delta(a, b, c)$ is the triangle determined by three vertices a, b, c, the corresponding circumcircle, circumcenter and circumradius are $\boldsymbol{o}(a, b, c)$, $o(a, b, c)$, $r(a, b, c)$, respectively.
3. $\pi(\delta)$ or $\pi(a, b, c)$ represents the plane determined by triangle δ or the three points a, b, c.
4. $O = O(o, r)$ represents a sphere with center o, radius r; $O(r)$ or $O(\epsilon)$ represent a rolling sphere with radius r or ϵ.
5. If $\Delta = \Delta(a, b, c, d)$ represents a tetrahedron with four vertices a, b, c, d, then $O(a, b, c, d)$ represents a circumscribed sphere determined by a, b, c, d, $o(a, b, c, d)$ represents the center of that circumscribed sphere, and $r(a, b, c, d)$ represents the radius of that circumscribed sphere.

13.1.3 Preliminary Analysis of the Morphology of Spatial Particle System

Let $\Omega = \Omega(A) = \Omega(A_0)$ be the convex hull of a spatial particle system A. A_0 is the set of all zero-depth points, and A_0^* is the set of vertices of $\Omega(A)$. These are determined by the method mentioned in Defintion 43. We first conduct a preliminary analysis of the morphology of the spatial particle system.

Basic Morphology of Protein

It is very useful to study the basic morphology of a protein, and its characteristic features include:

1. Long axis. We call ab the long axis of Ω, if $a, b \in A_0$, and for any $a', b' \in A_0$, such that $|a, b| \geq |a', b'|$ holds, where $|a, b|$ is the distance between points a, b. Then $\ell_z = |a, b|$ are the lengths of the particle system. We call the straight line determined by ab straight line OZ.
2. Middle axis. If $c, c' \in A_0$, we denote their projection to the long axis ab by o, o', and we call co the middle axis of Ω, if for any $c' \in A_0$, $|c, o| \geq |c', o'|$ holds. We call $\ell_y = |c, o|$ the width of the particle system. We call point o the center of the particle system. It must be on line segment ab, and the four points o, a, b, c are coplanar, with the plane denoted by π.
3. Normal axis and a natural coordinate system. The normal axis is a straight line through point o and perpendicular to plane π. With o as the origin; long axis, middle axis and normal axis as coordinate axes, and determined by right-hand rules, we find a coordinate system which is the natural coordinate system of the spatial particle system.

From the length and width of a particle system, we can roughly determine the basic morphology of the particle system. If the length and width are similar, this particle system is a sphere. If the length is clearly longer than the width, then the particle system is a cylinder. Upon further analysis, we can also analyze the projection or section of the convex closed hull $\Omega(A)$, which will not be discussed in detail.

Example 29. For the protein E. coli OmpF porin in Sect. 12.2, calculate the basic morphology of the protein:

1. Length of the long axis is $\ell_z = |a, b| = 62.969$, the coordinates of a, b are given as follows:

$$\begin{aligned} a &= (110.318, 53.037, 29.738)\,, \quad b = (55.901, 24.460, 16.054)\,, \\ \boldsymbol{k} &= (-0.864, -0.454, -0.217) \end{aligned}$$

2. Length of the middle axis is $\ell_y = |c, o| = 32.418$, the coordinates of c, o are given as follows:

$$\begin{aligned} c &= (85.685, 39.332, 56.668)\,, \quad o = (91.604, 43.209, 25.032)\,, \\ \mathbf{i} &= (-0.183, -0.120, 0.976)\,. \end{aligned}$$

   Then $\lambda = 0.344$ is the division of point o on line segment ab.
3. Normal axis is $\boldsymbol{j} = \boldsymbol{i} \times \boldsymbol{k} = (0.469, -0.883, -0.021)$.

Therefore, we find the natural coordinate system of the particle system: $\mathcal{E} = \{\boldsymbol{o}, \boldsymbol{i}, \boldsymbol{j}, \boldsymbol{k}\}$, in which, $\boldsymbol{o}$, $\boldsymbol{i}$, $\boldsymbol{j}$ and $\boldsymbol{k}$ are given by the above formulas.

Data Structure of the Circumcircle of the Surface

We have shown that the surface of convex closure in a spatial particle system is a convex hull, formed by a series of triangles. Each triangle has a circumcircle, and each circumcircle has an identified radius and center. These are the preliminary features of the morphology of a protein.

The data for the protein E. coli OmpF porin are listed in Tables 12.6 and 12.7, where the coordinates of the C_α atoms in the amino acids of the protein are listed in Table 12.6. Using this data, we can calculate their circumradii and the coordinates of the circumcenter. The result is shown as follows.

13.1.4 Example

In Table 12.6, we present the depth and level of each C_α atom in the protein E. coli OmpF porin. Then, the spatial particle system A is divided into eight levels, which we denote by A_k, $k = 0, 1, \cdots, 7$. Let $\Omega(A_k)$ denote the convex closure of A_k, then all boundary triangles of $\Omega(A_k)$, which we denote by $V_k^{(3)}$, form the convex hulls of different levels. We list them in Table 13.1.

C_α atoms in this protein can be divided into eight levels. The number of atoms in each level and the number of boundary triangles are given in Table 13.1.

Boundary triangles on each level are shown in Table 13.2.

Remark 23. Each lattice in Table 13.2 represents the data of a boundary triangle on the same level. (i, j, k) represents the serial number of the C_α atom of vertices of triangles on the convex hull, and the amino acids denoted by serial number (i, j, k) are given in Table 12.6.

Table 13.1. Distributed table of the number of atoms in the convex hull and the number of boundary triangles (NBT)

Level	0	1	2	3	4	5	6	7	Total
No. of atoms	51	66	69	65	43	20	17	9	340
NBT	98	128	134	126	86	36	30	12	646

NBT = number of boundary triangles

Table 13.2. Data structure of convex hulls on each level

No.	(i,j,k)	No.	(i,j,k)	No.	(i,j,k)	No.	(i,j,k)	No.	(i,j,k)
1	3,6,71	2	3,6,304	3	3,71,303	4	3,303,304	5	6,7,72
6	6,7,266	7	6,71,72	8	6,266,267	9	6,267,304	10	7,51,52
11	7,51,162	12	7,52,53	13	7,53,266	14	7,72,73	15	7,73,164
16	7,162,164	17	25,26,267	18	25,26,320	19	25,267,315	20	25,315,319
21	25,319,320	22	26,70,71	23	26,70,303	24	26,71,72	25	26,72,244
26	26,244,321	27	26,267,303	28	26,320,321	29	51,52,172	30	51,160,162
31	51,160,172	32	52,53,93	33	52,93,172	34	53,93,147	35	53,147,183
36	53,183,266	37	70,71,303	38	72,73,245	39	72,244,245	40	73,164,202
41	73,202,245	42	93,147,209	43	93,172,209	44	147,148,183	45	147,148,282
46	147,208,209	47	147,208,282	48	148,182,183	49	148,182,187	50	148,187,282
51	160,162,198	52	160,172,198	53	162,164,198	54	164,198,202	55	172,198,209
56	182,183,187	57	183,187,218	58	183,218,288	59	183,220,223	60	183,220,288
61	183,223,266	62	187,218,288	63	187,282,287	64	187,287,288	65	198,202,209
66	202,207,208	67	202,207,285	68	202,208,209	69	202,245,285	70	207,208,285
71	208,282,285	72	220,223,261	73	220,261,288	74	223,225,261	75	223,225,263
76	223,263,265	77	223,265,266	78	225,261,263	79	244,245,321	80	245,285,321
81	261,263,273	82	261,273,317	83	261,288,317	84	263,265,273	85	265,266,267
86	265,267,297	87	265,273,297	88	267,297,315	89	267,303,304	90	273,297,315
91	273,315,317	92	282,285,287	93	285,287,321	94	287,288,321	95	288,317,319
96	288,319,321	97	315,317,319	98	319,320,321				
1	3,6,7	2	3,6,19	3	3,7,25	4	3,25,162	5	3,162,69
6	3,19,23	7	3,23,24	8	3,24,69	9	6,7,285	10	6,285,315
11	6,315,19	12	7,25,147	13	7,72,93	14	7,72,147	15	7,73,93
16	7,73,183	17	7,183,187	18	7,187,282	19	7,282,285	20	25,147,209
21	25,162,209	22	26,71,148	23	26,71,57	24	26,148,69	25	26,24,57
26	26,24,69	27	51,52,53	28	51,52,317	29	51,53,70	30	51,70,71
31	51,71,57	32	51,317,9	33	51,9,57	34	52,53,54	35	52,317,55
36	52,54,55	37	53,70,71	38	53,71,148	39	53,148,160	40	53,160,297
41	53,297,50	42	53,37,50	43	53,37,54	44	72,93,147	45	73,93,182
46	73,182,183	47	93,147,207	48	93,164,182	49	93,164,218	50	93,202,207
51	93,202,218	52	147,207,208	53	147,208,209	54	148,160,162	55	148,162,69
56	160,162,209	57	160,209,265	58	160,265,297	59	164,182,218	60	172,182,183
61	172,182,273	62	172,183,187	63	172,187,198	64	172,198,223	65	172,223,245
66	172,245,273	67	182,218,220	68	182,220,261	69	182,261,273	70	187,198,223
71	187,223,244	72	187,225,244	73	187,225,282	74	202,207,218	75	207,208,263
76	207,218,261	77	207,261,263	78	208,209,263	79	209,263,265	80	218,220,261
81	223,244,245	82	225,244,282	83	244,245,303	84	244,282,285	85	244,285,287
86	244,287,288	87	244,288,303	88	245,273,321	89	245,303,321	90	261,263,266
91	261,266,273	92	263,265,266	93	265,266,267	94	265,267,297	95	266,267,4
96	266,273,4	97	267,297,4	98	273,321,4	99	285,287,8	100	285,315,317
101	285,317,319	102	285,319,8	103	287,288,304	104	287,304,320	105	287,320,8
106	288,303,304	107	297,4,5	108	297,5,50	109	303,304,320	110	303,320,321
111	315,317,9	112	315,9,57	113	315,19,23	114	315,23,24	115	315,24,57
116	317,319,55	117	319,8,55	118	320,321,5	119	320,5,8	120	321,4,5
121	5,8,36	122	5,36,37	123	5,37,50	124	8,27,36	125	8,27,54
126	8,54,55	127	27,36,54	128	36,37,54				

Table 13.2. (continued)

No.	(i,j,k)	No.	(i,j,k)	No.	(i,j,k)	No.	(i,j,k)	No.	(i,j,k)
1	3,6,7	2	3,6,164	3	3,7,25	4	3,25,93	5	3,26,93
6	3,26,90	7	3,164,319	8	3,319,27	9	3,27,90	10	6,7,72
11	6,72,164	12	7,25,148	13	7,72,148	14	25,93,148	15	26,93,147
16	26,147,37	17	26,36,37	18	26,36,90	19	51,52,71	20	51,52,75
21	51,71,147	22	51,147,37	23	51,37,75	24	52,53,70	25	52,53,147
26	52,70,69	27	52,71,147	28	52,69,74	29	52,74,75	30	53,70,288
31	53,147,148	32	53,148,288	33	70,288,57	34	70,57,69	35	72,73,162
36	72,73,164	37	72,148,160	38	72,160,162	39	73,162,164	40	93,147,148
41	148,160,182	42	148,182,202	43	148,202,297	44	148,288,297	45	160,162,182
46	162,164,207	47	162,182,207	48	164,172,183	49	164,172,207	50	164,183,187
51	164,187,319	52	172,183,208	53	172,198,208	54	172,198,223	55	172,207,223
56	182,202,207	57	183,187,208	58	187,208,265	59	187,209,218	60	187,209,267
61	187,218,265	62	187,267,319	63	198,208,220	64	198,220,223	65	202,207,244
66	202,244,245	67	202,245,297	68	207,223,225	69	207,225,245	70	207,244,245
71	208,220,287	72	208,265,285	73	208,285,304	74	208,287,304	75	209,218,267
76	218,265,282	77	218,266,267	78	218,266,282	79	220,223,287	80	223,225,287
81	225,245,263	82	225,263,287	83	245,261,297	84	245,261,5	85	245,263,5
86	261,297,5	87	263,287,303	88	263,303,5	89	265,282,285	90	266,267,273
91	266,273,317	92	266,282,315	93	266,315,317	94	267,273,319	95	273,317,319
96	282,285,304	97	282,304,8	98	282,315,8	99	287,303,304	100	288,297,5
101	288,5,55	102	288,55,57	103	303,304,5	104	304,5,8	105	315,317,4
106	315,4,9	107	315,8,9	108	317,319,321	109	317,321,19	110	317,4,19
111	319,320,24	112	319,320,27	113	319,321,23	114	319,23,24	115	320,24,27
116	321,19,23	117	4,9,19	118	5,8,57	119	5,55,57	120	8,9,69
121	8,57,69	122	9,19,69	123	19,23,54	124	19,54,69	125	23,24,54
126	24,27,37	127	24,37,50	128	24,50,54	129	27,36,37	130	27,36,90
131	37,50,75	132	50,54,74	133	50,74,75	134	54,69,74		
1	3,6,70	2	3,6,73	3	3,53,147	4	3,53,57	5	3,70,147
6	3,73,208	7	3,208,23	8	3,23,57	9	6,70,71	10	6,71,72
11	6,72,73	12	7,25,52	13	7,25,54	14	7,52,55	15	7,54,55
16	25,26,51	17	25,26,50	18	25,51,160	19	25,52,160	20	25,24,50
21	25,24,54	22	26,51,160	23	26,160,304	24	26,304,37	25	26,36,37
26	26,36,50	27	52,53,148	28	52,53,57	29	52,148,160	30	52,55,57
31	53,147,148	32	70,71,160	33	70,147,160	34	71,72,160	35	72,73,93
36	72,93,160	37	73,93,162	38	73,162,164	39	73,164,208	40	93,160,187
41	93,162,187	42	147,148,160	43	160,187,225	44	160,223,225	45	160,223,315
46	160,304,315	47	162,164,182	48	162,182,198	49	162,183,187	50	162,183,198
51	164,172,182	52	164,172,208	53	172,182,198	54	172,198,202	55	172,202,245
56	172,207,208	57	172,207,261	58	172,245,261	59	183,187,225	60	183,198,244
61	183,225,244	62	198,202,220	63	198,220,244	64	202,220,245	65	207,208,209
66	207,209,218	67	207,218,261	68	208,209,263	69	208,263,285	70	208,285,23
71	209,218,263	72	218,261,266	73	218,263,266	74	220,244,267	75	220,245,267
76	223,225,244	77	223,244,267	78	223,267,317	79	223,315,317	80	245,261,273
81	245,267,273	82	261,266,273	83	263,265,266	84	263,265,285	85	265,266,288
86	265,282,285	87	265,282,288	88	266,273,288	89	267,273,303	90	267,303,317

Table 13.2. (continued)

No.	(i,j,k)	No.	(i,j,k)	No.	(i,j,k)	No.	(i,j,k)	No.	(i,j,k)
91	273,288,297	92	273,297,320	93	273,303,320	94	282,285,287	95	282,287,288
96	285,287,5	97	285,321,4	98	285,321,23	99	285,4,5	100	287,288,5
101	288,297,320	102	288,320,5	103	303,304,317	104	303,304,319	105	303,319,8
106	303,320,8	107	304,315,317	108	304,319,8	109	304,8,9	110	304,9,37
111	320,5,8	112	321,4,19	113	321,19,36	114	321,23,36	115	4,5,9
116	4,9,19	117	5,8,9	118	9,19,37	119	19,36,37	120	23,24,27
121	23,24,55	122	23,27,36	123	23,55,57	124	24,27,50	125	24,54,55
126	27,36,50								
1	3,6,7	2	3,6,73	3	3,7,183	4	3,53,70	5	3,53,73
6	3,70,160	7	3,160,164	8	3,164,172	9	3,172,183	10	6,7,25
11	6,25,52	12	6,52,73	13	7,25,183	14	25,51,52	15	25,51,288
16	25,183,288	17	26,51,287	18	26,51,288	19	26,273,287	20	26,273,288
21	51,52,73	22	51,72,73	23	51,72,287	24	53,70,71	25	53,71,73
26	70,71,148	27	70,148,162	28	70,160,162	29	71,73,147	30	71,147,148
31	72,73,218	32	72,209,218	33	72,209,263	34	72,263,287	35	73,93,147
36	73,93,187	37	73,187,218	38	93,147,187	39	147,148,187	40	148,162,187
41	160,162,208	42	160,164,202	43	160,202,207	44	160,207,208	45	162,187,198
46	162,198,208	47	164,172,182	48	164,182,267	49	164,202,220	50	164,220,261
51	164,261,267	52	172,182,183	53	182,183,273	54	182,267,273	55	183,273,288
56	187,198,218	57	198,208,225	58	198,209,218	59	198,209,265	60	198,225,265
61	202,207,220	62	207,208,225	63	207,220,225	64	209,263,265	65	220,223,225
66	220,223,261	67	223,225,245	68	223,245,261	69	225,244,245	70	225,244,265
71	244,245,265	72	245,261,266	73	245,265,266	74	261,266,267	75	263,265,287
76	265,266,282	77	265,282,285	78	265,285,287	79	266,267,273	80	266,273,282
81	273,282,287	82	282,285,287						
1	3,6,51	2	3,6,93	3	3,51,52	4	3,52,172	5	3,93,182
6	3,172,182	7	6,7,26	8	6,7,93	9	6,26,52	10	6,51,52
11	7,25,26	12	7,25,160	13	7,71,72	14	7,71,160	15	7,72,73
16	7,73,93	17	25,26,52	18	25,52,53	19	25,53,160	20	52,53,147
21	52,147,172	22	53,147,148	23	53,148,160	24	70,72,160	25	70,72,162
26	70,160,162	27	71,72,160	28	72,73,182	29	72,162,182	30	73,93,182
31	147,148,162	32	147,162,164	33	147,164,172	34	148,160,162	35	162,164,172
36	162,172,182								
1	3,6,7	2	3,6,70	3	3,7,25	4	3,25,26	5	3,26,52
6	3,52,53	7	3,53,70	8	6,7,160	9	6,70,93	10	6,93,147
11	6,147,160	12	7,25,160	13	25,26,160	14	26,51,52	15	26,51,162
16	26,148,160	17	26,148,162	18	51,52,53	19	51,53,71	20	51,71,73
21	51,73,162	22	53,70,71	23	70,71,72	24	70,72,162	25	70,93,162
26	71,72,162	27	71,73,162	28	93,147,162	29	147,148,160	30	147,148,162
1	3,6,7	2	3,6,70	3	3,7,25	4	3,25,70	5	6,7,52
6	6,51,52	7	6,51,70	8	7,25,52	9	25,26,51	10	25,26,52
11	25,51,70	12	26,51,52						

13.2 Structural Analysis of Cavities and Channels in a Particle System

A cavity is one of the basic features of the morphology of a particle system, which we will discuss first. The main method of calculation is the rolling sphere method.

13.2.1 Definition, Classification and Calculation of Cavity

For a fixed spatial particle system A, its convex closure is $\Omega(A)$. We denote the sphere with center o and radius r by $O(o, r)$, and denote a rolling sphere with radius r by $O(r)$.

Definition of Empty Sphere

Definition 44. *1. If $O(r) \cap \Omega(A)$ is not an empty set, and $O(r)$ does not contain any points of A, then $O(r)$ is a hollow (or empty) sphere with radius r in particle system A.*

2. If a, b, c, d are four points in A, their circumscribed sphere is $O(a, b, c, d)$, If $O(a, b, c, d)$ is an empty sphere in particle system A, then we call $O(a, b, c, d)$ an empty sphere generated by the four points a, b, c, d.

Classification of Cavities

There are many methods for classifying the relationship between a cavity and the particle system A. First, we present the following four types of relationships between $O(r)$ and $\Omega(A)$:

1. Sphere $O(r)$ is entirely in $\Omega(A)$, we call it an I-0-class cavity of particle system A. In an I-0 class cavity, the cavity with the largest radius is the largest cavity in particle system A.
2. If sphere $O(r)$ is partly in $\Omega(A)$, then sphere $O(r)$ must intercept several boundary surfaces of $\Omega(A)$. If there exists a boundary surface π that intercepts most of the sphere $O(r)$ on the same side of $\Omega(A)$, then we call this cavity a I-1-class cavity of particle system A, otherwise, we call it a I-2-class cavity. I-1 and I-2 class cavities are also called spheroidal grooves of particle system A.
3. In addition, we can provide a classification using circumscribed sphere $O(a, b, c, d)$. Thus, the relationship between $a, b, c, d \in A$ and set A_0 is stated as follows: For $a, b, c, d \in A$, if there exist zero, one, two, three, or four points in set A_0, then the corresponding $O(a, b, c, d)$ is a II-0, II-1, II-2, II-3, or II-4 class sphere, respectively. Since we do not define the concept of II-4 class spheres in this chapter, we only discuss the first four types of cavities. For a II-0 class cavity, we can define the largest cavity of the particle system A similarly.

4. If sphere O is both a I-1-class and a II-3-class sphere, then a boundary triangle $\delta = \delta(a, b, c)$ of Ω must be on the surface of the sphere. We call O a I-1(r)-class sphere, where $r = r(\delta)$ is the circumradius of the boundary triangle δ.

Search and Calculation of a Cavity

Searching and calculating the cavity can be executed in two steps, namely:

1. For any four points a, b, c, d in set A, we can determine a circumscribed sphere $O(a, b, c, d)$ by the four-point-determined-sphere method, with the corresponding formulas given in Sect. 13.5. We denote the center and radius of the sphere by $o = o(a, b, c, d)$, $r = r(a, b, c, d)$, respectively.
2. Let $A(a, b, c, d) = A - \{a, b, c, d\}$, if for any $e \in A(a, b, c, d)$, e is not in sphere $O(a, b, c, d)$, then $O(a, b, c, d)$ is a cavity of A.
3. For a protein with length n, the computational complexity of finding all its cavities is n^5. Hence it is easily computed. For a protein with length 400, it takes 5 h to compute all the cavities on a 1 GHz PC. If we develop a proper algorithm, the computational complexity can be reduced. We do not discuss this further here.

Remark 24. The left object and the right object in Fig. 13.3 are the spatial structure of yeast hexokinase PII and E. coli OmpF porin, respectively. Each small sphere represents an atom; the two larger spheres are the "pocket" and the "center cavity."

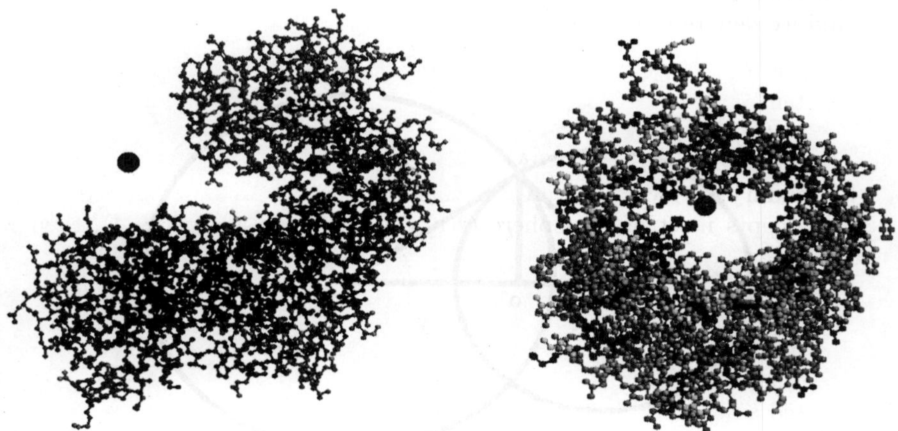

Fig. 13.3. A protein's "cavity" and the "pockets" of the surface

13.2.2 Relationship Between Cavities

Connected Cavities

If two known cavities $O(r)$, $O'(r')$ intersect, then they are called connected cavities. The intersected curve of the surfaces of connected cavities $O(r)$, $O'(r')$ must be a circle. Below, we provide the following discussion.

1. Let o, o' be the centers of cavities $O(r)$, $O'(r')$, respectively. We denote the distance between the two centers by $h = |oo'|$, and denote a point on the intersected curve of the surface of the cavities by a. Point a' is the projection of point a on line oo'. Thus, a' divides the line oo' into two parts: oa', $a'o'$, whose lengths are x, y, respectively. Their geometrical relationship and computational result are shown in Fig. 13.4.
2. x, y satisfy equations:

$$\begin{cases} x + y = h \,, \\ r^2 - x^2 = (r')^2 - y^2 \,. \end{cases}$$

Solving these equations, we have that

$$\begin{cases} x = \dfrac{h}{2} + \dfrac{r^2 - (r')^2}{2h} \,, \\ y = \dfrac{h}{2} - \dfrac{r^2 - (r')^2}{2h} \,. \end{cases} \tag{13.2}$$

Therefore, the values of x, y are independent of the selection of point a, and we denote a' by o''.

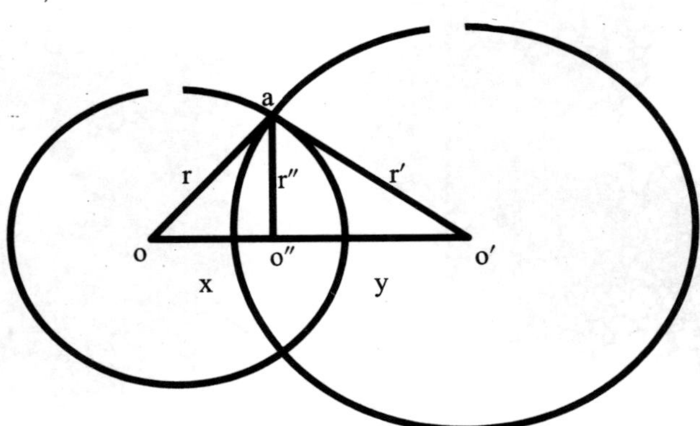

Fig. 13.4. Computation map of the radius of section between connected spheres

3. Let the intersected curve of the surface of $O(r)$, $O'(r')$ be L, then the distance between any point a on L and o'' is computed as follows:

$$
\begin{aligned}
r'' = (r^2 - x^2)^{1/2} &= \left[r^2 - \left(\frac{h}{2} + \frac{r^2 - (r')^2}{2h}\right)^2\right]^{1/2} \\
&= \frac{1}{2}\left\{2[r^2 + (r')^2] - h^2 \frac{r^4 - 2r^2(r')^2 + (r')^4}{h^2}\right\}^{1/2} . \qquad (13.3)
\end{aligned}
$$

Obviously, the values of r'' and point a are independent. So the intersected curve of the surface of $O(r)$, $O'(r')$, which we denoted by L, is a circle with center o'' and radius r''. The corresponding formulas are presented in (13.2) and (13.3). We call r'' in (13.3) the radius of the section between the two connected spheres.

Basic Properties of Connected Empty Spheres

If $O(r_1)$, $O'(r_2)$ are two connected empty spheres of set A, and r'' is their radius of section, then we say that the two empty spheres form a sphere pair, denoted by (r_1, r_2, r''). The following properties hold:

1. $r'' \leq \min\{r_1, r_2\}$.
2. For any sphere $O(r)$ such that $r < r''$ holds, the sphere can move freely within the connected empty spheres without encountering any points of set A.
3. If $O(a,b,c,d)$, $O'(a',b',c',d')$ are two connected empty spheres, then a,b,c,d,a',b',c',d' cannot be interior points of region

$$O(a,b,c,d) \cup O'(a',b',c',d') .$$

Connected Empty Spheres Map Generated by Particle System

For a fixed particle system A, we denote its entire connected empty spheres map by $\mathcal{G}_A = \{\mathcal{O}_A, \mathcal{V}_A, r_O, r_{(O,O')}\}$, where

$$
\begin{cases}
\mathcal{O}_A = \{O(a,b,c,d)\colon\ a,b,c,d \in A, \text{ and } O(a,b,c,d) \text{ is empty sphere of A}\} , \\
\mathcal{V}_A = \{(O,O')\colon\ O,\ O' \in \mathcal{Q}_A, \text{ and } (O,O') \text{ is connected sphere of A}\} ,
\end{cases} \qquad (13.4)
$$

where r_O is the radius of sphere O, and $r_{O,O'}$ is the radius of the section of connected empty spheres (O,O'). Then $\mathcal{G}_A$ is the connected empty sphere map generated by particle system A, where $\mathcal{O}_A$ contains all the vertices of the map, and $\mathcal{V}_A$ contains all the arcs of the map, while r_O, $r_{(O,O')}$ are the coloring functions of the vertices and arcs, respectively.

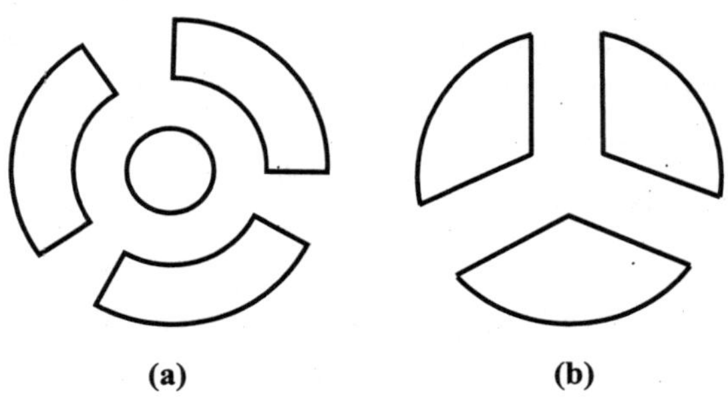

Fig. 13.5. Channel network map

The Threshold Subgraph

The threshold subgraph can be determined by the connected empty spheres map $\mathcal{G}_A$. If we let r^* be a fixed constant, we define the threshold subgraph of $\mathcal{G}_A$ by $\mathcal{G}_A(r^*) = \{\mathcal{O}_A(r^*), \mathcal{V}_A(r^*), r_O, r_{(O,O')}\}$, where

$$\begin{cases} \mathcal{O}_A(r^*) = \{O \in \mathcal{O}_A \colon r_O \geq r^*\} \ , \\ \mathcal{V}_A(r^*) = \{(O, O') \in \mathcal{V}_A \colon r_{O,O'} \geq r^*\} \ . \end{cases} \tag{13.5}$$

Therefore, in graph $\mathcal{G}_A(r^*)$, $r_O, r_{(O,O')} \geq r^*$ always holds.

Definition 45. *In the threshold subgraph of the connected empty sphere $\mathcal{O}_A(r^*)$, the path is a channel in particle system A.*

If L is a path in graph $\mathcal{G}_A(r^*)$, then for any sphere $O(r)$ such that $r < r''$ holds, the sphere can move freely through the path without encountering any points of set A.

To analyze a channel, we use the same definitions and symbols as in graph theory, such as the start vertex and end vertex, cycle, and so on. Several special definitions are listed as follows:

1. Channel L is connected with the exterior of particle system A, if the beginning sphere O of channel L is connected with another sphere O', the radius of sphere O' and the connected radius are $r_O, r_{OO'} \geq r^*$, and if sphere $O' = O(\delta)$ is a circumscribed sphere of a boundary triangle of $\Omega(A)$. Then we also say that one end of the channel is connected with the exterior of particle system A.
2. Channel L is penetrable according to particle system A, if both endpoints of channel L are connected with the exterior of particle system A.
3. If two channels L_1, L_2 have common points, then we say that they cross. Several crossed channels compose a channel network. See Fig. 13.5 for an illustration. Finding and characterizing a channel network is simply grid computing in graph theory, which we will not discuss here.

13.2.3 Example

We analyze the empty cavity and channel of protein E. coli OmpF porin. The data of the spatial structure of C_α atoms are given in Table 12.6. We continue our discussion based on these data:

1. In particle system A, the total number of cavities is 2033, where the radii range from 20.0 to 60.0, and the number of cavities whose four vertices are all in $\Omega(A)$ is 42. These are shown in Table 13.3.
2. Among these 42 cavities, let $\mathcal{G}_A(r^*) = \{\mathcal{O}_A(r^*), \mathcal{V}_A(r^*), r_O, r_{(O,O')}\}$ denote the channel network map constructed by the connected sphere pairs such that the radius of section $r^* > 20$ holds, where $\mathcal{O}_A(r^*), r_O$ are the parameters given in Table 13.3, and $\mathcal{V}(r^*)$, $r_{O,O'}$ are given in Table 13.4:
3. From Table 13.4, we can identify a series of channels as follows:

$$35 \to 26 \to 23 \to 24 \to 2 \to 25 \to 31 \to 22 \to 29$$
$$15 \to 24 \to 23 \to 26 \to 35 \to 22 \to 31 \to 25 \to 12 \ ,$$

   and so on. A small sphere with radius $r < 20$ can move through these channels freely, without encountering any points in set A.
   If these channels are connected with the circumscribed sphere of the boundary surface, then they are penetratable in convex closure $\Omega(A)$. We do not discuss this further at this point.

13.3 Analysis of γ-Accessible Radius in Spatial Particle System

13.3.1 Structural Analysis of a Directed Polyhedron

In order to calculate and analyze spatial particle systems, we will first introduce the theory of the structure of a directed polyhedron. In Sect. 13.1, we have introduced the hypergraph which is used to represent the spatial polyhedron; now we introduce the structure of the directed polyhedron.

In differential geometry, the theory of exterior differentiation is used to describe the direction of boundary surfaces of spatial regions. We do not use the theory here, but instead focus on the structure of the polyhedron. We explain using the structure of the directed polyhedron as follows.

Directed Triangle

The triangle δ which has been given a normal direction is a directed triangle, denoted by $\boldsymbol{\delta}$. It divides the space into two sides: inside and outside. The direction of δ of the boundary triangle on a convex polyhedron is defined: such that one side, in which set Ω lies, is the inside, and the other is the outside.

Table 13.3. Data table of cavities of orotein E. coli OmpF porin (vertices in convex hull)

No.	No. of 4 vertices	X, Y, Z; R	No.	No. of 4 vertices	X, Y, Z; R
1	4, 11, 13, 45	115.18, 32.45, 19.63; 20.54	2	4, 11, 45, 61	119.73, 30.42, 13.42; 28.04
3	4, 13, 44, 45	122.11, 26.09, 13.73; 31.34	4	4, 13, 44, 76	122.23, 25.96, 13.69; 31.50
5	4, 44, 45, 76	122.26, 26.01, 13.54; 31.58	6	4, 45, 61, 76	122.65, 26.15, 12.66; 32.26
7	9, 11, 59, 61	119.16, 33.88, 10.63; 28.08	8	9, 49, 59, 86	113.74, 42.72, 10.74; 21.32
9	13, 43, 44, 76	121.95, 25.89, 14.07; 31.10	10	21, 23, 36, 37	90.56, −1.11, 59.52; 29.57
11	21, 23, 334, 336	99.39, −15.77, 84.92; 59.45	12	43, 44, 61, 76	116.15, 26.83, 15.55; 25.49
13	43, 61, 63, 76	112.42, 27.25, 16.81; 21.86	14	44, 45, 61, 76	121.76, 26.22, 13.29; 31.24
15	59, 61, 84, 86	110.88, 35.85, 10.64; 21.22	16	61, 63, 76, 81	110.42, 27.58, 15.96; 20.51
17	61, 76, 81, 84	116.98, 27.51, 4.12; 31.87	18	61, 76, 84, 86	132.73, 26.16, −7.41; 50.87
19	76, 81, 84, 163	114.51, 26.82, −1.88; 34.12	20	81, 84, 99, 163	102.53, 30.54, 5.79; 20.33
21	84, 86, 99, 163	128.92, 25.99, −13.94; 52.18	22	86, 98, 99, 135	114.56, 42.74, −15.01; 40.44
23	88, 96, 98, 137	97.66, 56.69, −1.09; 22.53	24	88, 96, 137, 159	101.80, 71.54, −25.39; 50.32
25	88, 98, 137, 159	103.62, 60.16, −15.30; 37.95	26	144, 145, 155, 174	65.96, 80.47, −4.42; 39.16
27	144, 155, 174, 176	61.79, 74.35, 4.18; 32.12	28	144, 174, 176, 191	48.40, 82.52, −0.52; 47.19
29	144, 174, 191, 193	43.43, 87.07, −4.03; 54.65	30	144, 176, 178, 191	50.51, 79.98, 2.52; 42.90
31	173, 174, 195, 210	64.29, 65.08, −10.47; 32.56	32	189, 191, 210, 235	27.11, 79.97, 4.33; 57.59
33	189, 191, 214, 233	51.59, 58.96, 17.21; 23.43	34	189, 191, 233, 235	34.85, 72.24, 8.94; 45.97
35	189, 216, 231, 233	37.17, 65.04, 24.99; 36.92	36	189, 231, 233, 255	28.19, 70.63, 24.80; 47.39
37	191, 193, 210, 212	52.49, 60.41, 9.88; 25.65	38	191, 210, 212, 235	47.21, 62.85, 9.82; 30.96
39	191, 212, 233, 235	54.27, 55.35, 14.99; 20.03	40	216, 231, 232, 233	50.92, 55.71, 25.51; 20.55
41	233, 253, 255, 281	46.96, 49.57, 20.22; 20.41	42	301, 309, 311, 334	85.39, 9.94, 75.43; 34.40

The amino acids represented by the serial number of four vertices, as well as its coordinates, are shown in Table 12.6. $X, Y, Z; R$ the coordinates of the circumcenter and circumradius

Table 13.4. Data table of connected spheres of protein E. coli OmpF porin

Connected sphere	$r_O, r_{O'}, r_{O,O''}$	Connected sphere	$r_O, r_{O'}, r_{O,O''}$	Connected sphere	$r_O, r_{O'}, r_{O,O''}$
(2, 24)	28.04, 50.32, 23.73	(2, 25)	28.04, 37.95, 23.81	(3, 24)	31.34, 50.32, 24.60
(3, 25)	31.34, 37.95, 24.54	(4, 24)	31.50, 50.32, 24.69	(4, 25)	31.50, 37.95, 24.62
(5, 24)	31.58, 50.32, 24.82	(5, 25)	31.58, 37.95, 24.73	(6, 24)	32.26, 50.32, 25.69
(6, 25)	32.26, 37.95, 25.46	(7, 24)	28.08, 50.32, 25.58	(7, 25)	28.08, 37.95, 25.44
(9, 24)	31.10, 50.32, 24.23	(9, 25)	31.10, 37.95, 24.23	(12, 25)	25.49, 37.95, 20.49
(14, 24)	31.24, 50.32, 24.78	(14, 25)	31.24, 37.95, 24.70	(15, 24)	21.22, 50.32, 20.48
(17, 24)	31.87, 50.32, 28.72	(17, 25)	31.87, 37.95, 28.19	(18, 24)	50.87, 50.32, 41.53
(19, 23)	34.12, 22.53, 21.22	(19, 24)	34.12, 50.32, 31.58	(21, 24)	52.18, 50.32, 43.47
(22, 26)	40.44, 39.16, 24.68	(22, 29)	40.44, 54.65, 21.46	(22, 31)	40.44, 32.56, 23.65
(23, 26)	22.53, 39.16, 21.42	(23, 31)	22.53, 32.56, 20.15	(24, 28)	50.32, 47.19, 38.42
(24, 29)	50.32, 54.65, 41.48	(24, 30)	50.32, 42.90, 35.81	(24, 32)	50.32, 57.59, 35.61
(24, 34)	50.32, 45.97, 30.00	(24, 36)	50.32, 47.39, 20.06	(24, 37)	50.32, 25.65, 20.35
(24, 38)	50.32, 30.96, 22.97	(25, 26)	37.95, 39.16, 31.60	(25, 27)	37.95, 32.12, 25.21
(25, 28)	37.95, 47.19, 29.17	(25, 29)	37.95, 54.65, 31.00	(25, 30)	37.95, 42.90, 27.32
(25, 31)	37.95, 32.56, 28.79	(25, 32)	37.95, 57.59, 24.25	(26, 35)	39.16, 36.92, 31.00
(26, 36)	39.16, 47.39, 35.24	(31, 35)	32.56, 36.92, 26.49	(31, 36)	32.56, 47.39, 29.53

The spheres represented by the serial numbers of connected spheres are given in Table 13.3, and $r_O, r_{O'}, r_{O,O''}$ represent the radius and the connected radius, respectively

Definition 46. *1. Plane π is called a directed plane, if this plane divides the space into two sides: the inside and the outside. We denote the directed plane by $\boldsymbol{\pi}$.*

2. *Triangle δ is called a directed triangle, if the plane $\pi(\delta)$ determined by this triangle is a directed plane. We denote a directed triangle by $\boldsymbol{\delta}$. The direction of a directed triangle $\boldsymbol{\delta}$ is defined by the normal line through point $o(\delta)$, where its orientation is from outside to inside.*

Directed Polyhedron

In Sect. 13.1, we have introduced the bounded, closed, connective and nondegenerate properties of a spatial polyhedron Ω, and its representation using a hypergraph $G = \{A_0, V^{(3)}\}$, where A_0 contains all the vertices of the spatial polyhedron Ω, and $V^{(3)}$ contains all the boundary triangles. We now discuss the direction of the boundary triangle $\delta \in V^{(3)}$ as follows.

Theorem 42. *Let Ω be a spatial polyhedron, δ be one of its boundary triangles, and $o(\delta)$ its circumcenter. Let $\ell = \ell(\delta)$ be a line through point $o = o(\delta)$ and perpendicular to triangle δ; then point o divides line ℓ into two rays: ℓ_1 and ℓ_2, and we have the following propositions:*

1. *There exists a point a on ray ℓ_1 or ℓ_2 such that the tetrahedron $\Delta(\delta, a) \subset \Omega$.*
2. *Without loss of generality, we suppose that point a in proposition (1) is on ℓ_1; then for any points $b \neq o$ on ℓ_2, tetrahedron $\Delta(\delta, b)$ must not be in Ω.*

Proposition 1 can be proved by the nondegenerate condition of the polyhedron Ω, and proposition 2 can be proved by the condition that δ is a boundary triangle. We do not detail these proofs here.

Definition 47. *The spatial polyhedron Ω is a directed polyhedron, if each boundary triangle of $V^{(3)}$ is a directed triangle, and its direction is defined as follows: The inside is the side such that the tetrahedron $\Delta(\delta, a) \subset \Omega$ in Theorem 42, and the outside is then the other side. Here, we denote by $\boldsymbol{G} = \{A, \boldsymbol{V}^{(3)}\}$ the directed hypergraph, where each $\boldsymbol{\delta} \in \boldsymbol{V}^{(3)}$ is a directed triangle.*

If a directed polyhedron is a convex polyhedron, then we call it a convex directed polyhedron.

Envelope and Retraction of a Directed Polyhedron

Definition 48. *A directed polyhedron $\boldsymbol{G}' = \{A', \boldsymbol{U}^{(3)}\}$ is the envelope of another directed polyhedron $\boldsymbol{G} = \{A, \boldsymbol{V}^{(3)}\}$, if it satisfies the following:*

1. *Both $U^{(3)}, V^{(3)}$ are sets of boundary triangles, where the triangles in them are either completely coincident or have no common interior points of plane. We denote the set of coincident triangles by $V_0^{(3)}$.*

2. *If $\delta \in V^{(3)} - V_0^{(3)}$, a, b, c are three vertices of triangle δ, then there always exists a point $d \in A'$, such that the triangle $\delta_1 = \delta(a,b,d)$, $\delta_2 = \delta(a,c,d)$, $\delta_3 = \delta(b,c,d) \in U^{(3)}$. $\Delta = \Delta(\delta, d) = \Delta(a,b,c,d)$ forms a spatial tetrahedron such that triangles $\delta_1, \delta_2, \delta_3$ are in $U^{(3)}$, but not in $V^{(3)}$.*
3. *The directions of the triangles in $U^{(3)}$ and $V^{(3)}$ are defined as follows:*
   (a) *If $\delta \in V_0^{(3)}$, then the directions of the triangles in both $\{\boldsymbol{U}^{(3)}\}$ and $\boldsymbol{V}^{(3)}$ are consistent.*
   (b) *The direction of the four surfaces $\delta, \delta_1, \delta_2, \delta_3$ of tetrahedron $\Delta = \Delta(a,b,c,d)$ in condition 2 are defined as follows:*
      i. *For triangle δ, the side in which the tetrahedron Δ lies is the inside, and the other side is the outside.*
      ii. *For triangles $\delta_1, \delta_2, \delta_3$, the side in which tetrahedron Δ lies is the outside, the other side is the inside.*

The definitions for polyhedron $\boldsymbol{G}'$ and $\boldsymbol{G}$ are given as follows.

Definition 49. 1. *If the directed polyhedron $\boldsymbol{G}'$ is the envelope of another directed polyhedron $\boldsymbol{G}$, then the directed polyhedron $\boldsymbol{G}$ is the retraction of $\boldsymbol{G}'$.*
2. *Polyhedron $\Delta(\delta, d)$ is called the retracting polyhedron between directed polyhedron $\boldsymbol{G}$ and its envelope $\boldsymbol{G}'$. Then, δ is called the retracting triangle in $\Delta(\delta, d)$, and d is called the retracting point.*
3. *If the directed polyhedron $\boldsymbol{G}$ is the retraction of $\boldsymbol{G}'$, then we denote the entire retracting tetrahedrons by*

$$\tilde{\Delta}(\boldsymbol{G}, \boldsymbol{G}') = \{\boldsymbol{\Delta}_i = \boldsymbol{\Delta}(\delta_i, d_i),\ i = 1, 2, \cdots, k\} . \tag{13.6}$$

It called as the retracting set of $\boldsymbol{G}'$ about $\boldsymbol{G}$.

Definition 50. *If A is a spatial particle system, $\boldsymbol{G}$ is a directed polyhedron, and all the vertices of $\boldsymbol{G}$ are in A, then the directed polyhedron $\boldsymbol{G}$ is called the directed polyhedron of set A. If $\boldsymbol{G}$, $\boldsymbol{G}'$, and $\boldsymbol{G}''$ all are directed polyhedrons of set A, then we have the following definitions:*

1. *If $\boldsymbol{G}'$ is the envelope of $\boldsymbol{G}$, and for any retracting tetrahedron $\Delta(\delta, d)$ between $\boldsymbol{G}'$ and $\boldsymbol{G}$, it does not contain any points in A, then $\boldsymbol{G}'$ is called a simple envelope of $\boldsymbol{G}$ about set A, and $\boldsymbol{G}$ is called a simple retraction of $\boldsymbol{G}'$ about set A.*
2. *If $\boldsymbol{G}$ is a directed polyhedron, we denote all of its simple retractions by $\mathcal{G}_A$, called the retracting range of $\boldsymbol{G}$ on set A.*

The directed polyhedron and its envelope are shown in Fig. 13.6. Figure 13.6a shows a schematic diagram of the directed polyhedron, where the arrow is the direction of the directed triangle. Figure 13.6b shows the relation figure of the directed polyhedron, where graph $G = \{A, V\}$ is

$$A = \{1, 2, \cdots, 15\} , \quad V = \{(1,2), (2,3), \cdots, (14,15), (15,1)\} .$$

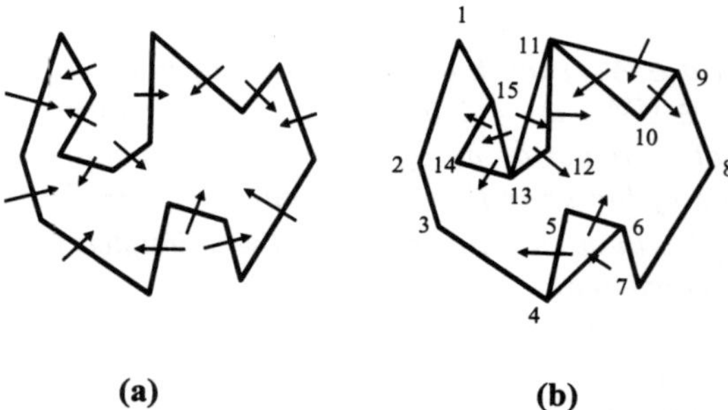

Fig. 13.6a,b. A directed polyhedron and its envelope

Graph $G' = \{A', V'\}$ is

$$\begin{cases} A' = \{1,2,3,4,6,7,8,9,11,13,15\} , \\ V' = \{(1,2),(2,3),(3,4),(4,6),(6,7),(7,8),(8,9),(9,11), \\ \qquad (11,13),(13,15),(15,1)\} . \end{cases}$$

The direction of each tetrahedron in the envelope is shown in Fig. 13.6.

Structural Analysis of a Directed Polyhedron

Theorem 43. *A directed polyhedron has the following basic properties:*

1. *A directed polyhedron is a convex directed polyhedron if and only if its envelope is itself.*
2. *An envelope of a directed polyhedron may not be convex, but after envelopments, it must become a convex directed polyhedron.*

Theorem 44. *Let the directed tetrahedron $\boldsymbol{G}'$, $\boldsymbol{G}''$ be the retraction of $\boldsymbol{G}$; their corresponding polyhedrons are $\Omega, \Omega', \Omega''$, and we denote their retracting sets by*

$$\begin{cases} \tilde{\Delta}(\boldsymbol{G}', \boldsymbol{G}) = \{\boldsymbol{\Delta}'_i = \boldsymbol{\Delta}(\delta'_i, d'_i), \ i = 1,2,\cdots,k'\} , \\ \tilde{\Delta}(\boldsymbol{G}'', \boldsymbol{G}) = \{\boldsymbol{\Delta}''_i = \boldsymbol{\Delta}(\delta''_i, d''_i), \ i = 1,2,\cdots,k''\} . \end{cases} \tag{13.7}$$

We then identify the following properties:

1. *If there are no common points between set $\{\boldsymbol{\Delta}'_i, \ i = 1,2,\cdots,k'\}$ and set $\{\boldsymbol{\Delta}''_i, \ i = 1,2,\cdots,k''\}$, the polyhedron is the retraction of $\Omega' \cap \Omega''$, and it is also the retraction of Ω, then its retracting set is*

$$\tilde{\Delta}(\boldsymbol{G}' \cap \boldsymbol{G}'', \boldsymbol{G}) = \{\boldsymbol{\Delta}'_i, \ i = 1,2,\cdots,k'\} \cup \{\boldsymbol{\Delta}''_j, \ j = 1,2,\cdots,k''\} . \tag{13.8}$$

2. *If there are no common points between set $\{\boldsymbol{\delta}'_i, \ i = 1,2,\cdots,k'\}$ and set $\{\boldsymbol{\delta}''_i, \ i = 1,2,\cdots,k''\}$, then the polyhedron $\Omega' \cap \Omega''$ is the second retraction of Ω, the first retraction is from Ω to Ω', and the second retraction is from Ω' to $\Omega' \cap \Omega''$.*

We do not prove this in detail here. In general, if both Ω' and Ω'' are retractions of Ω, then $\Omega' \cap \Omega''$ is still a directed polyhedron, but may not be the retraction of Ω.

As we have mentioned above, the biological definition of the surface of a protein is predicted on the basis of whether or not the amino acids of a protein are solvent accessible (especially by water). We now extend this definition using mathematical language by using the γ-accessible radius to describe it.

13.3.2 Definition and Basic Properties of γ-Accessible Radius

Definition of γ-Accessible Radius

If A is a spatial particle system, $\Omega(A)$ is its convex closure and a is a point in space, then $O(a, \gamma)$ denotes a small sphere with center a and radius γ, and $O(\gamma)$ denotes a rolling sphere with radius γ.

Definition 51. *For a fixed spatial particle system A and any point a in $\Omega(A)$, we say that point a is γ-accessible, if there is a broken line L which satisfies:*

1. *Point a is an endpoint of the broken line L, and the other endpoint b is outside of the region $\Omega(A)$.*
2. *Denote by d a moving point on broken line L; for any $d \in L$. There are no points of set A in the small sphere $O(d, \gamma)$.*

Then, we call the broken line L the γ-accessible path of the point a about A.

Definition 52. *For a fixed spatial particle system A and any point a in $\Omega(A)$, a point a has the γ-maximal accessible radius, if for all $\gamma' < \gamma$, point a is always γ'-accessible about A, and for any $\gamma'' > \gamma$, point a is always γ''-inaccessible.*

Therefore, the concept of a γ-maximal accessible radius is the maximal radius of a small sphere which can touch other molecules in the protein. By Definition 47, for a fixed spatial particle system A, the γ-maximal accessible radius of any point a in $\Omega(A)$ is uniquely determined. We denote it by $\gamma_A(a)$, and call it the γ-accessible radius or the γ-function of point a (about A).

Computational Terms of γ-Accessible Radius

For a fixed spatial particle system A, in order to compute the γ-function of each point in A, we introduce the following terms, which are used frequently in the following text:

1. For a fixed triangle δ or triangle $\delta(a, b, c)$, we denote the cirumcenter and the circumradius of this triangle by $o(\delta)$, $r(\delta)$, respectively. We denoted the circumcircle of this triangle by $\boldsymbol{o}(\delta)$, and denote the plane determined by this triangle by $\pi(\delta)$.

2. A small sphere $O(r)$ moves freely in region Σ. This means there are no points of set A in region Σ, and the small sphere $O(r)$ can move in region Σ without encountering any points of set A.
3. A small sphere $O(r)$ can access triangle δ, this means $\gamma[o(\delta)] \geq r$, where $\gamma[o(\delta)]$ is the γ-function of circumcenter $o(\delta)$ of triangle δ. We call $\gamma[o(\delta)]$ the accessible radius of triangle δ.
4. A small sphere $O(r)$ can traverse triangle δ, this means the small sphere $O(r)$ can access triangle δ, and $r(\delta) \geq r$, where $r(\delta)$ is the circumradius of triangle δ. We call $\min\{r(\delta), \gamma[o(\delta)]\}$ the traversable radius of triangle δ.
5. A small sphere $O(r)$ can access point a through triangle δ, meaning the small sphere $O(r)$ first accesses triangle δ, then continues moving and finally reaches the point a. In this moving process, we can choose an appropriate path without encountering any points of set A. Clearly, in later moving processes after the small sphere $O(r)$ reaches the triangle δ, it is possible that only a part of the small sphere traverses triangle δ and encounters the point a, while it is also possible that the whole small sphere has entirely traversed triangle δ and then encountered the point a.

For computation of the γ-function of each point in the spatial particle system A, we should choose an appropriate path for the small rolling sphere $O(r)$ in particle system A, such that the small sphere does not encounter any points of set A. Then, the small sphere may traverse the triangle formed by several points in the particle system and finally access the point a.

Basic Properties of the Calculation of γ-Accessible Radius

We formulate some basic properties of the γ-accessible radius as follows:

By the definition of the γ-accessible radius, we find a series of basic properties, such as if we denote $\gamma_0 = \min\{|ab| : a \neq b \in A\}$, then for any $d \in A$, $\gamma(d) \geq \gamma_0/2$ always holds.

Theorem 45. *$a \in A$ is a zero-depth point of A if and only if for any $\gamma > 0$, point a is always γ-accessible about set A. So for any point a on a convex hull $\Omega(A)$, $\gamma(a) = \infty$ always holds.*

The proof of this theorem is shown in Sect. 13.5.1.

13.3.3 Basic Principles and Methods of γ-Accessible Radius

Symbols and Terms in the Calculation of γ-Accessible Radius

As shown above, to calculate the γ-function for each point in set A, we must choose an appropriate path, such that a rolling sphere traverses the triangle formed by several points in the particle system A, and finally arrives at the point a. Therefore, we need to discuss the case of an arbitrary spatial particle system A_1 traversing a fixed triangle $\delta = \delta(a,b,c)$ and arriving at A_1. Its strict definition is presented as follows:

Definition 53. *For an arbitrary fixed spatial particle system A_1 and a fixed triangle $\delta = \delta(a,b,c)$, we say that r is the accessible radius of set A_1 about δ, if there exists a small rolling sphere $O(r)$, such that in the process of traversing triangle δ (namely, not encountering three points a,b,c) and finally arriving at set A_1 (namely arriving at a point d in set A_1), the small sphere does not encounter any points in A_1.*

The maximal accessible radius of set A_1 about δ is called the assessible radius of set A_1 about δ, in which, we denote by $\gamma_\delta(A_1)$.

To calculate $\gamma_\delta(A_1)$, we introduce the following symbols:

1. For a fixed triangle $\delta = \delta(a,b,c)$, we denote a sphere with center $o(\delta)$ and radius $r(\delta)$ by $O(\delta)$.
2. For a fixed triangle $\delta = \delta(a,b,c)$ and a point d, we denote by $O(\delta,d) = O(a,b,c,d)$ the circumscribed sphere determined by four points a,b,c,d, and the corresponding center and radius are denoted by $o(\delta,d)$, $r(\delta,d)$, respectively.
3. For a fixed set A_1 and a triangle δ, we can construct a series of spheres:
$$O(\delta,d)\ ,\ d \in A_1\ . \tag{13.9}$$
Here, $r(\delta,d) \geq r(\delta)$ always holds, and the equal sign is true if and only if point d is on the face of sphere $O(\delta)$.
4. For a fixed set A_1 and a triangle δ, let set $A_1(\delta) = A_1 \cap O(\delta)$. If $A_1(\delta)$ is not an empty set, then let d_0 be the point in set $A_1(\delta)$ such that $|o(\delta)o(\delta,d)|$, $d \in A_1(\delta)$ is the maximum, and d_0' be the point such that radius of sphere $r(\delta,d)$, $d \in A_1(\delta)$ is the maximum. Next, we will prove $d_0 = d_0'$.
5. Let $Q(\delta)$ be a cylinder, then the underside of this cylinder $\boldsymbol{o}(\delta)$ is the circumcircle of triangle δ and its long axis is perpendicular to plane $\pi(\delta)$. Let L_0 be a straight line through point $o(\delta)$ and perpendicular to plane $\pi(\delta)$; then $q(\delta)$ is the cylindrical face of the cylinder.
6. If $Q(\delta,A_1) = Q(\delta) \cap A_1$ is not an empty set, then for each $d \in Q(\delta,A_1)$, its projection on plane $\pi(\delta)$ is denoted by d'. Let d'' be the intersection point between line segment dd' and sphere $O(\delta)$, then calculate $|dd''|$ (the length of line segment dd''), and let d_1 be the point such that $|d_1d_1''|$ is the minimum of $|dd''|$, $d \in Q(\delta,A_1)$.
7. If $Q(\delta,A_1) = Q(\delta) \cap A_1$ is an empty set, then let $d_2 \in A_1$ be the point in A_1 such that the distance between d_2 and straight line L_0 is the minimum.

Relationships among d_0, d_1, d_2 and the small sphere $O(r)$, set A_1 are shown in Fig. 13.7.

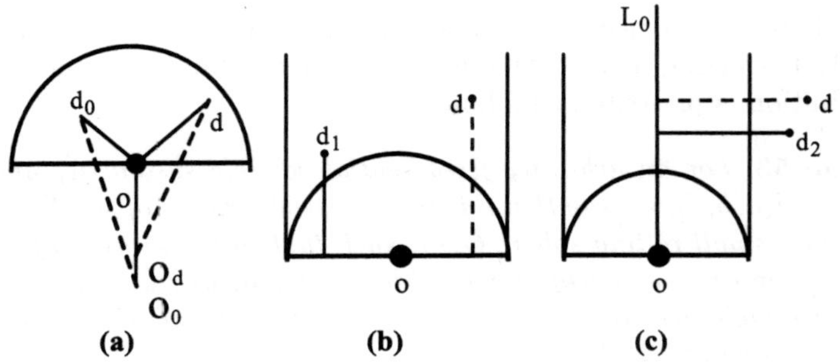

Fig. 13.7a–c. Relationship among d_0, d_1, d_2 and small sphere $O(r)$, set A_1

Thus, the definitions of d_0, d_0', d_1, d_2 are given as follows:

$$\begin{cases} |o(\delta)o(\delta, d_0)| = \max\{|o(\delta)o(\delta, d)| \colon d \in A_1(\delta)\} \,, \\ |o(\delta)d_0'| = \max\{r(\delta, d) \colon d \in A_1(\delta)\} \,, \\ |o(\delta)d_1| = \min\{|dd'| \colon d \in Q(\delta, A_1), \ d' \text{ is intersection point} \\ \qquad\qquad \text{of line segment } dd'' \text{ and sphere } O(\delta)\} \,, \\ |o(\delta)d_2| = \min\{|dd'| \colon d \in A_1, \ d' \text{ is projection of } d \text{ out to line } L_0\} \,, \end{cases} \tag{13.10}$$

where d'' is the projection of point d on plane $\pi(\delta)$.

Thus, we can distinguish the three cases to calculate the γ-function of points in set A_1, namely, where $A_1(\delta)$ is not an empty set; where $A_1(\delta)$ is an empty set, but $Q(\delta, A_1)$ is not an empty set; and where $Q(\delta, A_1)$ is an empty set.

Basic Theorem to Calculate γ-Accessible Radius

For a fixed particle system A_1 and triangle δ, let $r(\delta)$ be the circumradius of δ, and let $\gamma(\delta)$ be the γ-accessible function. The basic theorem is given as follows.

Theorem 46. *1. If $A_1(\delta)$ is not an empty set, then for d_0 and d_0' defined in (13.10), $d_0 = d_0'$ is always true, and the γ-function of point d_0 is $\gamma_\delta(A_1) = \min\{r(\delta), r(\delta, d_0)\}$, where the definitions of $r(\delta)$, $r(\delta, d_0)$ were given in Sect. 13.3.2.*

2. If $A_1(\delta)$ is an empty set and $Q(\delta, A_1)$ is not an empty set, then $\gamma_\delta(A_1) = r(\delta)$.

The proof of this theorem is given in Sect. 13.5.1.

From Fig. 13.8, we let the small sphere move upwards along the vertical direction if $Q(\delta, A_1)$ is an empty set, and then move along the horizontal direction. In the process of moving upwards along the vertical direction, the

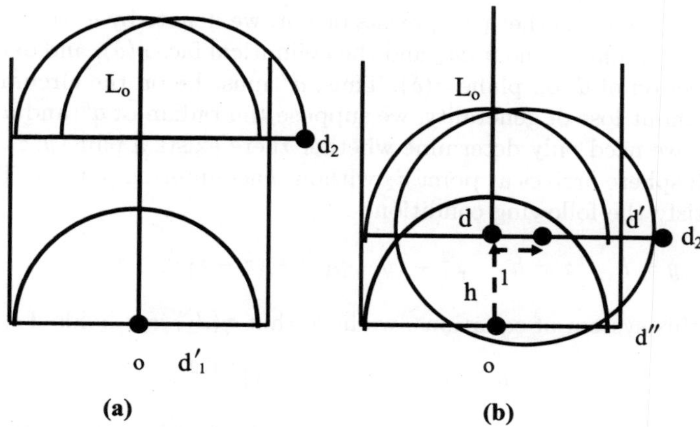

Fig. 13.8a,b. If $Q(\delta, A_1)$ is an empty set, this is an illustration of a changing figure of the small rolling sphere

small sphere will not encounter any points in set A; thus, we can only consider the case that the small sphere moves along the horizontal direction.

Let $O' = O(o', r(\delta))$ be a small sphere whose center moves upwards from point $o(\delta)$ along the vertical direction. If the moving distance $|oo'|$ is larger than $r(\delta)$, this small sphere can move along the horizontal direction without encountering any points of $\delta = \delta(a, b, c)$. Let d_2' be the projection of point d_2 on plane $\pi(\delta)$, if $h = |d_2 d_2'| \geq r(\delta)$, the small sphere $O(r)$, $r = \min\{r(\delta), \gamma(\delta)\}$ can move upwards along the vertical direction with distance h, and then move along the horizontal direction with distance $|d_2 d_2''| - r$ without encountering any points in A. Therefore, $\gamma(d_2) = r(\delta)$.

Similarly, for the small sphere $O' = O(o', r(\delta))$ which moves upwards along the vertical direction, if $|oo'|$ is small, then the small rolling sphere O' may encounter three vertices a, b, c of triangle δ in the process of moving along the horizontal direction, as shown in Fig. 13.7b, so further calculation is needed. We build a rectangular coordinate system $\mathcal{E}$, using $o(\delta)$ as the origin. We use line L_0 as the z-axis of $\mathcal{E}$, with its direction pointing to the side in which point d_2 lies; and we use $\overrightarrow{dd_2}$ as the Y-axis of $\mathcal{E}$, where d is the projection of d_2 on line L_0. Then, the x-axis of $\mathcal{E}$ is the normal vector of the YZ-plane which is through point o, and its direction can be determined by the right-hand screw rule. We denote the coordinates of the four points d_2, a, b, c by $\boldsymbol{r}_\tau = (x_\tau, y_\tau, z_\tau)$, $\tau = 0, 1, 2, 3$. Here, $x_0 = z_1 = z_2 = z_3 = 0$, let $y_0 - r = \ell$, $z_0 = h$.

In a rectangular coordinate system $\mathcal{E}$, if the rolling small sphere $O(r)$ moves upwards along z-axis a distance z, and then moves along the direction of d_2 a distance y, then if there exists a pair (y, z) such that the small sphere can arrive at point d_2 without encountering points a, b, c, $\gamma(d_2) = r$ holds.

To determine whether (y, z) exists or not, we denote by d' the intersection point between line segment dd_2 and the cylindrical face $q(\delta)$, and denote by d'' the projection of d' on plane $\pi(\delta)$. Thus, d'' must be on the circumference of $\boldsymbol{o}(\delta)$. Without loss of generality, we suppose the radian of d'' and a is closed, and then we need only determine whether there exists a pair (y, z) such that the small sphere arrives at point d_2 without encountering point a. Then (y, z) must satisfy the following condition:

$$y = \ell\,, \quad z = h\,, \quad x_1^2 + (y - y_1)^2 + (z - z_1)^2 \geq r^2\,. \tag{13.11}$$

Solving this system of equations, we have that $\gamma(d_2) = r$ holds if

$$h \geq [r^2 - x_1^2 - (\ell - y_1)^2]^{1/2}\,. \tag{13.12}$$

If there does not exist a corresponding value (h, ℓ) such that equations (13.11) hold for a fixed point d_2, we reduce the value of r to r_0 such that the equal sign of (13.12) holds. Here, k, ℓ, x_1, y_1 are constants, hence r_0 exists.

In conclusion, we state the following theorem.

Theorem 47. *If $Q(\delta, A_1)$ is an empty set:*

1. *If $h \geq r$, or h satisfies inequality (13.12), then $\gamma(d_2) = r$ holds.*
2. *If $h < r$, and h satisfies inequality (13.12), then $\gamma(d_2) \geq r_0$ holds, where r_0 is the solution when (13.11) is an equality.*

In proposition 2 of Theorem 47, the estimate of the lower bound of $\gamma(d_2)$ may not be optimal.

Definition 54. *Following from Theorems 44 and 45, we get the points d_0, d_1, $d_2 \in A_1$ which are the first-meeting points of set A_1 after traversing triangle δ under three cases: $A_1(\delta)$ is not empty; $A_1(\delta)$ is empty but $Q(\delta, A_1)$ is not empty; and $Q(\delta, A_1)$ is empty. We denote by $\gamma_\delta(d_\tau) = \gamma_\delta(A_1)$, $\tau = 0, 1, 2$ the accessible radius of d_τ about triangle δ.*

13.4 Recursive Algorithm of γ-Function

In the previous section, we have obtained the γ-function of each point a on the boundary surface of $\Omega(A)$. From calculations based on Theorem 44, we found the γ-function of some of the points of set A_1. Therefore, we continue discussing the γ function of the other points of set A_1, by introducing a recursive algorithm.

13.4.1 Calculation of the γ-Function Generated by 0-Level Convex Hull

In order to calculate the γ-function of the points of set A, we first realize that $\gamma(a) = \infty$ holds for any points in A_0, by Theorem 45. We now calculate the γ-function of the other points.

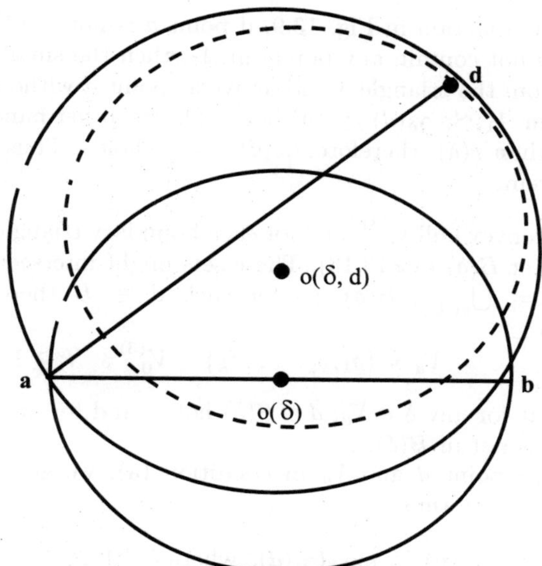

Fig. 13.9. Recursive calculation

We have presented the notation of $\boldsymbol{V}_0^{(3)}$ for the representation of a 0-level directed convex hull, where each $\boldsymbol{\delta} \in \boldsymbol{V}_0^{(3)}$ is a directed boundary triangle of convex closure $\Omega(A)$. We now perform the following calculations:

1. For a fixed directed triangle $\boldsymbol{\delta} = \boldsymbol{\delta}(a, b, c) \in \boldsymbol{V}_0^{(3)}$, let $A_1 = A - \{a, b, c\}$. For any $d \in A_1$, let $O(\delta, d)$ be a sphere and let

$$B(\delta) = \{d \in A_1, O(\delta, d) \text{ inside of spheroid containing no points of } A_1\} . \tag{13.13}$$

Set $B(\delta)$ must be a nonempty set, but may not be unique; there may be three cases for d, i.e., $d \in A_0$, or $d \in O(\delta)$, or d is not in the sphere $O(\delta)$.

Theorem 48. *For the γ-function of the points in set $B(\delta)$, we find the following properties.*

a) If $d \in A_0$, then $\gamma_\delta(d) = \infty$.

b) If $d \in O(\delta)$, then $\gamma_\delta(d) = r(\delta, d)$, where $r(\delta, d)$ is defined in Sect. 13.3.2. We find its properties via Theorem 46.

c) If point d is not in sphere $O(\delta)$, then $\gamma_\delta(d) = r(\delta)$, where $r(\delta)$ is defined in Sect. 13.3.2.

Proof. We can easily prove properties 1 and 2 by Theorems 45 and 46. The proof of property 3 is shown in Fig. 13.9.

Based on the diagram in Fig. 13.9, if point d is not in $O(\delta)$, and sphere $O(\delta, d)$ does not contain any points in A_1, then the small sphere $O[r(\delta)]$ can enter from the triangle δ and arrive at point d without encountering any points in A_1. So $\gamma_\delta(d) \geq r(\delta)$ holds. On the other hand, $\gamma_\delta(d)$ cannot be greater than $r(\delta)$, therefore, $\gamma_\delta(d) = r(\delta)$ holds. Thus ends the proof of the theorem.

2. For a fixed convex hull $V_0^{(3)}$, and for each boundary triangle with $\delta \in V_0^{(3)}$, we can get set $B(\delta)$ by (13.13). These sets might intersect, so:
   (a) Let $B = \bigcup_{\delta \in V_0^{(3)}} B(\delta)$, so for each $d \in B$ there always exists a group
   $$V_d = \{\delta_1, \delta_2, \cdots, \delta_k\} \subset V_0^{(3)}, \quad k \geq 1$$
   such that for any $\delta \in V_d$, $d \in S(\delta)$ holds, and for any $\delta \in V_0^{(3)} - V_d$, point d is not in $B(\delta)$.
   (b) Based on point d and V_d in condition (a), we get $\gamma_i(d) = \gamma_{\delta_i}(d)$, $i = 1, 2, \cdots, k$, thus
   $$\gamma(d) = \max\{\gamma_i(d), \text{ where } i = 1, 2, \cdots, k\} \tag{13.14}$$
   is the γ-function of point d.
   (c) By condition (b), for any point d in B, there exists a δ_d in (13.14), such that $d \in B(\delta_d)$, and $\gamma(d) = \gamma_{\delta_d}(d)$ holds. We call δ_d the approaching triangle of point d.

13.4.2 Recursive Calculation of γ-Function

In above section, we have calculated the γ-function of all points in set B by the 0-level convex hull $V_0^{(3)}$. To calculate the γ-function of other points, we introduce the following recursive steps:

1. We consider each triangle δ on a 0-level convex hull $V_0^{(3)}$ as a directed triangle. For each $d \in B$, we denote its approaching triangle by $\delta_d = \delta(a_d, b_d, c_d)$, and denote a tetrahedron with four vertices a_d, b_d, c_d, d by $\Sigma(a_d, b_d, c_d, d)$. Sphere $O(a_d, b_d, c_d, d)$ does not contain any points in $A - \{a_d, b_d, c_d, d\}$; this implies that $\Sigma(a_d, b_d, c_d, d)$ does not contain any points in $A - \{a_d, b_d, c_d, d\}$. Thus we can retract the directed polyhedron $\boldsymbol{V}_0^{(3)}$ according to (d, δ_d), $d \in B$, to get a new directed polyhedron $\boldsymbol{V}_1^{(3)}$.
2. For each boundary triangle $\boldsymbol{\delta} \in \boldsymbol{V}_1^{(3)}$ in the directed polyhedron $\boldsymbol{V}_1^{(3)}$, we can define set $B'(\delta)$ as in (13.13) and the γ-function of each point in this set about this triangle.
3. Similar to (13.14), we can calculate the γ-function of each point in set $B'(\delta)$. We repeat this to get the γ-function of each point in set A.

Table 13.5. Distribution of set A_1 in $O(\delta_i)$ interior

1	2	1	2	1	2	1	2	1	2	1	2	1	2	1	2
1	17.6747	2	28.0649	3	∞	4	134.445	5	56.1087	6	∞	7	∞	8	31.2947
9	34.3887	10	17.6422	11	33.5496	12	14.4538	13	31.5049	14	11.8489	15	19.0867	16	10.1349
17	35.2943	18	7.74091	19	310.286	20	6.89257	21	97.6989	22	5.92052	23	146.656	24	9.07193
25	∞	26	∞	27	183.222	28	5.22915	29	19.4598	30	15.7119	31	5.16211	32	8.29965
33	10.577	34	10.6558	35	8.04309	36	46.9502	37	171.338	38	8.77772	39	8.0484	40	8.59963
41	14.8524	42	9.84345	43	31.3584	44	31.5786	45	32.2711	46	14.1265	47	17.7637	48	15.7874
49	23.9796	50	13.0015	51	∞	52	∞	53	∞	54	110.071	55	62.6285	56	12.2717
57	22.635	58	10.1757	59	41.7513	60	10.0418	61	51.5585	62	10.8136	63	21.8588	64	8.88514
65	16.496	66	6.8638	67	16.8678	68	12.9889	69	10.7766	70	∞	71	∞	72	∞
73	∞	74	19.4956	75	18.6084	76	52.7615	77	13.7498	78	11.1552	79	11.9087	80	17.6727
81	34.1265	82	9.52151	83	9.66642	84	53.1736	85	18.096	86	94.8295	87	34.2828	88	56.8256
89	11.8952	90	20.7947	91	51.1955	92	29.9013	93	∞	94	6.47052	95	11.0615	96	50.5705
97	8.99533	98	63.0856	99	55.9255	100	9.11707	101	9.40965	102	10.549	103	4.98585	104	6.83729
105	4.84028	106	6.04947	107	12.3899	108	5.7534	109	4.99202	110	13.2468	111	10.0297	112	5.08576
113	10.0886	114	12.4667	115	5.66732	116	8.98487	117	9.30359	118	8.80155	119	5.55789	120	8.77183
121	8.51864	122	6.3084	123	6.8501	124	10.5035	125	9.8508	126	7.40003	127	5.22846	128	4.89868
129	4.62161	130	4.99202	131	8.07632	132	8.55018	133	6.12883	134	14.5463	135	60.1815	136	4.61216
137	50.3308	138	7.42384	139	19.628	140	11.3168	141	10.4905	142	15.084	143	9.01234	144	73.9075
145	318.287	146	36.0357	147	∞	148	∞	149	79.3614	150	15.9595	151	20.142	152	9.92854
153	17.0891	154	7.74252	155	39.2583	156	5.70074	157	63.0932	158	4.83674	159	325.22	160	∞
161	26.3831	162	∞	163	111.59	164	∞	165	15.038	166	7.25092	167	11.3238	168	9.01574
169	4.93283	170	4.83674	171	14.2906	172	∞	173	150.963	174	288.489	175	14.52	176	47.1928
177	7.85079	178	67.5936	179	7.95977	180	29.3243	181	14.1467	182	∞	183	∞	184	13.2111
185	32959.8	186	7.87724	187	∞	188	6.93986	189	286.936	190	8.31956	191	86.0665	192	13.1293
193	63.7394	194	12.0244	195	115.449	196	4.79312	197	15.0236	198	∞	199	64.5202	200	4.24766

Table 13.5. (continued)

1	2	1	2	1	2	1	2	1	2	1	2	1	2	1	2
201	16.5668	202	∞	203	14.6863	204	608.19	205	5.88999	206	5.19302	207	∞	208	∞
209	∞	210	483.284	211	4.75501	212	30.9646	213	5.81223	214	23.4438	215	10.2305	216	53.6861
217	6.62237	218	∞	219	8.30546	220	∞	221	14.3289	222	20.4224	223	∞	224	11.8745
225	∞	226	9.0402	227	457.683	228	7.35689	229	224.537	230	6.13938	231	158.361	232	20.5498
233	84.368	234	7.81338	235	84.225	236	4.62738	237	4.99202	238	5.94237	239	10.2674	240	10.0224
241	11.6428	242	5.90806	243	13.668	244	∞	245	∞	246	392.226	247	18.3927	248	10.7555
249	4.93617	250	4.24766	251	5.06028	252	10.7187	253	30.9887	254	5.42648	255	334.932	256	4.99202
257	132.451	258	4.98836	259	192.262	260	5.21359	261	∞	262	7.54876	263	∞	264	9.99908
265	∞	266	∞	267	∞	268	7.93322	269	82.7797	270	8.40605	271	59.3073	272	5.66904
273	∞	274	9.69061	275	275.624	276	5.50547	277	28.7123	278	5.26492	279	15.7718	280	5.06028
281	79.8257	282	∞	283	5.26597	284	26.0141	285	∞	286	97.967	287	∞	288	∞
289	10.1641	290	14.561	291	8.14075	292	7.47297	293	10.5766	294	9.1374	295	414.723	296	5.13234
297	∞	298	11.6084	299	28.409	300	7.83004	301	42.8329	302	8.8507	303	∞	304	∞
305	28.2231	306	46.4136	307	34.7497	308	8.635	309	50.7039	310	8.48863	311	68.7925	312	4.7766
313	35.915	314	5.55789	315	∞	316	8.40479	317	∞	318	8.84927	319	∞	320	∞
321	∞	322	8.1585	323	16.3929	324	5.51516	325	10.0118	326	8.88307	327	7.81688	328	6.54123
329	22.51	330	14.588	331	6.05049	332	24.096	333	6.09214	334	125.563	335	7.42444	336	1845.54
337	8.43237	338	281.116	339	10.5048	340	28.9558								

In the header *1* represents the serial number of the atom, and *2* represents the value of γ-function of the atom

13.4.3 Example

We still use the protein E. coli OmpF porin as an example. On the basis of Tables 12.6, 13.1, and 13.2, we calculate the γ-function of the C_α atoms in each amino acid, and finally find the values of the γ-function of all C_α atoms in this protein, which are shown in Table 13.5.

13.5 Proof of Relative Theorems and Reasoning of Computational Formulas

13.5.1 Proofs of Several Theorems

Proof of Theorem 43

If $a \in A$ is a zero-depth point in A, then there exists a plane Π through point a such that all points in A lie on the same side of Π. Draw a tangent sphere of plane Π through point a. If this tangent sphere lies on the other side of plane Π, and is outside of $\Omega(A)$ for any radius $\gamma > 0$, point a is always the γ-accessible radius point of A for any $\gamma > 0$.

Conversely, if point a is always the γ-accessible radius point of A for any $\gamma > 0$, then a must be the zero-depth point in A. We prove this using contradiction; if a is not the zero-depth point in A, then there always exist points of A on both sides of plane Π which is through point a. The following properties would then hold:

1. Let $\mathcal{E}$ be a rectangular coordinate system where origin a, $\boldsymbol{k}$ is the unit vector, its polar angle is (ψ, θ). Here (ψ, θ) changes by region:

$$\Psi \times \Theta = \{(\psi, \theta) : 0 \le \psi \le 2\pi,\ 0 \le \theta \le \pi\} . \tag{13.15}$$

   Denote a neighborhood of (ψ, θ) by

$$\boldsymbol{k}(\psi, \theta) = \{\boldsymbol{k}' = (\psi', \theta') : (\psi' - \psi)^2 + (\theta' - \theta)^2 < \epsilon(\boldsymbol{k}),\ \epsilon(\boldsymbol{k}) > 0\} . \tag{13.16}$$

2. Let $N_{\boldsymbol{k}}$ be a normal plane which is through point a, with $\boldsymbol{k}$ as its vector. $N_{\boldsymbol{k}}$ separates set A into sets, which we denote by $A_1(\boldsymbol{k}), A_2(\boldsymbol{k})$, respectively. Both sets are nonempty sets. Without loss of generality, we always adopt $1 \le ||A_1(\boldsymbol{k})|| \le ||A_1(\boldsymbol{k})||$.
   From geometric properties, we realize that there must exist a sufficiently small $\epsilon(\boldsymbol{k}) > \theta > 0$, for any $\boldsymbol{k}' \in \boldsymbol{k}(\psi, \theta)$, the two sets divided by its normal plane $N_{\boldsymbol{k}'}$ are denoted by $A_1(\boldsymbol{k}), A_2(\boldsymbol{k})$. There also exists a sufficiently large $\gamma(\boldsymbol{k}) > 0$ such that the normal plane $N_{\boldsymbol{k}}$ is a tangent plane and a is a tangent point. If the spheroid with radius $\gamma(\boldsymbol{k})$ is on the two sides of the normal plane $N_{\boldsymbol{k}}$, they contain set $A_1(\boldsymbol{k})$ and $A_2(\boldsymbol{k})$, respectively.

3. $\Psi \times \Theta$ is covered by all the small regions $\boldsymbol{k}(\psi, \theta)$ if $\boldsymbol{k}$ is changing in $\Psi \times \Theta$. Since $\Psi \times \Theta$ is a compact region in Euclidean space, it follows from the theorem of limited coverage in topological space that there must exist finite spheres as follows:

$$\boldsymbol{k}_1 = (\psi_1, \theta_1), \quad \boldsymbol{k}_2 = (\psi_2, \theta_2), \cdots, \boldsymbol{k}_m = (\psi_m, \theta_m), \tag{13.17}$$

such that corresponding neighborhoods

$$\boldsymbol{k}(\psi_1, \theta_1), \boldsymbol{k}(\psi_2, \theta_2), \cdots, \boldsymbol{k}(\psi_m, \theta_m) \tag{13.18}$$

cover the region $\Psi \times \Theta$.

4. By property 2, for each $\boldsymbol{k}_i$, $i = 1, 2, \cdots, m$, there exists a sufficiently large $\gamma(\boldsymbol{k}_i) > 0$ such that the sphere with radius $\gamma(\boldsymbol{k}_i)$ is tangent to the normal plane $N_{\boldsymbol{k}_i}$ at point a, and this sphere contains the set $A_1(\boldsymbol{k}_i)$. Let

$$\gamma_0 = \max\{\gamma(\boldsymbol{k}_1), \gamma(\boldsymbol{k}_2), \cdots, \gamma(\boldsymbol{k}_m)\} \tag{13.19}$$

5. For an arbitrary sphere through point a with radius $\gamma > \gamma_0$, we denote it by $O(a, \gamma)$. Let Π be its tangent plane through point a and let $\boldsymbol{k}$ be its normal vector. Then, there is a $i \in \{1, 2, \cdots, m\}$ such that $\Pi \in \boldsymbol{k}(\psi_i, \theta_i)$ and sphere $O(a, \gamma)$ contains set $A_1(\boldsymbol{k}_i)$ or set $A_2(\boldsymbol{k}_i)$. Therefore, point a could not be the γ-accessible radius point. Hence, the theorem holds.

Proof of Theorem 44

For the first proposition in Theorem 44, we consult Fig. 13.10 and prove it. In this figure, the bold straight line represents triangle δ, the bold circle represents the sphere $O(o(\delta), r(\delta))$ generated by triangle δ. d, d_0 are the points in $O(o(\delta), r(\delta))$, on the same side of the triangle δ, and o, o_0 represent the

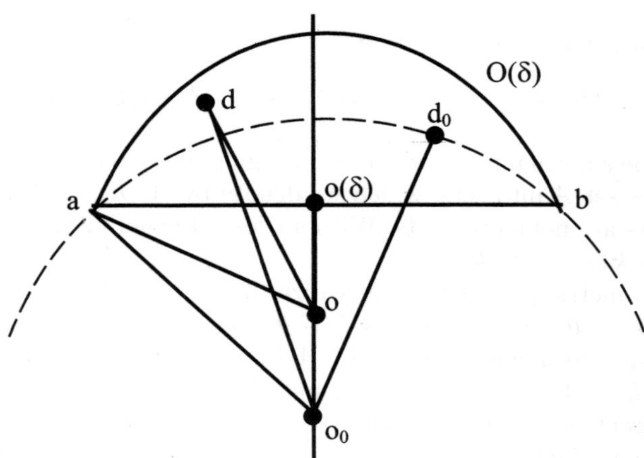

Fig. 13.10. Proof of Theorem 44, proposition 1

centers of $O(\delta, d)$, $O(\delta, d_0)$, respectively. To prove that $d_0 = d_0'$, we must prove:

1. If $|o_0 o(\delta)| \geq |oo(\delta)|$, then $|d_0 o_0| \geq |do|$ holds.
2. Conversely, if $|d_0 o_0| \geq |do|$, then $|o_0 o(\delta)| \geq |oo(\delta)|$ holds.

Without loss of generality, we may assume that points d and d_0 lie in two semispheres of $O(o(\delta), r(\delta))$. Following from the definition of o, o_0, we realize

$$|oa| = |ob| = |od| , \quad |o_0 a| = |o_0 b| = |o_0 d_0| .$$

We compare the two triangles $\delta(a, o(\delta), o_0)$ and $\delta(a, o(\delta), o)$. Both of them are right triangles, having the common edge $ao(\delta)$. The edge of sphere $o(\delta)o$ and the edge of $o(\delta)o_0$ are collinear, thus, $|o_0 o(\delta)| \geq |oo(\delta)|$ holds if and only if

$$|ad_0| = |d_0 o_0| \geq |ao| = |do| .$$

Therefore, $|o_0 o(\delta)| \geq |oo(\delta)|$ is equivalent to $|o_0 d_0| \geq |od|$; namely, the definitions of d_0 and d_0' are equivalent, and $d_0 = d_0'$ holds. The first proposition in Theorem 44 has been proven.

We consult Fig. 13.10 to prove the second proposition in Theorem 44. We compare the two triangles $\delta(a, o, o_0)$ and $\delta(d, o, o_0)$. Since oo_0 is their common edge, and $|od| = |oa|$ holds, it follows that these two triangles have two equivalent edges. On the other hand, following from the definition of $o(\delta), o, o_0$ and the inequality $\angle doo_0 > \angle aoo_0$, we have that $|do_0| > |ao_0|$ (in a triangle, a larger angle corresponds to a larger edge). Thus, point d must be outside of the sphere $O(a, b, cd_0)$. Proposition 1 in Theorem 44 is now proven.

To prove proposition 2 in Theorem 44 as shown in Fig. 13.11, we can project each point in $Q(\delta, A_1)$ to $\pi(\delta)$ if $A_1(\delta)$ is convex and $Q(\delta, A_1)$ is

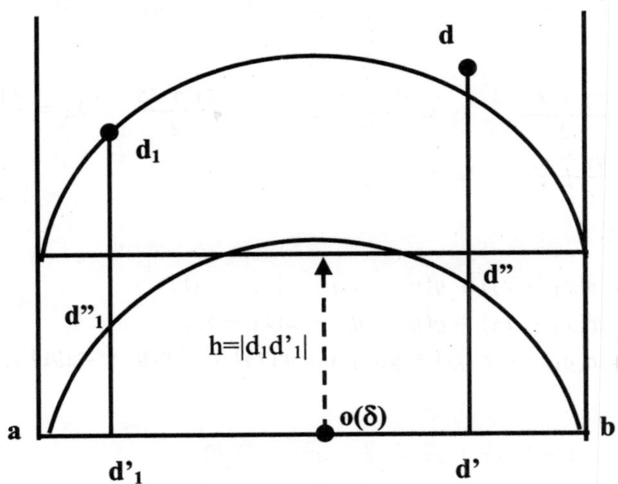

Fig. 13.11. Illustration of the proof of Theor. 44 (2)

a nonempty set. Furthermore, if d_1 is the point such that $|d_1 d_1''|$ is the minimum of $|dd''|$, $d \in Q(\delta, A_1)$, we only need to move the small sphere $O(\delta)$ vertically upwards. If the moving distance is $|d_1 d_1''|$, then the small sphere encounters point d_1 but does not encounter any points in $Q(\delta, A_1)$; thus, it does not encounter any points in A_1. Therefore, $\gamma(d_1) = r(\delta)$ holds, and the theorem has been proven.

13.5.2 Reasoning of Several Formulas

Circumcircle of Three Arbitrary Particles in Space

If a, b, c are three points in space, we now resolve their circumcenter and circumradius. We denote their coordinates by

$$\boldsymbol{r}_1 = (x_1, y_1, z_1)\,, \quad \boldsymbol{r}_2 = (x_2, y_2, z_2)\,, \quad \boldsymbol{r}_3 = (x_3, y_3, z_3)\,.$$

The circumcenter of three points a, b, c is the intersection point of plane Π_1, Π_2, Π_3, where Π_1 is the perpendicular plane intersecting midpoint of line segment AB, Π_2 is the perpendicular plane intersecting midpoint of line segment AC, and Π_3 is the plane determined by a, b, c. Hence, circumcenter o must be on the perpendicular plane intersecting the midpoint of line segment ab, ac. The four points a, b, c, o are on same circle, so the coordinate $\boldsymbol{r} = (x, y, z)$ of point o satisfies equations

$$\begin{cases} (x - \lambda_1)(x_1 - x_2) + (y - \lambda_2)(y_1 - y_2) + (z - \lambda_3)(z_1 - z_2) = 0\,, \\ (x - \lambda_4)(x_1 - x_3) + (y - \lambda_5)(y_1 - y_3) + (z - \lambda_6)(z_1 - z_3) = 0\,, \\ \begin{vmatrix} x - x_1 & y - y_1 & z - z_1 \\ x_2 - x_1 & y_2 - y_1 & z_2 - z_1 \\ x_3 - x_1 & y_3 - y_1 & z_3 - z_1 \end{vmatrix} = 0\,, \end{cases} \tag{13.20}$$

where

$$\lambda_1 = \frac{x_1 + x_2}{2}\,, \quad \lambda_2 = \frac{y_1 + y_2}{2}\,, \quad \lambda_3 = \frac{z_1 + z_2}{2}\,, \quad \lambda_4 = \frac{x_1 + x_3}{2}\,,$$
$$\lambda_5 = \frac{y_1 + y_3}{2}\,, \quad \lambda_6 = \frac{z_1 + z_3}{2}\,.$$

To simplify,

$$\begin{cases} \theta_1 = x(x_1 - x_2) + y(y_1 - y_2) + z(z_1 - z_2)\,, \\ \theta_2 = x(x_1 - x_3) + y(y_1 - y_3) + z(z_1 - z_3)\,, \\ \theta_3 = x(y_1 z_2 - z_1 y_2) + y(x_2 z_1 - x_1 z_2) + z(x_1 y_2 - x_2 y_1)\,, \end{cases} \tag{13.21}$$

in which

$$\begin{cases} \theta_1 = \left(x_1^2 + y_1^2 + z_1^2 - x_2^2 - y_2^2 - z_2^2\right)/2\,, \\ \theta_2 = \left(x_1^2 + y_1^2 + z_1^2 - x_3^2 - y_3^2 - z_3^2\right)/2\,, \\ \theta_3 = x_1(y_1 z_2 - z_1 y_2) + y_1(x_2 z_1 - x_1 z_2) + z_1(x_1 y_2 - x_2 y_1)\,. \end{cases}$$

Thus, the coordinate ρ_o of center $o_{a,b,c}$ can be computed as follows:

$$(x_o, y_o, z_o) = (\Delta_1/\Delta_0, \Delta_2/\Delta_0, \Delta_3/\Delta_0) \ , \tag{13.22}$$

in which

$$\Delta_0 = \begin{vmatrix} x_1 - x_2 & y_1 - y_2 & z_1 - z_2 \\ x_1 - x_3 & y_1 - y_3 & z_1 - z_3 \\ y_1 z_2 - z_1 y_2 & x_2 z_1 - x_1 z_2 & x_1 y_2 - x_2 y_1 \end{vmatrix} ,$$

$$\Delta_1 = \begin{vmatrix} \theta_1 & y_1 - y_2 & z_1 - z_2 \\ \theta_2 & y_1 - y_3 & z_1 - z_3 \\ \theta_3 & x_2 z_1 - x_1 z_2 & x_1 y_2 - x_2 y_1 \end{vmatrix} , \quad \Delta_2 = \begin{vmatrix} x_1 - x_2 & \theta_1 & z_1 - z_2 \\ x_1 - x_3 & \theta_2 & z_1 - z_3 \\ y_1 z_2 - z_1 y_2 & \theta_3 & x_1 y_2 - x_2 y_1 \end{vmatrix} ,$$

$$\Delta_3 = \begin{vmatrix} x_1 - x_2 & y_1 - y_2 & \theta_1 \\ x_1 - x_3 & y_1 - y_3 & \theta_2 \\ y_1 z_2 - z_1 y_2 & x_2 z_1 - x_1 z_2 & \theta_3 \end{vmatrix} .$$

The radius of the circle is then

$$\rho_{a,b,c} = |oa| = \sqrt{(x - x_1)^2 + (y - y_1)^2 + (z - z_1)^2} \ . \tag{13.23}$$

Let $O_{a,b,c}$ and $\boldsymbol{o}_{a,b,c}$ be the circle with center $o_{a,b,c}$, radius $\rho_{a,b,c}$, and it lies in the plane $\Pi(a, b, c)$.

Circumscribed Sphere of Four Arbitrary Particles in Space

Four points determine a sphere. If a, b, c, d are four noncoplanar points, then these four points determine a sphere, denoted by $O(a, b, c, d)$. If the rectangular coordinates of the four points a, b, c, d are (x_τ, y_τ, z_τ), where $\tau = 1, 2, 3, 4$, then the center of the sphere determined by the four points must be an intersection point of the perpendicular plane intersecting the midpoint of line segments ab, ac, ad, therefore, its coordinate satisfies the equations

$$\begin{cases} (x_2 - x_1)(x - \lambda_1) + (y_2 - y_1)(y - \lambda_2) + (z_2 - z_1)(z - \lambda_3) = 0 \, , \\ (x_3 - x_1)(x - \lambda_4) + (y_3 - y_1)(y - \lambda_5) + (z_3 - z_1)(z - \lambda_6) = 0 \, , \\ (x_4 - x_1)(x - \lambda_7) + (y_4 - y_1)(y - \lambda_8) + (z_4 - z_1)(z - \lambda_9) = 0 \ , \end{cases} \tag{13.24}$$

where $\lambda_1, \cdots, \lambda_6$ is given by (13.20),

$$\lambda_7 = \frac{x_1 + x_4}{2} , \quad \lambda_8 = \frac{y_1 + y_4}{2} , \quad \lambda_9 = \frac{z_1 + z_4}{2} .$$

If we denote the coordinates of the four points a, b, c, d by

$$o(a, b, c, d) = (x_0, y_0, z_0) \ ,$$

then the radius of the sphere is the solution of (13.24):

$$\begin{aligned} r(a, b, c, d) &= |ao| = |bo| = |co| = |do| \\ &= \left[(x_0 - x_1)^2 + (y_0 - y_1)^2 + (z_0 - z_1)^2\right]^{1/2} . \end{aligned}$$

Simplifying (13.24), we know it is equivalent to the equations

$$\begin{cases} x_5x + y_5y + z_5z = \theta_1 , \\ x_6x + y_6y + z_6z = \theta_2 , \\ x_7x + y_7y + z_7z = \theta_3 , \end{cases}$$

in which $x_{4+\tau} = x_{1+\tau} - x_1$, $y_{4+\tau} = y_{1+\tau} - y_1$, $z_{4+\tau} = z_{1+\tau} - z_1$, where $\tau = 1, 2, 3$ and

$$\begin{cases} \theta_1 = \frac{1}{2}\left(x_2^2 + y_2^2 + z_2^2 - x_1^2 - y_1^2 - z_1^2\right) , \\ \theta_2 = \frac{1}{2}\left(x_3^2 + y_3^2 + z_3^2 - x_1^2 - y_1^2 - z_1^2\right) , \\ \theta_3 = \frac{1}{2}\left(x_4^2 + y_4^2 + z_4^2 - x_1^2 - y_1^2 - z_1^2\right) . \end{cases}$$

The coordinates (x, y, z) of the radius of the sphere are $x = \Delta_1/\Delta_0$, $y = \Delta_2/\Delta_0$, $z = \Delta_3/\Delta_0$, in which

$$\Delta_0 = \begin{vmatrix} x_5 & y_5 & z_5 \\ x_6 & y_6 & z_6 \\ x_7 & y_7 & z_7 \end{vmatrix} , \quad \Delta_1 = \begin{vmatrix} \theta_1 & y_5 & z_5 \\ \theta_2 & y_6 & z_6 \\ \theta_3 & y_7 & z_7 \end{vmatrix} ,$$

$$\Delta_2 = \begin{vmatrix} x_5 & \theta_1 & z_5 \\ x_6 & \theta_2 & z_6 \\ x_7 & \theta_3 & z_7 \end{vmatrix} , \quad \Delta_3 = \begin{vmatrix} x_5 & y_5 & \theta_1 \\ x_6 & y_6 & \theta_2 \\ x_7 & y_7 & \theta_3 \end{vmatrix} .$$

Projection of a Point on a Plane Determined by Three Other Points

We resolve the coordinate of d', which is the projection of point d on the plane determined by three points a, b, c. We then denote the coordinates of d' by $\boldsymbol{r} = (x, y, z)$. The following properties hold true:

1. Let $\boldsymbol{r}_4 = \boldsymbol{r}_2 - \boldsymbol{r}_1$, $\boldsymbol{r}_5 = \boldsymbol{r}_3 - \boldsymbol{r}_1$; then the normal vector $\boldsymbol{n} = \boldsymbol{r}_4 \times \boldsymbol{r}_5 = (x_n, y_n, z_n)$ of plane $\Pi(a, b, c)$ is

$$\left(\begin{vmatrix} y_4 & z_4 \\ y_5 & z_5 \end{vmatrix} , - \begin{vmatrix} x_4 & z_4 \\ x_5 & z_5 \end{vmatrix} , \begin{vmatrix} x_4 & y_4 \\ x_5 & y_5 \end{vmatrix} \right) = (y_4z_5 - z_4y_5, x_5z_4 - x_4z_5, x_4y_5 - x_5y_4) . \tag{13.25}$$

2. Points a, b, c, d' are coplanar, satisfying

$$\langle \boldsymbol{r} - \boldsymbol{r}_1, \boldsymbol{r}_n \rangle = (x - x_1)x_n + (y - y_1)y_n + (z - z_1)z_n = 0 , \tag{13.26}$$

   where $\langle \boldsymbol{r}, \boldsymbol{r}' \rangle$ is the inner product of vector $\boldsymbol{r}$ and vector $\boldsymbol{r}'$.
3. Let $\boldsymbol{r}_0 = (x_0, y_0, z_0)$ be the coordinate of d, then $\boldsymbol{r} - \boldsymbol{r}_0$ is parallel to the normal vector $\boldsymbol{r}_n$. Thus, $\boldsymbol{r} - \boldsymbol{r}_0 = \lambda \boldsymbol{r}_n$ holds, where λ is an undetermined

coefficient. We obtain the equations

$$\begin{cases} x - x_0 = \lambda x_n = \lambda(y_4 z_5 - z_4 y_5)\,, \\ y - y_0 = \lambda y_n = \lambda(x_5 z_4 - x_4 z_5)\,, \\ z - z_0 = \lambda z_n = \lambda(x_4 y_5 - x_5 y_4)\,. \end{cases} \tag{13.27}$$

From (13.25) and (13.27), we find $\lambda = \theta_1/\theta_2$, where $\theta_1 = \langle \boldsymbol{r}_1 - \boldsymbol{r}_0, \boldsymbol{r}_n \rangle$, $\theta_2 = r_n^2 = \langle \boldsymbol{r}_n, \boldsymbol{r}_n \rangle$. So we obtain formula (13.27) to compute the coordinates $\boldsymbol{r} = (x, y, z)$ of the projection d'.

Intersection Point Between the Line Through Two Points and the Plane Through Three Points

We solve for the coordinates of the intersection point d'' between the line $L = dd'$ (determined by two points d, d') and the plane $\Pi(a, b, c)$ (determined by three points a, b, c). Denote the coordinates of d, d', d'' by $\boldsymbol{r} = (x, y, z)$, $\boldsymbol{r}' = (x', y', z')$, $\boldsymbol{r}'' = (x'', y'', z'')$, respectively, where $\boldsymbol{r}''$ is an undetermined coefficient. We have the following properties:

1. If the four points a, b, c, d'' are coplanar, then we obtain (13.26), where the undetermined variable (x, y, z) is replaced by (x'', y'', z'').
2. If three of the points d, d', d'' are collinear, then we get $\boldsymbol{r}'' - \boldsymbol{r} = \lambda(\boldsymbol{r}' - \boldsymbol{r})$, and $(x'' - x, y'' - y, z'' - z) = \lambda(x' - x, y' - y, z' - z)$ holds.
3. By properties 1 and 2, we get

$$\langle \boldsymbol{r}'' - \boldsymbol{r}_1, \boldsymbol{r}_n \rangle = \langle \boldsymbol{r}'' - \boldsymbol{r} + \boldsymbol{r} - \boldsymbol{r}_1, \boldsymbol{r}_n \rangle \langle \boldsymbol{r}'' - \boldsymbol{r}, \boldsymbol{r}_n \rangle + \langle \boldsymbol{r} - \boldsymbol{r}_1, \boldsymbol{r}_n \rangle = 0\,.$$

Therefore, $\lambda = \frac{\langle \boldsymbol{r}_1 - \boldsymbol{r}, \boldsymbol{r}_n \rangle}{\langle \boldsymbol{r}' - \boldsymbol{r}, \boldsymbol{r}_n \rangle}$. Since $\boldsymbol{r}, \boldsymbol{r}'$ are known vectors, we have that $\boldsymbol{r}''$ can be determined.

Projection of a Point on a Line Determined by Two Arbitrary Points in Space

To resolve the coordinates of the point c', which is the projection of point c onto the line $\ell(a, b)$ determined by two points a, b, we denote the coordinates of c' by $\boldsymbol{r} = (x, y, z)$. We find the following properties:

1. Let $\boldsymbol{r}_4 = \boldsymbol{r}_2 - \boldsymbol{r}_1$ be the coordinate of vector $\overrightarrow{ab}$; then, point $\boldsymbol{r} = (x, y, z)$ satisfies the equations

$$\begin{cases} x - x_1 = \lambda x_4 = \lambda(x_2 - x_1)\,, \\ y - y_1 = \lambda y_4 = \lambda(y_2 - y_1)\,, \\ z - z_1 = \lambda z_4 = \lambda(z_2 - z_1)\,, \end{cases} \tag{13.28}$$

and

$$(x - x_3)x_4 + (y - y_3)y_4 + (z - z_3)z_4 = 0\,. \tag{13.29}$$

2. To solve (13.28) and (13.29), substitute (13.29) into (13.28), so that

$$(x_1 - x_3 + \lambda x_4)x_4 + (y_1 - y_3 + \lambda y_4)y_4 + (z_1 - z_3 + \lambda z_4)z_4 = 0 \ .$$

   To solve

$$\lambda = \langle \boldsymbol{r}_5, \boldsymbol{r}_4 \rangle / r_4^2 \ , \tag{13.30}$$

   in which $\boldsymbol{r}_5 = \boldsymbol{r}_3 - \boldsymbol{r}_1$, we substitute (13.30) into (13.28), and the solution $\boldsymbol{r} = (x, y, z)$ is the coordinate of d'.

13.6 Exercises

Exercise 64. In the 2D case, generalize the main theorems and computational formulas presented in this chapter as follows:

1. Write down the definition and properties of the γ-accessible radius of the particles in the plane particle system A, and elaborate on Definitions 43 and 44, and prove Theorems 43–45 in the 2D case.
2. Write down the recursive algorithm of the γ-accessible radius in the plane particle system A.
3. Write down the computational algorithm for the cavity in the plane particle system A, and write down the definition and computational algorithm of channel network (including connected circle and connected radius).

Exercise 65. Let a, b, c be three points on a plane, and let their coordinates be $\boldsymbol{r}_s = (x_s, y_s)$, $s = 1, 2, 3$. Obtain formulas for the following objects:

1. Circumcenter and circumradius of three points a, b, c.
2. Coordinates of the projection of point c onto the line segment ab.
3. Rectangle such that the points a, b, c are on its edges.

Exercise 66. For the plane particle system A formed by the first 80 atoms in Table 12.6 on the XY-plane, calculate the following properties:

1. Depth function and level function of each particle in particle system A.
2. All boundary line segments and vertices of convex closure $\Omega(A)$.
3. γ-function of all points in set A.

Exercise 67. For the plane particle system A formed by the first 80 atoms in Table 12.6 on the XY-plane, calculate:

1. Depth function and level function of each particle in particle system A.
2. All boundary triangles, edges, and vertices of convex closure $\Omega(A)$.
3. γ-function of all points in set A.

14

Semantic Analysis for Protein Primary Structure

14.1 Semantic Analysis for Protein Primary Structure

14.1.1 The Definition of Semantic Analysis for Protein Primary Structure

Semantic analysis refers to analyzing grammar, syntax, and the construtction of words. Semantic analysis for common languages (such as English, Chinese, etc.) is what we are familiar with. If protein primary structure is considered to represent a special kind of biological language, then a protein structure database would be a kind of biological language library, and semantic analysis should aim to analyze the grammar, syntax, and semantics based on such a library.

Differences and Similarities Between Semantic Analysis for Protein Primary Structure and that for Human Languages

There are many similarities between protein primary structures and human languages. There are also significant differences. We list the similarities below:

1. Protein primary structures have the same language structure as human languages, especially English, French, and German. They are both composed of several basic symbols as building blocks. For example, English is composed of 26 letters, while proteins are composed of 20 common amino acids. A protein sequence can be considered to represent a sentence or a paragraph, and the set of all proteins can be considered to represent the whole language. Therefore, the semantic structure is similar to a language structure which goes from "letters" to "words," then to "sentences," to "chapters," "books," and finally to a "language library."
2. There are abundant language libraries to be studied and analyzed both for protein primary structures and human languages. Some of them have already been standardized, as in several kinds of language libraries in English, the PDB-Select database of protein, etc. These language libraries provide us with an information basis for the semantic analysis.

3. The goals of semantic analysis for protein primary structure and that for human languages are basically the same. That is, to find the basic words they are composed of, the meanings of these words and the role they play in the whole language system. It then goes on to the analysis of the structure of grammar, syntax and semantics.
4. Protein primary structure and human languages are both evolving. Two types of problems also occur in protein structures; problems when translating between different languages, and how the same language engenders differences between its regional dialects; adding difficulties to the semantic analysis for protein primary structure.

The vital difference between the semantic analysis for protein primary structure and that for human languages is that we are familiar with the semantics, grammar, syntax and word origin for human languages. The meanings of nouns, verbs and adjectives are clear, thus the semantic analysis for human languages is straightforward. Semantic analysis for protein primary structure is to find these kinds of words and to determine their meaning and grammar, syntax and word-building structures. Thus it is much harder to analyze the semantics of protein primary structure than it is to analyze the semantics of human languages.

Basic Methods of Semantic Analysis

The basic methods of semantic analysis for protein primary structure and those for human languages can be derived from information and statistics, together with methods based on permutation and combination operations. Statistical methods used in the semantic analysis for human languages are frequently used nowadays in computer semantic analysis. For example, the statistics-based analyses for single letters, double letters and multiple letters, those for letters and words, and those for the millennium evolution rate of important words are all results of statistical computation.

Information-based analysis methods rely on the information found in databases. For example, we use the relative entropy density theory to achieve systematic analysis and a series of results is obtained. Permutation and combination methods are commonly used in biology. This is called symbolic dynamics in some publications [7]. In this book, we use the cryptography analysis method for our computation, which has a definitive effect.

Moreover, in some commonly used small-molecule databases, such as the biological molecule and drug molecule libraries in combinatorial chemistry, peptide libraries and non-peptide small-molecule libraries (a general introduction of several libraries may be found in [86]), conformational libraries; in addition to the illustrations of their chemical composition, three-dimensional structures and biological functions [70] are given. They play an important role in drug design. A typical example from the area of biomedical materials is the synthesis of anti-HIV-1 protease inhibitors which was performed according to the molecular combination theory (a general introduction may be found in [20]).

Databases Used in Semantic Analysis

In semantic analysis for protein primary structures, the database we mainly use is the Swiss-Prot database. As of 19 September 2006 [8, 33], there were 232345 proteins (or polypeptide chains) and 85,424,566 amino acids stored in this database.

14.1.2 Information-Based Stochastic Models in Semantic Analysis

We begin with the introduction of the method using probability, statistics and information for our analysis. To give a clear description of semantics for protein primary structure sequences, we provide unified definitions for the related terms and elements.

Symbol, Vector, and Word

In human languages, symbols, words, and sentences are the basic elements of semantic analysis. They are also the basic elements of semantic analysis for biological sequences:

1. Symbols, also called single letters, are the basic unit of semantic structure analysis. The set of all possible symbols used is called the alphabet. In English, the basic alphabet is composed of 26 letters, while in a computer language, the alphabet of English text is the ASCII alphabet. In protein primary structure analysis, the 20 commonly occurring amino acids are considered to be single letter codes forming the alphabet. Their single letter codes were given in Chap. 8, and are

$$V_q = \{\text{A, R, N, D, C, Q, E, G, H, I, L, K, M, F, P, S, T, W, Y, V}\} .$$

   If each single letter code in V_q is denoted by a number, then the corresponding alphabet would be $V_q = \{1, 2, \cdots, q\}$, $q = 20$. In the PDB or Swiss-Prot databases, codes B, Z, X, and other amino acids that are not common sometimes do occur, but will not be introduced here.
2. Vectors and words. Several symbols arranged in a certain order are collectively called a vector, and we denote it by $b^{(k)} = (b_1, b_2, \cdots, b_k)$, $b_j \in V_q$, where each component represents one amino acid and k is the length of this vector, called the rank of the vector. The collection of vectors $b^{(k)}$ is denoted by $V_q^{(k)}$.

   In the English language, a vector is not usually a word, unless it has a specific meaning. Therefore the vectors of specific biological significance are called biological words. In protein primary structure sequences, we will discuss what kinds of combinations of amino acids can represents words, what their structures and meanings are, and how to determine their properties.

3. Local words. In the English language, besides their meaning, words have some special properties in the aspect of symbol structure relationships. For instance, in English words, the letter q must be followed by u, and the frequency of the three letters ordered to form the word "and" must be much higher than that of adn, nad, nda, dan, dna. Thus, vectors with these special statistical properties are called local words.
   Local words and biological words are different concepts, but many biological words may be or contain local words. For instance, "qu" has a special mathematical structure, hence it is a local word, but is not an English word. It can be a component of many English words (for example, "queen" contains local word "qu" and "and" is both a local word and an English word). Local words can be found by mathematical means, while biological words should be given definite biological content, which is the fundamental purpose of semantic analysis. Our point is that the analysis and search for local words will promote the search and discovery of biological words.
4. Phrases. They are composed of several words arranged in a certain order. They may be the superposition of several words or a new word. In mathematics, a phrase can be considered to be a compound vector composed of several vectors. Normally, idioms can be regarded as special words or phrases.

Databases and the Statistical Distribution of Vectors

Lexical analysis on biological sequences begins with statistical computation on a database of protein primary structures. We construct the following mathematical model for this purpose:

1. Mathematical description of the database. We denote Ω to be a database of protein primary structures, such as the Swiss-Prot database, etc. Here Ω consists of M proteins. For instance, in the Swiss-Prot database version 2000, $M = 107{,}618$. Thus, Ω can be denoted by a multiple sequence: $\Omega = \{A_s,\ s = 1, 2, \cdots, M\}$, where $A_s = (a_{s,1}, a_{s,2}, \cdots, a_{s,n_s})$ is the primary structure sequence of a single protein, its component $a_{s,i} \in V_q$ are amino acids, and n_s is the length of the protein sequence.
2. Frequency numbers and frequencies determined by a database. If Ω is given, the frequency numbers and frequencies of different vectors occurring in this database can be obtained.

In the following, we denote by $b^{(k)}$ the fixed vector of rank k in $V_q^{(k)}$. The number of times it occurs in the database Ω is the frequency number, denoted by $n(b^{(k)})$. Denote by n_0 the sum of the frequency numbers of all the vectors of rank k. Then $p(b^{(k)}) = \frac{n(b^{(k)})}{n_0}$ will be the normalized frequency or probability

Table 14.1. Frequency numbers and probabilities of the commonly occurring amino acids

	A	R	N	D	C	Q	E	G
$n(b)$	3010911	2049696	1707483	2067196	641622	1551221	2549016	2706769
$p(b)$	0.07629	0.05193	0.04326	0.05238	0.01626	0.03930	0.06458	0.06858
	H	I	L	K	M	F	P	S
$n(b)$	888197	2302697	3761963	2345725	932473	1612406	1928666	2780793
$p(b)$	0.02250	0.05834	0.09532	0.05943	0.02363	0.04085	0.04887	0.07046
	T	W	Y	V				
$n(b)$	2190245	476445	1240969	2608996				
$p(b)$	0.05549	0.01207	0.03144	0.06610				

with which vector $b^{(k)}$ occurs in database Ω. For instance, when $k = 1$, in the Swiss-Prot database, the frequency numbers and probabilities of the 20 common amino acids are as given in Table 14.1.

$n_0 = 39{,}467{,}963$ is the total number of the amino acids (except B, Z, X) occurring in database Ω, and $n(b)$ is the frequency number of amino acid b occurring in Ω. Then, $p(b) = \frac{n(b)}{n_0}$ is the probability of amino acid b occurring in Ω.

We can also obtain the frequency number and frequency distribution of $k = 2, 3, 4$ ranked vectors in database Ω. See [99].

Characteristic Analysis for Words

In human languages, if vector $b^{(k)}$ is a word, then normally it has the following characteristics:

1. The occurrence of each component in vector $b^{(k)}$ has certain dependence on the context. For instance, in English words, q must be followed by u, and the possibility of letter f, or t following an i must be higher than that of any other two-letter combinations among these three. This statistical characteristic of words is called the statistical correlation between letters.
2. Changing the order of the components in vector $b^{(k)}$ may influence their frequency of occurrence. For instance, in English, the frequency of the three letters "and" is much higher than that of "nad," "nda," "nad," or "dan." This statistical characteristic of words is called the statistical order preservation between letters.
   Statistical correlation and statistical order preservation are the two characteristics of word-building. Hence, they are an important factor in the determination of local words.

14.1.3 Determination of Local Words Using Informational and Statistical Means and the Relative Entropy Density Function for the Second and Third Ranked Words

Principle of the Analysis of Local Words Using Informational and Statistical Means

To determine local words, the information-based and statistics-based approaches can be used. In this analysis we use the probability distribution of the vectors and their relative entropy density function. In informatics, relative entropy is also called Kullback–Leibler entropy. Its definition is as follows:

1. If $p(b^{(k)})$ and $q(b^{(k)})$ are two frequency distributions of $V_q^{(k)}$, then we define
$$k\left(p,q;b^{(k)}\right) = \log \frac{p\left(b^{(k)}\right)}{q\left(b^{(k)}\right)} \tag{14.1}$$
to be the relative entropy density function of the frequency distribution q concerning p, or the relative entropy density function for short, where $b^{(k)} \in V_q^{(k)}$ is the independent variable of the function. The relative entropy density function is the measurement of the differences between distribution functions p and q. In informatics, many discussions are presented on the properties of the Kullback–Leibler entropy [23, 88]). In this book, we mainly discuss the relative entropy density function defined by formula (14.1), as it possesses a much deeper meaning than that of Kullback–Leibler entropy.
2. The mean (or expectation), variance and standard deviation of the relative entropy density function are denoted respectively by: $\mu(p,q)$, $\sigma^2(p,q)$ and $\sigma(p,q) = \sqrt{\sigma^2(p,q)}$. Here
$$\mu(p,q) = \sum_{b^{(k)} \in V_q^{(k)}} p\left(b^{(k)}\right) k\left(p,q;b^{(k)}\right)$$
is the Kullback–Leibler entropy which is common in informatics. In informatics, it has been proved that $\mu(p,q) \geq 0$ always holds, and the equality holds if and only if $p(b^{(k)}) \equiv q(b^{(k)})$ holds.
3. The frequency number distribution of the relative entropy density function. In addition to the eigenvalue, the relative entropy density function has another property, namely the frequency distribution function. We denote by $F_{(p,q)}(x)$ the set $\{a^{(k)} \in V_q^{(k)}: \ k(p,q;a^{(k)}) \leq x\}$. Then, from $F_{(p,q)}(x)$, a histogram can be drawn. This is a common method in statistics.

Relative Entropy Density Function

When using the relative entropy density function $k(p,q;b^{(k)})$ in formula (14.1), frequency distributions p, q can be chosen by different methods. The value of the relative entropy density function $k(p, q; b^{(k)})$ then reflects different aspects of the different internal structure characteristics of vector $b^{(k)}$ in database Ω. Then p, q can be chosen as follows:

1. Let $q(b^{(k)}) = p(b_1)p(b_2)\cdots p(b_k)$. The value of the relative entropy density function $k(p, q; b^{(k)})$ then reflects the "affinity" of each amino acid in vector $b^{(k)} = (b_1, b_2, \cdots ; b_k)$. The concept of "affinity" reflects "correlation" or "cohesion" and "attraction" between the amino acids in vector $b^{(k)}$. That is, $k(p, q; b^{(k)}) > 0$, $k(p, q; b^{(k)}) < 0$, or $k(p, q; b^{(k)}) \sim 0$, express that each amino acid in vector $b^{(k)}$ is in a relation of cohesion, exclusion or neutral reaction (in which case it is neither cohesion nor exclusion), respectively.
2. If γ represents a permutation in set $\{1, 2, \cdots, n\}$, we denote as

$$\gamma\left(b^{(k)}\right) = \left(b_{\gamma(1)}, b_{\gamma(2)}, \cdots, b_{\gamma(n)}\right) , \tag{14.2}$$

the permutation on the position of each amino acid in vector $b^{(k)}$. If we take $q(b^{(k)}) = p[\gamma(b^{(k)})]$, then the value of the relative entropy density function $k(p, q; b^{(k)})$ reflects the "order orientation" of the order of each amino acid in vector $b^{(k)}$.
3. "Conditional affinity." We take $k = 3$ as an example. Here we denote $b^{(3)} = (a, b, c)$, and take

$$k(p, q; a, c|b) = \log \frac{p(a, b, c)p(b)}{p(a, b)p(b, c)} = \log \frac{p(a, c|b)}{p(a|b)p(c|b)} , \tag{14.3}$$

which denotes the affinity of a and c on the condition when b is given. Here $k(p, q; a, b|c) > 0$, $k(p, q; a, b|c) < 0$, and $k(p, q; a, b|c) \sim 0$, denote, respectively that amino acids a, c are of cohesion, exclusion or neutral reaction on the condition of the given b. If amino acids a, c are of neutral reaction on the condition of the given b, then we say the tripeptide (a, b, c) is a Markov chain.

The three different types of relative entropy density functions above reflect the stability of the protein primary structure in three different aspects. Normally, if $q(b^{(k)})$ is considered to be the replacement of the polypeptide chain structure $p(b^{(k)})$, then $k(p, q; b^{(k)}) > 0$, expresses the fact that the stability of the original polypeptide chain $b^{(k)}$ is higher than that of the other replacements.

The Generation of Local Words

It has been shown in the above text that there are eigenvalues and distribution functions of the relative entropy density function $k(p, q; b^{(k)})$. For different choices of p, q, the relative entropy density function may reflect the stability of the protein primary structure in different aspects. Thus we refer to the conditions of $k(p, q; b^{(k)}) \gg 0$, or $k(p, q; b^{(k)}) \ll 0$ as the "strong stability" state or "strong nonstability" state, respectively. The "strong stability" state or "strong unstability" state is often assessed by $\tau\sigma$ determination in statistics. That is,

$$\begin{cases} k\left(p,q;b^{(k)}\right) - \mu(p,q) \geq \tau_k\sigma(p,q)\,, & b^{(k)} \text{ is said to be } \tau_k \text{ stable}\,, \\ k\left(p,q;b^{(k)}\right) - \mu(p,q) \leq -\tau_k\sigma(p,q)\,, & b^{(k)} \text{ is said to be } \tau_k \text{ nonstable}\,, \end{cases} \tag{14.4}$$

where $b^{(k)}$ is a polypeptide chain, τ_k is taken to be constant and $\mu(p,q)$, $\sigma(p,q)$ are the mean and standard deviation of $k(p,q;b^{(k)})$, respectively. The determination of local words using the formula (14.4) is called the (p,q,τ) determination, or (τ,σ) determination for short. It is a common method in statistics, while the theories of data analysis and data mining do not commonly employ several types of relative entropy density functions.

14.1.4 Semantic Analysis for Protein Primary Structure

As stated above, protein primary structures are the sequence structures composed of amino acids. Their structure characteristics are similar to human languages we use, thus the research on protein primary structures can be considered a branch of research on semantic structures. We now use the determination of local words given above and the Swiss-Prot database to construct a table of relative entropy density functions of protein primary structures and the local words of second, third, and fourth rank. The computation process is described below.

Table of Relative Entropy Density Functions of Dipeptides

Tables of relative entropy density functions of dipeptides can come in one of two types. When $k = 2$, for the fixed distribution $p(a,b)$, $a,b \in V_q$, we take $q_1(a,b) = p(a)p(b)$, $q_2(a,b) = p(b,a)$, and the corresponding relative entropy density functions are, respectively,

$$k_0(a,b) = \log\frac{p(a,b)}{p(a)p(b)}\,, \quad k_1(a,b) = \log\frac{p(a,b)}{p(b,a)}\,. \tag{14.5}$$

In the Swiss-Prot database, the computational results of $k_0(a,b)$, $k_1(a,b)$ are listed as in Tables 14.2, 14.3, 14.4, and 14.5.

From Tables 14.4 and 14.5, the eigenvalues of k_0, k_1 are obtained, and are shown in Table 14.6.

Searching for the Second Ranked Local Words

When $k = 2$, from Tables 14.4, 14.5, and 14.6 we can find the second ranked local words. For example, when we take $\tau = 2$, the dipeptides of the second ranked local words and their types can be obtained, and are listed in Table 14.7.

In Table 14.7, type 1 is the affinitive type, where the dipeptides are attractive. Type 2 is the repulsive type, where the dipeptides are repulsive. Type 3 is the ordered type, where the probability of the dipeptides occurring is obviously related to the order.

$k_1(p,q;a,b) = -k_1(p,q;b,a)$ is antisymmetric; for example, if

$$k_1(p,q;M,W) = 0.3535 ,$$

then

$$k_1(p,q;W,M) = -0.3535 .$$

Thus the probability of dipeptide MW occurring is much higher than that of WM.

Determination of the Third Ranked Local Words

The determination of the third ranked local words is generally the same as that for the second ranked ones. This is discussed as follows.

From database Ω, we can determine the frequency distribution $p(a,b,c)$ of the third ranked vector $b^{(3)} = (a,b,c)$, $a,b,c \in V_q$, and compute its first and second ranked marginal distributions $p(a,b)$, $p(b,c)$, $p(a)$, $p(b)$, and $p(c)$. From this, eight types of relative entropy density functions can be generated:

$$\begin{aligned} k_0(a,b,c) &= \log \frac{p(a,b,c)}{p(a)p(b)p(c)} , \quad & k_1(a,b,c) &= \log \frac{p(a,b,c)}{p(a)p(c,b)} , \\ k_2(a,b,c) &= \log \frac{p(a,b,c)}{p(b)p(a,c)} , \quad & k_3(a,b,c) &= \log \frac{p(a,b,c)}{p(c)p(a,b)} , \\ k_4(a,b,c) &= \log \frac{p(a,b,c)}{p(b,a,c)} , \quad & k_5(a,b,c) &= \log \frac{p(a,b,c)}{p(c,b,a)} , \\ k_6(a,b,c) &= \log \frac{p(a,b,c)}{p(a,c,b)} , \quad & k_7(a,b,c) &= \log \frac{p(a,b,c)}{p(b,c,a)} , \\ k_8(a,b,c) &= \log \frac{p(a,b,c)}{p(c,a,b)} . \end{aligned} \tag{14.6}$$

Their eigenvalues are given in Table 14.8, where the number of the local words of type I refers to the number of tripeptide chains following $k_\tau(a,b,c) > \mu_\tau + 3.5\sigma_\tau$, and the number of the local words of type II refers to the number of tripeptide chains which follow $k_\tau(a,b,c) < \mu_\tau - 3.5\sigma_\tau$.

We comment on the following aspects for Table 14.9:

1. In this table, the capital letter is the tripeptide and the number is function k_3. When $k_3 > 0$, it is a chain of peptides of type I, which attract each other, and when $k_3 < 0$, it is a chain of peptides of type II, which repel each other. There are in total 54 local words in Table 14.9.
2. Data in this table are obtained on the condition of $\tau = 3.5$. If we lower the value of τ, there will be an increase in the number of local words.
3. For each type of dynamic function $k_0 - k_8$, we can also obtain the vocabulary of local words in tripeptide chains, which will not be listed here.

Table 14.2. Joint probability distribution of dipeptides ($100.0 \times p(a, b)$)

One-letter code	A	R	N	D	C	Q	E	G	H	I
A	0.76537	0.39096	0.28178	0.37277	0.11765	0.29835	0.48523	0.55710	0.15979	0.43803
R	0.38201	0.36240	0.21778	0.27674	0.08674	0.21473	0.34813	0.33846	0.12562	0.30029
N	0.29388	0.20178	0.23192	0.21355	0.07119	0.16826	0.25486	0.30345	0.09457	0.28579
D	0.38088	0.25462	0.21765	0.30138	0.08013	0.17400	0.37399	0.36371	0.10670	0.34148
C	0.10631	0.08940	0.06923	0.08561	0.04366	0.06543	0.08956	0.12952	0.04489	0.08495
Q	0.31392	0.22679	0.17286	0.18411	0.06314	0.24227	0.26160	0.24338	0.09281	0.22321
E	0.50501	0.35878	0.30708	0.36572	0.09053	0.26458	0.57415	0.38987	0.13739	0.40062
G	0.51538	0.36239	0.28144	0.35612	0.11160	0.25190	0.41022	0.56177	0.15732	0.41570
H	0.15288	0.12318	0.09023	0.09877	0.04571	0.09565	0.11834	0.15805	0.07984	0.13083
I	0.44105	0.28904	0.27924	0.33045	0.10024	0.21637	0.36631	0.37594	0.13426	0.36708
L	0.76506	0.52405	0.41123	0.49989	0.15044	0.39975	0.60975	0.62532	0.21746	0.51781
K	0.43623	0.31970	0.29318	0.32832	0.08609	0.22728	0.45619	0.35329	0.12500	0.37411
M	0.21003	0.12093	0.10774	0.12880	0.03391	0.08653	0.15529	0.15487	0.05190	0.13134
F	0.27427	0.19538	0.18369	0.23432	0.07325	0.15431	0.23798	0.29506	0.09590	0.24714
P	0.37621	0.23453	0.19209	0.25548	0.07115	0.19636	0.35396	0.37597	0.11270	0.24024
S	0.49778	0.35150	0.29734	0.35345	0.12012	0.27642	0.39911	0.52118	0.15754	0.39264
T	0.43252	0.25090	0.22992	0.27473	0.09673	0.20075	0.32007	0.40910	0.12336	0.32815
W	0.08342	0.06732	0.05894	0.06386	0.02053	0.05210	0.06772	0.08240	0.02972	0.07087
Y	0.20324	0.17052	0.14609	0.17386	0.05816	0.13009	0.18654	0.22707	0.07535	0.18135
V	0.52862	0.32743	0.28030	0.36796	0.11328	0.23625	0.42456	0.42717	0.14048	0.39462

Table 14.2. (continued)

One-letter code	L	K	M	F	P	S	T	W	Y	V
A	0.77063	0.42322	0.17673	0.29664	0.34644	0.51443	0.41759	0.08361	0.21116	0.54317
R	0.49873	0.31285	0.11488	0.21685	0.23865	0.33371	0.25895	0.06568	0.17469	0.33757
N	0.40347	0.26204	0.09420	0.18662	0.24167	0.30913	0.23280	0.05646	0.15246	0.27861
D	0.51897	0.29412	0.11104	0.23785	0.26769	0.34573	0.26183	0.07013	0.19042	0.36284
C	0.15066	0.08859	0.03041	0.06904	0.08920	0.12561	0.08722	0.02347	0.05353	0.10264
Q	0.38655	0.23968	0.08844	0.14327	0.19435	0.23981	0.20613	0.04958	0.11866	0.25011
E	0.60379	0.47997	0.14805	0.23310	0.23898	0.36464	0.33464	0.07261	0.18995	0.41787
G	0.61366	0.42341	0.15387	0.29662	0.29043	0.49632	0.38941	0.09033	0.23280	0.47256
H	0.22857	0.10892	0.04632	0.10621	0.13912	0.15904	0.12174	0.03050	0.08222	0.13916
I	0.54553	0.35519	0.11761	0.24300	0.29385	0.42543	0.34235	0.06332	0.18572	0.38177
L	0.97237	0.58216	0.19742	0.38382	0.48733	0.70084	0.53970	0.10246	0.27004	0.60292
K	0.53766	0.47457	0.12613	0.20728	0.26782	0.36416	0.33120	0.06473	0.19653	0.38002
M	0.21546	0.15529	0.06536	0.09118	0.10903	0.17159	0.13605	0.02380	0.06373	0.15826
F	0.40538	0.21864	0.08318	0.19305	0.18031	0.32761	0.23687	0.05348	0.14181	0.26289
P	0.43470	0.25264	0.09039	0.19595	0.32115	0.37428	0.27757	0.06108	0.15024	0.33794
S	0.68411	0.38268	0.14252	0.30264	0.36642	0.65234	0.40922	0.08769	0.21847	0.45245
T	0.54884	0.28878	0.11458	0.23100	0.31747	0.40661	0.35595	0.07044	0.16896	0.40420
W	0.12401	0.07429	0.03041	0.05196	0.04582	0.08136	0.06791	0.02176	0.04053	0.07504
Y	0.30093	0.17524	0.06603	0.14935	0.14852	0.22264	0.17620	0.04275	0.12169	0.19739
V	0.64004	0.38336	0.14311	0.27200	0.32753	0.46504	0.39540	0.07986	0.19772	0.48693

Table 14.3. Probability distribution of $100.0 \times q_0(a,b) = 100.0 \times p(a)p(b)$

One-letter code	A	R	N	D	C	Q	E	G	H	I
A	0.58635	0.39948	0.33278	0.40287	0.12503	0.30231	0.49680	0.52733	0.17310	0.44880
R	0.39895	0.27181	0.22642	0.27411	0.08507	0.20569	0.33802	0.35880	0.11778	0.30536
N	0.33237	0.22645	0.18864	0.22837	0.07087	0.17136	0.28161	0.29892	0.09812	0.25440
D	0.40276	0.27440	0.22859	0.27673	0.08588	0.20765	0.34125	0.36222	0.11890	0.30828
C	0.12484	0.08506	0.07085	0.08578	0.02662	0.06436	0.10577	0.11228	0.03686	0.09556
Q	0.30202	0.20576	0.17141	0.20751	0.06440	0.15571	0.25589	0.27162	0.08916	0.23117
E	0.49643	0.33822	0.28175	0.34109	0.10586	0.25594	0.42061	0.44646	0.14656	0.37998
G	0.52754	0.35941	0.29940	0.36246	0.11249	0.27198	0.44697	0.47444	0.15574	0.40379
H	0.17285	0.11776	0.09810	0.11876	0.03686	0.08911	0.14645	0.15545	0.05103	0.13230
I	0.44864	0.30566	0.25462	0.30825	0.09566	0.23130	0.38012	0.40348	0.13245	0.34339
L	0.73267	0.49917	0.41583	0.50341	0.15623	0.37774	0.62077	0.65893	0.21630	0.56080
K	0.45597	0.31066	0.25879	0.31329	0.09723	0.23509	0.38633	0.41008	0.13461	0.34901
M	0.18172	0.12381	0.10314	0.12486	0.03875	0.09369	0.15397	0.16343	0.05365	0.13909
F	0.31381	0.21380	0.17810	0.21561	0.06691	0.16179	0.26588	0.28222	0.09264	0.24019
P	0.37589	0.25610	0.21334	0.25827	0.08015	0.19380	0.31848	0.33806	0.11097	0.28772
S	0.54152	0.36894	0.30734	0.37207	0.11547	0.27919	0.45881	0.48701	0.15987	0.41449
T	0.42712	0.29100	0.24241	0.29347	0.09108	0.22021	0.36189	0.38413	0.12610	0.32693
W	0.09273	0.06318	0.05263	0.06372	0.01977	0.04781	0.07857	0.08340	0.02738	0.07098
Y	0.24165	0.16464	0.13715	0.16603	0.05153	0.12459	0.20474	0.21733	0.07134	0.18496
V	0.50825	0.34628	0.28846	0.34921	0.10838	0.26204	0.43063	0.45710	0.15005	0.38903

Table 14.3. (continued)

One-letter code	L	K	M	F	P	S	T	W	Y	V
A	0.73324	0.45717	0.16378	0.31425	0.37578	0.54169	0.42681	0.09286	0.24186	0.50833
R	0.49889	0.31106	0.11143	0.21381	0.25568	0.36856	0.29040	0.06318	0.16456	0.34587
N	0.41563	0.25915	0.09284	0.17813	0.21301	0.30705	0.24193	0.05264	0.13710	0.28815
D	0.50366	0.31403	0.11250	0.21585	0.25812	0.37208	0.29317	0.06378	0.16613	0.34917
C	0.15612	0.09734	0.03487	0.06691	0.08001	0.11533	0.09087	0.01977	0.05150	0.10823
Q	0.37768	0.23548	0.08436	0.16186	0.19356	0.27901	0.21984	0.04783	0.12458	0.26183
E	0.62079	0.38706	0.13866	0.26605	0.31815	0.45862	0.36135	0.07862	0.20477	0.43038
G	0.65969	0.41132	0.14735	0.28273	0.33809	0.48736	0.38400	0.08354	0.21760	0.45735
H	0.21615	0.13477	0.04828	0.09263	0.11077	0.15968	0.12582	0.02737	0.07130	0.14985
I	0.56103	0.34980	0.12531	0.24044	0.28752	0.41446	0.32657	0.07105	0.18506	0.38894
L	0.91622	0.57126	0.20465	0.39266	0.46956	0.67686	0.53332	0.11603	0.30222	0.63519
K	0.57020	0.35552	0.12736	0.24437	0.29223	0.42124	0.33191	0.07221	0.18808	0.39531
M	0.22725	0.14169	0.05076	0.09739	0.11646	0.16788	0.13228	0.02878	0.07496	0.15754
F	0.39242	0.24467	0.08765	0.16818	0.20111	0.28990	0.22842	0.04970	0.12944	0.27205
P	0.47006	0.29308	0.10499	0.20145	0.24090	0.34726	0.27362	0.05953	0.15505	0.32588
S	0.67717	0.42221	0.15125	0.29022	0.34705	0.50027	0.39417	0.08576	0.22337	0.46947
T	0.53413	0.33302	0.11930	0.22891	0.27374	0.39459	0.31091	0.06764	0.17618	0.37029
W	0.11596	0.07230	0.02590	0.04970	0.05943	0.08567	0.06750	0.01469	0.03825	0.08039
Y	0.30219	0.18841	0.06750	0.12951	0.15487	0.22324	0.17590	0.03827	0.09968	0.20950
V	0.63558	0.39628	0.14196	0.27239	0.32573	0.46954	0.36996	0.08049	0.20965	0.44063

Table 14.4. The value of relative entropy density function $k_0(a,b) = \log \frac{p(a,b)}{p(a)p(b)}$

One-letter code	A	R	N	D	C	Q	E	G	H	I
A	0.38439	−0.03112	−0.24002	−0.11206	−0.08779	−0.01899	−0.03400	0.07922	−0.11549	−0.03505
R	−0.06261	0.41499	−0.05614	0.01374	0.02799	0.06209	0.04254	−0.08417	0.09301	−0.02416
N	−0.17757	−0.16636	0.29805	−0.09678	0.00654	−0.02635	−0.14396	0.02172	−0.05321	0.16783
D	−0.08060	−0.10796	−0.07074	0.12312	−0.10004	−0.25510	0.13219	0.00593	−0.15618	0.14757
C	−0.23176	0.07182	−0.03349	−0.00285	0.71381	0.02375	−0.23999	0.20613	0.28448	−0.16978
Q	0.05580	0.14037	0.01220	−0.17265	−0.02857	0.63776	0.03187	−0.15838	0.05780	−0.05057
E	0.02473	0.08513	0.12421	0.10061	−0.22562	0.04785	0.44895	−0.19553	−0.09316	0.07633
G	−0.03363	0.01189	−0.08924	−0.02547	−0.01149	−0.11068	−0.12376	0.24376	0.01455	0.04193
H	−0.17707	0.06488	−0.12055	−0.26589	0.31066	0.10206	−0.30746	0.02396	0.64579	−0.01614
I	−0.02460	−0.08066	0.13313	0.10034	0.06735	−0.09630	−0.05339	−0.10200	0.01965	0.09622
L	0.06240	0.07016	−0.01605	−0.01012	−0.05447	0.08169	−0.02584	−0.07553	0.00771	−0.11506
K	−0.06387	0.04140	0.18002	0.06759	−0.17548	−0.04872	0.23978	−0.21505	−0.10692	0.10017
M	0.20889	−0.03390	0.06299	0.04480	−0.19228	−0.11467	0.01229	−0.07767	−0.04774	−0.08275
F	−0.19429	−0.12999	0.04458	0.12002	0.13044	−0.06829	−0.15996	0.06416	0.04989	0.04113
P	0.00123	−0.12694	−0.15136	−0.01570	−0.17187	0.01892	0.15235	0.15333	0.02226	−0.26020
S	−0.12148	−0.06987	−0.04771	−0.07407	0.05691	−0.01438	−0.20113	0.09784	−0.02114	−0.07811
T	0.01812	−0.21393	−0.07635	−0.09519	0.08691	−0.13348	−0.17717	0.09084	−0.03166	0.00538
W	−0.15276	0.09156	0.16336	0.00329	0.05386	0.12398	−0.21447	−0.01736	0.11836	−0.00231
Y	−0.24973	0.05066	0.09117	0.06646	0.17474	0.06230	−0.13434	0.06329	0.07895	−0.02844
V	0.05669	−0.08072	−0.04138	0.07542	0.06381	−0.14946	−0.02049	−0.09770	−0.09505	0.02060

Table 14.4. (continued)

One-letter code	L	K	M	F	P	S	T	W	Y	V
A	0.07175	−0.11132	0.10986	−0.08316	−0.11728	−0.07449	−0.03150	−0.15141	−0.19587	0.09562
R	−0.00047	0.00829	0.04391	0.02038	−0.09948	−0.14330	−0.16536	0.05587	0.08616	−0.03504
N	−0.04286	0.01600	0.02109	0.06717	0.18213	0.00973	−0.05549	0.10119	0.15325	−0.04856
D	0.04320	−0.09449	−0.01882	0.14000	0.05252	−0.10599	−0.16309	0.13693	0.19685	0.05540
C	−0.05132	−0.13592	−0.19742	0.04520	0.15688	0.12312	−0.05923	0.24724	0.05578	−0.07645
Q	0.03349	0.02550	0.06816	−0.17598	0.00593	−0.21842	−0.09293	0.05195	−0.07023	−0.06608
E	−0.04006	0.31037	0.09453	−0.19076	−0.41284	−0.33082	−0.11078	−0.11461	−0.10839	−0.04253
G	−0.10436	0.04180	0.06252	0.06920	−0.21922	0.02630	0.02018	0.11260	0.09740	0.04721
H	0.08060	−0.30725	−0.05962	0.19733	0.32872	−0.00576	−0.04756	0.15590	0.20558	−0.10673
I	−0.04042	0.02205	−0.09146	0.01530	0.03142	0.03768	0.06809	−0.16609	0.00520	−0.02686
L	0.08581	0.02727	−0.05189	−0.03287	0.05360	0.05022	0.01716	−0.17947	−0.16239	−0.07522
K	−0.08477	0.41671	−0.01404	−0.23747	−0.12581	−0.21008	−0.00309	−0.15785	0.06341	−0.05690
M	−0.07681	0.13222	0.36472	−0.09511	−0.09517	0.03153	0.04060	−0.27387	−0.23408	0.00652
F	0.04687	−0.16231	−0.07545	0.19899	−0.15750	0.17641	0.05240	0.10577	0.13168	−0.04942
P	−0.11282	−0.21423	−0.21600	−0.03995	0.41477	0.10808	0.02070	0.03718	−0.04549	0.05242
S	0.01471	−0.14183	−0.08578	0.06049	0.07837	0.38292	0.05403	0.03212	−0.03197	−0.05326
T	0.03921	−0.20565	−0.05828	0.01311	0.21385	0.04328	0.19521	0.05842	−0.06041	0.12641
W	0.09683	0.03918	0.23165	0.06407	−0.37534	−0.07451	0.00878	0.56754	0.08348	−0.09934
Y	−0.00601	−0.10458	−0.03176	0.20565	−0.06045	−0.00389	0.00250	0.15978	0.28785	−0.08587
V	0.01008	−0.04784	0.01159	−0.00206	0.00794	−0.01391	0.09595	−0.01138	−0.08453	0.14414

Table 14.5. The value of relative entropy density function $k_1(a,b) = \log \frac{p(a,b)}{p(b,a)}$

One-letter code	A	R	N	D	C	Q	E	G	H	I
A	0.00000	0.03342	−0.06067	−0.03105	0.14615	−0.07340	−0.05766	0.11228	0.06371	−0.00991
R	−0.03342	0.00000	0.11007	0.12018	−0.04358	−0.07882	−0.04345	−0.09855	0.02834	0.05510
N	0.06067	−0.11007	0.00000	−0.02742	0.04043	−0.03895	−0.26887	0.10863	0.06769	0.03346
D	0.03105	−0.12018	0.02742	0.00000	−0.09542	−0.08147	0.03226	0.03044	0.11145	0.04736
C	−0.14615	0.04358	−0.04043	0.09542	0.00000	0.05153	−0.01547	0.21489	−0.02621	−0.23877
Q	0.07340	0.07882	0.03895	0.08147	−0.05153	0.00000	−0.01629	−0.04965	−0.04351	0.04488
E	0.05766	0.04345	0.26887	−0.03226	0.01547	0.01629	0.00000	−0.07340	0.21537	0.12918
G	−0.11228	0.09855	−0.10863	−0.03044	−0.21489	0.04965	0.07340	0.00000	−0.00671	0.14502
H	−0.06371	−0.02834	−0.06769	−0.11145	0.02621	0.04351	−0.21537	0.00671	0.00000	−0.03739
I	0.00991	−0.05510	−0.03346	−0.04736	0.23877	−0.04488	−0.12918	−0.14502	0.03739	0.00000
L	−0.01047	0.07144	0.02748	−0.05403	−0.00210	0.04846	0.01418	0.02715	−0.07187	−0.07522
K	0.04367	0.03126	0.16202	0.15870	−0.04117	−0.07662	−0.07330	−0.26120	0.19869	0.07487
M	0.24904	0.07413	0.19370	0.21404	0.15732	−0.03143	0.06885	0.00926	0.16403	0.15926
F	−0.11315	−0.15046	−0.02283	−0.02160	0.08539	0.10706	0.02986	−0.00762	−0.14731	0.02435
P	0.11893	−0.02511	−0.33128	−0.06739	−0.32616	0.01479	0.56669	0.37242	−0.30390	−0.29066
S	−0.04745	0.07490	−0.05612	0.03187	−0.06449	0.20496	0.13030	0.07052	−0.01370	−0.11572
T	0.05070	−0.04557	−0.01800	0.06938	0.14938	−0.03810	−0.06425	0.07116	0.01911	−0.06110
W	−0.00330	0.03566	0.06200	−0.13519	−0.19316	0.07146	−0.10074	−0.13248	−0.03737	0.16236
Y	−0.05512	−0.03483	−0.06155	−0.13124	0.11988	0.13267	−0.02613	−0.03593	−0.12574	−0.03435
V	−0.03916	−0.04399	0.00874	0.02020	0.14221	−0.08223	0.02288	−0.14570	0.01359	0.04776

Table 14.5. (continued)

One-letter code	L	K	M	F	P	S	T	W	Y	V
A	0.01047	−0.04367	−0.24904	0.11315	−0.11893	0.04745	−0.05070	0.00330	0.05512	0.03916
R	−0.07144	−0.03126	−0.07413	0.15046	0.02511	−0.07490	0.04557	−0.03566	0.03483	0.04399
N	−0.02748	−0.16202	−0.19370	0.02283	0.33128	0.05612	0.01800	−0.06200	0.06155	−0.00874
D	0.05403	−0.15870	−0.21404	0.02160	0.06739	−0.03187	−0.06938	0.13519	0.13124	−0.02020
C	0.00210	0.04117	−0.15732	−0.08539	0.32616	0.06449	−0.14938	0.19316	−0.11988	−0.14221
Q	−0.04846	0.07662	0.03143	−0.10706	−0.01479	−0.20496	0.03810	−0.07146	−0.13267	0.08223
E	−0.01418	0.07330	−0.06885	−0.02986	−0.56669	−0.13030	0.06425	0.10074	0.02613	−0.02288
G	−0.02715	0.26120	−0.00926	0.00762	−0.37242	−0.07052	−0.07116	0.13248	0.03593	0.14570
H	0.07187	−0.19869	−0.16403	0.14731	0.30390	0.01370	−0.01911	0.03737	0.12574	−0.01359
I	0.07522	−0.07487	−0.15926	−0.02435	0.29066	0.11572	0.06110	−0.16236	0.03435	−0.04776
L	0.00000	0.11471	−0.12622	−0.07884	0.16488	0.03485	−0.02423	−0.27546	−0.15625	−0.08618
K	−0.11471	0.00000	−0.30005	−0.07692	0.08421	−0.07158	0.19771	−0.19886	0.16546	−0.01262
M	0.12622	0.30005	0.00000	0.13238	0.27042	0.26778	0.24783	−0.35355	−0.05106	0.14517
F	0.07884	0.07692	−0.13238	0.00000	−0.11999	0.11436	0.03619	0.04164	−0.07474	−0.04915
P	−0.16488	−0.08421	−0.27042	0.11999	0.00000	0.03059	−0.19379	0.41490	0.01664	0.04514
S	−0.03485	0.07158	−0.26778	−0.11436	−0.03059	0.00000	0.00923	0.10813	−0.02729	−0.03958
T	0.02423	−0.19771	−0.24783	−0.03619	0.19379	−0.00923	0.00000	0.05267	−0.06059	0.03175
W	0.27546	0.19886	0.35355	−0.04164	−0.41490	−0.10813	−0.05267	0.00000	−0.07700	−0.08969
Y	0.15625	−0.16546	0.05106	0.07474	−0.01664	0.02729	0.06059	0.07700	0.00000	−0.00236
V	0.08618	0.01262	−0.14517	0.04915	−0.04514	0.03958	−0.03175	0.08969	0.00236	0.00000

Table 14.6. The eigenvalues of relative entropy density function k_0, and k_1

	Mean (μ)	Variance (σ^2)	Standard deviation (σ)	Maximum (μ_M)	Minimum (μ_m)	$\mu - 2\sigma$	$\mu + 2\sigma$
k_0	−0.00633	0.01879	0.13708	0.71381	−0.41384	0.28049	−0.26783
k_1	0.00441	0.01274	0.11289	0.56669	−0.56669	0.23019	−0.22137

Determination of Fourth and Higher Ranked Local Words

The search for fourth ranked local words is similar to that of the second and third ranked ones, while their relative entropy density functions are more complex. For example, there are 11+18 = 29 relative entropy density functions of fourth rank. We can also find local words by choosing a proper value of τ. Because of the amount of statistical data, local words with a rank higher than fourth cannot be processed by the statistical methods used for the second and third ranked ones. They must be processed by permutation and combination methods or the method of combining lower ranked words. Permutation and combination methods will be discussed in detail in the next section.

14.2 Permutation and Combination Methods for Semantic Structures

In the previous section, informational and statistical analysis methods for the semantic structure of protein primary structure have been given. In using this method, we notice that the basis of the informational and statistical methods is the computation of frequency numbers and probabilities for polypeptide chains. However, if the lengths of the polypeptide chain vectors increase, the number of combinations (20^n of the polypeptide chain vector $b^{(k)}$) will increase rapidly. For example, $20^6 = 6.4 \times 10^7$. This exceeds the total number of amino acids in the Swiss-Prot database (version 2000); hence the informational and statistical methods no longer work. Therefore, other methods must be used to analyze the semantic structures of higher ranked words.

In this section, we continue to analyze the semantic structure of protein primary structures with combinatorial graph theory methods, and give the definition of the key words and core words as well as their characteristic parameters for protein primary structures in the Swiss-Prot database. The concept of the key words and core words refers to a special type of polypeptide chains, which exist uniquely in a protein database (i.e., in nature). Hence the key words and core words are actually special kinds of biological signatures [91].

The concept of a biological signature occurs commonly in biological semantic analysis. Besides the small-molecule library and conformation theory mentioned above, many research institutions are building their own annotated

Table 14.7. Parameters of the second ranked local words determined by k_0, k_1

Dipeptide	AA	RR	NN	CC	CH	QQ	EE	EK	HC	HH	HP	KK	MM
Type	1	1	1	1	1	1	1	1	1	1	1	1	1
Relative entropy	0.3843	0.4149	0.2980	0.7138	0.2844	0.6377	0.4489	0.3103	0.3106	0.6457	0.3287	0.4167	0.3647
Dipeptide	PP	SS	WW	YY	EP	ES	HE	HK	MW	NP	CP	EN	GK
Type	1	1	1	1	2	2	2	2	2	3	3	3	3
Relative entropy	0.4147	0.3829	0.5675	0.2878	−0.4128	−0.3308	−0.3074	−0.3072	−0.2738	0.3312	0.3261	0.2688	0.2612
Dipeptide	HP	IC	IP	MA	MK	MP	MS	MT	PE	PG	PW	WL	WM
Type	3	3	3	3	3	3	3	3	3	3	3	3	3
Relative entropy	0.3039	0.2387	0.2906	0.2490	0.3000	0.2704	0.2677	0.2478	0.5666	0.3724	0.4148	0.2754	0.3535

Table 14.8. The eigenvalues of the relative entropy density function of the third rank

Function type	Mean (μ)	Variance (σ^2)	Standard deviation (σ)	Maximum (μ_M)	Minimum (μ_m)	$\mu + 3.5\sigma$	$\mu - 3.5\sigma$	Number of local words of type I	Number of local words of type II
k_0	0.02239	0.06933	0.26331	2.82880	−0.98377	0.94399	−0.89920	50	1
k_1	0.01590	0.04803	0.21916	2.18053	−0.92031	0.78295	−0.75114	53	6
k_2	0.01652	0.04983	0.22323	2.23581	−1.13892	0.79782	−0.76478	49	9
k_3	0.01592	0.04823	0.21962	2.18089	−0.90762	0.78458	−0.75274	54	2
k_4	0.01778	0.05152	0.22697	1.37851	−1.37851	0.81218	−0.77663	36	48
k_5	0.01466	0.04256	0.20630	1.58496	−1.58496	0.73673	−0.70740	48	57
k_6	0.01726	0.05002	0.22364	1.50198	−1.50198	0.80001	−0.76549	37	40
k_7	0.01400	0.04054	0.20134	1.70828	−1.58496	0.71869	−0.69069	31	46
k_8	0.01398	0.04040	0.20100	1.58496	−1.70828	0.71748	−0.68951	39	39

Table 14.9. Vocabulary of local words in tripeptide chains obtained by function k_3 where $\tau = 3.5$

Tripeptide	Function k_3	Tripeptide	Function k_3	Tripeptide	Function k_3	Tripeptide	Function k_3
AAR	0.8800	AWY	0.9401	RRN	1.0088	RCQ	0.8608
NND	1.1113	DWY	1.0809	CCN	0.9025	CCQ	1.1099
CCH	0.9238	CCI	0.9867	CCP	1.1000	CCS	0.9419
CCT	1.1079	CCY	1.0179	CCV	0.9288	CGM	0.9367
CHW	0.8036	CWF	0.9154	CWY	0.9028	QQE	2.1651
QQI	0.9102	QWY	0.8563	EEG	1.0708	HCQ	1.1058
HQE	0.8404	HHE	1.0121	HHI	2.2358	HHF	0.8031
HHS	1.2020	HPG	0.8862	HPY	0.8606	HWV	0.9085
HYQ	0.8992	LCQ	0.8312	PPH	0.8159	PPS	1.3695
SCQ	1.1518	SST	1.0226	SWY	0.9231	TCQ	0.8391
TWY	0.9172	WCQ	1.0100	WHI	1.0138	WHF	1.3881
WWD	1.7609	WWV	0.9166	WYP	0.8767	WYY	1.0541
YWY	1.0033	CMQ	−1.1389	CMS	−0.7809	EPF	−0.8306
HEI	−0.7943	HES	−0.7871	FKY	−0.8506	WPQ	−0.8086
WPM	−0.7852	YPQ	−0.7881				

databases, such as PROSITE, Pratt, EMOTIF, etc. [43, 45, 47, 48, 94]. They use different methods; for example, Pratt and EMOTIF are index databases analyzing the structures and functions of the polypeptide chains directly, and PROSITE is an index database of homologous protein classes obtained by the alignment of homologous proteins using PSI-BLAST [4]. Therefore, it is meaningful to compare the relationships between these different types of databases.

The core of the combinational graph theory method is the vector segment of the sequence data. When its length reaches a certain level, data structures of the long sequences may form a recursive relation. This characteristic is widely used in the analysis of shift register sequences and codes. Theoretically, its scope makes it applicable by means of the complexity theory of sequence data and the theory of Boolean function or the de Bruijn–Good graph of data structures. Therefore, many theories and tools used in the analysis of shift register sequences and codes can be brought in. However, the purpose of the research on biological sequences is different from that of the code analysis. The former aims to find the relationship between words and the language of biological sequences, while the latter aims to construct the pseudorandomicity of the sequences. In-depth discussion on combinatorial graph theory can be found in the literature [35]. Combinatorial graph theory methods can also be used to discuss the complexity, classification, cutting and regulation of databases. In this book, we only discuss the use of core words for the classification and prediction of homologous proteins.

14.2.1 Notation Used in Combinatorial Graph Theory

The mathematical model and involved definitions and notation used for the protein primary structures can readily be found in the literature [89], and hence it will not be repeated in detail. The theory of Boolean functions and that of the de Bruijn–Good graph can also be found in the literature [35].

Combination Space and Database

Let $V_q = \{1, 2, \cdots, q\}$ be a set of integers, which represents an alphabet for biological sequences. In the database of protein primary structures, we take $q = 20$ to denote the 20 commonly occurring amino acids. For the sake of convenience, in this book we set $q = 23$, and take 21 to be the zero element. Here V_q is a finite field, in which addition and multiplication are integer operations modulo 23.

Let $V_q^{(k)}$ be the kth ranked product space of V_q, whose element $b^{(k)} = (b_1, b_2, \cdots, b_k) \in V_q^{(k)}$ is the kth ranked vector on V_q. $V_q^{(k)}$ is also called the kth ranked combination space of V_q.

As mentioned in Sect. 14.1.2, Ω is a database of protein primary structures, which is composed of M proteins. Here

$$\Omega = \{C_s,\ s = 1, 2, \cdots, M\} , \tag{14.7}$$

where $C_s = (c_{s,1}, c_{s,2}, \cdots, c_{s,n_s})$ is the primary structure sequence of a single protein and n_s is the length of the sth protein sequence, whose components $c_{s,i} \in V_q$ are the commonly occurring amino acids. If C_s is the primary structure sequence of a protein, then we denote

$$c_s^{[i,j]} = (c_{s,i}, c_{s,i+1}, \cdots, c_{s,j}), \quad 1 \le i \le j \le n_s , \tag{14.8}$$

to represent a polypeptide chain of protein C_s, whose length is $k = j - i + 1$. Here $c_s^{[i,j]} \in V_q^{(k)}$ is a vector in kth ranked combination space.

Boolean Functions on a Combination Space

If $V_q^{(k)}$ is a combination space and $f(b^{(k)})$ is a single-valued mapping on $V_q^{(k)} \to V_q$, then we say f is a kth ranked Boolean function with q elements on V_q. In mathematics, the Boolean functions can have several representations, such as the listing representation, combination representation, function representation, graph representation, etc. They are described in detail in the literature [35], and here we only introduce the related notation.

1. Combination representation. The listing representation is what we are familiar with. The combination representation can be represented by a group of subsets of $V_q^{(k)}$

$$\mathcal{A}_f = \{\mathbf{A}_{f,1}, \mathbf{A}_{f,2}, \cdots, \mathbf{A}_{f,q}\} , \tag{14.9}$$

where

$$\mathbf{A}_{f,j} = \left\{b^{(k)} \in V_q^{(k)} : f\left(b^{(k)}\right) = j\right\} , \quad j \in V_q , \tag{14.10}$$

then $\mathcal{A}_f$ is called the combination representation of the Boolean function. Here $\mathcal{A}_f$ is a division of $V_q^{(k)}$.

2. Function representation. If f is a mapping on $V_q^{(k)} \to V_q$, then f is a function whose domain is $V_q^{(k)}$ and takes values in V_q. If V_q is a finite field, the Boolean function can be calculated by addition and multiplication operations on a finite field. The formula is

$$f\left(b^{(k)}\right) = \sum_{j_1, j_2, \cdots, j_k = 0}^{q-1} \left(\alpha_{j_1, j_2, \cdots, j_k} \prod_{i=1}^{k} b_i^{j_i}\right) , \tag{14.11}$$

where $\alpha_{j_1, j_2, \cdots, j_k} \in V_q$, and the addition, multiplication and power operations in formula (14.11) are operations on field V_q.

3. Graph representation. The definition of graph is given in [12, 35]. In this section, we denote it by $G = \{A, V\}$, where A is a vertex set, V is the dual point set of A, which is called the edge set in graph theory. In the following, we denote the vertices in A as a, b, c, etc., and the edges in V as

$(a, b), (a, c), (b, c)$, etc. The definitions of graphs fall into the categories of finite graphs, directed graphs, undirected graphs, subgraphs, supergraphs, plot graphs, etc. In this section we only discuss finite graphs and directed graphs.

The Theory of the de Bruijn–Good Graph

One of the important graphs is the de Bruijn–Good graph (which is called the DG graph for short). We denote a kth ranked DG graph by $G_{q,k} = \{B_{q,k}, V_{q,k}\}$, where

$$B_{q,k} = V_q^{(k)} , \quad V_{q,k} = V_q^{(k+1)} . \tag{14.12}$$

Here the elements in $B_{q,k}, V_{q,k}$ can be denoted respectively by

$$\begin{cases} b^{(k)} = (b_1, b_2, \cdots, b_k) , \\ b^{(k+1)} = ((b_1, b_2, \cdots, b_k), (b_2, b_3, \cdots, b_k, b_{k+1})) . \end{cases} \tag{14.13}$$

Then, $b^{(k+1)}$ can be considered to be dual points in $B_{q,k}$, or edges in $G_{q,k}$. When q is fixed, we denote $G_{q,k} = \{B_{q,k}, V_{q,k}\}$ as $G_k = \{B_k, V_k\}$ for short.

In the following, the subgraph of a DG graph is also called a DG graph. The Boolean graph is an important DG graph. If f is the Boolean function on $V_q^{(k)} \to V_q$, then we call $G_f = \{B, V_f\}$ the Boolean graph determined by f, where

$$B = V_q^{(k)} , \quad V_f = \left\{(b^{(k)}, f(b^{(k)})),\ b^{(k)} \in B\right\} . \tag{14.14}$$

Here G_f is a subgraph of graph G_k. Boolean graphs and Boolean functions determine each other. In graph theory, DG graphs can have several specific representations, which will not be introduced here.

Properties of the Boolean Graph

The definitions of edge, path and tree in graph theory have been given in Chap. 6. Detailed properties of the Boolean graph in the DG graph can be found in the literature [35]. In this section, we only introduce the basic properties. We know from the definition of a Boolean graph that a DG graph is a Boolean graph if and only if there is at most one outer-edge coming from each vertex. From this we arrive at the following conclusions:

1. There must be several cycles in a Boolean graph. We call the cycle with only one vertex a trivial cycle. There is no common vertex in different cycles. We denote all the cycles in a Boolean graph G_f by $\mathcal{O}_G = \{O_1, O_2, \cdots, O_m\}$, where each cycle is

$$O_s = \{b_{s,0} \to b_{s,1} \to \cdots \to b_{s,k_s-1} \to b_{s,k_s}\} , \quad s = 1, 2, \cdots, m , \tag{14.15}$$

where $b_{s,0} = b_{s,k_s}$.

2. In the vertex set B of a Boolean graph, each vertex b arrives at a cycle O_s in the end. In B, we denote all vertices arriving at cycle O_s by W_s.
3. Sets $W_1, W_2, \cdots, W_m$ are disjoint with each other, and their combination is set B.

DG Boolean Graph Generated by Sequences

Let $C(c_1, c_2, \cdots, c_n)$ be a sequence of length n in V_q, where $c_j \in V_q$. The subvector of sequence C is denoted as

$$c_i^{(k)} = (c_i, c_{i+1}, \cdots, c_{i+k-1}), \quad 1 \le i \le n-k+1 . \tag{14.16}$$

Let

$$B_k(C) = \left\{ c_i^{(k)} = (c_i, c_{i+1}, \cdots, c_{i+k-1}), \quad i = 1, 2, \cdots, n-k+1 \right\}, \tag{14.17}$$

then we call $B_k(C)$ the kth ranked vector family determined by sequence C.

Definition 55. *If C is a sequence in V_q, for any positive integer k, we define the kth ranked DG Boolean graph determined by sequence C as follows:*

$$G_k(C) = \{B_k(C), B_{k+1}(C)\} , \tag{14.18}$$

where the elements in $B_{k+1}(C)$ are

$$c_i^{(k+1)} = (c_i, c_{i+1}, \cdots, c_{i+k}) = ((c_i, \cdots, c_{i+k-1}), (c_{i+1}, \cdots, c_{i+k})) ,$$

which are dual points in $B_k(C)$. Thus $G_k(C)$ is a DG Boolean graph determined by C.

Graph $G_k(C)$ has the following properties:

1. Sequence C is a path in graph $G_k(C)$, and its terminus is $c_{n-k+1}^{(k)}$.
2. For the fixed sequence C, if vertices in $B_k(C)$ are different from each other, we denote $C' = (c_1, c_2, \cdots, c_n, c_1)$, where vector $(c_1, c_2, \cdots, c_n)$ is given in C. Then, the graph $G_k(C')$ is a Boolean graph, and sequence C' comprises a maximum cycle of $G_k(C')$, which traverses each vertex in $B_k(C)$ once and only once. We call it the Boolean cycle. Here graph $G_k(C)$ is a trunk tree (this tree does not have inner nodes).
3. If sequence C can generate a Boolean function, then we take

$$f(c_i, c_{i+1}, \cdots, c_{i+k-1}) = c_{i+k} , \quad i = 1, 2, \cdots, n-k+1 , \tag{14.19}$$

to be the Boolean function which generates sequence C. This will be called the generating function of sequence C for short in the following. In general, the solution of formula (14.19) is not unique. We denote by $\mathbf{F}^{(k)}(C)$ all the solutions that hold for formula (14.19), which are called the Boolean function family that generates C.

4. If there is a vertex occurring in $\mathbf{A}_k(C)$ more than once, then graph $G_k(C)$ contains a cycle. If $c_i^{(k)} = c_j^{(k)}$, $i < j$, then

$$c_i^{(k)} \to c_{i+1}^{(k)} \to \cdots \to c_{j-1}^{(k)} \to c_j^{(k)} = c_i^{(k)} \tag{14.20}$$

comprise a cycle. To have graph $G_k(C)$ contain no cycle, the value of k must be increased. This is to be solved by nonlinear complexity theory.

14.2.2 The Complexity of Databases

To discuss under what condition $G_k(C)$ would be a Boolean graph, we examine the problem of the sequence complexity in cryptography.

The Complexity of a Sequence

In sequence analysis used in cryptography, complexity can be associated with three different definitions: linear complexity, nonlinear complexity and nonsingular complexity. These concepts are frequently cited in the combinatorial analysis of semantics, and we begin with these definitions.

Definition 56. *If C is a given sequence, several definitions of complexity can be formulated as follows:*

1. *We call $k = K_N(C)$ the nonlinear complexity of C, if DG graph $G_k(C)$ is a Boolean graph while $G_{k-1}(C)$ is not a Boolean graph.*
2. *We call $k = K_L(C)$ the linear complexity of C, if $G_k(C)$ is a Boolean graph while its generating function f is a linear function on $V_q^{(k)} \to V_q$.*
3. *We call $k = K_S(C)$ the nonsingular complexity of C, if the generating function f of C is a nonsingular function on $V_q^{(k)} \to V_q$.*

Nonsingular functions are an important class of Boolean function. We define them as follows: in formula (14.19), when $(c_{i+1}, \cdots, c_{i+k-1})$ is fixed, each c_i corresponds to c_{i+k}.

In the literature [87], a series of properties and formulas for these three complexities are given. For instance, they follow formula $K_S(C) \le K_N(C) \le K_L(C)$ etc., which will not be discussed here in detail.

The Complexity of a Database

In Definition 55, the related definitions of the graph generated by a sequence and the complexity can be expanded to those of a database. In the database of protein primary structures given in Sect. 14.1.2, graphs generated by each protein primary structure sequence C_s, are defined in formula (14.18) to be

$$G_k(C_s) = \{B_k(C_s), V_k(C_s)\}, \quad s = 1, 2, \cdots, m. \tag{14.21}$$

We define $\mathcal{G}_k(\Omega) = \{\mathcal{A}_k(\Omega), \mathcal{V}_k(\Omega)\}$ as the graph generated by database Ω, where

$$\mathcal{A}_k(\Omega) = \bigcup_{s=1}^{m} B_k(C_s), \quad \mathcal{V}_k(\Omega) = \bigcup_{s=1}^{m} V_k(C_s).$$

From graph $\mathcal{G}_k(\Omega)$, the linear complexity, nonlinear complexity and nonsingular complexity of database Ω can be defined similarly. They follow from Definition 56, and are not repeated here.

The Biological Significance of Complexity

The essential significance of the sequence complexity is to discuss under what conditions in the same sequence, segments of different vectors lead to recursive relations and how the recursive relations express themselves and change. Thus, it is closely related to the concepts of the regulation and splicing of biological sequences.

We find in the calculation of biological sequences that, the computation of sequence complexity is effective for single protein sequences, but not as effective for the analysis of databases Ω.

Example 30. Trichosanthin is a kind of pharmaceutical protein extracted from Chinese herbs. It was an abortion-inducing drug [105], and in recent years, it was found to have an inhibition effect on several types of cancer and AIDS; attracting much attention. In the Swiss-Prot database, two homologous proteins RISA-CHLPN and RISA-CHLTR have the primary structures, given in Fig. 14.1.

We denote these two sequences by C, D. Their nonlinear complexity and nonsingular complexity are found to be, respectively,

$$\begin{aligned} K_N(C) = K_N(D) = 3, \quad K_N(C, D) = 94, \\ K_S(C) = K_S(D) = 4, \quad K_S(C, D) \geq 95. \end{aligned} \tag{14.22}$$

It shows that the nonlinear complexity of database Ω can become very high (in the Swiss-Prot database, the nonlinear complexity can be higher than 3000). The reason for this increase is that there are many mutually homologous sequences and self-homologous sequences in database Ω. Therefore, complexity analysis is useful in the database searches, mutation predictions and the general analysis of homologous sequences.

14.2.3 Key Words and Core Words in a Database

The Definitions of Key Words and Core Words

In order to analyze the database of protein primary structures efficiently using a combinatorial method, we set up the theory of key words and core words in a database.

RISA-CHLPN:

```
MIRFLVFSLLILTLFLTAPAVEGDVSFRLSGATSSSYGVFISNLRKALPYERKLYDIPLLRSTLPGSQRYALIHLTNYADETISVAIDVTNVYVMG
YRAGDTSYFFNEASATEAAKYVFKDAKRKVTLPYSGNYERLQIAAGKIRENIPLGLPALDSAITTLFYYNANSAASALMVLIQSTSEAARYKFIEQ
QIGKRVDKTFLPSLAIISLENSWSALSKQIQIASTNNGQFETPVVLINAQNQRVTITNVDAGVVTSNIALLLNRNNMAAIDDDVPMAQSFGCGSYAI
```

RISA-CHLTR:

```
MIRFLVLSLLILTLFLTTPAVEGDVSFRLSGATSSSYGVFISNLRKALPNERKLYDIPLLRSSLPGSQRYALIHLTNYADETISVAIDVTNVYIMG
YRAGDTSYFFNEASATEAAKYVFKDAMRKVTLPYSGNYERLQTAAGKIRENIPLGLPALDSAITTLFYYNANSAASALMVLIQSTSEAARYKFIEQ
QIGKRVDKTFLPSLAIISLENSWSALSKQIQIASTNNGQFESPVVLINAQNQRVTITNVDAGVVTSNIALLLNRNNMAAMDDDVPMTQSFGCGSYAI
```

Fig. 14.1. Primary structures of RISA-CHLPN and RISA-CHLTR

Definition 57. *1. We call vector $b^{(k)}$ the τth ranked key word in database Ω, if the frequency number $n_\Omega(b^{(k)})$ of $b^{(k)}$ occurring in Ω follows $n_\Omega(b^{(k)}) = \tau$, where $n_\Omega(b^{(k)})$ denotes the number of times vector $b^{(k)}$ occurs in database Ω.*

2. We call $b^{(k)}$ the τth ranked core word in database Ω, if $b^{(k)}$ is the τth ranked key word in Ω, and $n_\Omega(b^{(k-1)}) > \tau$, $n_\Omega(b_+^{(k-1)}) > \tau$ both hold, where

$$b^{(k-1)} = (b_1, b_2, \cdots, b_{k-1}), \quad b_+^{(k-1)} = (b_2, b_3, \cdots, b_k)$$

are subvectors with $(k-1)$ elements before or after $b^{(k)}$, respectively. The first ranked key word and core word are called for short the key word and core word, respectively.

Key words and core words are "labels" for protein primary structure sequences. That is, if $b^{(k)}$ is a core word in Ω, then there is one and only one sequence C_s in Ω that contains this vector.

Key words and core words can also serve as a "classification" method for proteins. If $b^{(k)}$ is the τth ranked key word in Ω, contained in proteins C_{s_i}, $i = 1, 2, \cdots, k$, then proteins $s_1, s_2, \cdots, s_k$ contain the same key word $b^{(k)}$. They can be considered to be homologous (or locally homologous) proteins.

A protein C_s may contain several core words, thus protein primary structure sequences have multiple "labels" or characteristics.

Example 31. In the trichosanthin RISA-CHLPN and RISA-CHLTR given in Example 30, the core words of length 6 are

```
C:  RFLVFS, PYERKL, YERKLY, TLPGSQ, TNVYVM, NVYVMG, VYVMGY,
    YVMGYR, VMGYRA, NGQFET, GQFETP, FETPVV, NMAAID, MAAIDD,
    AAIDDD, DVPMAQ, VPMAQS, PMAQSF, MAQSFG, AQSFGC.

D:  IRFLVL, TLFLTT, LPNERK, YIMGYR, VFKDAM, FKDAMR, KDAMRK,
    NNMAAM, NMAAMD, MAAMDD, VPMTQS, PMTQSF, MTQSFG, TQSFGC.
```

Properties of Key Words and Core Words

If $b^{(k)}$, $c^{(k')}$ are two vectors, and there exist $1 \le i < j \le k'$, and $j - i + 1 = k$ such that $b^{(k)} = c^{[i,j]}$ holds, then we say that vector $c^{(k')}$ contains $b^{(k)}$, and $c^{(k')}$ is an extension of $b^{(k)}$, while vector $b^{(k)}$ is a contraction of $c^{(k')}$. Key words and core words have the following properties:

1. If $b^{(k)}$ is a key word in the protein sequence C_s, then any extension $b^{(k')}$ in protein C_s is a key word and its frequency number recursively declines such that $n_\Omega(b^{(k)}) \ge n_\Omega(b^{(k')})$.

2. In database Ω, different core words do not contain each other. They occur once and only once in database Ω.
3. When the vector of a core word contracts, this core word changes into the τth ranked key word and the τth ranked core word, thereby generating the homologous structure tree of the protein. The homologous structure tree is significant for long core words.

Searching for the Core Words in a Database

Searching for core words in a database is similar to that in a sequence. Here the database is considered a long sequence of joined proteins and each protein is distinguished by list separators. Thus the definition of core words in a database requires the absence of the list separators in the core words. Hence we denote by

$$\Omega = \{c_1, c_2, \cdots, c_n\}, \quad c_i \in V_{q+1}, \tag{14.23}$$

the sequence generated by the database, where the list separator 0 is added to set $V_q = \{1, 2, \cdots, 20\}$, giving V_{q+1}. The search for the core words in a database is implemented by the following recursive computation.

Step 14.2.1 Take a positive integer k_0, which follows the conditions below:

1. Any vector $b^{(k_0)}$ in $V_q^{(k_0)}$ occurs at least twice in database Ω. That is, $n_\Omega(b^{(k_0)}) \geq 2$ always holds.
2. In $V_q^{(k_0+1)}$, there is a vector $b^{(k_0+1)}$ that occurs in database Ω only once. That is, $n_\Omega(b^{(k_0+1)}) = 1$ holds.

We call k_0 the original rank of the recursive computation.

Step 14.2.2 Take $k = k_0 + 1, k_0 + 2, \cdots$, and compute recursively. Here

$$c_i^{(k)} = (c_i, c_{i+1}, \cdots, c_{i+k-1}), \quad i = 0, 1, \cdots, n - k + 1. \tag{14.24}$$

From this, integers in the set $N_k = \{0, 1, 2, \cdots, n - k + 1\}$ are classified into two classes by $c_i^{(k)}$. They are:

1. We define

$$N_{k,1} = \left\{i \in N_k : n_\Omega\left(c_i^{(k)}\right) = 1\right\},$$

which is the subscript set of $c_i^{(k)}$ which occurs only once in Ω.

2. We define

$$N_{k,2} = \left\{i \in N_k : n_\Omega\left(c_i^{(k)}\right) > 1\right\},$$

which is the subscript set of $c_i^{(k)}$ which occurs more than once in Ω.

Then set

$$\Omega_k = \left\{c_i^{(k)},\ i \in N_{k,1},\ i \in N_{k-1,2}\right\} \tag{14.25}$$

will be the collection of the kth ranked core words in database Ω.

Step 14.2.3 The subscript set of set $N_{k,2}$ given in Step 14.2.2 can also be classified. Here, we take $N_{k,2,i_1}, N_{k,2,i_2}, \cdots, N_{k,2,i_p}$ to be a group of subsets of $N_{k,2}$. We denote the set $D_k = \{i_1, i_2, \cdots, i_p\}$, which follows the conditions below:

1. If $i_j \neq i_{j'}$, then $c_{i_j}^{(k)} \neq c_{i_{j'}}^{(k)}$ must hold.
2. If $i' \in N_{k,2,i_j}$, then $c_{i'}^{(k)} = c_{i_j}^{(k)}$ must hold. Therefore, sets $N_{k,2,i_1}$, $N_{k,2,i_2}, \cdots, N_{k,2,i_p}$ are disjoint with each other.
3. $\bigcup_{j=1}^{p} N_{k,2,i_j} = N_{k,2}$. Thus, sets $N_{k,2,i_1}, N_{k,2,i_2}, \cdots, N_{k,2,i_p}$ are a division of set $N_{k,2}$.

From the classification $N_{k,2,i_1}, N_{k,2,i_2}, \cdots, N_{k,2,i_p}$ of $N_{k,2}$, the recursive computation can be continued as follows:

4. With every subscript in set $N_{k,2}$ as a starting point, construct $k+1$ dimensional vectors. That is, take vectors $c_i^{(k+1)}$, $i \in N_{k,2}$ and repeat Step 14.2.2, to obtain two subsets $N_{k+1,1}$ and $N_{k+1,2}$ of set $N_{k,2}$, whose definitions are the same as that of $N_{k,1}$ and $N_{k,2}$ in Step 14.2.2, classifications 1 and 2, respectively. At this time, $\Omega_{k+1} = \{c_i^{(k+1)}, i \in N_{k+1,1}\}$ is the collection of the $(k+1)$th ranked core words in database Ω. For the elements in $N_{k+1,2}$, we can repeat the computation in Steps 14.2.2 and 14.2.3 to construct the set $N_{k+2,1}$ and $N_{k+2,2}$. Recurring in this way, the collection of the core words in database Ω can be obtained.
5. In the previous computation (4), the comparison of vectors $c_i^{(k+1)}$, $i \in N_{k,2}$ can only be implemented in the divided set $N_{k,2,i_1}, N_{k,2,i_2}, \cdots$, $N_{k,2,i_p}$ of $N_{k,2}$. Here, each $N_{k,2,i_j}$ is divided into two sets $N_{k+1,1,i_j}$, $N_{k+1,2,i_j}$, where the vectors $c_i^{(k+1)}$ corresponding to subscript i in $N_{k+1,1,i_j}$ occur only once in $c_{i'}^{(k+1)}$, $i' \in N_{k,2,i_j}$, while the vectors $c_i^{(k+1)}$ corresponding to subscript i in $N_{k+1,2,i_j}$ occur more than once in $c_{i'}^{(k+1)}$, $i' \in N_{k,2,i_j}$.

The amount of computation can be greatly reduced in this way, which makes the complexity of searching and computation for all core words in database Ω a linear function of $|\Omega|$. We refer to the above algorithm as recursive computation for nonlinear complexity and core words of a database.

With the discussion on the nonlinear complexity of databases, key words and core words, proteins can be classified and aligned. In this book, we will not discuss this problem; instead we focus on some special problems in the next section.

14.2.4 Applications of Combinatorial Analysis

Besides what was mentioned in Sect. 14.1 (the application of analysis on protein structure using information and statistics, and the application of relative entropy density dual theory), there are many applications of protein structure analysis using combinatorial analysis. We discuss these as follows.

Run Analysis on Databases of Protein Primary Sequences

We see from Sect. 14.1 that, several amino acids, such as A, R, N, etc., are of very high affinity. Thus we assume that, vectors consisting of single letters may form long polypeptide chains. If we name vectors consisting of single letters as runs, then analysis on these special polypeptide chains will then be called run analysis.

Suppose $c^{(s,t)}$ is a local sequence of sequence C. If $c_s = c_{s+1} = \cdots = c_t = c \in V_q$, $c^{(s,t)}$ is to be called a run vector and $t - s + 1$ is to be called the run length of this vector. We do statistical analysis on these run vectors in the database Ω, and some results are summarized as follows:

1. The differences between the maximum run lengths of different amino acids are quite significant. In the Swiss-Prot database, the maximum run lengths of different amino acids are shown in Table 14.10.
2. We see from the computation in the Swiss-Prot database that I and W are two special amino acids. The affinity of peptide chains with double I is neutral, $k(\mathrm{I},\mathrm{I}) = \log \frac{p(\mathrm{I},\mathrm{I})}{p(\mathrm{I})p(\mathrm{I})} = 0.0964$, while the affinity of peptide chains with triple I is repulsive, $k(\mathrm{I},\mathrm{I},\mathrm{I}) = \log \frac{p(\mathrm{I},\mathrm{I},\mathrm{I})}{p(\mathrm{I},\mathrm{I})p(\mathrm{I})} = -0.0926$. The affinity of peptide chains with double W or triple W is attractive, while the affinity of peptide chains with W of four runs is repulsive. Here

$$\begin{cases} k(\mathrm{W},\mathrm{W}) = \log \dfrac{p(\mathrm{W},\mathrm{W})}{p(\mathrm{W})p(\mathrm{W})} = 0.5663\,, \\ k(\mathrm{W},\mathrm{W},\mathrm{W}) = \log \dfrac{p(\mathrm{W},\mathrm{W},\mathrm{W})}{p(\mathrm{W},\mathrm{W})p(\mathrm{W})} = 0.5075\,, \\ k(\mathrm{W},\mathrm{W},\mathrm{W},\mathrm{W}) = \log \dfrac{p(\mathrm{W},\mathrm{W},\mathrm{W},\mathrm{W})}{p(\mathrm{W},\mathrm{W},\mathrm{W})p(\mathrm{W})} = -0.8316\,. \end{cases}$$

3. Amino acids with long runs are often of high ranked Markovity. Here

$$\begin{aligned} & k(c_1, c_2, \cdots, c_{k+2}) - k(c_1, c_2, \cdots, c_{k+1}) \\ &= \log \frac{p(c_1, c_2, \cdots, c_{k+2})}{p(c_1, c_2, \cdots, c_k)} - \log \frac{p(c_1, c_2, \cdots, c_{k+1})}{p(c_1, c_2, \cdots, c_k)} \\ &= \log \frac{p(c_1, c_2, \cdots, c_k)p(c_1, c_2, \cdots, c_{k+2})}{p(c_1, c_2, \cdots, c_{k+1})p(c_1, c_2, \cdots, c_k)} \\ &= \log \frac{p(c_1, c_{k+2}|c_2, \cdots, c_{k+1})}{p(c_1|c_2, \cdots, c_{k+1})p(c_{k+2}|c_2, \cdots, c_{k+1})} \sim 0\,. \end{aligned}$$

Table 14.10. The maximum run lengths of different amino acids

One-letter code	A	R	N	D	C	Q	E	G	H	I
Maximum run length	21	14	50	44	11	40	31	24	14	7
One-letter code	L	K	M	F	P	S	T	W	Y	V
Maximum run length	19	10	7	10	50	42	18	4	12	8

4. We see from the computation in the Swiss-Prot database that, for run sequences of different amino acids, when the run lengths reach a certain number, their transferences have a great propensity for orientation. For example, when the run length of amino acid A reaches 12, the possibility of its state transferring to A, G, S, V is very high, higher than 85%. This characteristic is similar to the English word structure. For instance, in English, what follows "qu" will always be one of the four letters a, e, i, o.
5. We can see from the fact that different amino acids have relatively long runs, that the lexical structure of protein primary structure is quite different from that of English and Chinese. In English or Chinese vocabularies, the maximum run of single letters are often only 2, while that of single letters in protein primary structure sequences may reach 50 (e.g., amino acid P).

Search and Alignment of Homologous Proteins

We have stated in the above text that, if $b^{(k)}$ is a core word, it occurs in the database Ω once, and only once. If vector $b^{(k)}$ is contracted back and forth, then key words of different ranks can be formed in database Ω. Proteins with these key words consist of the same peptide chains in considerably longer segments, where the corresponding homologous proteins and their stable regions can be found.

Example 32. In Swiss-Prot version 2000, we have a core word with length 29:

$$\text{GREFSLRRGDRLLLFPFLSPQKDPEIYTE}\ .$$

The protein it locates and its starting site are CAML-MOUSE, 373. Its segments with length 7 and the serial number and sites of the segments occurring in other proteins are

1 Segment	Serial no.	2	Serial no.	2	Serial no.	2	Serial no.	2
4 RGDRLLL	CAMG-MOUSE	379	CAML-HOMAN	379	CAML-MOUSE	380	CAML-RAT	380
4 GDRLLLF	CAMG-MOUSE	380	CAML-HOMAN	380	CAML-MOUSE	381	CAML-RAT	381
4 LLLFPFL	CAMG-MOUSE	383	CAML-HOMAN	383	CAML-MOUSE	384	CAML-RAT	384
4 LLFPFLS	CAMG-MOUSE	384	CAML-HOMAN	384	CAML-MOUSE	385	CAML-RAT	385
4 LFPFLSP	CAMG-MOUSE	385	CAML-HOMAN	385	CAML-MOUSE	386	CAML-RAT	386
3 FSLRRGD	CAMG-MOUSE	375	CAML-MOUSE	376	CAML-RAT	376		
3 PQKDPEI	CAMG-MOUSE	391	CAML-MOUSE	392	CAML-RAT	392		
3 SLRRGDR	CAMG-MOUSE	376	CAML-MOUSE	377	CAML-RAT	377		

Therefore, "1" is frequency number and "2" is location. The vector segment GREFSLRRGDRLLLFPFLSPQKDPEIYTE is a marker of the homology of gene CAML-MOUSE, CAMG-MOUSE, and CAML-RAT. We can also search for several core words in the same protein, and thereby obtain its homologous proteins.

The Cutting of Protein Sequences and the Prediction of Homologous Proteins

We analyze the cutting of protein sequences and the prediction of homologous proteins by using protein RISA-CHLPN and RISA-CHLTR in Example 30. We denote these two protein sequences as C and D respectively. Their nonlinear complexity and nonsingular complexity are both 4. Thus, the fourth ranked DG graph $G_4(C), G_4(D)$ generated by C and D are antitrunk trees with two branches. We can then discuss the cutting of these proteins:

1. We compare the primary structure of proteins C and D given in Example 30. The lengths of these two sequences are both $n = 289$, where the amino acids differ in ten sites and stay the same in the remaining sites. The differing sites are 7, 18, 50, 63, 94, 123, 139, 234, 272, and 280. We call these cutting sites.
2. As is shown in Fig. 14.2, we draw proteins C and D into two parallel lines. In these parallel lines, the lines present the same amino acids, while the sites corresponding to the hollow points and solid points present different amino acids. If we hold the lines presenting the same amino acids and randomly choose amino acids from sequence C or D to the sites where the amino acids are different in C and D, this operation is called the cutting operation of homologous proteins. This operation actually divides two homologous proteins into several segments and then combines them in the original order. If the original proteins C and D are denoted by

$$C = S_{1,1}0S_{1,2}0S_{1,3}0S_{1,4}0S_{1,5}0S_{1,6}0S_{1,7}0S_{1,8}0S_{1,9}0S_{1,10}0S_{1,11} \ ,$$

and

$$D = S_{2,1}1S_{2,2}1S_{2,3}1S_{2,4}1S_{2,5}1S_{2,6}1S_{2,7}1S_{2,8}1S_{2,9}1S_{2,10}1S_{2,11} \ ,$$

then the cutting sequence generated by this is

$$C' = S_{1,1}0S_{1,2}1S_{2,3}1S_{2,4}1S_{2,5}0S_{1,6}0S_{1,7}0S_{1,8}1S_{2,9}1S_{2,10}0S_{1,11} \ .$$

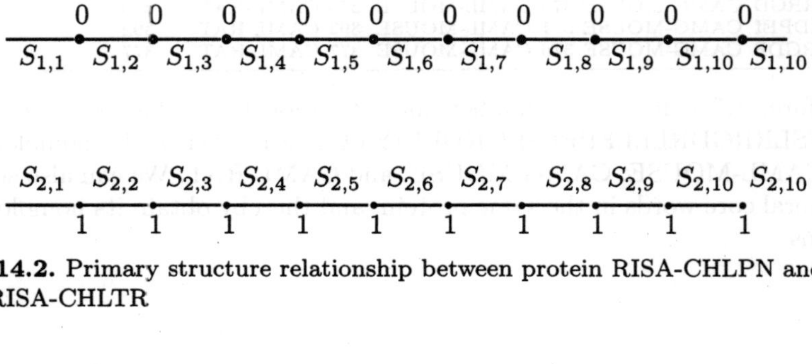

0 0 0 0 0 0 0 0 0 0
$S_{1,1}$ $S_{1,2}$ $S_{1,3}$ $S_{1,4}$ $S_{1,5}$ $S_{1,6}$ $S_{1,7}$ $S_{1,8}$ $S_{1,9}$ $S_{1,10}$ $S_{1,10}$

$S_{2,1}$ $S_{2,2}$ $S_{2,3}$ $S_{2,4}$ $S_{2,5}$ $S_{2,6}$ $S_{2,7}$ $S_{2,8}$ $S_{2,9}$ $S_{2,10}$ $S_{2,10}$
1 1 1 1 1 1 1 1 1 1

Fig. 14.2. Primary structure relationship between protein RISA-CHLPN and protein RISA-CHLTR

3. We reach the following conclusion: If there are k cutting points in two proteins, then there would be 2^k different combinations for this cutting procedure. All of these combinations of cutting may generate new homologous proteins.

From this we can predict that, there may be $2^{10} \sim 1000$ kinds of homologous trichosanthin structures. Biological experimentation is required to demonstrate and illustrate whether these homologous proteins exist and what their function and characteristics are. One particularly significant role of bioinformatics is to provide the scope, content and direction of experiments for biologists, thereby greatly decreasing the number of the possible experiments. The cutting of sequences provides a tool for this.

Semantic analysis for biological sequences is an important problem that combines mathematics and biology. We see from the above discussion that the corresponding analysis is related to the essential problems of the life sciences. Thus, a close relationship exists between the life sciences and biological engineering. Both the depth and breadth of the discussions in this book are preliminary. If the theories and methods in biology, mathematics and biological computation are further combined and different types of databases, such as databases of protein three-dimensional structures, protein functions, special (such as enzyme, antibody, virus, etc.) databases, GenBank, cDNA database, genome database are synthetically analyzed, its development and applications will lead to significant progress in biology and the related sciences. These include the core problems in genometics, proteomics theory, the applications of biological engineering and drug design, etc. We hope this book can stimulate further research using a synergistic combination of mathematics and biology.

14.3 Exercises, Analyses, and Computation

Exercise 68. Explain the biological significance of "local words," "key words," and "core words."

Exercise 69. Explain the meaning of "relative entropy density function" in protein primary structure.

Exercise 70. Perform the following computation for bacteria or archaebacteria (see the data given in [99]):

1. Take $k = 9$. Compute the probability distribution $p(b^{(9)})$ of $b^{(9)} \in V_4^{(9)}$, and the marginal distribution $p_1(b^{(3)}, p_0(b_1))$.

2. For

$$q_0\left(b^{(9)}\right) = p_0(b_1)p_0(b_2)p_0(b_3)p_0(b_4)p_0(b_5)p(b_6)p_0(b_7)p_0(b_8)p(b_0) ,$$
$$q_1\left(b^{(9)}\right) = p_0(b_1, b_2, b_3)p_0(b_4, b_5, b_6)p_0(b_7, b_8, b_0) ,$$
$$q_2\left(b^{(9)}\right) = p_0(b_4, b_5, b_6)p_0(b_7, b_8, b_9)p_0(b_1, b_2, b_3) ,$$
$$q_3\left(b^{(9)}\right) = p_0(b_7, b_8, b_9)p_0(b_1, b_2, b_3)p_0(b_4, b_5, b_6) ,$$

compute the relative entropy density functions

$$k_s\left(b^{(9)}\right) = \log_2 \frac{p\left(b^{(9)}\right)}{q_s\left(b^{(9)}\right)}, \quad s = 0, 1, 2, 3 ,$$

and give the table of their relative entropy density functions.
3. Based on task 2, compute the mean, variance, and standard deviation of $k_s(b^{(9)})$.
4. For $\tau = 2.5$, find the "local words" generated by $k_s(b^{(9)})$.

Exercise 71. Find the "key words" and "core words" generated by these bacteria or archaebacteria.

Epilogue

In this book, we introduced the stochastic models for DNA (or RNA) mutations, the theory of modulus structure used for gene recombination, gene mutation and gene alignment, the fast alignment algorithms for pairwise and multiple sequences, and the topological and graphical structures on the set of outputs induced from multiple alignments, respectively. These new concepts illustrate the large number of approaches for analyzing the structures of biosequences, many of which strongly rely on mathematics. We hope that the theory of the modulus structure will lead to new advances in algebra theory and will play an important role in the study of gene recombination and mutation.

On the other hand, this book is only the first step in what we expect will become a long history of applying mathematics to improve both the depth and meaning within the study of biology. Rather than developing the mathematical theory and methods in a vacuum, it is far more interesting when the mathematics evolves in order to solve problems in biology, biomedicine, and biomedical engineering.

Currently, we are in an ideal period for advancement in the life sciences, and many scientists from disciplines other than biology are facing the challenge of working in this field. Incorporating mathematics into the life sciences can only enhance the quality and accuracy of research in the life sciences. The question of how to ideally incorporate these two disciplines is of great importance, and it is our hope that this book will contribute towards this goal. Although every attempt was made to ensure correctness and accuracy in this edition before publication, the authors welcome comments and suggestions so that any remaining errors or omissions may be addressed in future editions.

References

1. Allen, F.H. The Cambridge Structural Database: a quarter of a million crystal structures and rising, Acta Crystallogr. Sect. B 58, 380–388 (2002)
2. Altekar, G., Dwarkadas, S., Huelsenbeck, J.P., and Ronquist, F. Parallel Metropolis coupled Markov chain Monte Carlo for Bayesian phylogenetic inference, Bioinformatics 20, 407–415 (2004)
3. Altschul, S.F., Gish, W., Miller, W., Myers, E.W., and Lipman, D.J. Basic local alignment search tool, J. Mol. Biol. 215, 403–410 (1990)
4. Altschul, S.F., Madden, T.L., Schaffer, A.A., Zhang, J., Zhang, Z., Miller, W., and Lipman, J.D. Gapped BLAST and PSI-BLAST: a new generation of protein database search programs, Nucleic Acids Res. 25, 3389–3402 (1997)
5. Andreeva, A., Howorth, D., Brenner, S.E., Hubbard, T.J.P., Chothia, C., Murzin, A.G. SCOP database in 2004: refinements integrate structure and sequence family data, Nucleic Acids Res. 32, 226–229 (2004)
6. Anfinsen, C.B. Principles that govern the folding of protein chains, Science 181:223–230 (1973)
7. Attwood, T.K. and Parry-Smith, D.J. Introduction to bioinformatics (Longman, Harlow, 1999)
8. Bairoch, A., Boeckmann, B., Ferro, S., and Gasteiger, E. Swiss-Prot: juggling between evolution and stability, Brief. Bioinform. 5, 39–55 (2004)
9. Barker, W.C., Garavelli, J.S., Haft, D.H., Hunt, L.T., Marzec, C.R., Orcutt, B.C., Srinivasarao, G.Y., Yeh, L.S.L., Ledley, R.S., Mewes, H.W., Pfeiffer, F., and Tsugita, A. The PIR—international protein sequence database, Nucleic Acids Res. 26, 27–32 (1998)
10. Benson, D.A., Karsch-Mizrachi, I., Lipman, D.J., Ostell, J., and Wheeler, D.L. GenBank, Nucleic Acids Res. 31, 23–27 (2003)
11. Benson, D.A., Karsch-Mizrachi, I., Lipman, D.J., Ostell, J., and Wheeler, D.L. GenBank: update, Nucleic Acids Res. 32, 23–26 (2004)
12. Berge, C. Hypergraphs–combinatorics of finite sets (North-Holland, Amsterdam, 1989)
13. Berman, H.M., Westbrook, J., Feng, Z., and Bourne, P.E. The Protein Data Bank, Nucleic Acids Res. 28, 235–242 (2000)
14. Chapman, M.S. Mapping the surface properties of macromolecules, Protein Sci. 2, 459–469 (1993)

15. Chan, S.C., Wong, A.K.C., and Chiu, D.K.Y. A survey of multiple sequence comparision methods, Bull. Math. Biol. 54(4), 563–598 (1992)
16. Chao, K.-M., Pearson, W.R., and Miller, W. Aligning two sequences within a specified diagonal band, CABIOS 8, 481–487 (1992)
17. Chao, K.-M., Hardison, R.C., and Miller, W. Recent developments in linear-space aligning methods: a survey, Comput. Appl. Biosci. 1(4), 271–291 (1994)
18. Chao, K.-M., Ostell, J., and Miller, W. A local aligning tool for very long DNA sequences, Comput. Appl. Biosci. 11(2), 147–153 (1995)
19. Chao, K.-M., Zhang, J., Ostell, J., and Miller, W. A tool aligning very similar DNA sequences, Comput. Appl. Biosci. 13(1), 75–80 (1997)
20. Chen, K.X., Jiang, H.L., and Ji, R.Y. Computer-aided drug design (Shanghai Scientific & Technical, Shanghai, 2000)
21. Chew, L.P., Kedem, K., Huttenlocher, D.P., and Kleinberg, J. Fast detection of geometric substructure in proteins, J. Comput. Biol. 6(3–4), 313–325 (1999)
22. CLUSTAL-W. http://www.ebi.ac.uk/Tools/clustalw/index.html (2007)
23. Cover, T.M. and Thomas, J.A. Elements of information theory (Wiley, New York, 1991)
24. Dayhoff, M.O., Schwartz, R.M., Orcutt, B.C. A model of evolutionary change in proteins. In: Dayhoff, M.O. (ed.) Atlas of protein sequence and structure, vol. 5, suppl. 2, pp. 345–352 (National Biomedical Research Foundation, Washington, 1978)
25. Doob, J.L. Stochastic processes (Wiley, New York, 1953)
26. Eisenberg, D., Schwarz, E., Komaromy, M., and Wall, R. Analysis of membrane and surface protein sequences with the hydrophobic moment plot, J. Mol. Biol. 179, 125–142 (1984)
27. Elshorst, B., Hennig, M., Forsterling, H., Diener, A., Maurer, M. et al. NMR solution structure of a complex of calmodulin with a binding peptide of the Ca^{2+} pump, Biochemistry 38(38), 12320–12332 (1999)
28. Felsenstein, J. Evolutionary trees from DNA sequences: a maximum likelihood approach, J. Mol. Evol. 17, 368–376 (1981)
29. Felsenstein, J. PHYLIP (the PHYLogeny inference package), version 3.66 (University of Washington, Seattle, 2007)
30. Fitch, W.M. Toward defining the course of evolution: minimum change for a specified tree topology, Syst. Zool. 20, 406–416 (1971)
31. Gallo, G., Longo, G., Nguyen, S., and Pallottino, S. Directed hypergraphs and applications, Discrete Appl. Math. 42, 177–201 (1993)
32. Garavelli, J.S., Hou, Z., Pattabiraman, N., and Stephens, R.M. The RESID database of protein structure modifications and the NRL-3D sequence-structure database, Nucleic Acids Res. 29(1), 199–201 (2001)
33. Gasteiger, E., Jung, E., and Bairoch, A. SWISS-PROT: Connecting biological knowledge via a protein database, Curr. Issues Mol. Biol. 3, 47–55 (2001)
34. Goldman, N. and Yang, Z. A codon-based model of nucleotide substitution for protein-coding DNA sequences, Mol. Biol. Evol. 11, 725–735 (1994)
35. Golomb, S.W. Shift register sequences (Holden-Day, San Francisco, 1967)
36. Goonet, G.H., Korostensky, C., and Benner, S. Evaluation measures of multiple sequence alignments, J. Comput. Biol. 7, 261–276 (2000)
37. Green, P.J. Reversible jump Markov chain Monte Carlo computation and Bayesian model determination, Biometrika 82, 711–732 (1995)
38. Hasegawa, M., Kishino, H., and Yano, T. Dating of the human-ape splitting by a molecular clock of mitochondrial DNA, J. Mol. Evol. 22, 160–174 (1985)

39. Hastings, W.K. Monte Carlo sampling methods using Markov chains and their applications, Biometrika 57, 97–109 (1970)
40. Henikoff, S., and Henikoff, J.G. Amino acid substitution matrices from protein blocks, Proc. Natl. Acad. Sci. USA 89, 10915–10919 (1992)
41. Hobohm, U. and Sander, C. Enlarged representative set of protein structures, Protein Sci. 3, 522–524 (1994)
42. Holm, L. and Sander, C. Mapping the protein universe, Science 273, 595–603 (1996)
43. Huang, J.Y. and Brutlag, D.L. The EMOTIF database, Nucleic Acids Res. 29(1), 202–204 (2001)
44. Huelsenbeck, J.P., Ronquist, F. MRBAYES: Bayesian inference of phylogenetic trees, Bioinformatics 17(8), 754–755 (2001)
45. Hulo, N., Bairoch, A., Bulliard, V., Cerutti, L., De Castro, E., Langendijk-Genevaux, P.S., Pagni, M., and Sigrist, C.J.A. The PROSITE database, Nucleic Acids Res. 34, 227–230 (2006)
46. Jiang, T., Kearney, P., and Ming, L. Some open problems in computational molecular biology, J. Algorithms 34, 194–201 (2000)
47. Jonassen, I. Efficient discovery of conserved patterns using a pattern graph, Comput. Appl. Biosci. 13(5), 509–522 (1997)
48. Jonassen, I., Collins, J.F., and Higgins, D. Finding flexible patterns in unaligned protein sequences, Protein Sci. 4(8), 1587–1595 (1995)
49. Jukes, T.H. and Cantor, C.R. Evolution of protein molecules. In: Munro, H.N. (ed.) Mammalian protein metabolism, pp. 21–132 (Academic, New York, 1969)
50. Kabsch, W. and Sander, C. Dictionary of protein secondary structure: Pattern recognition of hydrogen-bond and geometrical features, Biopolymers 22, 2577–2637 (1983)
51. Kedem, K., Chew, L.P., and Elber, R. Unit-vector RMS (URMS) as a tool to analyze molecular dynamics trajectories, Proteins Struct. Funct. Genet. 37, 554–564 (1999)
52. Kimura, M. A simple method for estimating evolutionary rates of base substitutions through comparative studies of nucleotide sequences, J. Mol. Evol. 6, 111–120 (1980)
53. Kyte, J. and Doolittle, R.F. A simple method for displaying the hydropathic character of a protein, J. Mol. Biol. 157, 105–132 (1982)
54. Lee, B. and Richards, F.M. The interpretation of protein structures: estimation of static accessibility, J. Mol. Biol. 55, 379–400 (1971)
55. Levenshtein, V.I. Binary coded capable of correcting deletion, insertions and reversals, Sov. Phys. Dokl. 10(8), 707–710 (1965)
56. Lipman, D.J., Altschul, S.F., and Kececioglu, J.D. A tool for multiple sequence alignment, Proc. Natl. Acad. Sci. USA 86, 4412–4415 (1989)
57. Liu, R.Y. On a notion of data depth based on random simplices, Ann. Stat. 18, 405–414 (1990)
58. Liu, R.Y. and Singh, K. A quality index based on data depth and multivariate rank tests, J. Am. Stat. Assoc. 88, 252–260 (1993)
59. Lo Conte, L., Ailey, B., Hubbard, T.J.P., Brenner, S.E., Murzin, A.G., and Chothia, C. SCOP: a structural classification of proteins database. Nucleic Acids Res. 27, 254–256 (2000)
60. Lo Conte, L., Brenner, S.E., Hubbard, T.J.P., Chothia, C., and Murzin, A. SCOP database in 2002: refinements accommodate structural genomics, Nucleic Acids Res. 30(1), 264–267 (2002)

61. Meek, J.L. Prediction of peptide retention times in high-pressure liquid chromatography on the basis of amino acid composition, Proc. Natl. Acad. Sci. USA 77, 1632–1636 (1980)
62. Metropolis, N., Rosenbluth, A.W., Rosenbluth, M.N., Teller, A.H., and Teller, E. Equations of state calculations by fast computing machines, J. Chem. Phys. 21, 1087–1091 (1953)
63. Michener, C.D., Sokal, R.R. A quantitative approach to a problem in classification, Evolution 11, 130–162 (1957)
64. Mount, D.W. Bioinformatics—sequence and genome analysis (Cold Spring Harbor Laboratory Press, Cold Spring Harbor, 2001)
65. Murray, R.K, Granner, D.K, Mayes, P.A., and Rodwell, V.W. Harper's biochemistry (McGraw-Hill, New York, 2000)
66. Murzin, A.G., Brenner, S.E., Hubbard, T., and Chothia, C. SCOP: a structural classification of proteins atabase for the investigation of sequences and structures, J. Mol. Biol. 247, 536–540 (1995)
67. National Center for Biotechnology Information. BLAST guide, http://0-www.ncbi.nlm.nih.gov.catalog.llu.edu/BLAST/ (2000)
68. Navarro, G. Gided tour to approximate string matching, ACM Comput. Surv. 33(1), 31–88 (2001)
69. Needleman, S.B. and Wunsch, C.D. A general method applicable to the search for similarities in the amino acid sequence of two proteins, J. Mol. Biol. 48, 443–453 (1970)
70. Nelson, D.L. and Cox, M.M. Lehninger principles of biochemistry, 3rd. edn. (Freeman, New York, 2000)
71. O'Donovan, C., Martin, M.J., Gattiker, A., Gasteiger, E., Bairoch, A., and Apweiler, R. High-quality protein knowledge resource: SWISS-PROT and TrEMBL, Brief. Bioinform. 3, 275–284 (2002)
72. Oja, H. Descriptive statistics for multivariate distributions, Stat. Probab. Lett. 1, 327–333 (1983)
73. Pearson, W.R. FASTA/TFASTA/FASTX/TFASTX users manual, http://www.ebi.ac.uk/fasta/ (1998)
74. Pearson, W.R. and Lipman, D.J. Improved tools for biological sequence comparison, Proc. Natl. Acad. Sci. USA 85(8), 2444–2448 (1988)
75. Pearson, W.R., Wood, T., Zhang, Z., and Miller, W. Comparison of DNA sequences with protein sequences, Genomics 46(1), 24–36 (1997)
76. RasMol. http://www.openrasmol.org/ (2007)
77. RDP service: Maidak, B.L., Cole, J.R., Parker, C.T. Jr., Garrity, G.M., Larsen, N., Li, B., Lilburn, T.G., McCaughey, M.J., Olsen, G.J., Overbeek, R., Pramanik, S., Schmidt, T.M., Tiedje, J.M., and Woese, C.R. A new version of the RDP (ribosomal database project), Nucleic Acids Res. 27, 171–173 (1999)
78. Richard, D., Sean, E., Anders, K., and Graeme, M. Biological sequence analysis: probabilistic models of proteins and nucleic acids (Cambridge University Press, Cambridge, 1998)
79. Rost, B. Review: protein secondary structure prediction continues to rise, J. Struct. Biol. 134(2–3), 204–218 (2001)
80. Rousseeuw, P.J. and Struyf, A. Computing location depth and regression depth in higher dimensions, Stat. Comput. 8, 193–202 (1998)
81. Saitou, N. and Nei, M. The neighbor-joining method: a new method for reconstructing phylogenetic trees, Mol. Biol. Evol. 4(4), 406–425 (1987)

82. Sankoff, D.D. Minimal mutation trees of sequences. SIAM J. Appl. Math. 28, 35-42 (1975)
83. Sayle, R.A. and Milner-White, E.J. RasMol: biomolecular graphics for all, Trends Biochem. Sci. 20, 374–376 (1995)
84. Schumacher, M.A., Crum, M., and Miller, M.C. Crystal structures of apocalmodulin and an apocalmodulin/SK potassium channel gating domain complex, Structure (Camb.) 12(5), 849–860 (2004)
85. Sellers, P.H. On the theory and computation of evolutionary distances, SIAM J. Appl. Math. 26(4), 787–793 (1974)
86. Sewald, N. and Jakubke, H.D., Peptides: chemistry and biology (Wiley-VCH, Weinheim, 2002)
87. Shen, S.Y. Combinatorial cryptography (Zhejiang Publishing House of Science and Technology, 1992)
88. Shen, S.Y. Information theory and coding theory (Science, Beijing, 2002)
89. Shen, S.Y. Structure analysis of biological sequence mutation and alignment (Science, Beijing, 2004)
90. Shen, S.Y, Yang, J., Yao, A., and Hwang, P.I. Super pairwise alignment (SPA): an efficient approach to global alignment for homologous sequences, J. Comput. Biol. 9, 477–486 (2002)
91. Shen, S.Y., Kai, B., Ruan, J., Huzil, J.T., Carpenter, E., and Tuszynski, J.A. Probabilistic analysis of the frequencies of amino acid pairs within characterized protein sequences, Physica A 370(2) 651–662 (2006)
92. Shen, S.Y., Hu, G., and Tuszynski, J.A. Analysis of protein three-dimension structure using amino acids depths, Protein J. 26, 183–192 (2007)
93. Shi, F. CAGD and non-uniformly rational B-sample (Higher Education, Beijing, 2001)
94. Sigrist, C.J.A., Cerutti, L., Hulo, N., Gattiker, A., Falquet, L., Pagni, M., Bairoch, A., and Bucher, P. PROSITE: a documented database using patterns and profiles as motif descriptors, Brief. Bioinform. 3, 265–274 (2002)
95. Smith, T.F., Waterman, M.S., and Fitch, W.M. Comparative biosequence metrics, J. Mol. Evol. 18, 38-46 (1981)
96. Sneath, P.H.A. and Sokal, R.R. Numerical taxonomy (Freeman, San Francisco, 1973)
97. Srinivasarao, G.Y., Yeh, L.-S., Marzec, C.R., Orcutt, B.C., and Barker, W.C. PIR-ALN: a database of protein sequence alignments, Bioinformatics 15, 382–390 (1999)
98. Su, B. and Liu, D. Computational geometry (Shanghai Scientific & Technical, Shanghai, 1981)
99. Super Alignment. http://mathbio.nankai.edu.cn (2007)
100. Tavaré, S. Some probabilistic and statistical problems in the analysis of DNA sequences. In: Miura, R.M. (ed.) Lectures in mathematics in the life sciences, vol. 17, pp. 57–86 (American Mathematical Society, Providence, 1986)
101. The Chinese SARS Molecular Epidemiology Consortium. Molecular evolution of the SARS coronavirus during the course of the SARS epidemic in China, Science 303, 1666–1669 (2004)
102. Thompson, J.D., Higgins, D.G., and Gibson, T.J. CLUSTAL W: improving the sensitivity of progressive multiple sequence alignment through sequence weighting, position-specific gap penalties and weight matrix choice, Nucleic Acids Res. 22, 4673–4680 (1994)

103. Tukey, J.W. Mathematics and picturing data. In: James, R.D. (ed.) Proceedings of the international congress on mathematics, vol. 2, pp. 523–531 (Canadian Mathematics Congress, 1975)
104. Wang, L. and Jiang, T. On the complexity of multiple sequence alignment, J. Comput. Biol. 1, 337–348 (1994)
105. Wang, Y. Trichosanthin, 2nd edn. (Science, Beijing, 2000)
106. Winfried, J. Computational complexity of multiple sequence alignment with SP-score, J. Comput. Biol. 8(6), 615–623 (2001)
107. Wolfenden, R., Andersson, L., Cullis, P., and Southgate, C. Affinities of amino acid side chains for solvent water, Biochemistry 20, 849–855 (1981)
108. Yang, Z. Maximum-likelihood-estimation of phylogeny from DNA sequences when substitution rates differ over sites, Mol. Biol. Evol 10, 1396–1401 (1993)
109. Yang, Z. Estimating the pattern of nucleotide substitution. J. Mol. Evol. 39, 105–111 (1994)
110. Yang, Z. Maximum-likelihood phylogenetic estimation from DNA sequences with variable rates over sites-approximate methods, J. Mol. Evol. 39, 306–314 (1994)
111. Yang, Z. PAML: a program package for phylogenetic analysis by maximum likelihood, Comput. Appl. Biosci. 13(5), 555–556 (1997)
112. Yang, Z. and Rannala, B. Bayesian phylogenetic inference using DNA sequences: a Markov chain Monte Carlo method, Mol. Biol. Evol. 14, 717–724 (1997)
113. Yap, K.L., Yuan, T., Mal, T.K., Vogel, H.J., and Ikura, M. Structural basis for simultaneous binding of two carboxy- terminal peptides of plant glutamate decarboxylase to calmodulin, J. Mol. Biol. 328(1), 193–204 (2003)
114. Zuo, Y. Multivariate monotone location estimators, Sankhyã, Ser. A 62(2), 161–177 (2000)
115. Zuo, Y. A note on finite sample breakdown points of projection based multivariate location and scatter statistics, Metrika, 51(3), 259–265 (2000)
116. Zuo, Y. and Serling, R. General notions of statistical depth function, Ann. Stat. 28(2), 461–482 (2000)

Further Reading

Altschul, S.F. A protein alignment scoring system sensitive at all evolutionary distances, J. Mol. Evol. 36, 290–300 (1993)

Altschul, S.F. and Lipman, D. Trees, stars and multiple biological sequence alignment, SIAM J. Appl. Math. 49, 197–209 (1989)

Baldi, P. and Brunak, S. Bioinformatics (MIT Press, London, 1998)

Baldi, P. and Brunak, S. Bioinformatics—the machine learning approach (MIT Press, Cambridge, 2001)

Baxevanis, A.D. and Francis, B.F. Bioinformatics (Wiley, New York, 1998)

Carrillo, H. and Lipman, D. The multiple sequence alignment problem in biology, SIAM J. Appl. Math. 48, 1073–1082 (1988)

Comparison. http://www.accelrys.com/products/cg/comparison.html

Durbin, R., Eddy, S., Krogh, A., and Mitchison, G. Biological sequence analysis—probabilistic models of proteins and nucleic acids (Cambridge University Press, Cambridge, 1998)

Fuellen, G. A gentle guide to multiple alignment, version 2.03 (1997)

Giegerich, R. and Wheeler, D. Pairwise sequence alignments, version 2.01 (1996)

Gusfield, D. Efficient methods for multiple alignment with guaranteed error bounds, Bull. Math. Biol 55, 141–145 (1993)

Hu, S., Xue, Q. Data analysis of genomes (Hangzhou Zhejiang University Press, 2003)

Huang, X. A context dependent method for comparing sequences. In: Proceedings of the 5th symposium on combinatorial pattern matching. Lecture notes in computer science 807, pp. 54–63 (Springer, Berlin, 1995)

Kirkpatrick, S., Gelatt, C.D. Jr., and Vecchi, M.P. Optimization by simulated annealing, Science 220, 671–680 (1983)

Loeve, M. Probability theory (Van Nostrand Reinhold, New York, 1960)

Minoru, K. Post-genome informatics (Oxford University Press, Oxford, 2000)

Morgenstern, B., Dress, A., and Werner, T., Multiple DNA and protein sequence alignment based on segment-to-segment comparison, Proc. Natl. Acad. Sci. USA 93, 12098–12103 (1996)

Morgenstern, B., Atchley, W.R., Hahn, K., and Dress, A. Segment-based scores for pairwise and multiple sequence alignments. In: Proceedings of the 6th international conference on intelligent systems for molecular biology (ISMB 98) (1998)

Morgenstern, B., Frech, K., Dress, A., and Werner, T. DIALIGN: finding local similarities by multiple sequence alignment, Bioinformatics 14(3), 290–294 (1998)

Pearson, W.R. Rapid and sensitive sequence comparison with FASTP and FASTA, Methods Enzymol. 183, 63–98 (1990)

Shen, S.Y. Optimal principle of information measurement for the mutiple sequence alignment, J. Eng. Math. 19, 3 (2002)

The Human Genome Program. The Human Genome Project information, http://www.ornl.gov/sci/techresources/Human_Genome/home.shtml (1998)

Wang, L. and Gusfield, D. Improved approximation algorithms for tree alignment, J. Algorithms 25, 255–273 (1997)

Waterman, M.S. Efficient sequence alignment algorithms, J. Theor. Biol. 108–333 (1984)

Waterman, M.S. Sequence alignment. In: Waterman, M.S. (ed) Mathematical methods for DNA sequences (CRC, Boca Raton, 1989)

Waterman, M.S. Introduction to computational biology (Chapman and Hall, London, 1995)

Zhao, G. et al. Bioinformatics (Science, Beijing, 2002)

Zhang, C.G. and He, F. Bioinformatics—method and practice (Science, Beijing, 2002)

Index